IET CONTROL, ROBOTICS AND SENSORS SERIES 116

Imaging Sensor Technologies and Applications

IET International Book Series on Sensors–Call for Authors

The use of sensors has increased dramatically in all industries. They are fundamental in a wide range of applications from communication to monitoring, remote operation, process control, precision and safety, and robotics and automation. These developments have brought new challenges such as demands for robustness and reliability in networks, security in the communications interface, and close management of energy consumption. This Book Series covers the research and applications of sensor technologies in the fields of ICTs, security, tracking, detection, monitoring, control and automation, robotics, machine learning, smart technologies, production and manufacturing, photonics, environment, energy, and transport. Book Series Editorial Board

Proposals for coherently integrated international multi-authored edited or co-authored handbooks and research monographs will be considered for this Book Series. Each proposal will be reviewed by the IET Book Series Editorial Board members with additional external reviews from independent reviewers. Please email your book proposal to: vmoliere@theiet.org or author_support@theiet.org.

Imaging Sensor Technologies and Applications

Edited by
Wuqiang Yang

The Institution of Engineering and Technology

Published by The Institution of Engineering and Technology, London, United Kingdom

The Institution of Engineering and Technology is registered as a Charity in England & Wales (no. 211014) and Scotland (no. SC038698).

First published 2020

The Institution of Engineering and Technology
Michael Faraday House
Six Hills Way, Stevenage Herts,
SG1 2AY, United Kingdom

www.theiet.org

British Library Cataloguing in Publication Data
A catalogue record for this product is available from the British Library

ISBN 978-1-78561-497-2 (Hardback)
ISBN 978-1-78561-498-9 (PDF)

Typeset in India by MPS Limited

Contents

Preface

Imaging has become an attractive research topic and has found many applications. Imaging sensors are the basis of generating images and hence the basis of imaging technology. This book covers imaging sensors in a wide frequency range, that is different sensing modalities, including X-ray beam imaging sensors, optical scattering sensors, smart visual sensors in robotic systems, CCD and CMOS sensors and their applications in flame measurement, tunable diode Laser absorption spectroscopy (TDLAS) sensors, light detection and ranging (LiDAR) sensors, microwave imaging (MWI) sensors, ultra-wideband (UWB) sensors, synthetic aperture radar (SAR), electrical resistance tomography (ERT) sensors and applications, electrical tomography for medical applications, electro-magnetic tomography (EMT) sensors, micro-sensors for cell and blood imaging and ultrasonic imaging sensors. This book consists of 14 chapters. All the authors, who contributed to this book, are experienced researchers in their specific field, that is sensing modality and hence they present the state-of-the-art technologies.

Chapter 1 introduces X-ray beam imaging sensors. At the high-frequency side of the electro-magnetic spectrum, the instrumentation described in this chapter is used for in situ characterisation of X-ray beams. The fully transparent method to localise and measure of intensity distribution of highly energetic beams is remarkable because the method only extracts 0.0001% of the incident beam.

Chapter 2 introduces sensors for particle measurements based on optics and acoustics. It is divided into four parts: static light scattering, dynamic light scattering, RGB multi-wavelengths light extinction and ultrasonic scattering.

Chapter 3 is about smart visual sensors in robotic systems. It provides an overview of camera models and calibration processes, and includes both position and image-based visual servoing, using neural and fuzzy networks, Kohonen self-organising map and other learning models. Several techniques, such as 3D reference object-based calibration, 2D plane based-calibration and self-calibration are reviewed. Experimental results in some interesting applications, including warehouse automation and implementation issues of visual sensing due to cameras, image processing and feature extraction are given.

Chapter 4 is about CCD- and CMOS-based imaging sensors and their applications in flame measurement. It covers the architecture and characteristics of CCD and CMOS sensors, CCD/CMOS cameras, special cameras, camera interfaces and a general guide for selection of cameras for imaging systems. Vision-based technologies for flame measurement and characterisation are introduced, including typical 2D and 3D imaging techniques for measuring the characteristic parameters of flames in combustion systems. The basic principle, advantages and disadvantages of each technique are also discussed.

Chapter 5 is about TLDAS sensors, which are used for non-invasive temperature measurement in a wide temperature range. TDLAS can be used in many industrial cases that include absorbable gas species, such as oxygen and water. A tomographic TDLAS sensor enables extraction of temperature distribution. In this chapter, TDLAS sensors are introduced, including principle, hardware design, performance evaluation, image reconstruction as well as two case studies for combustion diagnostics.

Chapter 6 is about LiDAR sensors. In this chapter, a terrestrial Laser scanner (TLS) is introduced as a typical time-of-flight LiDAR sensor, including sections on TLS instruments, error analysis and applications. By the design of TLS instruments to obtain the ranging and cloud points data and deduct angle and ranging errors and its self-calibration to ensure that the survey returns the best quality data, accurate 3D point data can be processed for 3D imaging of various scenes, such as urban objects, rock or ice outcrops, trees and vegetation.

Chapter 7 discusses MWI sensors. First, the microwave imaging basics are briefly reviewed and the common architecture of MWI systems is introduced. Subsequently, antennas adopted in different applicative areas are detailed and discussed.

Chapter 8 is about UWB sensors. UWB signals can travel through most dielectric obstacles, maintaining separation between a target and background noise. Because of non-radiation and low power consumption, such a system is ideally suited for solving various imaging problems. This chapter discusses the development of a UWB prototype, detailing the requirements for software and sensor hardware, and focusing on breast cancer detection.

Chapter 9 introduces the basic theory and processing algorithm of SAR. First, the principles of both synthetic aperture and azimuth high resolution in Radar imaging are presented. Then, the classical range-Doppler algorithm is derived. Finally, a unified focusing algorithm based on the concept of the rotation-centre range is given for multi-mode SAR.

Chapter 10 is about ERT sensors and applications. This chapter mainly introduces the principle of ERT and system design, including the excitation and measurement strategies. Typical forward and inverse problems in ERT image reconstruction are also described. Finally, some typical application examples of ERT are given.

Chapter 11 reviews the use of electrical tomography (ET) for medical applications. It consists of four parts: (1) early cancer detection (breast, cervical, prostate and colorectal cancer), (2) lung imaging, (3) brain imaging and (4) image-navigated surgery. The current status, limitations and prospects of ET techniques for those medical applications are also introduced in this chapter.

Chapter 12 is about EMT sensors. This chapter presents the principle, simulation and application of several different EMT sensors. It also briefly describes the system hardware and a scaling relationship in an EMT sensor.

Chapter 13 is about micro-sensors for cell and blood imaging. With development of MEMS, the size of bio-impedance sensors is reduced from millimetre scale for organ detection to micron scale for cell detection. This chapter introduces various micro-sensors with the emerging technology and their applications to biomedical measurement, such as micro-electrical impedance tomography (EIT) sensors for cell imaging in a blood flow, micro-fluidic cytometer for electrical properties detection in a single cell analysis and real-time quantitative imaging of 3D cell culture.

Chapter 14 introduces ultrasound imaging sensors. It covers the fundamentals of acoustics, a variety of acoustic sensing technologies and typical applications for medical diagnosis. The acoustic sensors include piezoelectric transducer/array, capacitive micro-machined ultrasound transducer and optical ultrasound sensors. The imaging modalities include amplitude mode (A-mode), brightness mode (B-mode), motion mode (M-mode), intravascular ultrasound, Doppler imaging and harmonic imaging.

This book is useful for both researchers and engineers working in the field related to imaging sensors. It is also useful for postgraduate students to understand different sensing modalities for imaging.

I would like to take this opportunity to thank all the chapter authors, who have contributed to the book. Without their tremendous effort, it would not be possible to have this book published. I would also like to thank Olivia Wilkins, Valerie Moliere and other team members at IET Publishing for their encouragement, guidance, management and support during the whole process.

Wuqiang Yang
The University of Manchester, UK
August 2019

Imaging sensor technologies and applications

Imaging has become an attractive research topic and has found many applications. Imaging sensors are the basis of generating images and hence the basis of imaging technology. This book covers imaging sensors in a wide frequency range, that is different sensing modalities, including X-ray beam imaging sensors, optical scattering sensors, smart visual sensors in robotic systems, CCD and CMOS sensors and their applications in flame measurement, tunable diode Laser absorption spectroscopy (TDLAS) sensors, light detection and ranging (LiDAR) sensors, microwave imaging sensors, ultra-wideband (UWB) sensors, synthetic aperture radar (SAR), electrical resistance tomography (ERT) sensors and applications, electrical tomography for medical applications, electro-magnetic tomography (EMT) sensors, micro-sensors for cell and blood imaging and ultrasonic imaging sensors. This book consists of 14 chapters. All the authors, who contributed to this book, are experienced researchers in their specific field, that is sensing modality and hence they present the state-of-the-art technologies.

About the editor

Wuqiang Yang is a Fellow of the IEEE, Fellow of the IET and Fellow of the Institute of Measurement and Control. He spent 13 years at Tsinghua University in Beijing, as an undergraduate, MSc and PhD student and then as a lecturer. Since 1991, he has been working with UMIST and University of Manchester in the UK. He became a Professor in the Department of Electrical and Electronic Engineering in 2005. His main research interests include electrical capacitance tomography (ECT), instrumentation and multiphase flow measurement. He has published 400 papers and edited a book '*Sensor array*', which is published by Intech. He is an Associate Editor of *IEEE Trans. on Instrum. and Meas.*. and *IET Sci., Meas. and Technol.*, editorial board member of several other journals (including *Meas. Sci. and Technol.* and *Sensor Review*), guest editor of many journal special issues and visiting professor at several other universities. He is one of the key organisers of the annual *IEEE International Conference on Imaging Systems and Techniques*. He was an IEEE IMS Distinguished Lecturer (2010–2016) and a JSPS Invitation Fellow (2016) and is a JSPS Bridge Fellow (2020). His biography has been included in Who's Who in the World since 2002.

Chapter 1

X-ray beam imaging sensors

Roelof van Silfhout[1]

List of abbreviations

ASIC	application-specific integrated circuits
CCD	charge coupled device
CMOS	complementary metal oxide semiconductor
CSPAD	Cornell-SLAC pixel array detector
CVD	chemical vapour deposition
DLS	Diamond Light Source
HPD	hybrid pixel detector
LCLS	linear coherent light source
MAPS	monolithic active pixel sensor
PID	proportional, integral and derivative
PSD	position sensitive device
PSF	point spread function
QD	quadrant photodiode
SNR	signal-to-noise ratio

1.1 Introduction

Measurements of the position and profile of high energetic particle beams, such as X-rays, are important during experiments, in which these beams are used to precisely map or pinpoint features of interest. In medical application, such measurements are crucial during radiotherapy sessions to target tumours without damaging healthy tissue. Only with suitable transparent or non-destructive means of localising the beam accurately and precisely, a system can be developed to stabilise beam delivery. This chapter focuses on solutions that have been developed and exploited at synchrotron radiation sources, where microscopic beams are used to study

[1]Department of Electrical and Electronic Engineering, The University of Manchester, Manchester, United Kingdom

similarly sized samples. Various methods of accomplishing transparent X-ray beam imaging will be described and a detailed description of a complete system that consists of sensors, image acquisition, data processing and feedback control is given. The techniques and sensing methods discussed are applicable to many other precision measurement techniques where a light beam is used to gauge the position of an object.

Conducting experiments with synchrotron radiation often involve in working with samples that are illuminated by a probing X-ray beam at a safe distance and often enclosed by a lead-lined hutch. As a consequence the task of alignment of optical elements, mono-chromators and sample with the incident beam must be done remotely and often involves in the use of fluorescent screens, special X-ray sensitive paper or X-ray cameras. Traditionally such alignments would be done at the start of the experiment and settings would then be frozen for the duration of the measurement process. With the arrival of third-generation light sources and the ability to focus their beams down to sub-micrometre level, experimenters can no longer assume that the alignment of both beam and sample remains constant even over the duration of a short time period for experiment. To illustrate the issue of misalignment, an often over-looked effect is referred, which may be caused by the thermal expansion coefficient of materials. For example, a small temperature difference of 0.5 °C between 1.4 m high steel supports results in a length difference of about 10 μm. Although it is technically feasible to stabilise the temperature to values smaller than 0.5 °C, it becomes prohibitively expensive to do so over the full length of a typical X-ray optical setup (beamline), which is typically 50 m long.

Even if facility builders could cancel thermal variations there are many other sources of misalignment related to either beamline components or source instabilities – both static and dynamic in nature. Long beamlines in particular will suffer, for example, from the beam pointing instability, which for a typical third-generation source, is approximately 0.1 μrad. Beam position changes may be caused by many sources, including refilling of the storage ring, thermal cycles of the mono-chromators, mirrors and other temperature, pressure and mechanical related factors, which lead to loss of focus on the sample or result in the beam missing the sample all together. Synchrotron beamlines seldom operate without optical elements that perform beam defining, focusing and energy selecting functions. In most cases all these three types of elements are combined to give a well-defined small beam size with a set energy at the sample position. In turn, each of these components influences the beam position and/or beam shape.

Over the last decade, several groups have developed a range of methods to monitor the X-ray beam without blocking it with highly absorbing media in such a way that researchers have access to real-time information regarding the position of the incident beam during their experiments. Traditionally, the beam centre of mass and beam intensity measurements are performed with either quadrant photodiodes or lateral effect devices. More recent work has added the capability to record the detailed cross-sectional intensity distribution of the beam as a function of time. Several of these designs are fast enough to establish active feedback loops, to provide a constant illumination of microscopic samples.

Capturing a precise beam profile and position would ideally be performed in a single operation without a mechanical scanning step so that all information is gathered at the same moment in time and repeated at a high (frame) rate to measure beam displacements and vibrations at relevant frequencies. Practically, such measurements are taken with imaging sensors, which feature a large amount of small individual sensors or pixels tiled in a large array.

With beam profile monitors charge coupled devices (CCDs) have been the choice of preference for various studies and also for general use in operational beamlines. More recently complementary metal oxide semiconductor (CMOS) image sensors have been used in either indirect, direct or hybrid imaging mode. For beam imaging, a high frame rate, a high signal-to-noise ratio (SNR) and a high pixel density are required.

Recording high-resolution images of the intensity distribution across the beam is but one aspect of the challenge to locate and maintain the beam position. To direct the beam to the right part of the sample and keep it there requires fast sensor readout and data processing engines that can cope with high-bandwidth data flows whilst performing various algorithms needed to extract key information regarding the beam profile and position at sufficiently high update rates.

After reviewing the main methods of non-destructive X-ray beam monitoring, suitable detector systems will be discussed before a method of transparent X-ray beam imaging is introduced, which has been developed at the University of Manchester. A flexible and integrated readout, processing and control system that can interface to a large variety of suitable sensors, autonomously process the image data at high frame rates and integrate a multi-channel proportional, integral and derivative (PID) controller for feedback control to suitable actuators that allow stabilising of the X-ray beam position will also be introduced. Finally, measurements taken from this instrument will be presented.

1.2 Non-destructive X-ray beam monitoring

Interactions of X-ray with solids or gas provide designers with several options for performing X-ray beam imaging. Figure 1.1 shows a schematic overview of these options. Devices based on all these options have been reported in the literature. An incident X-ray beam onto solid matter or gas causes a weak interaction with the electrons giving rise to various processes that lead to the emission of electrons, elastic scattered X-rays or fluorescence that spans much of the electromagnetic spectrum. Furthermore, X-ray excited electrons can be collected from the conduction band by applying suitable electric fields inside semiconducting materials.

In general, two distinct approaches have been followed most for non-destructive X-ray beam diagnostics. The first one infers only the beam centre position ('centre-of-mass') by intercepting the fringes of the beam profile using four thin metal plates that can record the photo current due to the photo-electric effect from the X-rays at the tail end of the intensity distribution. An alternative approach takes four readings from the blades of an entrance slit that is used to

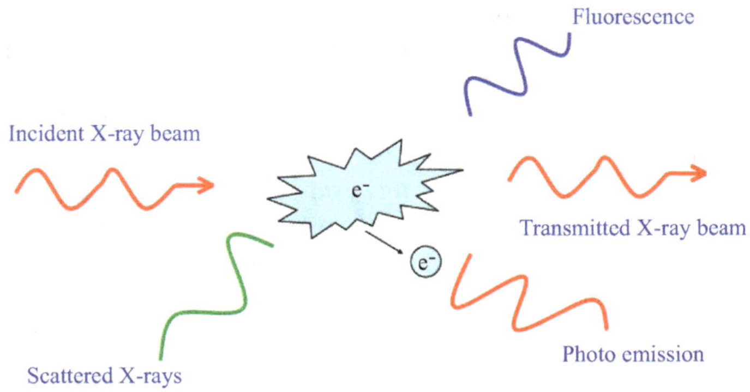

Figure 1.1 Interactions of X-rays with matter

define the X-ray beam. Alternatively, researchers have equipped individual slit blades with fluorescent material, so that the beam position can be inferred visually. Researchers have also used quadrant photodiodes with a central hole. The benefit of this approach is that the interaction between the beam and the sensor is minimal and that the centre of the intensity distribution passes freely. The accuracy of this method, however, relies on the fact that the beam profile remains unchanged. Any slight changes in the beam profile (even though its centre of mass does not change) would result in a recorded change in beam position.

The second method of transparent beam imaging provides a more comprehensive diagnostic of the incident beam, in which not only the beam position but also its profile is measured. Generally, such methods use either a gas or a solid in the form of a thin film or membrane that is traversed by the X-ray beam. The following is an overview of the various methods that are reported in the literature. Generally, this method offers the capability of recording the X-ray beam profile for X-ray beam energies above 5 keV. Below this energy foils and gases absorb a significant part of the incident beam rendering it no longer as a transparent or non-destructive beam imaging method.

Destructive X-ray beam profile measurement techniques place items in the beam path that are greatly or completely absorbing the beam to collect a measurement. Examples of these are wire scanners, blade scanners and cameras that either directly take the X-ray (often with attenuators) or measure the fluorescence from a specially prepared scintillating material.

1.2.1 Gas scattering

For a gas as a scatter medium, one can opt for two basic approaches. The first one uses the fluorescence from the atoms in the gas. For example, at the Cornell synchrotron (CHESS) a small section of the X-ray beam path traverses a chamber filled with Helium gas at atmospheric pressure [1]. The visible fluorescence light is observed with two cameras placed outside the chamber that receive the fluorescent light through two viewports positioned at right angles with the direction of the

X-ray beam, one placed above the beam to measure the integrated beam profile in the horizontal plane and the other placed in the horizontal plane to measure the vertical beam profile and position. Using standard charge coupled device (CCD) cameras and frame grabbers, a resolution of measurement of the beam position for intense synchrotron X-ray beams as emitted by the source (white or polychromatic) was reported to be smaller than 1 μm [2].

The second approach collects the photo-electrons created by the white beam passing through a vacuum tube with a residual amount of gas [3,4]. To collect the electrons or ions, electric fields are used and micro-channel plates are inserted to record two integrated beam profiles from the side and top of the beam path to determine the beam position. For this rest gas X-ray beam position monitor, a position resolution of about 5 μm was reported. Designers have the option of collecting the photo-electrons or ions resulting from the X-ray beam interaction with the residual gas. A simplified scheme, which only measures the beam centre, works without the micro-channel plates and uses a split electrode ion chamber [5]. Using normalised recording of the difference between the two split electrodes, a reading is obtained of the beam position. Gas-type beam position monitors are particularly suitable for high-intensity ('white') X-ray beams.

1.2.2 Foil scattering

Placing a thin foil in the X-ray beam provides several options for imaging the X-ray beam non-destructively. Again one can collect fluorescent radiation (both in the visible [2] and X-ray region of the spectrum [6]), photo electrons or the elastically scattered X-rays. By using thin foils of low-density materials, the transmitted beam is largely unaffected rendering the device transparent to the incident beam. Most of the methods published measure beam position and some add an option of beam profiling but few provide a truly cross-sectional intensity distribution (beam imaging) measurement.

The use of a thin metallic foil placed in the X-ray beam allows a determination of the beam position by measuring the fluorescent X-rays with four photodiodes placed in a back scattering geometry with respect to the foil and the incoming X-ray beam [6]. In this geometry, the photo-diodes are placed well away from the incident beam avoiding any beam-induced damage. Each diode covers a quadrant and thus by comparing the photocurrents as measured by the photodiodes the position of the beam centre-of-mass in both horizontal and vertical directions is obtained. Typically, the foil consists of a very thin low-density metal foil so that a negligible beam attenuation is caused. Because elements, such as Ti or Cr, have K-edges at low X-ray energies one can use these devices with most X-ray energies of interest to synchrotron users. Typically, this scheme works well for monochromatic radiation where a typical resolution of 1 μm can be achieved. Due to the fragile nature of the thin foils, this method does not work with intense polychromatic beams, which would readily evaporate the thin metal layer. Also, for spectroscopic beamlines such an approach, in which the transmission of the foil shows transient behaviour near absorption edges of the foil material, is unwanted.

Thin foils of a semi-conductor material can also be used. Several authors have developed devices based on Diamond and Silicon membranes. These act like solid state ion chambers that collect charge created by the energetic X-rays inside the foils. Typically, Diamond-based devices have proven to work well when placed in monochromatic beams without evidence of beam-induced damage [7,8].

Diamond foils can also be used as scintillators: a thin membrane without any further metalised contacts is placed in the beam and a camera system records the visible fluorescent light when placed in the X-ray beam. When placed into the white beam researchers have reported 'burn-in' of the foils, showing a reduced fluorescent light yield from locations that had been exposed to the X-ray beam previously.

Finally, one can imagine the X-ray beam by recording elastically scattered X-rays either directly or through a collimator or a pin-hole using a camera [9,10]. With this method it is possible to select a suitable geometry to achieve the highest resolution and sensitivity whilst keeping the image sensor well away from the incident beam. This method works well with a range of foils and in the literature researchers have used polyimide (Kapton^{TM}), Beryllium, quartz glass, glassy carbon, chemical vapour deposition (CVD) diamond and other foils made from low-density materials that are radiation hard (not susceptible to damage).

1.3 Sensors

Figure 1.2 shows sensors for light beam position measurements. For quadrant photodiodes (QD) and position sensitive device (PSD) sensors lateral positions are readily derived using the simple differences between electrode readings that are normalised. Camera sensors provide position and beam spot size information. For the camera the lateral position is given as a centre-of-mass calculation over all pixels with charge Q_{ij}.

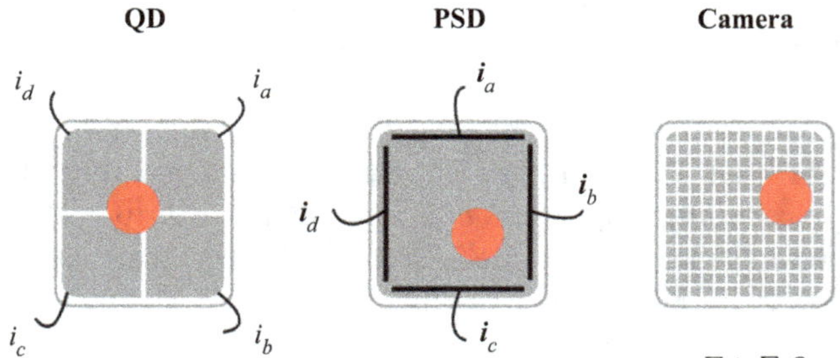

$$x_{QD} \propto \frac{(i_a + i_b) - (i_c + i_d)}{(i_a + i_b + i_c + i_d)} \qquad x_{PSD} \propto \frac{i_b - i_d}{i_b + i_d} \qquad x_{Cam} \propto \frac{\sum_i i \sum_j Q_{ij}}{\sum_{i,j} Q_{ij}}$$

Figure 1.2 Sensors for light beam position measurements

1.3.1 Quadrant photodiode

In many of the above-listed schemes for localising the beam centre a combination of four sensing elements, each taking up one of four quadrants, is used. Through a simple comparison between their readings, a 'centre of mass' measurement is obtained. Such devices are generally called QD. The use of discrete photo-diodes has the additional benefit that devices can be engineered with a centre hole to give a clear passage to the centre of the beam using collection of the fringes of the incident beam only [11]. By integrating the four diodes onto a single semiconductor wafer, the gap between each photodiode can be kept small so that it is possible to track small beams.

Such sensing methods provide, to a good approximation, linear response to displacements of the beam provided excursions of the beam are kept to a relatively small size as compared to the beam size itself (see Figure 1.2). Disadvantage of this type of device has a limited range: the moment one of the sensing elements is not exposed or the recorded signal becomes close to the noise level of the sensor, no meaningful recording of displacements can be made.

A relative comparison of QD and lateral effect PSD performance

Optical position-sensitive detectors are widely used in commercial and industrial applications where low-cost or high-speed position sensing is needed. For example, quad diodes are often used in atomic force microscopes to detect the position of the scanning tip above the sample. Lateral effect diodes are mostly used in optical distance meters and autofocus capable cameras.

Compared to PSDs, QDs have the disadvantage that the measurement range is limited by the beam size. PSDs have the benefit of a continuous sensitive surface which lies between the two (four) electrodes. For a QD beam size d not only determines the measurement range but is also a key parameter in the precision of the measurement σ that can be attained. For a QD the four segments of the noise contribution are not correlated and the noise sensitivity is the same for all noise sources [12]: $\sigma_{\mathrm{QD}} \approx \frac{\pi}{4}\frac{i_n\sqrt{B}}{P_r S}d$, where i_n is the root-mean-square value of the current noise density at the input of a single receiver channel, B is the noise equivalent bandwidth, P_r is the total signal power received and S is the responsivity of the photodiode quadrants. The precision for a PSD is approximated by $\sigma_{\mathrm{PSD}} \approx \frac{k_n}{\mathrm{SNR}}d$.

In the assumption of negligible background, the ratio between the achievable precision of the QD and PSD devices becomes: $\frac{\sigma_{\mathrm{QD}}}{\sigma_{\mathrm{PSD}}} \approx \frac{\pi}{4}\sqrt{\frac{R_i}{R_f}}$, where R_f is the feedback resistor in the current-to-voltage amplifier of the QD and R_i is the inter-electrode resistance of the PSD. With typical values of 10 MΩ for R_f and 10 kΩ for R_i one concludes that a QD offers a significant better precision.

1.3.2 Lateral effect photodiode

The term PSD was first used by Wallmark in a 1957 publication, describing the lateral photoelectric effect [13]. In the literature the term lateral effect sensor is also used when PSDs are discussed. Lateral effect PSDs are non-segmented or continuous photo-sensors (like photo-diode or photo-transistor) that can differentiate between the positions where incident light strikes the sensing surface. There are uniaxial sensors, which can only distinguish position along a single axis (x), and duo-lateral or tetra-lateral sensors, which can distinguish position along two axes (x, y). For the 2D devices there are two distinct ways of achieving measurements in X and Y. The first, called a duo-lateral PSD, uses two uniaxial sensors implemented on either side of a single silicon die with their axis rotated by 90°. The second, called a tetra-lateral PSD, uses four rather than two electrodes at the surface of the device to provide four individual measurements, from which the X- and Y-coordinate of the light spot can be inferred (see Figure 1.2). The functional principle is a differential measurement of the photo-diode surface resistance by comparing the photo-induced currents at a split anode and cathode in the duo-lateral or a fourfold split anode in the tetra-lateral detectors [14]. The use of PSD devices has been limited and can be attributed to the fact that these devices are nonlinear, and perhaps more importantly, offer a precision that is not as good as can be achieved by discrete photo-diodes [12].

1.3.3 Image (pixel array) sensors

Large arrays of multiple sensing elements (diodes) constitute a very appealing alternative to the traditional sensors listed earlier. So far their use has been limited because in many cases the SNR per sensing element (pixel) is not as high as that for large area sensing devices, such as quad diodes. If special techniques, such as on chip pixel binning (noise free addition of signals of neighbouring pixels) are used, these types of sensors become interesting. Ultimately, one could simply simulate a quad diode element by summing all pixel values in four quadrants through binning. Moreover, the multi-pixel sampling with flexible integration timing allows very high-resolution beam position determination analogous to a highly oversampled data acquisition system.

Such highly over-sampled measurements using image sensors give experimenters an opportunity to fit the dataset with a Gaussian intensity distribution. By doing so it has been shown that the position of the beam centre can be obtained with a sub-pixel resolution [15].

Image sensors for X-ray radiation are not as ubiquitous as those for visible light for the simple reason that the base material, Silicon, is rather transparent for X-rays and very few X-ray photons are captured in the shallow active area of such a photo-diode array. Good reviews of the available technology are published elsewhere [16,17]. The following briefly discusses the state-of-the-art technologies.

There are three distinct ways to overcome the problem of the transparency to hard X-rays of silicon. The first uses a special diode structure with a thicker depleted layer so that more X-ray can be captured. Recently, such monolithic active pixel arrays sensors (MAPS) using CMOS technology have been developed [18].

The second type of detector uses a photon energy conversion process in which X-ray photons produce photons in the visible range using a scintillator. This scintillator is either directly deposited on a traditional image sensor or its visible light output is coupled to the photo-diode array through a fibre optic bundle (fibre optic face plate), see Figure 1.3.

A suitable scintillator creates large number of visible photons for each X-ray photon at a wavelength for which the image sensing elements have the highest quantum efficiency. A simple procedure [19] allows researchers to glue fibre optics to commercial sensors after removal of the glass window that protects the sensor die. The key requirement is that the glue layer is uniform and as thin as possible. The use of a fibre-optic coupling to a CCD or CMOS image sensor is the most practical solution to guide the visible light to the sensor because it is often not possible to deposit the scintillator material directly on to the sensor die. In addition the fibre optic protects the semiconductor against radiation damage stemming from radiation that is not captured by the scintillator.

The third sensing method is a variation of the first method, in which a dedicated sensing layer consisting of an array of diodes only. These diodes have a depletion layer equivalent to the thickness of the wafer (often 0.3 to 5 mm thick), which is created by a high reverse bias voltage. The signal captured by the diode array is then processed by a second silicon die that is mounted using solder balls under the sensing layer using flip-chip mounting techniques [20]. Such a sensor is often referred to as a hybrid pixel detector (HPD). The benefits of the hybrid are two-fold. First, one can easily change the sensing layer material to suit the energy range of the X-ray photons. Hybrid sensors are not limited to silicon and can use different sensing materials, allowing for a higher detection efficiency at different X-ray energies. Another advantage of HPDs is that they are more resilient to radiation effects than MAPS, which is a necessity, specifically for experiments with large dynamic ranges where radiation damage to the detectors is likely. Different

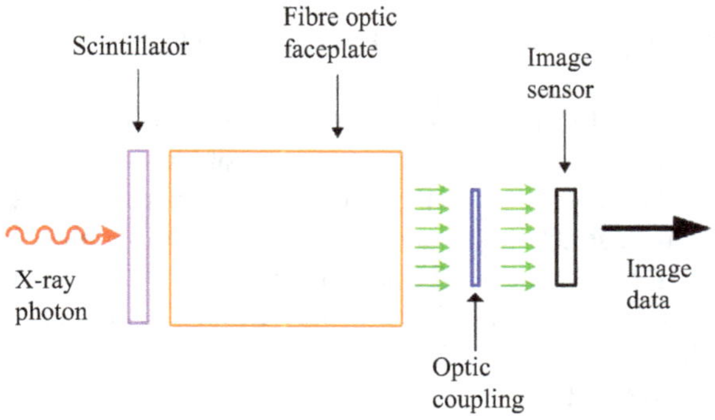

Figure 1.3 A fibre-optically coupled indirect X-ray image sensor. The image sensor can be selected from a wide range of commercial devices

materials have been tested for the sensing layer of HPDs, incrementing the range of energies that can be detected. For example, Germanium has been proven to be suitable to detect X-rays at 50 keV [21], while Gallium Arsenide has been tested for a range of up to 70 keV [22].

There are two types of HPDs: X-ray photon-integrating and X-ray photon counting. Photon-integrating HPDs integrate the total charge collected during a user-defined period. This method enables the handling of large instantaneous fluxes. The disadvantage of photon-integrating detectors is that it is difficult to discern between the dark current generated noise and photons, or between photons of different energies. An example of a photon-integrating HPD is the Cornell-SLAC pixel array detector (CSPAD) developed by the linear coherent light source (LCLS) [23].

Photon-counting HPDs count each photon individually. The signal is recorded as a photon only if it passes a threshold criterion. Due to the separation of sensing and processing circuitry, there is more chip space available and one can introduce signal thresholding, processing and dedicated logic that simply count charge pulses created in the sensing layer corresponding to absorption of a single X-ray photon of a particular X-ray energy. Because the number of charge carriers (holes or electrons) generated per absorbed X-ray is large (it only takes about 3.6 eV to create an electron-hole pair for silicon whereas an X-ray photon corresponds to 10 keV or more), this process of charge pulse counting is not affected by electronic noise so that every absorbed X-ray photon is registered. The threshold comparison function grants the ability to discern between different photon energies and the dark current generated noise, allowing a practically zero-noise performance. A disadvantage of the photon counting HPDs is that they cannot handle large fluxes of X-ray photons. In such cases the charge packets from individual X-ray photons overlap and cannot be distinguished by the thresholding logic.

Several direct X-ray detector application-specific integrated circuits (ASICs) have been developed. Some examples of these ASICs are: Pilatus [24], XPAD [25], ChromAIX [26] and MEDIPIX [27]. These circuits have similar characteristics, with pixel pitches ranging from 55 to 175 μm. The Medipix ASIC offers the smallest pixel size of 55 μm per side. The latest version of the MEDIPIX ASIC is the 3RX hybrid X-ray sensor developed at CERN and funded by a large international consortium.

Figure 1.4 shows the cross-section of a typical HPD. The sensor layer is interconnected to the readout ASIC by using a solder process. When an X-ray photon impinges the sensor material, its energy is transferred into electron-hole pairs via ionisation. Due to the influence of an externally applied electric field, a depletion region is generated on the sensor layer. The electric field forces the electrons and holes to drift to opposite directions of the sensor material. The drift of electrons and holes under the electric field induces an electric current on the charge sensing nodes bump-bonded to the ASIC. The generated current is then processed by the ASIC electronics.

A typical detection channel of a photon-counting HPD is shown in Figure 1.5 (modified from [28]). The signal induced in the sensor layer is read out and

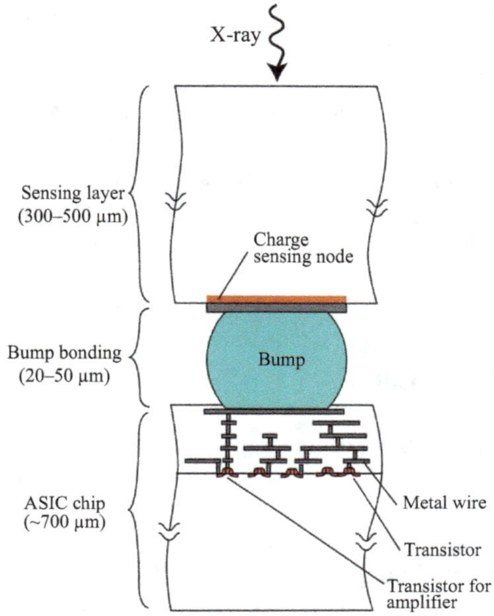

Figure 1.4 Cross-sectional view of one pixel of a hybrid pixel detector. The sensing layer and the ASIC chip are connected by a bump. The signal generated on the sensing layer flows to the ASIC where it is processed. Figure modified from [17]

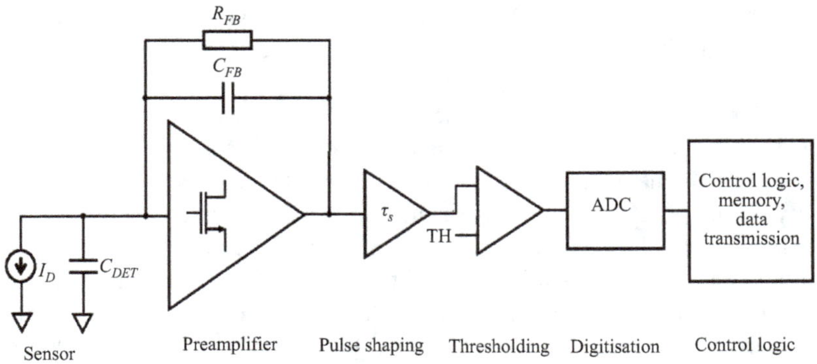

Figure 1.5 One pixel showing circuitry for a typical hybrid pixel detector

amplified using a charge sensitive amplifier. A band-pass filter is used to improve the SNR. Because this filter changes the time shape of the signal, it is commonly referred to as a 'Shaper'. The filtered output is compared against a threshold value in a comparator, the threshold value is set by the user. The output of the comparator is a pulse that is passed to the digital processing section of the pixel circuitry.

HPDs can have many functionalities embedded in ASIC, from multiple comparators to smart reconstructions of the signal, but this comes at a cost of a more complex electronic design.

An electronic design with more features means more space to implement these circuits is required, increasing the minimum size of each pixel. More features also mean a higher power consumption. An increase in power consumption is already limited by the use of long metal wires, with a significant resistance, connected to the periphery of the chip. This comes as a trade-off of functionalities versus power consumption and pixel size.

1.4 X-ray beam imaging

A key requirement for comprehensive diagnostics of the incident beam is to record more than just the centre of the intensity distribution. Especially in experiments where this beam is focused, it is important to know its size, or better still, obtain a full intensity distribution across the beam. Ideally, a device would consist of a high-density pixel array created on a thin membrane of low-density material. Attempts have been made to create a multi-pixel CVD diamond diode array but unfortunately with a rather small number of pixels. Quad diodes based on this idea have been successful and are available commercially [7,8].

In Manchester, research on developing X-ray beam imaging based on the elastic scattering of X-rays by foils was conducted. Unlike with recording fluorescent light there is no deterioration or 'burn in' of the foils. Key demands of such a system are that the devices would work reliably for long periods and allow the incident beam to pass freely. The first demand has proven to be hard with most silicon-based devices suffering from increased dark currents or failure after relatively short periods of operation.

To address these demands and measure a true cross-sectional intensity measurement, a device geometry was chosen, in which the sensing element is placed well away from the incident X-ray beam. This choice was inspired by beam intensity monitors at the former SRS synchrotron (Daresbury, UK), which use an inclined polymer foil placed in the X-ray beam. A measurement from the elastically scattered radiation from this foil was used to normalise measurements because the photon flux from the source is not constant over time. In this case, the scattered radiation from the inclined foil is intercepted by a set of parallel thin foils from a highly X-ray absorbing (high-density) material such as Tungsten or Molybdenum placed at a distance from each other, which corresponds the dimension of the individual photodiodes in the array, which is used to collect the transmitted X-rays. The foil stack is a traditional method of collecting only those X-rays from a scattering or diffraction experiment, which have a direction of travel that is parallel to the foils, and in the literature such a device is often called a Soller slit [29] (see Figure 1.6).

By recording the distribution of photo-diode current readings a histogramme-like intensity distribution that represents the footprint of the beam on the scatter foil

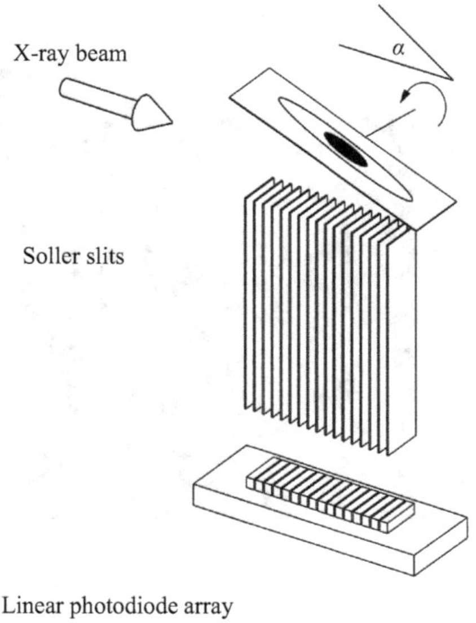

Figure 1.6 1D X-ray beam profiling detector. Radiation scattered from a foil is collimated by a Soller slit and recorded by a linear photodiode array

is obtained. Fitting this distribution with an (assumed) Gaussian beam shape gave surprisingly precise beam positions that could be refreshed at sufficiently high rates to determine beam vibrations.

In a 2007 paper [30], a 2D Soller slit in combination with a commercial CMOS sensor was introduced. The CMOS sensor was fibre-optically coupled to a Terbium doped Gadolinium oxide sulphide scintillator (Gd_2O_2S:Tb or Gadox), see Figure 1.7.

The obtained images allow researchers to measure the beam intensity distribution but with limited spatial resolution. In the same paper, the removal of the Soller slit was investigated in order to see whether the much higher signal levels would allow the determination of beam position and size using the fact that the scattered radiation intensity will obey an inverse square dependency on path length (see Figure 1.8). The integrated intensity collected by an individual sensor pixel at a large scattering angle (around 90° in this case) is approximated by

$$I_{\text{int}} = \frac{\Delta\Omega}{4\pi}\Delta t\frac{I_0 r_e^2}{R^2}\sum_{a=C,H,N,O} N_a\left[Z_a^2 f_a^2 + Z_a\left(1 - f_a^2\right)\right]. \tag{1.1}$$

The setup without collimator provides a straightforward measurement of the incident X-ray flux by plotting the intensity of each pixel against the inverse square of its distance to the centre of the footprint of the incident beam on the scatter foil. The total flux of the incoming X-ray beam is derived from the gradient of the best-fit line to the data in Figure 1.8, using (1.1). The CMOS sensor has a conversion

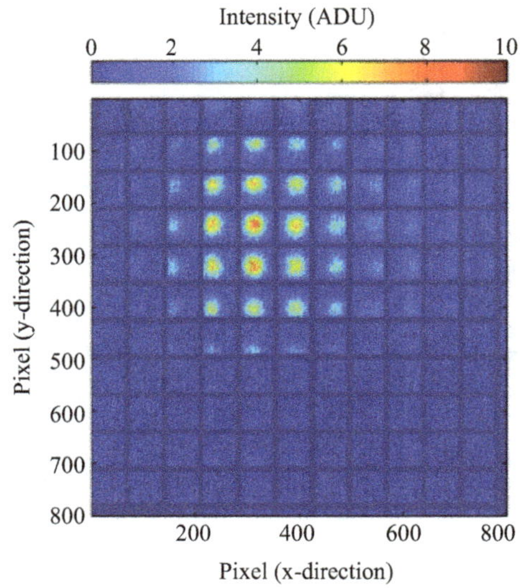

Figure 1.7 Recording of X-ray scattered from a polyimide foil after passing through a 2D collimator

constant of 0.83 ADU per incident X-ray photon at an energy of 12.7 keV. The data presented in Figure 1.8 also give a measure of the beam size. If the X-ray beam had infinitesimal size the scatter foil would behave as a point source and all the data points would fall on the best-fit line. For the width of this distribution the averaged size of the footprint of the incident beam is easily determined.

More recently, a dedicated pinhole imaging setup was introduced [10]. With hindsight such a setup represents an obvious choice but one should remember that the scattering yield is very low, typically seven orders of magnitude below that of the incident X-ray beam. Hence, such pinhole camera recordings were initially deemed to be unfeasible because the resultant signal levels per pixel are rather low (often less than an X-ray photon per pixel on average). For a suitable polymer such as polyimide (Kapton^{TM}) the scatter efficiency of a 25 μm thick foil as a function of X-ray energy is shown in Figure 1.9.

Dedicated signal and image processing and the use of coded apertures (rather than simple pinholes) have proven to give high-resolution beam images, from which centre-of-mass measurements are obtained with sub-pixel size precision at suitable frequency. For example, careful background image correction using the average of a large number of images taken without X-ray beam has proven to be essential when using CCD or CMOS image sensors. Even at very short exposures, where most sensor pixels will not have recorded an X-ray event, one can still collect two 1D-profiles by adding sensor pixel values row- and column-wise, due to the large number of pixels. Typically, these profiles feature a good SNRs from which the beam position is determined with high precision.

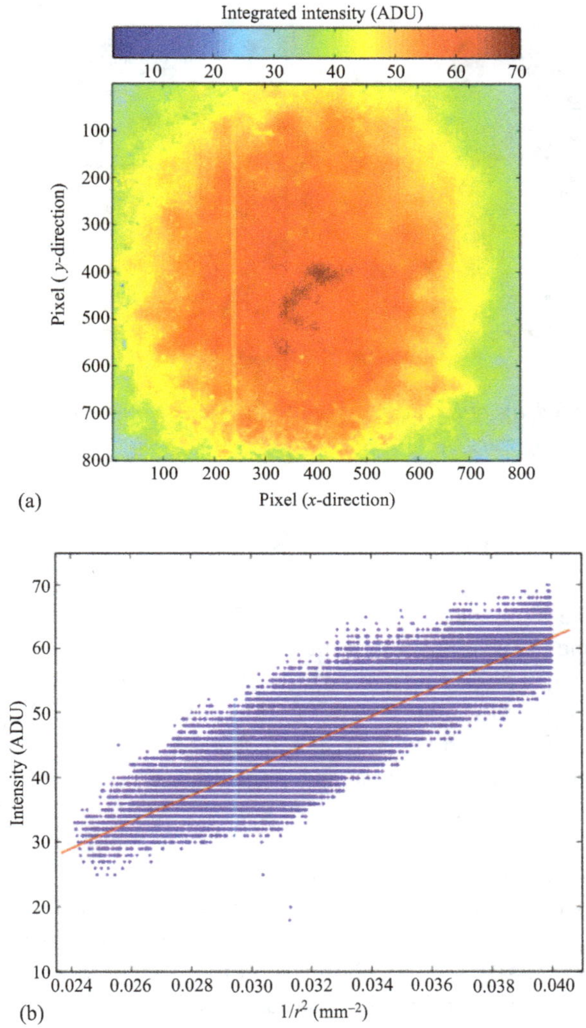

Figure 1.8 Recording of scattered X-rays without a collimator (top). The intensity distribution is due to the $1/r^2$ fall off as shown in the plot (bottom) in which for each pixel its intensity in arbitrary units (ADU) is plotted against the inverse square distance to the centre of the scatter foil

One of the great benefits of the pinhole setup is that beam movements and beam size can be magnified at the image sensor simply by adjusting relative distances between scatter foil, pinhole and sensor. This feature is particularly helpful to tailor the measurement range and the precision, with which the beam has to be recorded. A detailed theoretical model for the system was developed to predict its performance as a function of beam intensity, detector pixel size and scatter foil properties.

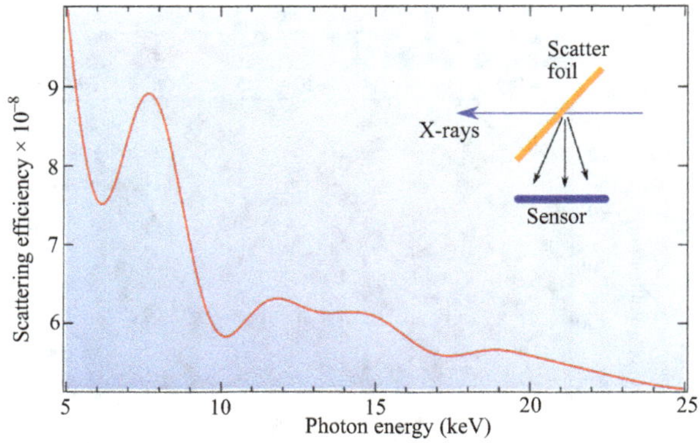

Figure 1.9 Scattering efficiency of X-rays from a 25 μm polyimide foil versus photon energy

1.5 Pinhole/coded aperture X-ray camera

A diagram of the X-ray beam imaging setup [31] is shown in Figure 1.10. A negligible part of the incident beam is scattered by a thin amorphous sheet of a low-Z material in all directions. Using a pinhole an enlarged image of the footprint of the incident beam is recorded by a camera which is placed underneath the foil.

The centre of the recorded image is a representative measure of the beam position along the horizontal (x-axis) and vertical (z-axis) directions. If we assume that the sensor is set up so that its sides are parallel to the x- and y-axes, the respective coordinates of the image centre u_c and v_c can be inferred from two image profiles obtained by summing image rows and columns, respectively. Reducing 2D image data to two 1D profiles improves the measured SNR and significantly simplifies and speeds up subsequent signal processing. Because most beam profiles closely match a Gaussian shape each image profile is fitted to a Gaussian function whose centre gives the corresponding coordinate of the image position. After the image position is determined, it can be related to the actual beam position if the geometry of the device is known.

In a typical application, the user is interested in keeping the beam in a predefined position. By detecting any deviation from that position, compensating control signals can be generated and used to drive appropriate actuators to return the beam to the desired position. Therefore, the linearity and precision of beam displacement measurements that can be achieved are of primary interest. Below the main results are presented that are published elsewhere [10,33].

1.5.1 Linearity

To investigate the expected behaviour in terms of linearity an X-ray beam with width w and height h is propagated along the y-axis of the Cartesian coordinate system whose origin is at the centre of the aperture (see Figure 1.10) [33]. The

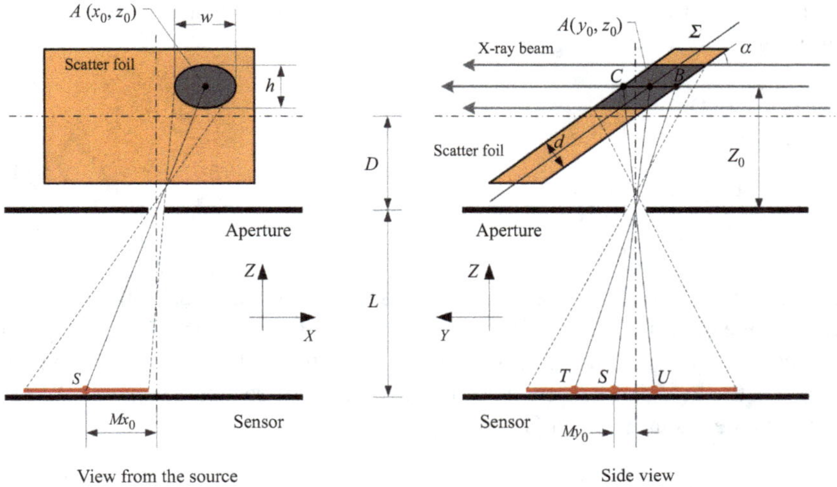

Figure 1.10 Beam view and side view of the X-ray beam imager (not to scale) showing key parameters of the setup used in the proposed model [32]

beam passes through the foil of thickness d so that the centre of its footprint on the front face of the foil has coordinates (x_0, y_0, z_0). Due to the foil tilt, the y- and z- coordinates of the footprint are linearly coupled: $z = -y \tan \alpha + D$, where α is the foil tilt angle and D is the distance between the foil and the aperture. The y-coordinate can be omitted when describing beam position, for example the beam centre is located at point (x_0, z_0). The radiation scattered from the foil passes through the aperture and creates an image in the sensor plane at distance L below the aperture plane. It is assumed that the aperture can have any shape that exhibits symmetry relative to the x- and y-axes (e.g. circular, square and rectangular).

The geometrical centre of the image of the footprint as recorded by the sensor, that is the coordinates of the centre of its orthogonal projections to the x- and y-axes (u_c and v_c, respectively), is determined by

$$\begin{aligned} u_c &= -\frac{L}{z_0}x_0 \\ v_c &= -\frac{L}{\tan \alpha}\left(\frac{D + \dfrac{d}{2\cos\alpha}}{z_0} - 1\right) \end{aligned} \qquad (1.2)$$

Let's assume that the beam is kept near the centre of the foil so that z_0 is close to $D + d/(2\cos\alpha)$. Due to the geometry of the actual device, D is at least a factor of 10 larger than h, d and b (e.g. $D \approx 5$ mm, $h \leq 0.5$ mm, $b \leq 0.5$ mm and $d \leq 0.125$ mm).

Small beam movements ($x_0 + \Delta x$, $z_0 + \Delta z$) are readily measured from shifts in position of the recorded image (Δu, Δv) using (1.2):

$$\Delta x = -\frac{D + \dfrac{d}{2\cos\alpha}}{L}\Delta u$$
$$\Delta z = \frac{\tan\alpha\left(D + \dfrac{d}{2\cos\alpha}\right)}{L}\Delta v \quad (1.3)$$

Therefore, the device exhibits a linear response to small beam position fluctuations. Note that since D is usually smaller than L that changes in Δx and Δz are magnified by a factor M. Any movement in the vertical direction achieves an extra magnification due to the foil tilt. For an infinitesimally thin scatter foil ($d \approx 0$), (1.3) reproduces the result given in [10].

1.5.2 Precision

The precision of the system is determined by the smallest beam position fluctuation (i.e. beam displacement or movement) that can reliably be detected by the instrument. The ability of the instrument to measure small beam movements is limited by errors in beam position measurements introduced due to fundamental photon shot noise and by the instrument itself. The precision is therefore taken to be equal to the standard deviation of beam displacement measurements when all beam parameters, such as position, shape, size and intensity, are kept constant.

Let σ_u and σ_v be the statistical errors of measuring the image x- and y-coordinates u_c and v_c, respectively, and assume that σ_u and σ_v do not depend on u_c and v_c. Then, by applying the uncertainty propagation rules to (1.3), the errors in beam displacement measurements along the horizontal, $\sigma_{\Delta x}$, and vertical, $\sigma_{\Delta z}$, directions are expressed via σ_u and σ_v:

$$\sigma_{\Delta x} = \sqrt{2}\frac{D + \dfrac{d}{2\cos\alpha}}{L}\sigma_u$$
$$\sigma_{\Delta z} = \sqrt{2}\tan\alpha\frac{D + \dfrac{d}{2\cos\alpha}}{L}\sigma_v \quad (1.4)$$

where a$\sqrt{2}$ multiplier accounts for the error increase due to coordinates subtraction when image shift (Δu, Δv) is calculated.

For the given SNR of the image profiles data and the size of sensor, pixel and beam, respectively, the quantities σ_u and σ_v can be estimated using a method outlined in [34], which considers χ^2 statistics of the difference between the model instrument response and the observed data. The standard deviation of coordinate measurements is estimated by

$$\sigma_v = \Gamma_v\left(\frac{\Delta_{\chi^2}}{A_v^2 N_{\Gamma_v} t F_v(t)}\right)^{\frac{1}{2}}, \quad (1.5)$$

where Δ_{χ^2} is the maximum variation of χ^2 for a given confidence level and in this particular case (four parameter fit, '1σ' confidence level) is equal to a numeric value of 4.7, N_{Γ_v} is the number of pixels per length Γ_v and $tF_v(t)$ is an integral term, which includes the noise in the measured signal, the image profile parameters and sensor size and sampling rate:

$$tF_v(t) = \int_{-t}^{t} \frac{\xi^2 e^{-\xi^2}}{M\left(B + \sigma_r^2\right) + A_v e^{-\frac{\xi^2}{2}}} d\xi \tag{1.6}$$

where $t = N\delta_p/(2\Gamma_v)$ is the parameter describing which fraction of the image profile is measured by the sensor.

From inspection of (1.5) and (1.6), it is clear that for the highest precision (i.e. lowest value of σ_v) one should choose the magnification such that the beam image entirely fits in the active area of the sensor. The integrand's denominator in (1.6) describes the contribution of sensor shot and read noise. Equations similar to (1.4)–(1.6) are readily obtained for σ_u by evaluating the measured beam profile along the x-axis.

For a 2D Gaussian, the amplitude A_v, as presented in (1.6), is a function of the integrated intensity I_{int}:

$$A_v = \frac{I_{\text{int}}}{\sqrt{2\pi} N_{\Gamma_v}} \tag{1.7}$$

I_{int} is a function of the flux and energy of the X-ray beam, detector response, the material and thickness of the scatter foil, and the solid angle subtended by the aperture of the pinhole camera. For the error in image coordinate measurements we obtain:

$$\sigma_v = \frac{\Gamma_v}{I_{int}} \left(\frac{2\pi \Delta_{\chi^2} N_{\Gamma_v}}{tF_v(t)} \right)^{\frac{1}{2}} \tag{1.8}$$

The resolution of beam movement measurements $\sigma_{\Delta z}$ is determined by the combined effect of (1.8) and the terms preceding σ_v in the right-hand part of the corresponding equation in (1.4). These terms, which are collectively referred to as a precision factor, are a function of the pinhole camera magnification. Figure 1.11 shows how σ_v and the corresponding resolution factor change with respect to the pinhole camera magnification. Point-by-point multiplication of these curves gives the resolution of beam displacement measurements as shown in the inset of Figure 1.11. This precision curve has a distinct minimum that marks the optimal pinhole camera magnification for a given set of instrument and beam parameters. The magnification, at which the minimum occurs, mostly depends on the slope of the image position precision curve, and hence largely determined by the integrated intensity of the scattered radiation registered by the detector, I_{int} [see (1.8)].

When I_{int} increases, the minimum moves towards higher magnification values up to the certain value, for which the beam image size becomes equal or larger than the detector. In this case, the gain in precision due to higher magnification is

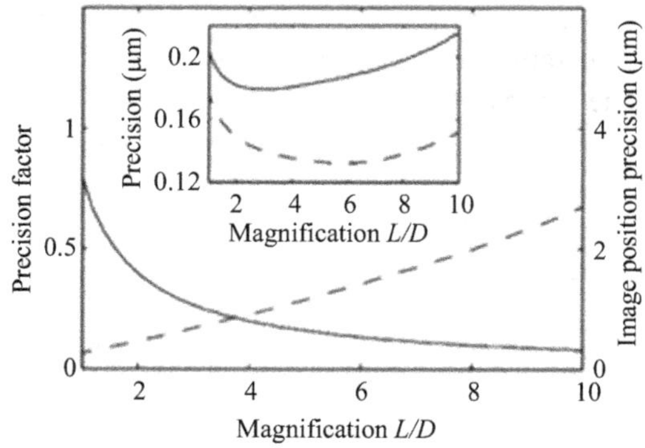

Figure 1.11 Precision factor (solid line) and precision of image position measurements σ_v (dashed line) as a function of camera magnification. Their combined effect determines the precision of beam displacement measurements (solid line in the inset). Parameters: $\alpha = 29$, $I_{int} = 10^7$, $D = 5$ mm, $d = 125$ μm, 200×200 μm^2 beam size, pinhole diameter 500 μm. The dashed line in the inset shows the use of a 'noiseless' detector [33]

negated by decreasing values of t in the integral term $tF_v(t)$. On the other hand, the lower I_{int}, the lower is the optimal magnification, so that for a particularly weak signal the device has to be operated at the lowest attainable, and often suboptimal, magnification. The optimal magnification can be different for X- and Y- image directions due to the effect of the foil tilt and thickness and differing detector and pixel sizes along these directions.

The inverse proportional relation between the precision and integrated intensity advocates the use of special variants of coded apertures which overcomes the limitation of the integrated flux available through a pinhole. For very high beam position measurements a coded aperture in the form of a '+' or crossed slit shape is very attractive. When the cross-shaped aperture is carefully aligned with the rows and columns of the image sensor, such a coded aperture significantly increases the recorded amplitude A_u (and integrated intensity I_{int}), without creating an enlarged image (i.e. $\Gamma_v =$ constant) and the usual complication of having to de-convolve the recorded image in order to find a beam position.

1.5.3 Beam imaging

A detailed imaging model for the above X-ray beam imaging system was described by Kachatkou *et al.* [32]. An image $i(x, y)$ created by a practical system is obtained by convolution of image $s(x, y)$ with its impulse response $h(x, y)$:

$$i(x,y) = s(x,y) * h(x,y). \tag{1.9}$$

This impulse response is a function of three factors: aperture size and shape, foil thickness and detector point spread function (PSF). In the geometry with the inclined foil in combination with a pinhole of arbitrary shape (slit, crossed slits and pinhole) the impulse response function is not trivial. To calculate the foil contribution to the impulse response, an infinitesimally thin beam is assumed that propagates along the y-axis and the use of a minuscule pinhole. The centre of the beam has coordinates (x_0, z_0) and the intensity distribution across its cross-section at the XZ-plane is given by the Dirac delta function $\delta(x - x_0, z - z_0)$. The corresponding intensity distribution is given by

$$i_f(x,y) \propto \delta\left(\frac{x}{M} + x_0\right)\exp\left[-\mu\left(-\frac{y}{M} - \left(y_0 - \frac{d}{2\sin\alpha}\right)\right)\right] \tag{1.10}$$

where $y \in \left[-M\left(y_0 + \frac{d}{2\sin\alpha}\right), -M\left(y_0 - \frac{d}{2\sin\alpha}\right)\right]$ and μ is the linear absorption coefficient of the scattering foil material.

The foil impulse response, $h_f(x, y)$, which describes image blur caused by the finite thickness of the foil, is given by (1.10) for $x_0 = 0$ and $y_0 = 0$. The absolute value of $i_f(x, y)$ and $h_f(x, y)$ is determined by the dependence of the scattered intensity on the direction of scattering. Assuming that the distance between the foil and the pinhole is significantly larger than the thickness of the foil ($D >> d$), this dependence is negligible and the proportionality sign in (1.10) can be replaced with the equality sign if the right-hand side is multiplied by a constant intensity factor. Because this work is focused on the imaging performance of the system and not on performing absolute intensity measurements, the intensity factor is taken to be equal to one in the subsequent discussion. The combined contribution of the foil and an arbitrary shaped aperture defined by $a(x, y)$ into the impulse response is given by

$$\begin{aligned} h_{fa}(x,y) &= h_f(x,y) * a\left(\frac{x}{M+1}, \frac{y}{M+1}\right) \\ &= \int_{-M\frac{d}{2\sin\alpha}}^{M\frac{d}{2\sin\alpha}} \exp\left[-\mu\left(-\frac{v}{M} + \frac{d}{2\sin\alpha}\right)\right] a\left(\frac{x}{M+1}, \frac{y-v}{M+1}\right) dv \end{aligned} \tag{1.11}$$

The impulse response $h(x, y)$ is obtained by convolution of $h_{fa}(x, y)$ and the detector's PSF. Note that the magnification factor M in (1.11) depends on z_0 and, consequently, on y_0. Therefore, $h(x, y)$ varies for different locations along the image Y-direction. However, in practice M does not vary much (small beam movements compared to D) and hence can be deemed to be constant: $M = L/(D + d/(2\cos\alpha)) \approx L/D$. For this approximation, the impulse response is spatially invariant. The exponential term in (1.11), which accounts for the foil thickness, results in the impulse response being elongated along the direction of the foil tilt.

To reconstruct the cross-sectional image of the beam from the recorded image requires knowledge of the instrument impulse response, which is determined by the device and can be calculated from the known device geometry and foil thickness. The PSF of the detector can be measured separately [19]. The image of the beam is obtained through the use of well-established deconvolution techniques, such as the

Lucy-Richardson algorithm [35]. When images are reconstructed one also needs to factor in the effect of detector noise. The detector noise primarily consists of two components: photon-counting noise and readout noise. The photon-counting noise has a Poisson distribution and is implicitly accounted for in the derivation of the Lucy-Richardson iteration. The readout noise is typically described as additive noise with a Gaussian distribution. As shown in [35], the Lucy-Richardson algorithm can be modified to accommodate this type of noise. Other types of noise, such as impulse noise, bad pixels and background signal (pixel dark current, X-ray air scattering, etc.), can also be accounted for during Lucy-Richardson iteration. However, it is found that it is sometimes more convenient and efficient to correct for these distortions during the pre-processing step using pixel interpolation and dedicated filters such as median and threshold filters.

1.6 System implementation and description

Throughout the development of indirect CMOS, CCD and HPD detectors for the X-ray beam imaging instruments use was made of an in-house designed heterogeneous processing platform consisting of traditional sequential, graphics processor and parallel processing hardware with an optimised memory layout dedicated for signal acquisition and processing (ARMflash). The combination of an ARM-based embedded processor (Texas Instruments, AM335X) with a memory-mapped field programmable gate array (FPGA, Cyclone V) allows for a flexible sensor interface so that data can be collected from a wide range of image sensors (see Figure 1.12).

Image processing tasks, such as histogramming, 1D profiling and background correction, are fully implemented in the FPGA, which due to its massive parallel operation capability can perform all operations at the sensor frame rate. Extensive use of the Avalon video streaming protocol is made to have a modular structure, in which it is easy to make changes and introduce new functionalities. The FPGA is equipped with its own high bandwidth memory bank to allow for sufficient local storage to complete a series of image processing tasks and storing of subsequent frames.

The ARM processor runs an embedded version of the Linux operation system, which provides networking capability through a TCP/IP stack using a 1 Gb/s link and runs server code to interface to hosts. The processor can also implement various house-keeping functions, such as firmware upgrade, sensor temperature control and others. Programming of the FPGA is done under processor control so updating of its functionality is easy with firmware upgrades. Access to image data from the FPGA is available through a dedicated parallel data bus, which allows transfer speeds of 140 MBytes/s under processor direct memory access control. Control over the various functions of the FPGA is performed through a bank of registers. In a similar way the processor can obtain status information from the FPGA.

A commercial version of the system based on an indirect CMOS image sensor is available from FMB Oxford Ltd. The NanoBPM device relies on a standard 1 Mpixel sensor from ON Semiconductor which is fibre-optically coupled to a

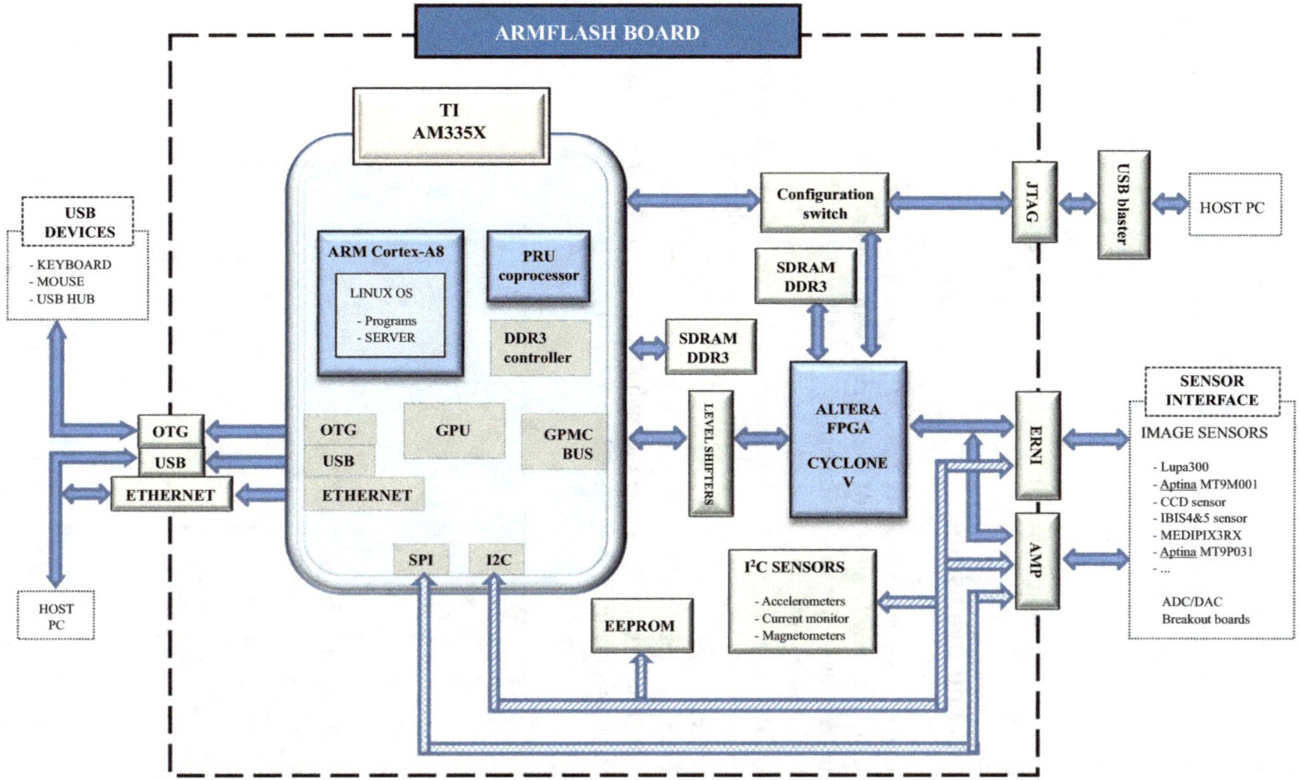

Figure 1.12 Heterogeneous processing platform used for sensor control, image processing and communication with host systems. The platform combines three different types of data processing engines

Gadox scintillator which is deposited directly on the fibre faceplate that is composed of individual fibres with a diameter of 6 μm. The CMOS imager has square pixels of 7 μm size [36].

Since the development of the commercial version a switch was made to direct imaging sensors and a new system was created (LANCELOT) based on the Medipix3RX ASIC. Although the pixel size of this imager (55 μm) is significantly larger the fact that this sensor does not suffer from noise and background; the achievable SNR is significantly higher. Together with the higher frame rate this system is ideally suited to implement feedback control so that beam position changes are removed actively. Two different systems were built: one portable and one that is suited for ultra-high vacuum environments [37,38].

1.7 Results and discussion

The instrument has proven to be very reliable and accurate. The NanoBPM has been available commercially since 2014 and devices are used at most synchrotron radiation sources around the world. Experience has shown that these devices are mainly used for observing (and correcting) beam position and beam size (shape). As such they operate continuously and autonomously whilst providing operational and measurements to synchrotron beam line operators for monitoring and trouble-shooting purposes.

Here a couple of results obtained with the proposed type of X-ray beam imaging devices are presented. To validate the pinhole camera model, the precision of the system at various magnifications (see Figure 1.13) was measured.

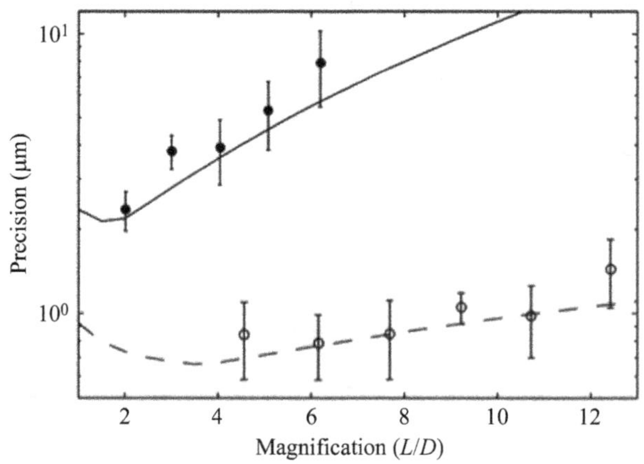

Figure 1.13 Precision measurements in vertical (open circles) and horizontal (closed circles) directions measured at different magnification settings of the instrument. The model predictions for the instrument taking into account the flux as produced by a bending magnet beamline (DLS, B16)

Changing magnification was achieved by varying the sensor distance to the aperture plate whilst keeping the aperture scatter foil distance constant. The experiment was conducted at the Diamond Light Source (DLS) facility using bending magnet beamline B16, which has a reduced photon flux compared to regular insertion device beamlines. The resultant measurements of the obtained precision for horizontal and vertical displacement is shown in Figure 1.14. The measured values for the precision of beam displacements fit the model well with the given device geometry, sensor specifications and available photon flux.

Figure 1.14 shows an example of a several hour-long recording of the vertical beam position measured at beamline I15 at DLS. The precision of the measurements was determined to be 0.21 μm. It is clear that during the recording various large source movements were seen on top of smaller, more regular displacements due to vibrations.

The LANCELOT system developed for DLS can record about 680 beam profiles per second. With this capability it is straightforward to analyse characteristic frequencies of the vibrations in the two orthogonal directions and an example is shown of the measured spectrum calculated from data recorded at beamline I19 at DLS [37] (see Figure 1.15).

The unique feature of the instrument is that it provides detailed images from the impinging beam during experiments due to its transparent or non-destructive nature. It allows researchers to conduct experiments knowing where the beam is located and what its shape is. Any issues can thus be detected at an early stage and corrected either manually or using automatic feedback.

Figure 1.16 shows a beam image that was taken with a 100 μm diameter pinhole and compared to the image obtained by placing a Gafchromix™ film in the

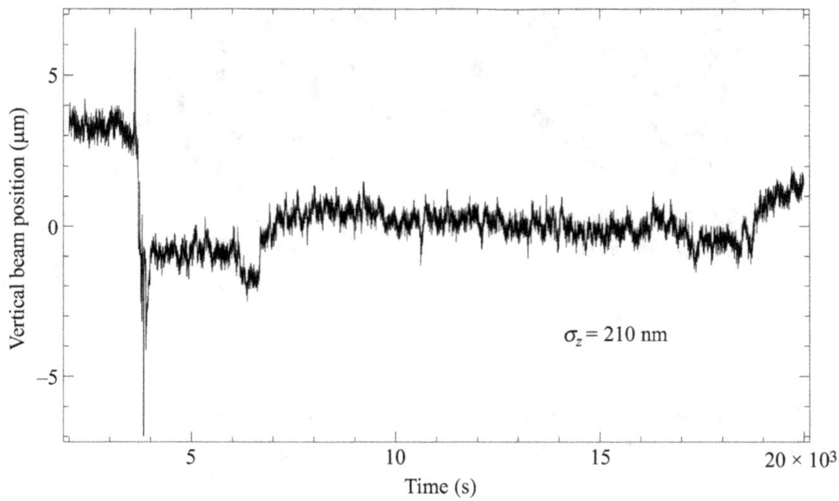

Figure 1.14 Time scan of vertical component of X-ray beam position. The precision of the measurement is 210 nm

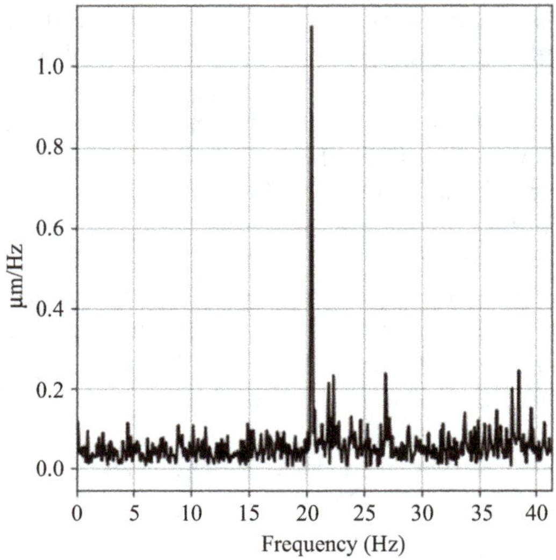

Figure 1.15 Fast Fourier spectrum calculated from vertical beam displacements measured at beamline I19 showing several prominent peaks (e.g. 21 Hz) which are probably associated with vacuum pumps. Beamline staff had previously identified similar resonant frequencies

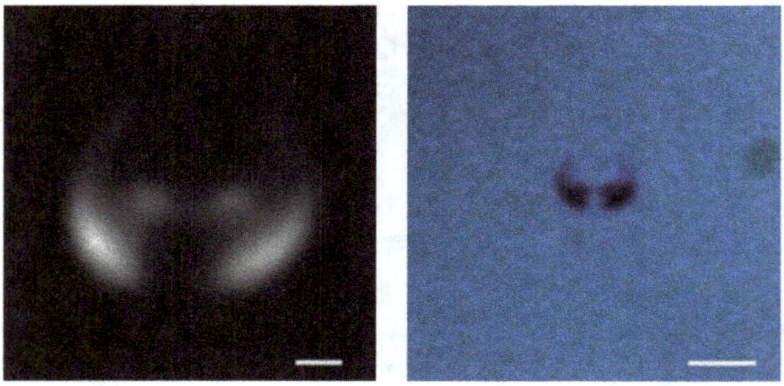

Figure 1.16 Focused beam image captured with the instrument (left) at DLS beamline B16. The scale bar in the transparently imaged beam on the left is 200 µm long whereas the scale bar in the Gafchromix^{TM} image is 1 mm long

X-ray beam at the same position for a direct comparison with the image obtained from the instrument through deconvolution. The image provided by the pinhole camera gives beamline users a very detailed view of issues, in this case with a toroidal mirror deformations. Figure 1.17 shows that even in the case of a coded aperture it is possible to recover the profile of the X-ray beam using the previously outlined deconvolution procedure. As discussed earlier it is beneficial to make use

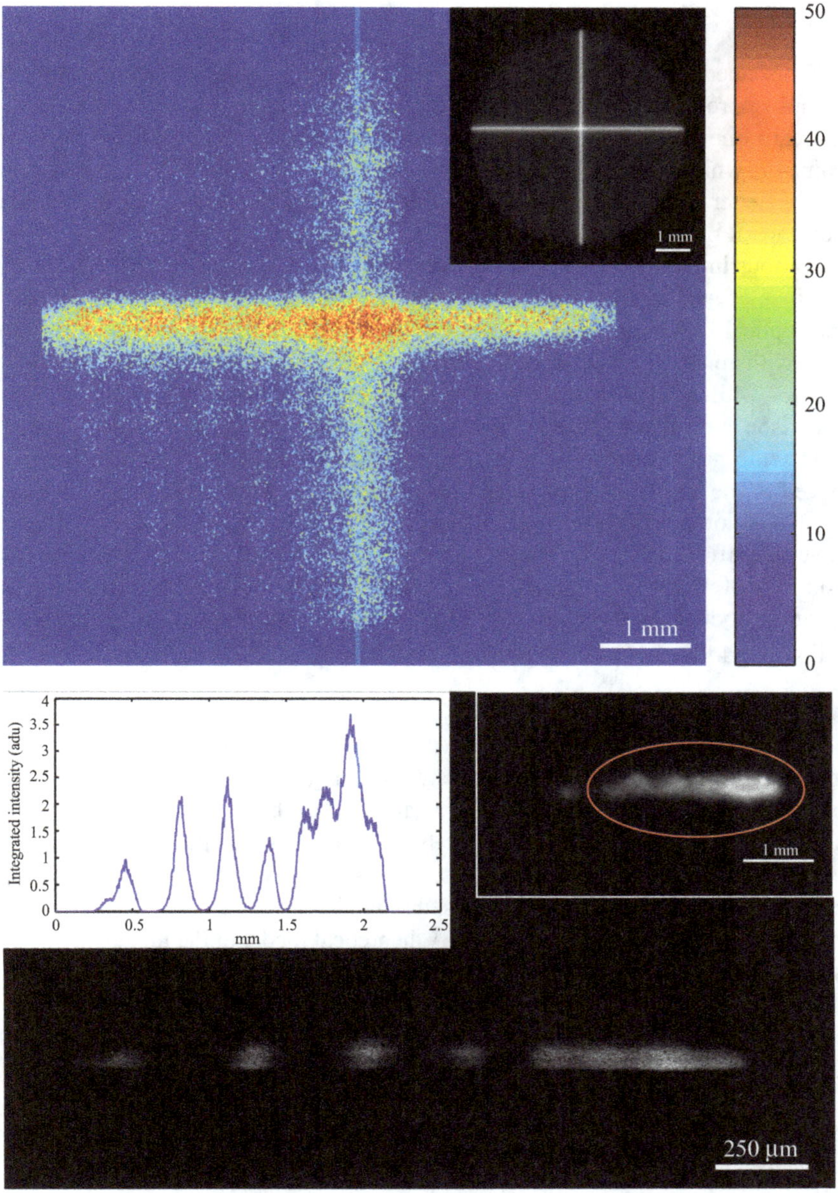

Figure 1.17 *Images of a focused beam at beamline BM26A taken using a cross-shaped aperture (3000 × 25 μm² slits cut in a 13 μm thick stainless steel foil) placed on a 0.2 mm-thick Molybdenum foil with a hole of about 2 mm in diameter. Top: the image obtained by averaging 10 raw images corrected by background subtraction and the calculated instrument impulse response (inset). Bottom: the reconstructed cross-sectional image with the corresponding horizontal image profile (left inset) and the image obtained by exposing the X-ray camera directly to the beam (right inset). Instrument parameters: L/D = 2, 50 μm scatter foil, α = 18.7, CMOS detector with 7 μm pixels fibre-optically coupled with a Gd_2O_2S:Tb scintillator, integration time 0.25 s [32]*

of cross-shaped apertures so that the measured image has a significantly better SNR and thus improved precision in beam position measurements.

Beam line BM26A [39] was equipped with a ribbed sagitally focusing monochromator, which was not correctly bent. The resultant image is very detailed and actually better than achieved with another camera that was placed in the beam absorbing all photons.

Various implementations of the patented technology are in operation at various beamlines at synchrotron radiation sources and often provide active feedback to X-ray optical components to provide stable beams for use in demanding experiments with micro-focused beams. The use of noise-less hybrid photon-counting detectors with small pixel size has been of real benefit to this application. In comparison with imaging detectors that are directly placed in the beam the presented setup has two large benefits. In the proposed geometry, the detector is not exposed to the high fluxes that could lead to damage in the medium to long term. From a position precision and spatial resolution point of view the setup allows for a simple amplification of the image which provides images with high spatial resolution. This feature is particularly important because modern sources and optics regularly produce beams with cross-sectional dimensions that are significantly smaller than pixel sizes of available image sensors.

1.8 Summary

A non-destructive method of continuously recording the position and intensity distribution of high-intensity synchrotron radiation X-ray beams is described in detail. The method is based on collecting diffusely scattered radiation from a thin membrane using a pinhole camera consisting of state-of-the-art hybrid pixel X-ray detector. The use of specially coded apertures provides, through iterative deconvolution procedures, very detailed images of the X-ray beam. A theoretical model of the imaging and beam position localisation is presented that provides a way to calculate the device performance based on simple geometric parameters of the system and the available photon flux. The model predictions were verified by experiments. The architecture of the embedded processing platform based on heterogeneous computation that is optimised for real time, high frame rate image processing tasks is outlined. Besides sensor control, image processing and network communication tasks the system also provides four independent feedback control signals that can be used to stabilise beam position and intensity. Some key results obtained at two synchrotrons are presented which show that the proposed instrument is capable of achieving a sub-micrometre precision although sensor pixel size is significantly larger. The system is in operation at several synchrotrons around the world.

References

[1] Revesz P, White J A, 'An X-ray beam position monitor based on the photoluminescence of helium gas', *Nucl. Instrum. Methods Phys. Res. Sect. Accel. Spectrometers Detect. Assoc. Equip.*, 2005, **540** (2), pp 470–479

[2] Bunk O, Pfeiffer F, Stampanoni M, Patterson B D, Schulze-Briese C, David C, 'X-ray beam-position monitoring in the sub-micrometre and sub-second regime', *J. Synchrotron Radiat.*, 2005, **12** (6), pp 795–799

[3] Ilinski P, Hahn U, Schulte-Schrepping H, Degenhardt M, 'Residual gas X-ray beam position monitor development for PETRA III', *AIP Conf. Proc.*, 2007, **879** (1), pp 782–785

[4] Artemiev A N, Latushkin S, Mikhailov V, Rezvov V, Yudin L, 'Position-sensitive ionization detectors for ionizing beams diagnostics', *Nucl. Instrum. Methods Phys. Res. A*, 2002, **477**, pp 335–339

[5] Schildkamp W, 'Nondestructive position monitor for synchrotron radiation', *Nucl. Instrum. Methods Phys. Res. Sect. Accel. Spectrometers Detect. Assoc. Equip.*, 1987, **258** (2), pp 275–280

[6] Alkire R W, Rosenbaum G, Evans G, 'Design of a vacuum-compatible high-precision monochromatic beam-position monitor for use with synchrotron radiation from 5 to 25 keV', *J. Synchrotron Radiat.*, 2000, **7** (2), pp 61–68

[7] Morse J, Salomé M, Berdermann E, Pomorski M, Cunningham W, Grant J, 'Single crystal CVD diamond as an X-ray beam monitor', *Diam. Relat. Mater.*, 2007, **16** (4), pp 1049–1052

[8] Desjardins K, Pomorski M, Morse J, 'Ultra-thin optical grade scCVD diamond as X-ray beam position monitor', *J. Synchrotron Radiat.*, 2014, **21** (Pt 6), pp 1217–1223

[9] Kyele N R, Decanniere K, van Silfhout R G, 'A transparent two-dimensional in situ beam-position and profile monitor for synchrotron X-ray beamlines', *J. Synchrotron Radiat.*, 2005, **12** (6), pp 800–806

[10] van Silfhout R, Kachatkou A, Kyele N, Scott P, Martin T, Nikitenko S, 'High-resolution transparent x-ray beam location and imaging', *Opt. Lett.*, 2011, **36** (4), pp 570–572

[11] van Silfhout R G, 'A beam tracking optical table for synchrotron X-ray beamlines', *Nucl. Instrum. Methods Phys. Res. Sect. Accel. Spectrometers Detect. Assoc. Equip.*, 1998, **403** (1), pp 153–160

[12] Makynen A J, Kostamovaara J T, Myllyla R A, 'A high-resolution lateral displacement sensing method using active illumination of a cooperative target and a focused four-quadrant position-sensitive detector', *IEEE Trans. Instrum. Meas.*, 1995, **44** (1), pp 46–52

[13] Wallmark J T, 'A new semiconductor photocell using lateral photoeffect', *Proc. IRE*, 1957, **45** (4), pp 474–483

[14] Fuchs M R, Holldack K, Bullough M, *et al.*, 'Transmissive x-ray beam position monitors with submicron position- and submillisecond time resolution', *Rev. Sci. Instrum.*, 2008, **79**, 063103

[15] Scott P F I, Kachatkou A S, Kyele N R C, van Silfhout R G, 'Real-time photon beam localization methods using high-resolution imagers and parallel processing using a reconfigurable system', *Opt. Eng.*, 2009, **48** (7), 073601

[16] Gruner S M, 'X-ray imaging detectors', *Phys. Today*, 2012, **65** (12), pp 29–34

[17] Hatsui T, Graafsma H, 'X-ray imaging detectors for synchrotron and XFEL sources', *IUCrJ*, 2015, **2** (3), pp 371–383

[18] Chen W, De Geronimo G, Li Z, *et al.*, 'Active pixel sensors on high-resistivity silicon and their readout', *IEEE Trans. Nucl. Sci.*, 2002, **49** (3), pp 1006–1011

[19] van Silfhout R G, Kachatkou A S, 'Fibre-optic coupling to high-resolution CCD and CMOS image sensors', *Nucl. Instrum. Methods Phys. Res. Sect. Accel. Spectrometers Detect. Assoc. Equip.*, 2008, **597** (2–3), pp 266–269

[20] Tong H M, Lai Y S, Wong C P, *Advanced Flip Chip Packaging*, Springer Science & Business Media, 2013

[21] Pennicard D, Struth B, Hirsemann H, *et al.*, 'A germanium hybrid pixel detector with 55 um pixel size and 65,000 channels', *J. Instrum.*, 2014, **9** (12), P12003

[22] Kozhevnikov D, Chelkov G, Demichev M, Gridin A, Smolyanskiy P, Zhemchugov A, 'Performance and applications of GaAs:Cr-based Medipix detector in X-ray CT', *J. Instrum.*, 2017, **12** (1), C01005

[23] Herrmann S, Boutet S, Duda B, *et al.*, 'CSPAD-140k: A versatile detector for LCLS experiments', *Nucl. Instrum. Methods Phys. Res. Sect. Accel. Spectrometers Detect. Assoc. Equip.*, 2013, **718**, pp 550–553

[24] Loeliger T, Brönnimann C, Donath T, Schneebeli M, Schnyder R, Trüb P, 'The new PILATUS3 ASIC with instant retrigger capability', *Proceedings of IEEE Nuclear Science Symposium and Medical Imaging Conference Record*, Anaheim, California, USA, 27 October–3 November 2012, pp 610–615

[25] Le Bourlot C, Landois P, Djaziri S, *et al.*, 'Synchrotron X-ray diffraction experiments with a prototype hybrid pixel detector', *J. Appl. Crystallogr.*, 2012, **45** (1), pp 38–47

[26] Steadman R, Herrmann C, Mülhens O, Maeding D G, 'ChromAIX: Fast photon-counting ASIC for spectral computed tomography', *Nucl. Instrum. Methods Phys. Res. Sect. Accel. Spectrometers Detect. Assoc. Equip.*, 2011, **648**, pp S211–S215

[27] Ballabriga R, Campbell M, Heijne E H M, Llopart X, Tlustos L, 'The Medipix3 prototype, a pixel readout chip working in single photon counting mode with improved spectrometric performance', *IEEE Trans. Nucl. Sci.*, 2007, **54** (5), pp 1824–1829

[28] Ballabriga R, Alozy J, Campbell M, *et al.*, 'Review of hybrid pixel detector readout ASICs for spectroscopic X-ray imaging', *J. Instrum.*, 2016, **11**, P01007

[29] van Silfhout R G, 'A high-precision X-ray beam-position and profile monitor for synchrotron beamlines', *J. Synchrotron Radiat.*, 1999, **6** (6), pp 1071–1075

[30] Kyele N R, van Silfhout R G, Manolopoulos S, Nikitenko S, 'In situ synchrotron x-ray photon beam characterization', *J. Appl. Phys.*, 2007, **101** (6), 064901

[31] Kyele N, van Silfhout R, 'Beam sensing', Patent WO/2010/004258, 14–January 2010

[32] Kachatkou A, Kyele N, Scott P, van Silfhout R, 'In situ X-ray beam imaging using an off-axis magnifying coded aperture camera system', *J. Synchrotron Radiat.*, 2013, **20** (4), pp 596–602
[33] Kachatkou A, van Silfhout R, 'On the resolution and linearity of lensless in situ X-ray beam diagnostics using pixelated sensors', *Opt. Express*, 2013, **21** (4), pp 4291–4302
[34] Bobroff N, 'Position measurement with a resolution and noise-limited instrument', *Rev. Sci. Instrum.*, 1986, **57** (6), pp 1152–1157
[35] Hanisch R J, White R L, Gilliland R L, *Deconvolution of Images and Spectra*, 2nd edn, Mineola, NY, Dover Publications Inc., 2009
[36] Nano beam position monitor, FMB Oxford Ltd
[37] Chagani H, Garcia-Nathan T B, Jiang C, *et al.*, 'Performance of the Lancelot beam position monitor at the diamond light source', *J. Instrum.*, 2017, **12** (12), C12044
[38] Garcia-Nathan T B, Kachatkou A, Jiang C, *et al.*, 'Compact and portable X-ray imager system using Medipix3RX', *J. Instrum.*, 2017, **12** (10), C10011
[39] Borsboom M, Bras W, Cerjak I, *et al.*, 'The Dutch–Belgian beamline at the ESRF', *J. Synchrotron Radiat.*, 1998, **5** (3), pp 518–520

Roelof van Silfhout obtained his PhD from the University of Leiden, The Netherlands. After post-doctoral work at Edinburgh University on structural changes of semiconductors at high pressures, he became staff scientist at the European Synchrotron Radiation Facility in Grenoble, France and at the EMBL Hamburg Outstation, Germany. In 2001, he accepted a post as lecturer in Manchester. His research interests include the application of X-rays in structural studies of materials at the atomic scale, optical characterisation methods at the meso- and macroscopic scale and associated high precision instrumentation. Roelof has a strong track record in developing innovative instrumentation and to exploit their application in close collaboration with dedicated users groups both in the UK, Europe and the USA.

Chapter 2

Optical scattering sensors

Wu Zhou[1], Haitao Yu[2], Mingxu Su[1] and Xiaoshu Cai[1]

List of abbreviations

ACF	auto-correlation function
BEM	boundary element method
BFGS	Broyden–Fletcher–Goldfarb–Shanno
CCD	charge-coupled device
CMOS	complementary metal–oxide–semiconductor
DFP	Davidon–Fletcher–Powell
DLS	dynamic light scattering
DOP	degree of polarisation
EC	Epstein–Carhart
ECAH	Epstein–Carhart–Allegra–Hawley
FFT	fast Fourier transform
GLMT	generalised Lorenz–Mie theory
GRT	global rainbow technique
IDLS	image-based dynamic light scattering
IGA3	improved genetic algorithm 3
LM	Levenberg–Marquard
MIA	microscope image analysis
NNLS	non-negative least squares
RMS	root mean square
SGA	standard genetic algorithm
SNR	signal-to-noise ratio
SRT	standard rainbow technique
UAS	ultrasonic attenuation spectroscopy
VRT	vector ray tracing

[1]School of Energy and Power Engineering, University of Shanghai for Science and Technology, Shanghai, China
[2]College of Science, University of Shanghai for Science and Technology, Shanghai, China

2.1 Introduction

Optical and acoustic scatterings are general physical processes where light and sound are forced to deviate from a straight trajectory due to non-uniformities in the medium. The non-uniformities are known as scatters, such as particles or bubbles in fluid. The scattered light or sound intensity, as well as the spatial distribution of them, are strongly affected by characteristics of different scatters. These phenomena are used for measurement, especially for size and concentration measurement of particles and two-phase flows. This chapter will cover four of the commonly used methodologies for particle measurement, that is static light scattering, dynamic light scattering, RGB multi-wavelength light extinction and ultrasonic scattering. For each of these methodologies, the measurement principle, data processing or inversion algorithm, experimental system and some measurement results are included.

2.2 Static light scattering

Optical method is one of the most important methods for particle characterisation, such as size, size distribution, temperature and velocity. The particle can be a solid particle, droplet or bubble. Measurement techniques for optical particle characterisation can be classified into direct imaging, intensity or intensity ratio, interferometry, time shift, pulse delay and Raman scattering [1]. For further advancement and improvement of these optical techniques, interpretation of light scattering from a particle is a pre-requisite. Here, only Lorenz–Mie theory and one technique for droplet measurement are given in detail.

2.2.1 Lorenz–Mie theory

Plane-wave scattering by a homogeneous sphere is always referred to Mie theory [2]. It is also called as Lorenz–Mie theory or Lorenz–Mie–Debye theory [3]. Suppose that a particle is positioned at the origin of the coordinate and a plane wave propagates along z-axis (see Figure 2.1).

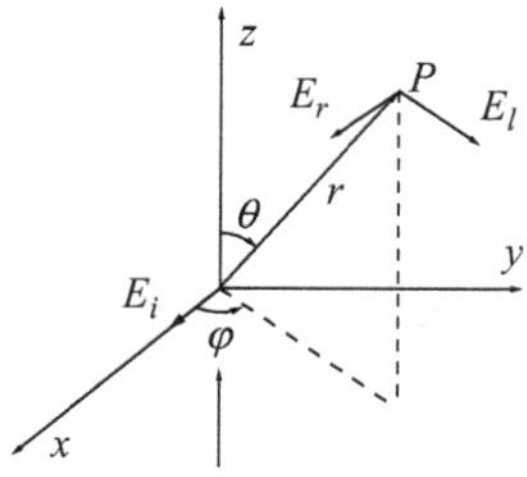

Figure 2.1 Light scattering by a particle

The scattering amplitudes for the perpendicular and parallel to the scattering plane are given by

$$\begin{aligned} E_r &= \frac{i}{kr} S_1(\alpha, m, \theta) E_{r0} \\ E_l &= \frac{i}{kr} S_2(\alpha, m, \theta) E_{l0} \end{aligned} \tag{2.1}$$

where $k = 2\pi/\lambda$ is the wave number of incident light and r is the distance from the particle centre to the observer. $\alpha = \pi a/\lambda$ is the dimensionless particle size parameter, where a is the diameter of spherical particle and λ is the wavelength of the incident light in the surrounding medium. The relative refractive index $m = m_2/m_1$ is the ratio of the refractive index of the particle m_2 to that of the surrounding medium m_1. E_{r0} and E_{l0} are incident electric field components of two polarised directions, respectively. $S_1(\alpha, m, \theta)$ and $S_2(\alpha, m, \theta)$ are the dimensionless scattering amplitudes for the perpendicular and parallel to the scattering plane, which are given by [3,4]:

$$\begin{aligned} S_1(\alpha, m, \theta) &= \sum_{n=1}^{\infty} \frac{2n+1}{n(n+1)} [a_n(\alpha, m)\pi_n(\cos\theta) + b_n(\alpha, m)\tau_n(\cos\theta)] \\ S_2(\alpha, m, \theta) &= \sum_{n=1}^{\infty} \frac{2n+1}{n(n+1)} [a_n(\alpha, m)\tau_n(\cos\theta) + b_n(\alpha, m)\pi_n(\cos\theta)] \end{aligned} \tag{2.2}$$

where the functions π_n and τ_n are related to Legendre functions as follows:

$$\begin{aligned} \pi_n(\cos\theta) &= \frac{P_n^{(1)}(\cos\theta)}{\sin\theta} = \frac{dP_n(\cos\theta)}{d(\cos\theta)} \\ \tau_n(\cos\theta) &= \frac{d}{d\theta} P_n^{(1)}(\cos\theta) = -P_n^{(1)'}(\cos\theta)\sin\theta \end{aligned} \tag{2.3}$$

where a_n and b_n are Mie coefficients given as

$$\begin{aligned} a_n &= \frac{\psi_n(\alpha)\psi_n'(m\alpha) - m\psi_n'(\alpha)\psi_n(m\alpha)}{\xi_n^{(1)}(\alpha)\psi_n'(m\alpha) - m\xi_n^{(1)'}(\alpha)\psi_n(\beta)} \\ b_n &= \frac{m\psi_n(\alpha)\psi_n'(m\alpha) - \psi_n'(\alpha)\psi_n(m\alpha)}{m\xi_n^{(1)}(\alpha)\psi_n'(m\alpha) - \xi_n^{(1)'}(\alpha)\psi_n(m\alpha)} \end{aligned} \tag{2.4}$$

where $\psi_n(\alpha)$ and $\xi_n^{(1)}(\alpha)$ are the Riccati–Bessel functions given by

$$\begin{aligned} \psi_n(\alpha) &= \left(\frac{\pi\alpha}{2}\right)^{1/2} J_{n+\frac{1}{2}}(\alpha) \\ \xi_n^{(1)}(\alpha) &= \left(\frac{\pi\alpha}{2}\right)^{1/2} H_{n+\frac{1}{2}}^{(1)}(\alpha) \end{aligned} \tag{2.5}$$

where $J_{n+\frac{1}{2}}(\alpha)$ is the Bessel function of the first kind, and $H_{n+\frac{1}{2}}^{(1)}(\alpha)$ is the Hankel function of the first kind.

The Hankel function of the first kind is used because it makes $\boldsymbol{E}\exp(-i\omega t + i\mathbf{k}\cdot\mathbf{r})$ to an outgoing wave at larger distance from the particle, as the scattered wave is supposed to be. The Hankel function of the second kind can be used if one adopts the convention of $\boldsymbol{E}\exp(i\omega t - i\mathbf{k}\cdot\mathbf{r})$.

According to (2.1), the far-field intensities for two polarised directions can be obtained:

$$\begin{aligned} I_r &= |E_r|^2 = \frac{\lambda^2}{8\pi^2 r^2} i_1(\alpha, m, \theta)|E_{r0}|^2 \\ I_l &= |E_l|^2 = \frac{\lambda^2}{8\pi^2 r^2} i_2(\alpha, m, \theta)|E_{l0}|^2 \end{aligned} \tag{2.6}$$

Generally, the dimensionless far-field intensities $i_1(\alpha, m, \theta)$ and $i_2(\alpha, m, \theta)$ are used, which are given by

$$\begin{aligned} i_1(\alpha, m, \theta) &= |S_1(\alpha, m, \theta)|^2 \\ i_2(\alpha, m, \theta) &= |S_2(\alpha, m, \theta)|^2 \end{aligned} \tag{2.7}$$

The total intensity of scattered light $i(\alpha, m, \theta)$ is simply obtained by

$$i(\alpha, m, \theta) = i_1(\alpha, m, \theta) + i_2(\alpha, m, \theta) \tag{2.8}$$

For the scattered light from a particle, the degree of polarisation (DOP) is defined as

$$P = \frac{i_1(\theta) - i_2(\theta)}{i_1(\theta) + i_2(\theta)} \tag{2.9}$$

It can be seen that the scattering amplitudes (S_1 and S_2) given by (2.2) are expressed in terms of an analytical infinite series. For simulation, the criterion of the terminating order has to be chosen. For $\alpha \in (0, 100{,}000)$, a criterion is given by the following equation [5]:

$$n_{\text{stop}} = \begin{cases} \alpha + 7.5\alpha^{0.34} + 2 & |A_n(\alpha)| < e^{-18} \\ \alpha + 6\alpha^{1/3} + 2 & |A_n(\alpha)| < e^{-12} \\ \alpha + 4.88\alpha^{0.31} & |A_n(\alpha)| < e^{-5} \end{cases} \tag{2.10}$$

The definition of $A_n(\alpha)$ can be found in [5]. The Lorenz–Mie theory gives the analytical solution of the interaction between electromagnetic radiation and a homogeneous spherical particle. It has to be noted that the above solutions are given for the fields at far distance outside of the particle. Of course, the solutions can be obtained at any point inside the particle. Due to the development of computation, it can be used for simulation of the scattered light from larger size particle.

Based on the Lorenz–Mie theory, the scattering intensity distributions are given for particles with relative refractive index $m = 1.333$, but with different dimensionless particle size parameter (i.e. $\alpha = 10, 100, 1{,}000, 10{,}000$). The intensity distributions are related to the size parameter of the particle. The intensity close to $\theta = 0^\circ$ (i.e. the forward direction) is higher than that at other angle (see Figure 2.2). The intensity in the forward direction can be calculated by the diffraction theory. It can be seen from Figure 2.2 that the primary and secondary rainbows are obvious for large particle. Alexander's dark band is between the primary rainbow and the secondary rainbow.

The intensity distribution is dependent on the absorption of the particle. A comparison of the scattering intensity distributions for a non-absorbing particle and an absorbing particle is shown in Figure 2.3.

For the non-absorbing particle with $m = 1.333$, the light intensity oscillates in all scattering angles. However, for the absorbing particle with $m = 1.333 + i0.01$, the oscillation of the intensity distribution is only observed in the forward directions (i.e. small scattering angles). In the viewpoint of geometric optics, the oscillation is caused by the interference of diffraction, reflection and refraction rays. For an absorbing particle, the refraction ray is absorbed inside the particle. Diffraction plays important role in forward scattering angles. Therefore, the intensity is mainly brought by reflection in large scattering angles and no oscillation is observed.

The Lorenz–Mie theory describes the scattering by a spherical homogeneous dielectric particle illuminated by an incident plane wave. However, the Lorenz–Mie theory fails to accurately describe the scattering characteristics of particles illuminated by a shaped beam. For this situation, Gouesbet *et al.* [6] developed the generalised Lorenz–Mie theory (GLMT), which can rigorously describe the interaction between an arbitrarily shaped beam and a regularly shaped particle. During the past three decades, GLMT for spherical particles has been developed to a state of maturity [6–13] and extensions to GLMT allow prediction of light from spheroidal particles [14–18]. For scattering from spherical, non-spherical or coated particles, GLMT can be complemented by computations using the Debye series

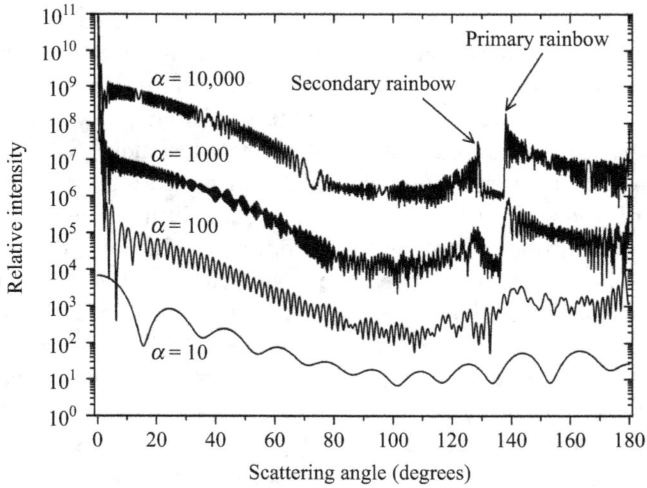

Figure 2.2 Light scattering intensity distributions for different particles

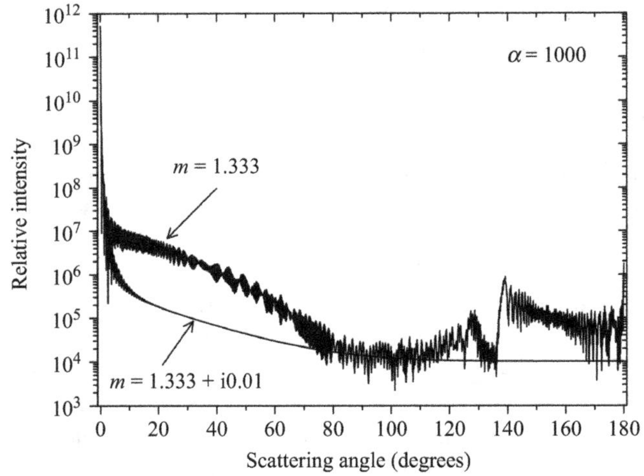

Figure 2.3 Light scattering intensity distributions for a non-absorbing particle and an absorbing particle

[19–25]. The Debye series also can give insight into either the scattering process, or changes in the pattern resulting from changes in shape, refractive index or profile of the incident wave fronts. For more information about light scattering theories, one can refer to the literatures [3,4,26,27].

2.2.2 Rainbow technique

The scattering distributions in rainbow region (see Figure 2.2) are related to droplet parameters (i.e. refractive index, temperature and size). The rainbow phenomena from a single droplet or from a cloud of droplets can be used for droplet diagnostics, that is rainbow refractometry, sometimes simply called the rainbow technique. It was first proposed by Roth *et al.* [28] for measuring the size and refractive index of a spherical droplet, which is called the standard rainbow technique (SRT). Then the SRT was generalised to a global rainbow technique (GRT) to measure clouds of droplets in sprays [29]. The GRT can measure temperature and size distribution of spray droplets [30–32], temperature distributions inside burning droplets [33], and mixing ratio of multiphase flows [34,35]. A one-dimensional (1D) rainbow refractometry [36,37], phase rainbow refractometry [38] and rainbow technique with Gaussian beam illumination [39] were developed for droplet diagnostics. However, SRT can only measure spherical droplets. GRT can measure clouds of spray droplets, but non-sphericity of droplets is not explicitly evaluated. For a spheroidal droplet, a generalised rainbow technique was developed to measure the refractive index, size and shape of droplets [40,41]. Nevertheless, the generalised rainbow technique was developed only for spheroidal droplets with symmetric illumination, without considering the tilted illumination condition. As an example of rainbow technique, the spheroidal droplet measurement is shown in the following.

Light scattered from the spheroidal water droplet in the vicinity of the primary rainbow region is shown in Figure 2.4. The droplet diameters are denoted by a in the horizontal equatorial plane, and by b along the vertical axis of the rotational symmetry. The droplet shape is defined by aspect ratio (i.e. a/b). Figure 2.4(a) shows the scattering pattern from a near-spherical droplet having $a = 1.58$ mm and $a/b = 1.02$. The fringes are almost straight in the viewed region. Figure 2.4(b) shows the scattering pattern for an oblate spheroidal droplet with $a = 1.77$ mm and $a/b = 1.15$. For an oblate spheroidal droplet illuminated by a parallel beam of light, the diffraction catastrophes in the primary rainbow region were first revealed by Marston and Trinh [42]. The observed scattering patterns are not expected for the light scattering from a spherical droplet. They are called 'generalised' rainbow patterns to distinguish them from those for a spherical droplet. These generalised rainbow patterns include transverse cusp caustic, hyperbolic umbilic and other catastrophes in the language of catastrophe optics.

An oblate spheroid is a surface of revolution obtained by rotating an ellipse about its minor axis. Therefore, the scattering patterns are symmetric about the horizontal equatorial plane. In the horizontal equatorial plane the section of the oblate spheroid is a circle. Therefore, the rays transmitting on the equatorial plane will be confined to this same plane. In this plane, the intensity distribution can be described by the Airy theory [4]. The angle for the first extreme value of the Airy approximation is denoted θ_1. The angle for the second maximum in the Airy approximation is denoted θ_2. To obtain the value θ_1 and θ_2 from the experimental data, the intensity curve is first filtered to remove the ripple structure. A low-pass Gaussian filter was used for this purpose. The filtered experimental result is shown in Figure 2.5.

It is obvious that the angular positions of first two extreme values agree very well. The values of θ_1 and θ_2 found experimentally can then be used to calculate the relative refractive index and size parameter. According to the Airy theory, the relation between the rainbow angles θ_{rg}, θ_1 and θ_2 is given by [4]

$$\theta_{rg} = (\theta_1 - C\theta_2)/(1 - C) \tag{2.11}$$

where C is a constant defined by $C = \alpha_1/\alpha_2$, $\alpha_1 = 1.0874$, $\alpha_2 = 3.4668$.

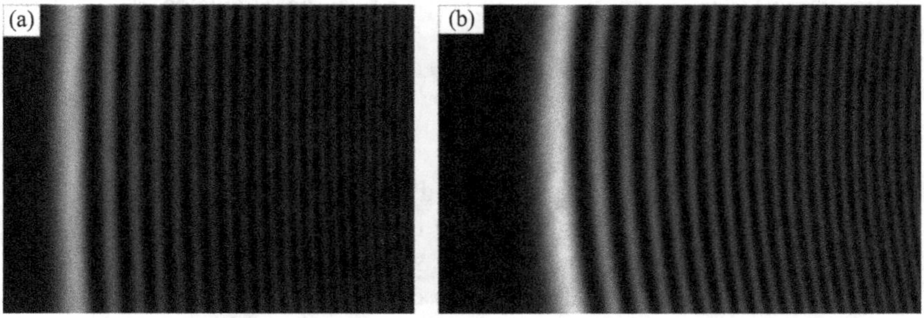

Figure 2.4 Generalised rainbow patterns from spheroidal water droplets: (a) a/b = 1.02 and (b) a/b = 1.15

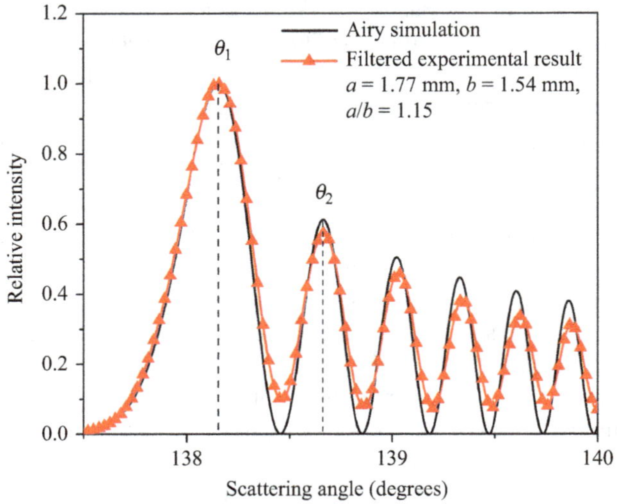

Figure 2.5 Comparison of intensity distribution calculated by Airy theory with the filtered experimental result for a water droplet with a = 1.77 mm and a/b = 1.15

The constants α_1 and α_2 are calculated by the Airy integral. θ_1 and θ_2 are measured from the experiment. After knowing the rainbow angle, the relative refractive index can be calculated. According to geometric optics, the relation between the rainbow angle and the relative refractive index is given by

$$\theta_{rg} = \pi + 2\cos^{-1}\left(\frac{m_{\text{rainbow}}^2 - 1}{3}\right)^{1/2} - 4\cos^{-1}\left(\frac{4\left(m_{\text{rainbow}}^2 - 1\right)}{3m_{\text{rainbow}}^2}\right)^{1/2} \tag{2.12}$$

To differentiate the relative refractive index of a pure water droplet and that calculated from the rainbow pattern, we denote them as m and m_{rainbow}, respectively.

The diameter D_{rainbow} in the equatorial plane can be calculated by

$$D_{\text{rainbow}} = \frac{\lambda}{4}\left[\left(\frac{3\left(4 - m_{\text{rainbow}}^2\right)^{1/2}}{\left(m_{\text{rainbow}}^2 - 1\right)^{3/2}}\right)^{1/2}\left(\frac{\alpha_1 - \alpha_2}{\theta_1 - \theta_2}\right)^{3/2}\right] \tag{2.13}$$

where α_1, α_2, θ_1, θ_2 and λ have been defined above.

Based on (2.11) and (2.12), the relative refractive indices calculated from the relevant generalised rainbow patterns are shown in Figure 2.6, in which the refractive indices are shown with error bars corresponding to one standard deviation. It can be seen that the inversion results of refractive index are consistent with that of pure water ($m = 1.333$ at 632.8 nm). Water droplets have been studied only in the range 0.80–2.00 mm and aspect ratios smaller than 1.23, because larger and

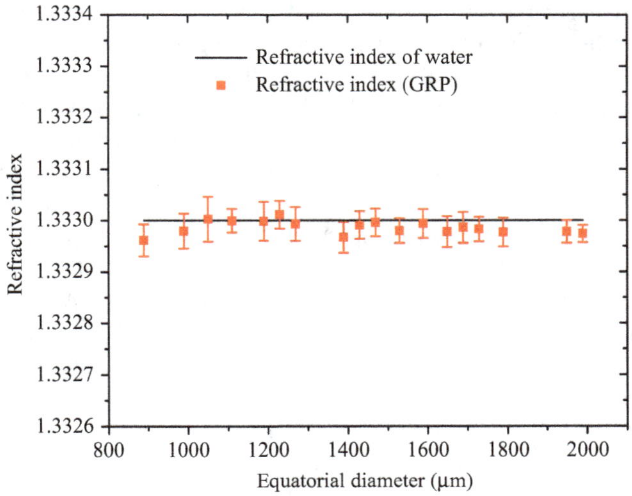

Figure 2.6 Refractive index of pure water droplet and that calculated from corresponding generalised rainbow patterns

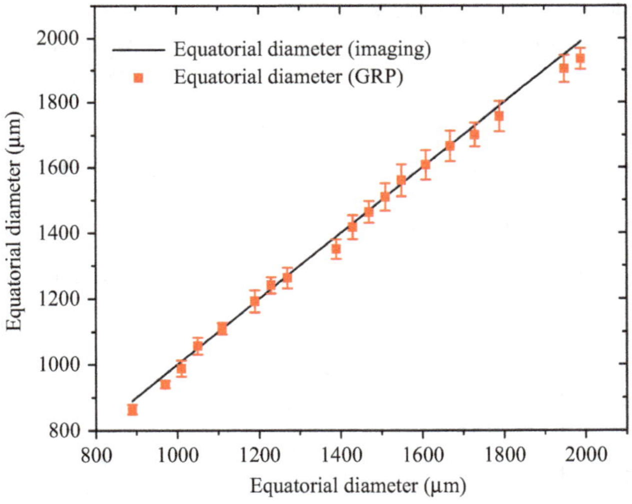

Figure 2.7 Water droplet diameters in equatorial plane measured by imaging and that calculated from corresponding generalised rainbow patterns

smaller droplets are not held stable in the acoustic levitator. The absolute error of relative refractive index inverted from the corresponding generalised rainbow patterns is smaller than 0.5×10^{-4} [40].

Figure 2.7 shows the water droplet diameters in the equatorial plane calculated by (2.13), in which the droplet diameters are also shown with error bars relevant to one standard deviation. The equatorial diameter calculated from the corresponding rainbow pattern matches well the values obtained by imaging with the camera. The absolute relative error is less than 5%, and hence very reliable droplet diameter estimation can be obtained from the corresponding generalised rainbow patterns.

As shown by Figure 2.4, the fringes and the fringe curvature of generalised rainbow patterns are dependent on the shape of droplet. However, there is no analytic solution for this relation. Therefore, the fringe is simulated based on vector ray tracing (VRT). The curvature of the rainbow fringe can be calculated (see Figure 2.8). A comparison of the curvature of rainbow fringes obtained from a VRT simulation and that from experiments for water droplets is given in [41]. The agreement between the VRT model and measurement is excellent at most aspect ratios. Most deviations are within the experimental uncertainty. Attention has to be paid to the fact that for large aspect ratio (e.g. $a/c > 1.23$), the droplet vibrates significantly in the levitator and the relevant generalised rainbow pattern becomes highly instable and blurry.

Based on the relation between the aspect ratio and the curvature of rainbow fringe as shown in Figure 2.8, the aspect ratio of the water droplet can be inverted from the corresponding generalised rainbow patterns and is compared with observed value from the directly recorded droplet image. Figure 2.9 gives such a comparison. The aspect ratios inverted from the generalised rainbow patterns are also shown with error bars relevant to one standard deviation. It reveals that the inverted droplet aspect ratios agree well with that from direct imaging. The relative errors lie between –1% and 1%, which indicates good measurement accuracy of droplet non-sphericity (in terms of aspect ratio) from the corresponding generalised rainbow patterns.

Here, only the measurement of a spheroidal droplet is introduced in brief. For measurement of size distribution and temperature of clouds of droplets in sprays, mixing ratio of multiphase flows and so on, one can refer to literatures [30–38].

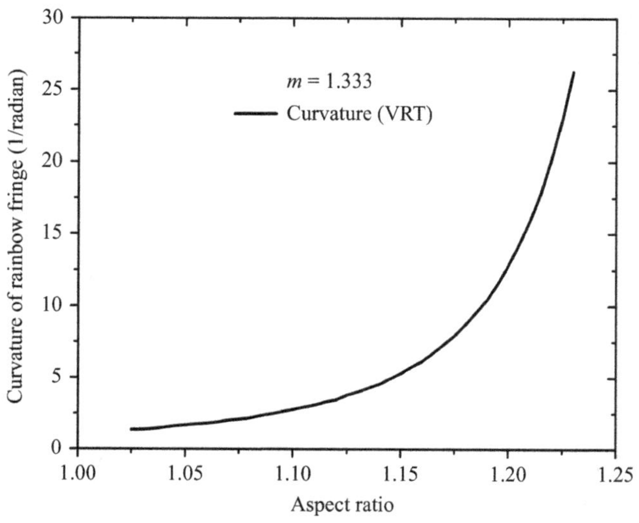

Figure 2.8 Curvatures of rainbow fringes calculated from VRT simulations

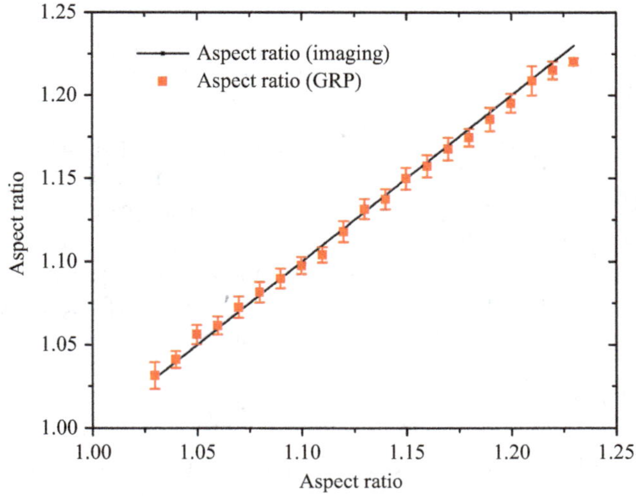

Figure 2.9 Comparison of water droplet aspect ratios observed from direct imaging and that inverted from the corresponding generalised rainbow patterns

Recently, the boundary model is proposed to calculate the optical caustics for a tilted spheroidal droplet, which can be used for droplet measurement [43]. More information about the measurement techniques can be found in [1].

2.3 Dynamic light scattering

Dynamic light scattering (DLS) has also been called as photon correlation spectroscopy, quasi-elastic light scattering or intensity fluctuation spectroscopy in the past. When in solution, nanoparticles are buffeted by the solvent molecules. This leads to a random motion of the nanoparticles called Brownian motion [44–46]. As light scatters from the moving nanoparticles, this motion imparts a randomness to the phase of the scattered light, so that when the scattered light from two or more particles is added together, there will be a changing destructive or constructive interference. This leads to time-dependent fluctuation in the intensity of the scattered light. The fluctuation is directly related to the rate of diffusion of the particles through the solvent, which is related in turn to the particles' hydrodynamic radii. Smaller particles move faster and the fluctuation of scattered light intensity is also faster. It was first used for the particle sizing in the early 1970s [47,48] and then got a booming development in the applications, including concentration and Zeta potential measurement.

2.3.1 Principle of DLS

For spherical nanoparticles dispersed in solutions, the Stokes–Einstein equation reveals the relation between particle hydrodynamic diameter D and the diffusion coefficient D_T of the Brownian motion as

$$D_T = \frac{K_B T}{3\pi\mu D} \tag{2.14}$$

where K_B is the Boltzmann constant, T is the absolute temperature of the solvent and μ is the dynamic viscosity of the dispersion medium. D_T is determined using DLS technology by analysing the scattering signal from nanoparticles, which is illuminated using coherent light beam.

Assuming that the 1D fluctuation information of scattered light intensity is represented by $I(t)$, the intensity auto-correlation function will be fitted according to (2.15) as the conventional DLS data processing method:

$$G(\tau) = \langle I(t)I(t+\tau)\rangle = \frac{1}{N}\sum_{i=1}^{N} I(t_i)I(t_i+\tau) \tag{2.15}$$

where τ is the time delay between signals and N is the total number of intensity data points.

For mono-disperse system, the following auto-correlation function [44] can also characterise such a steady fluctuation:

$$G(\tau) = g^{(2)}(\tau) - 1 = \beta\exp(-2\Gamma\tau) \tag{2.16}$$

where β is an instrumental constant between 0 and 1, and the decay constant Γ can be obtained by curve fitting from (2.16). Γ is related to the translational diffusion coefficient D_T by

$$\Gamma = D_T q^2 \tag{2.17}$$

where q is the norm of scattering vector, which indicates the difference between wave vector of the scattered light and the incident light. It is related to the refractive index of the dispersion medium n, the scattering angle θ and the wavelength λ of light by

$$q = \frac{4\pi n}{\lambda}\sin\left(\frac{\theta}{2}\right) \tag{2.18}$$

Thus, the particle diameter is derived from DLS measurements. Take three different sizes (79/482/948 nm) of standard polystyrene particles for example. The theoretical time-dependent auto-correlation functions with $\beta = 1$ under certain conditions can be calculated and shown in Figure 2.10. It shows that the auto-correlation function values decrease from 1 to nearly 0 in approximately 2 ms, 20 ms and 40 ms, for the respective three sizes of particles.

2.3.2 Particle sizing system based on DLS

A particle sizing system based on conventional DLS method is shown in Figure 2.11. The nano-particles are dispersed in solutions in the sample cell, which is illuminated by a focused laser beam from the left side. A photo-multiplier tube is

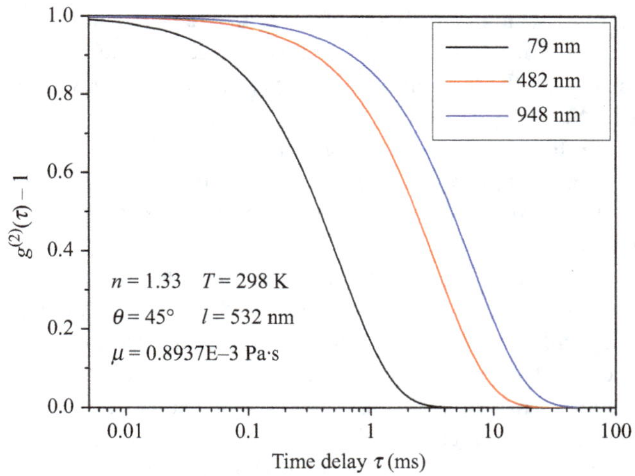

Figure 2.10 Theoretical ACF for nanoparticles of different sizes in water [49]

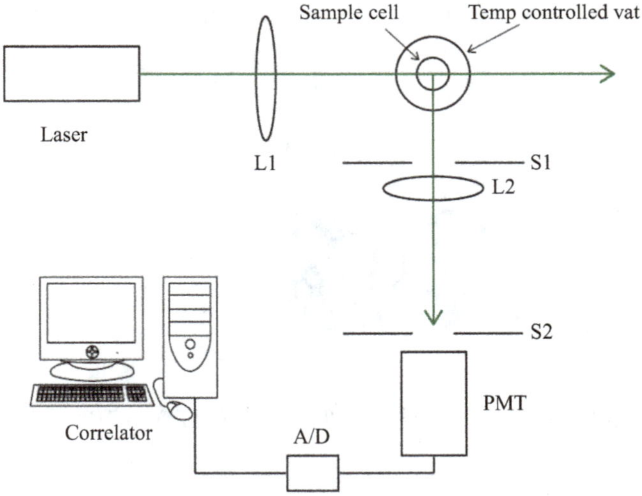

Figure 2.11 Conventional DLS experimental setup

used to receive the scattered light from nano-particles at a certain scattering angle, and a digital correlator is utilised to process the 1D time-series signal to obtain the auto-correlation function of the intensity fluctuation.

Commercial instrument based on DLS has become popular in off-line laboratory analysis. However, to obtain reliable and accurate result, averaging was adopted on a large number of time-series data, because the fluctuation goes in a random way. It needs a long time for both data acquisition and data processing. The typical measurement time for the current conventional DLS particle sizer is generally 1–2 min, which is difficult to meet the requirements of real-time and online measurement for rapidly changing process. Because the Brownian motion of particles is sensitive to the temperature, a temperature control device is necessary in conventional DLS.

To overcome the above disadvantages, image-based dynamic light scattering (IDLS) is proposed to capture the information of Brownian motion using an image sensor [50,51]. One pixel or several pixels of the sensor is equivalent to one photomultiplier tube. There are generally tens of or hundreds of thousands of pixels in one image sensor. As shown in Figure 2.12, the image is divided into $m \times n$ grids, which is equivalent to $m \times n$ photomultiplier tubes. $m \times n$ time-series of light intensity signals will be obtained simultaneously. Each series can be deal with as conventional data. This takes the place of conventional averaging on autocorrelation functions from a long time-series signal, and so the data acquisition time can be reduced dramatically.

Schematic drawing of the IDLS system is shown in Figure 2.13, including light source, lenses, sample cell, diaphragm iris, high-speed camera and computer. The laser beam was focused in the sample cell, just as other commercial DLS nanoparticle sizer does. The scattered light from the nano-particles was collected by a lens and an adjustable diaphragm was applied to define the measured region.

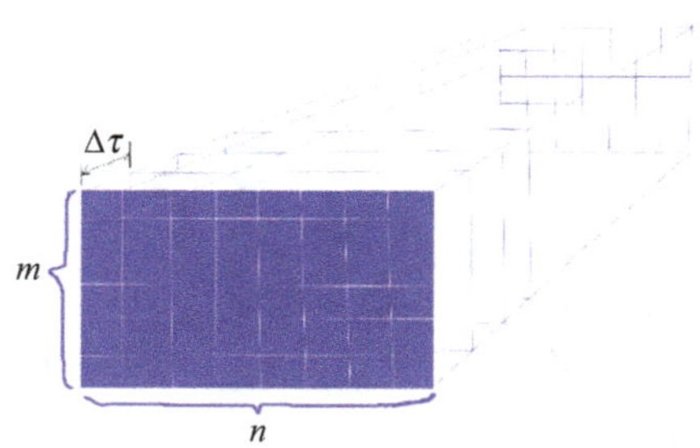

Figure 2.12 Division of images for data processing [50]

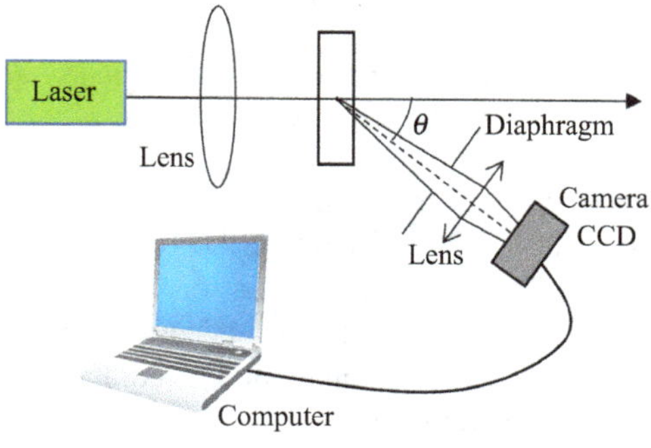

Figure 2.13 IDLS experimental setup

Instead of a photo-multiplier tube in conventional DLS, the limited scattered light was recorded by a high-speed camera at angle θ with the incident light.

As long as the frame rate of the camera is high enough to characterise the light intensity fluctuation, the data acquisition time will be shortened to the characteristic time of theoretical auto-correlation function (ACF) curve, that is about 2 ms, 20 ms and 40 ms in Figure 2.10. For example, if the images are captured in the frame rate of 10,000 frames per second, the number of images required to cover the typical time-scales in auto-correlation function curves should be about 20, 200 and 400 frames.

The data processing time can be further decreased by using image cross-correlation [49,52] as (2.19):

$$G(\tau) = R(A(i, j), B(i, j)) = \frac{\sum_i \sum_j (A_{i,j} - \bar{A})(B_{i,j} - \bar{B})}{\sqrt{\left(\sum_i \sum_j (A_{i,j} - \bar{A})^2\right)\left(\sum_i \sum_j (B_{i,j} - \bar{B})^2\right)}} \tag{2.19}$$

where τ is the time interval (equal to the value of time delay) between two images A and B, which are captured by the camera. i and j represent pixel row and column positions in the images. $\bar{A}$ and $\bar{B}$ are the pixel-averaged intensity value of the corresponding images A and B.

As shown in Figure 2.10, under the same time delay, the 2D ACF is a monotonically decreasing function of particle size. Lager particles lead to greater correlation coefficient. Therefore, the particle size may be determined by one correlation coefficient, which can be obtained by cross-correlation of two-dimensional light intensity signals in a time interval. Figure 2.14 shows the

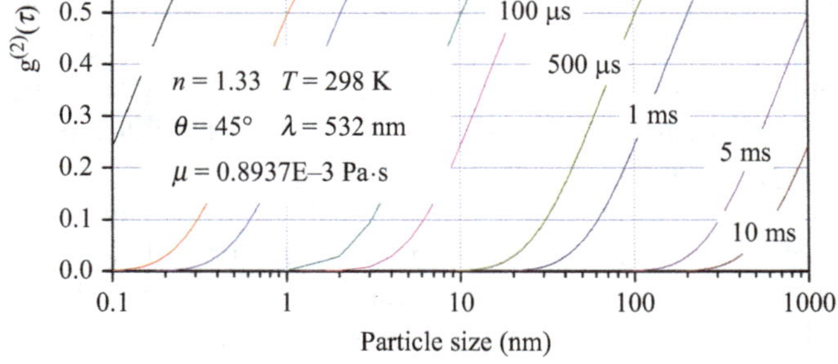

Figure 2.14 Variation of theoretical ACF value as particle size at different delayed times

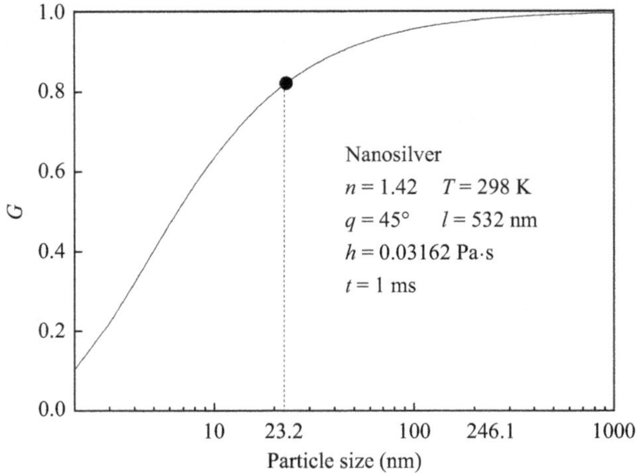

Figure 2.15 Measured result of the nano-silver by IDLS method

variation of correlation coefficient, or theoretical ACF value, as particle size under different time delays (from 1 μs to 10 ms). It indicates that the data acquisition time can be shortened to the acquisition time of two images, and the data processing time is shortened to the time needed for cross-correlation of these two images, that is several milliseconds.

Figure 2.15 shows an example of the experimental results on a sample of nano-silver. The experimental conditions are shown in the figure and the measured mean diameter 23.2 nm shows good agreements with the data 20 ± 2 nm provided by the manufacturer.

2.4 RGB multi-wavelengths light extinction

2.4.1 Measuring fine particle size with RGB three-wavelength bands light extinction

Measurement of particle size distribution by multi-wavelength light extinction method has been widely used [53–56]. In recent years, colour digital camera has been more often used in image method to measure particle size distribution [57]. However, the measurement ranges of these two methods are different. Due to the limitation of light scattering theory, the measurement range of multi-wavelength light extinction method is about 0.1–5 μm [58]. While the image measurement is limited by camera resolution, ranging from about 10 μm to hundreds of microns, and the measurement error will increase greatly when the range is less than 10 μm. In many applications for online measurement, the particle size distribution may span the measurement range of these two methods, ranging from submicron to hundreds of microns. For example, in the study of wet steam of steam turbine, a large number of fine water droplets whose sizes are about 0.5–2 μm are generated by spontaneous condensation, along with a small number of course water droplets ranging from tens of microns to hundreds of microns [59]. Thus, it is necessary to develop a method to measure the size of these particles.

The image method has the characteristics of 'seeing is believing' and can measure the morphology of the measured particles. In recent years, an image method has developed rapidly with the rapid development of charge-coupled device (CCD)/complementary metal–oxide–semiconductor (CMOS). Especially for the colour image method, it can provide more information about particle morphology, which promotes the development of image method for measuring particle size. The basic principle of particle size measurement by multi-wavelength light extinction method is that particles will scatter and absorb incident light at different wavelengths. The attenuation (extinction) of transmitted light is related to particle size distribution and concentration, which can be described by Mie's light scattering theory. The transmitted light signal received by colour CCD/CMOS is also light extinction signal, which is only light extinction signal of RGB three-wavelength bands light. By using Mie's light scattering theory, it is possible to obtain the size distribution of fine particles.

2.4.2 Measurement principle of RGB three-wavelength bands light extinction method

The principle of RGB three-wavelength bands light extinction method is based on multi-wavelength light extinction method and CCD colour signal measurement. Before proposing RGB three-wavelength bands light extinction method, the light extinction method and the principle of CCD colour signal formation are introduced briefly.

2.4.2.1 Principle of particle measurement by light extinction method

The basic principle of light extinction method is that the incidence of light into poly-disperse system has intensity of I_0 and wavelength of λ in medium, and the intensity of the transmitted light is I, as shown in Figure 2.16.

According to the Lambert–Beer theorem, the intensity attenuation can be obtained by

$$\ln \frac{I}{I_0} = -\frac{\pi}{4} L \int_a^b N(D) D^2 K_{\text{ext}} dD \tag{2.20}$$

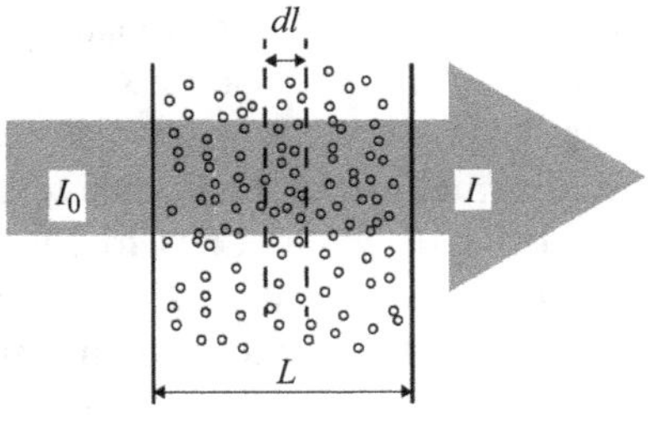

Figure 2.16 Light extinction measurement

where L is the optical path, $N(D)$ is the particle size distribution function, D is the particle diameter and K_{ext} is the light extinction coefficient, which can be calculated by Mie's light scattering theory.

Multi-wavelength light extinction method is used to measure the incident light intensity I_0 and transmitted light intensity I on the same sample, and the following equations are then obtained:

$$\ln(I/I_0)_{\lambda_i} = -\frac{\pi}{4}L\int_a^b N(D)D^2 k_{m\lambda_i} dD \quad i = 1, 2, \ldots m \tag{2.21}$$

In practical measurement, if the particle size distribution is not needed and only the average particle size is measured, the Sauter mean diameter D_{32} can be used instead of the particle size distribution function. In this case, the above formula is simplified as follows:

$$\ln(I/I_0)_{\lambda_i} = -\frac{\pi}{4}LND_{32}^2 k_{m\lambda_i} \qquad i = 1, 2, \ldots, m \tag{2.22}$$

For the same measured particle system, the values of D_{32} and N in (2.20) should be the same in different wavelength measurement. By comparing measurements at multiple wavelengths, it is obtained that

$$\frac{\ln(I/I_0)_{\lambda_i}}{\ln(I/I_0)_{\lambda_j}} = \frac{k_{m\lambda_i}}{k_{m\lambda_j}} \qquad i \neq j; \qquad i,j = 1, 2, \ldots, m \tag{2.23}$$

The left side of (2.23) is the measured value. When the refractive index and the measured wavelength are given, the light extinction coefficient of the right side of (2.23) is only a function of D_{32} to be evaluated. In this way, the particle size can be inversely calculated by measuring the intensities of incident and transmitted light at multiple wavelengths.

2.4.2.2 RGB signal of CCD camera

A color CCD (CMOS) sensor is generally a single-chip array element. RGB tri-colour separation is achieved by installing a basic colour filter arranged according to Bayer's law on the surface of each CCD pixel. Each pixel sensitises only one primary colour, and gets the colour image by interpolating with the other two primary colour signals around it. Figure 2.17 shows RGB pixels of CCD arranged according to Bayer's law [60]. At present, colour image acquisition by an area-array CCD is widely used in digital cameras and remote sensing imaging equipment. The spectral characteristics of colour CCD tri-colour filters from various companies are different. Figure 2.18 shows the response curve of IMx-720G CCD array used in three-wavelength bands light extinction measurement experiment.

Figure 2.17 CCD RGB pixel arrangement diagram according to Bayer's law

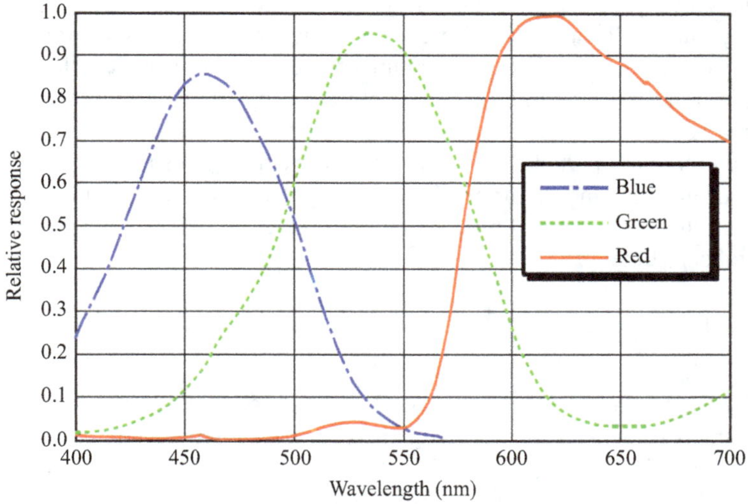

Figure 2.18 CCD RGB response curves

2.4.2.3 RGB three-wavelength bands light extinction method

The measurement principle of RGB three-wavelength bands light extinction method is similar to that of multi-wavelength bands light extinction method, as shown in Figure 2.19.

When there are no particles in the measuring area, light emitted by a white light forms parallel incident light through a lens, and then illuminate the CCD camera through the measuring area. The response charge number is generated by the camera at each pixel, which is related to the light source spectrum and the response curve of CCD, and the related equation is

$$I_0 = \int_{\lambda_1}^{\lambda_2} f(\lambda) I_\lambda d\lambda \tag{2.24}$$

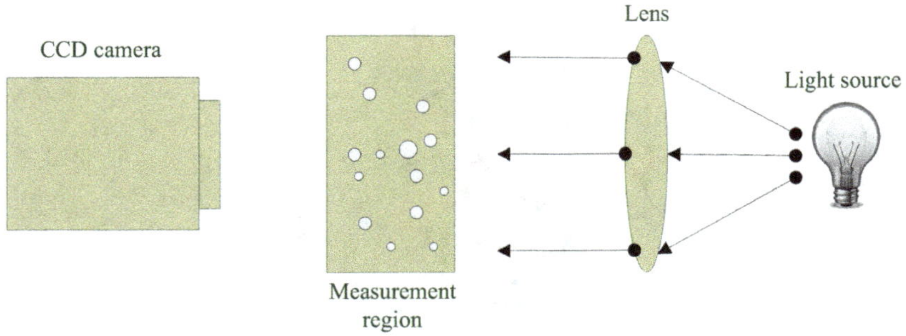

Figure 2.19 Measurement principle of RGB three-wavelength bands light extinction method

where I_0 is the light intensity obtained by CCD, λ_1 is the lower limit of light source wavelength and CCD quantum efficiency curve wavelength; λ_2 is the upper limit of light source wavelength and CCD quantum efficiency curve wavelength; $f(\lambda)$ is the response curve of CCD and I_λ is the light intensity of light source at different wavelengths.

When there are particles, the light emitted by the light source passes through the measuring area, and the original incident light will attenuate. At this time, the number of charges generated by each pixel of the CCD will be reduced under the same exposure time:

$$I = \int_{\lambda_1}^{\lambda_2} f(\lambda) I_\lambda d\lambda \tag{2.25}$$

where I is the transmitted light intensity after particle scattering obtained by CCD, I_λ is the intensity of transmitted light with different wavelengths scattered by particles.

The extinction coefficient of particle scattering can be calculated by Mie's scattering theory [27]. With the Lambert-Beer Formula, the relationship between the attenuation of transmitted light band and particle size and concentration can be obtained by (2.21) and (2.22):

$$\ln \frac{I}{I_0} = -\frac{\pi}{4} D^2 \int_{\Delta\lambda} f(\lambda) I(\lambda) E(\lambda, D, m) d\lambda \tag{2.26}$$

For colour CCD, three formulas can be obtained from the response curves of R, G and B:

$$\begin{cases} \ln\left(\dfrac{I}{I_0}\right)_R = -\dfrac{\pi}{4} D^2 \int_{\Delta\lambda} f_R(\lambda) I(\lambda) E(\lambda, D, m) d\lambda \\ \ln\left(\dfrac{I}{I_0}\right)_G = -\dfrac{\pi}{4} D^2 \int_{\Delta\lambda} f_G(\lambda) I(\lambda) E(\lambda, D, m) d\lambda \\ \ln\left(\dfrac{I}{I_0}\right)_B = -\dfrac{\pi}{4} D^2 \int_{\Delta\lambda} f_B(\lambda) I(\lambda) E(\lambda, D, m) d\lambda \end{cases} \tag{2.27}$$

While for particle groups with distributed particle sizes, there is

$$\ln\left(\frac{I}{I_0}\right) = -\frac{\pi}{4}L\int_{\Delta D} D^2N(D)\int_{\Delta\lambda} f(\lambda)I(\lambda)E(\lambda, D, m)d\lambda dD \tag{2.28}$$

The discretisation of this formula is obtained:

$$\ln\left(\frac{I}{I_0}\right) = -\frac{\pi}{4}L\sum_{\Delta D}\left[D^2N(D)\sum_{\Delta\lambda} f(\lambda)I(\lambda)E(\lambda, D, m)\right] \tag{2.29}$$

Similarly, for colour CCD, three discretisation equations can be obtained from the response curves of R, G and B:

$$\begin{cases} \ln\left(\frac{I}{I_0}\right)_R = -\frac{\pi}{4}\sum_{\Delta D}\left[D^2N(D)\sum_{\Delta\lambda} f(\lambda_R)I(\lambda)E(\lambda, D, m)\right] \\ \ln\left(\frac{I}{I_0}\right)_G = -\frac{\pi}{4}\sum_{\Delta D}\left[D^2N(D)\sum_{\Delta\lambda} f(\lambda_G)I(\lambda)E(\lambda, D, m)\right] \\ \ln\left(\frac{I}{I_0}\right)_B = -\frac{\pi}{4}\sum_{\Delta D}\left[D^2N(D)\sum_{\Delta\lambda} f(\lambda_B)I(\lambda)E(\lambda, D, m)\right] \end{cases} \tag{2.30}$$

By dividing the formulas of (2.30) separately, the following equations are then obtained:

$$\begin{cases} \dfrac{\ln\left(\frac{I}{I_0}\right)_R}{\ln\left(\frac{I}{I_0}\right)_G} = \dfrac{-\frac{\pi}{4}\Sigma_{\Delta D}\left[D^2N(D)\Sigma_{\Delta\lambda} f(\lambda_R)I(\lambda)E(\lambda, D, m)\right]}{-\frac{\pi}{4}\Sigma_{\Delta D}\left[D^2N(D)\Sigma_{\Delta\lambda} f(\lambda_G)I(\lambda)E(\lambda, D, m)\right]} \\ \dfrac{\ln\left(\frac{I}{I_0}\right)_R}{\ln\left(\frac{I}{I_0}\right)_B} = \dfrac{-\frac{\pi}{4}\Sigma_{\Delta D}\left[D^2N(D)\Sigma_{\Delta\lambda} f(\lambda_R)I(\lambda)E(\lambda, D, m)\right]}{-\frac{\pi}{4}\Sigma_{\Delta D}\left[D^2N(D)\Sigma_{\Delta\lambda} f(\lambda_B)I(\lambda)E(\lambda, D, m)\right]} \\ \dfrac{\ln\left(\frac{I}{I_0}\right)_G}{\ln\left(\frac{I}{I_0}\right)_B} = \dfrac{-\frac{\pi}{4}\Sigma_{\Delta D}\left[D^2N(D)\Sigma_{\Delta\lambda} f(\lambda_G)I(\lambda)E(\lambda, D, m)\right]}{-\frac{\pi}{4}\Sigma_{\Delta D}\left[D^2N(D)\Sigma_{\Delta\lambda} f(\lambda_B)I(\lambda)E(\lambda, D, m)\right]} \end{cases} \tag{2.31}$$

The particle size D satisfying all these three equations simultaneously is the particle size of measured particles. This establishes the relationship between RGB three-wavelength bands extinction signal and particle size. The colour CCD can be used as a sensor, and the particle size can be obtained by inverse algorithm based on the ratio of the measured light intensity in three bands.

2.4.2.4 Simulation and analysis of three-wavelength bands extinction method

First, the theory of RGB three-wavelength bands extinction method is numerically simulated to simulate the attenuation of light intensity $\ln(I/I_0)$ measured by CCD at different particle sizes. The IMX-720G 12-bit colour camera and the ICX274 AL/AQ CCD produced by IMI Tech Company are selected. The corresponding response curve is shown in Figure 2.18. The spectral response range is between 400 nm and 700 nm. A white LED is used as the light source, the spectral curve of which is presented in Figure 2.20.

The standard polystyrene particles were selected as the tested particles, with the refractive index of 1.59 and the relative refractive index of 1.195 in water. By simulating different particle sizes ranging from 0.1 μm to 5 μm under the above white LED light source, the analogue signal of colour CCD with $\frac{\ln(I/I_0)_R}{\ln(I/I_0)_G}$, $\frac{\ln(I/I_0)_R}{\ln(I/I_0)_B}$, $\frac{\ln(I/I_0)_G}{\ln(I/I_0)_B}$ are then obtained, as shown in Figure 2.21.

In Figure 2.21, there is curve oscillation at less than 0.5 μm, which indicates that there will be multiple solutions in this region. Therefore, the lower limit of RGB three-wavelength bands extinction method is determined to be 0.5 μm. In the region of 0.5–2.8 μm, the curves are clearly distinguished, and the particle size can be easily calculated by inversion of the measured values. In the region of 2.8–3.6 μm, when the curve distribution is concentrated, more precise measurements are needed to retrieve the exact particle size. In the region of 3.6–5 μm, the curve distribution is similar to region of 0.6–2 μm, and the inversion results of measurement results can easily fall into the range of 0.6–2 μm, and the wrong results will be obtained. In summary, RGB three-wavelength bands extinction method can measure standard particles in the range of 0.5–2.8 μm. If the measurement accuracy is further improved, the measuring range can be expanded to about 0.5–3.6 μm. If additional constraints are added, the measurement range can be extended to 5 μm.

2.4.2.5 Inversion simulation of RGB three-wavelength bands extinction method

For the inversion calculation of RGB three-band extinction method, the non-negative least squares (NNLS) algorithm is adopted to solve the inversion (2.34).

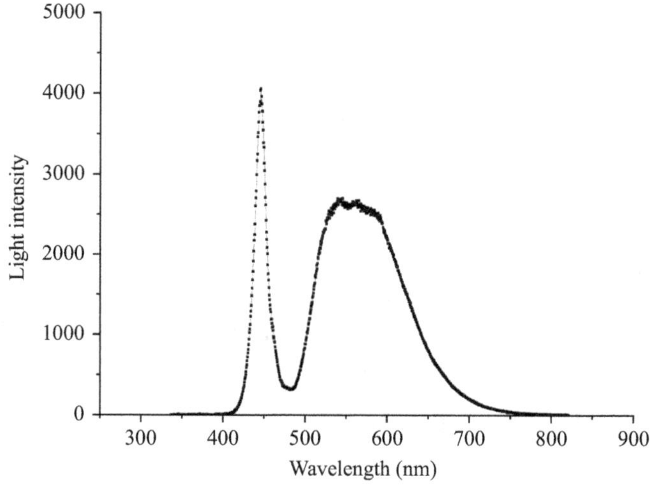

Figure 2.20 White LED spectrum

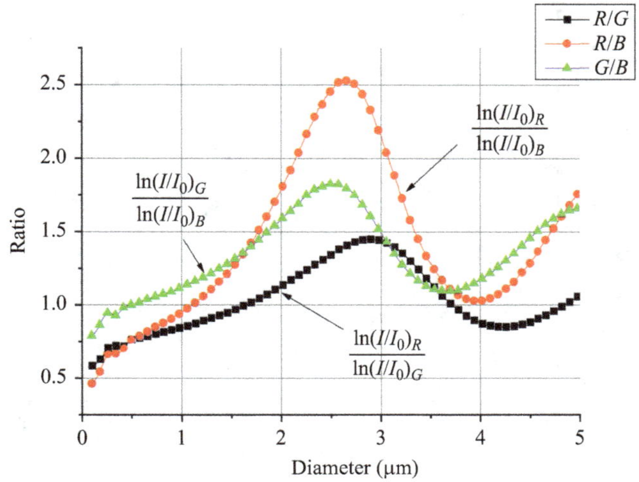

Figure 2.21 CCD responding charge number ratio of natural logarithm of single particle from 0.1 μm to 5 μm

This issue can be reduced to solving the matrix equation when the following conditions apply simultaneously:

$$\begin{aligned} &||TW - E|| = \min \\ &W \geq 0 \end{aligned} \tag{2.32}$$

where $W \geq 0$ represents the non-negative number solution, namely, the particle size must be greater than or equal to zero.

For monodisperse particles, the inversion formulas using non-negative least squares method are as follows:

$$\sum_{\substack{i,j=1 \\ i \neq j}}^{3} \left\{ \frac{\ln\left(\frac{I}{I_0}\right)_i}{\ln\left(\frac{I}{I_0}\right)_j} - \frac{-\frac{\pi}{4} L D^2 \sum_{\Delta\lambda} f(\lambda_i) I(\lambda) E(\lambda, D, m)}{-\frac{\pi}{4} L D^2 \sum_{\Delta\lambda} f(\lambda_j) I(\lambda) E(\lambda, D, m)} \right\}^2 = \min \tag{2.33}$$

For distributed particle swarms, the inversion formulas of non-negative least squares method are as follows:

$$\sum_{\substack{i,j=1 \\ i \neq j}}^{3} \left\{ \frac{\ln\left(\frac{I}{I_0}\right)_i}{\ln\left(\frac{I}{I_0}\right)_j} - \frac{\sum_{\Delta D} \left[D^2 N(D) \sum_{\Delta\lambda} f(\lambda_i) I(\lambda) E(\lambda, D, m) \right]}{\sum_{\Delta D} \left[D^2 N(D) \sum_{\Delta\lambda} f(\lambda_j) I(\lambda) E(\lambda, D, m) \right]} \right\}^2 = \min \tag{2.34}$$

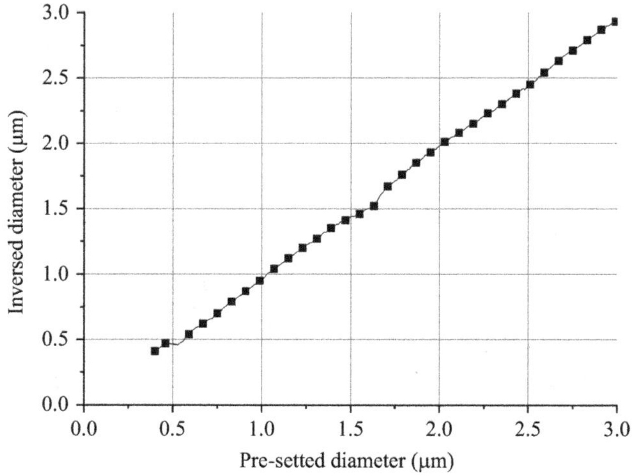

Figure 2.22 Particle inversion results

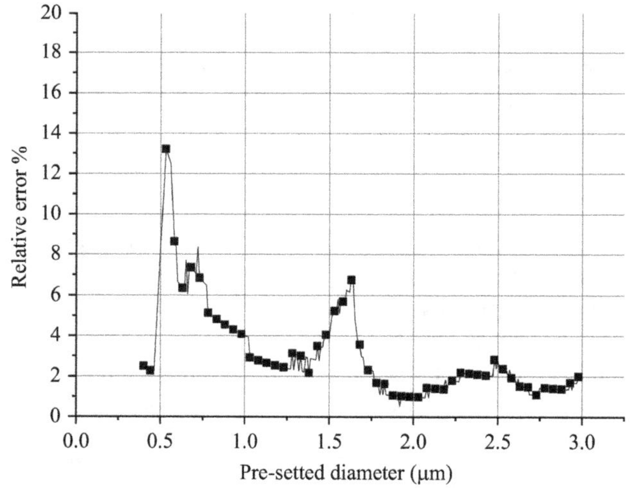

Figure 2.23 Particle inversion relative error

The CCD response values of 0.4–3.0 μm particles with different particle sizes were simulated and the inversion calculation based on NNLS algorithm is adopted. The calculated results are shown in Figures 2.22 and 2.23.

From Figure 2.23, the largest relative error exceeding 10% is reached when the particle size is 0.5 μm. The relative error can be controlled within 15% in the range of 0.4–3 μm, and the relative error in particle size range of 1–3 μm is less than 10%. In the measurement of wet steam of steam turbine, the diameter of fine water droplets concentrates around 1–2 μm, so the measurement error is less than 10%, which can meet the requirements.

2.4.3 Experimental verification

In the experiment, 2.1 μm and 960 nm polystyrene standard particles, 3 W white LED light source and 12-bit colour IMX-720G camera produced by IMI Tech Company were selected, and the exposure time was 20 μs. Tables 2.1 and 2.2 are part of the data of CCD response matrix obtained in the 2.1 μm standard particle experiment without and with adding particles in the measurement area. Tables 2.3 and 2.4 are the partial data of CCD response matrix obtained in 960 nm standard particle experiment without and with particles in the measurement area, respectively.

Table 2.1 CCD response matrix (partial data) without adding 2.1 μm standard particles

2,886	947	2,888	2,969	2,938	1,210	2,615	2,981
2,975	1,073	2,773	2,899	3,275	1,407	2,662	2,954
2,952	953	2,855	2,980	3,115	1,177	2,615	2,965
3,054	1,045	2,769	2,930	3,334	1,363	2,619	3,022
2,813	983	2,812	2,927	3,043	1,175	2,645	3,014
3,077	1,083	2,795	2,951	3,364	1,411	2,686	3,075
2,926	1,005	2,908	2,911	2,953	1,157	2,722	3,039
3,112	1,094	2,827	2,957	3,317	1,490	2,555	3,045

Table 2.2 CCD response matrix (partial data) with adding 2.1 μm standard particles

2,376	751	2,300	2,602	2,500	998	2,151	2,475
2,470	848	2,208	2,348	2,761	1,148	2,144	2,467
2,405	818	2,373	2,484	2,509	931	2,061	2,459
2,548	862	2,251	2,443	2,716	1,176	2,131	2,457
2,321	820	2,292	2,517	2,492	998	2,119	2,454
2,591	925	2,277	2,419	2,712	1,183	2,143	2,430
2,302	832	2,345	2,518	2,487	1,014	2,072	2,495
2,562	946	2,263	2,403	2,751	1,130	2,090	2,514

Table 2.3 CCD response matrix (partial data) without adding 960 nm standard particles

2,813	1,322	2,777	2,905	2,893	1,485	2,464	2,919
2,966	1,329	2,676	2,848	3,179	1,713	2,517	2,865
2,824	1,314	2,708	2,845	2,933	1,523	2,488	2,807
2,990	1,369	2,607	2,821	3,206	1,784	2,476	2,793
2,887	1,314	2,603	2,916	2,896	1,592	2,369	2,923
3,023	1,374	2,570	2,837	3,246	1,756	2,453	2,869
2,881	1,286	2,700	2,926	2,989	1,511	2,528	2,863
2,968	1,513	2,667	2,859	3,091	1,785	2,414	2,896

Table 2.4 CCD response matrix (partial data) with adding 960 nm standard particles

2,237	1,061	2,106	2,335	2,312	1,187	2,011	2,326
2,455	1,098	2,066	2,266	2,585	1,416	2,002	2,273
2,325	1,048	2,140	2,413	2,328	1,233	1,939	2,376
2,438	1,140	2,030	2,285	2,567	1,481	1,942	2,251
2,315	1,088	2,157	2,371	2,385	1,211	2,017	2,320
2,413	1,127	2,016	2,312	2,520	1,462	1,935	2,272
2,256	1,117	2,133	2,291	2,342	1,271	2,042	2,362
2,422	1,167	2,068	2,287	2,563	1,446	2,017	2,307

Table 2.5 Experimental results

Particle size	**Inversion particle size**	**Error**
2.1 μm	1.9 μm	9.50%
960 nm	1,044 nm	8.75%

From the CCD response measured above, it can be seen that the number of responding charges per pixel (i.e. light intensity signal) decreases due to the addition of standard particles. For these light intensity signals, the NNLS inversion method is used to calculate, and the results are obtained as shown in Table 2.5. The measurement errors of the two methods are less than 10%.

2.4.4 Conclusion

In recent years, the development of CCD sensor technology has provided a very convenient and intuitive measurement method for particle measurement, which can obtain particle size and morphology characteristics. However, for particles less than 3 μm, accurate measurement is quite difficult because of the resolution of CCD sensor. Based on the principle of multi-wavelength extinction method and the RGB imaging principle of colour CCD, a three-wavelength bands RGB extinction method using colour CCD sensor to measure the particle size of fine particles is proposed. The feasibility of this method is verified by numerical simulation and experiment. The measurement range of the three-wavelength bands RGB extinction method is from 0.5 μm to 5 μm, which extends the measurement range of particle size by the image method, and the particle size measured by colour camera from submicron to hundreds of microns is realised.

2.5 Ultrasonic scattering

In recent years, ultrasound technique is widely used in particle sizing. Compared to optical and electrical methods, ultrasound technique has merits on its availability to optically opaque, undiluted and high/non-conducting dispersions [61]. The ultrasonic attenuation spectroscopy (UAS) method is especially considered to be very promising for nano-particle suspension system as a rapid on-line measurement technology, which has advantages of simple structure, non-invasive and on-line measurement and being available to relatively dense suspensions. UAS for particle sizing mainly includes three aspects: establishment of mathematical model, measurement of ultrasound attenuation spectroscopy and the retrieval of particle size distribution. The accuracy of the retrieval results would be influenced by selection of inappropriate inversion algorithms and unreasonable parameters.

2.5.1 Principle of ultrasonic scattering

The interaction between ultrasonic beams and particulate two-phase flow is a quite complicated process, in which the sound scattering phenomena must be taken into account in the first place. By assuming small particles as elastic objects, it can be analogous to the light scattering to reach some similar deduction, and certainly the shape of particles, the types of ultrasonic beams and incident angles, as well as the ratio of sound wavelength to particle size can be significant for the description of sound scattering. The analytical solution can be obtained for those regular-shaped particles like a sphere, while for more irregular particles the analytical solution is rarely available up to now. Numerical investigation was carried out [62] on acoustic scattering for non-spherical objects from gas to weak fluid and rigid spheroids, as well as the multiple scattering problems from an ensemble of scatterers [63]. However, for more cases, a numerical solving method must be relied on. For example, the boundary element method (BEM) can be employed for some special cases like the scattering sound pressure calculation for one or several irregular particles, which can give useful estimation for the shape effects on sound scattering directive properties and scattering cross section. Except scattering, there are still several important interaction behaviour, relevant with the viscous and thermal dissipations. A further observation of the aforementioned phenomena can help find obvious intrinsic differences in that the sound scattering merely causes the redirection of acoustic energy away from the incident beam, while the dissipations essentially give rise to the conversion from acoustic energy to heat. Dukhin and Goetz [64] pointed out that in fact the scattering and dissipations can be divided into two separated components and analysed independently, which actually facilitated the understanding of acoustic loss and simplified the mathematical modelling. Typically, a complete mathematical model is still necessary for describing all these physical phenomena and establishing the rigorous relationship between the particle size and the ultrasonic attenuation spectrum.

When modelling the interaction between particles and the surrounding sound waves, Epstein and Carhart [65] focused on acoustic scattering and attenuation by droplets suspended in the air. Allegra and Hawley [66] further extended EC's model to solid particles and unified the formulisation of ultrasound scattering for particle sizing by fulfilling the ECAH model. As a typical linear scattering theory, their theoretical model fully accounted for the influence on acoustic waves of mixture viscous and thermal dissipation, which coupled to the scattering in the form of shear waves and thermal waves, respectively. By solving the complex wave number, the macroscopic phenomena like ultrasonic attenuation and phase shift are judiciously connected with the microscopic process like the interaction between a single particle and surrounding waves.

For a poly-dispersed particle system, the attenuation coefficient α_s can be evaluated from the algebraic superposition of contribution from every single particle, namely:

$$\alpha_s = \frac{3\varphi}{16k^2}\sum_{j=1}^{N}\frac{q_j}{D_j^3}\sum_{n=0}^{\infty}(2n+1)\mathrm{Re}(A_n(D_j,f)) \tag{2.35}$$

where φ denotes the volume fraction of particles, k is the wavenumber of incident compression wave, D is the diameter of the particle and q_i is the frequency distribution of particles with radii between D_j and D_{j+1}. The summation of series of the compression wave-scattering coefficients A_n (a function of particle diameter D and ultrasonic frequency f) plays the key role for the calculation of the frequency-dependent attenuation coefficients. The solution of A_n finally reduces to a six-order complex linear equations, whose coefficient matrix elements are composed of spherical Bessel and Hankel functions with complex variables as arguments.

Figure 2.24 shows the numerical results from simulation of acoustic propagation in 0.5% v/v silver aqueous suspension [67]. It manifests that calculation range extends to the ultrasonic frequency 1–100 MHz with the particle size 1 nm–1000 μm, which indeed covers the major particle size range under potential investigation, although particles outside this range have been successfully measured. At the same time, it can be observed that the attenuations grow gradually with the increasing frequency. For a fairly small frequency like 1 MHz, a sensitive peak can be found for particles in the submicron range (i.e. 100 nm–1,000 nm) due to dissipation mechanism, which indicates the ultrasonic signal would vary with the particle size seriously. With the increase in frequency, the peak moves towards the smaller particles direction. Thus, a broader frequency range can undoubtedly contain more information about the particle size, accordingly in favour of the improvement of resolution of particle sizing.

2.5.2 Data inversion for particle sizing

The inversion algorithms to determine particle size distribution have been generally classified into two types: dependent and independent modes. The dependent mode algorithm assumes that the particle system to be measured conforms to one or two given size distribution functions, that is lognormal or Rosin–Rammler [68], and two (unimodal distribution assumption) or five (bimodal distribution assumption) characteristic parameters need to be determined by optimisation. So far, the Broyden–Fletcher–Goldfarb–Shanno (BFGS) and Davidon–Fletcher–Powell (DFP) [69] are two significant algorithms to perform the optimisation process to determine the particle size distribution, although they both possess the drawback that optimisation may be trapped into the local optimised solutions instead of the global one.

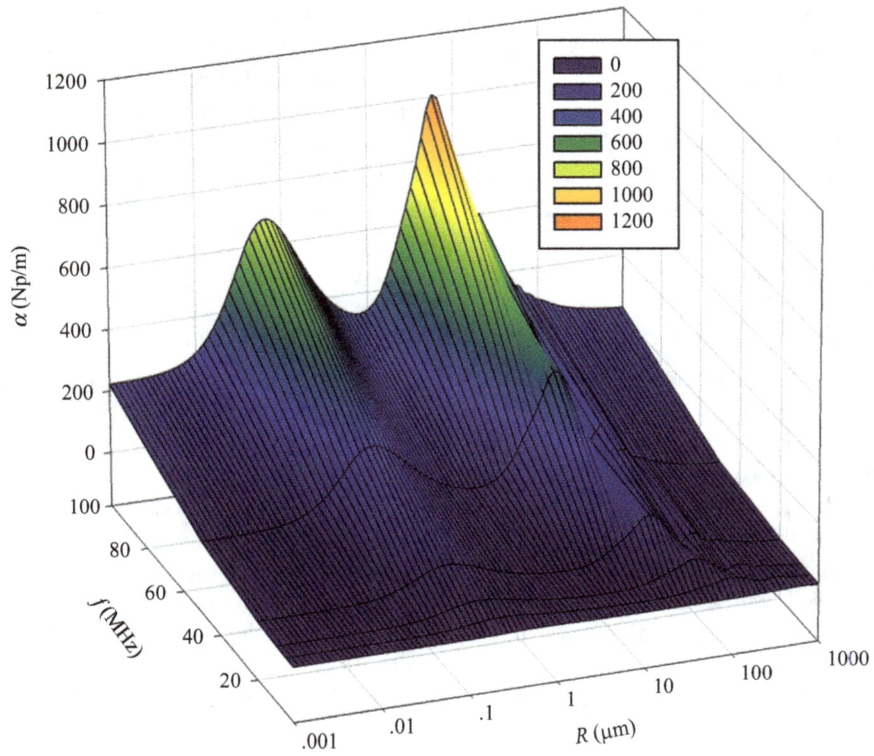

Figure 2.24 Predicted ultrasonic attenuation coefficients against particle diameter and ultrasonic frequency (0.5% v/v silver aqueous suspension)

Unlike the competitor, the independent mode algorithm does not assume a priori particle size distribution. The particle size solution is obtained directly by minimising the root mean square (RMS) error describing the difference between the simulated and the measured spectra:

$$E_{\mathrm{RMS}} = \sqrt{\frac{1}{N}\sum\nolimits_{j=1}^{N}\left(\frac{a_s - a_{\mathrm{meas},j}}{a_{\mathrm{meas},j}}\right)^2} \tag{2.36}$$

where $\alpha_{\mathrm{meas},j}$ is the measured spectrum, N is the number of selected frequencies in all the ultrasound attenuation spectrum.

Dependent model algorithms, such as BFGS algorithm, DFP algorithm and LM algorithm are usually employed to solve (2.36). However, when the initial value or the particle size range is inappropriately chosen, it would lead to an inaccurate particle size distribution. This situation can be avoided with the global optimisation genetic algorithm or other evolutionary algorithms (EA) because multiple search points are used to search information simultaneously.

Figure 2.25 shows the simulated particle size distribution with four kinds of algorithms (BFGS, DFP, LM and Standard Genetic-Algorithm (SGA)) for the particles with radius 5 μm, the particle range was assumed between 0 and 100 μm, and initial radius was 50 μm for local optimization algorithms. The objective function is minimised with four different optimisation algorithms and the particle size distribution can be then obtained. As shown in Figure 2.25, the median particle radius determined by BFGS, DFP and LM algorithm were 3.23 μm, 37.59 μm and 41.41 μm, respectively. The radius calculated by SGA was 5.72 μm, which was in good agreement with given parameter.

2.5.3 Measurement system for ultrasonic attenuation

A measurement setup configuration and its working process are shown in Figure 2.26.

A pair of ultrasonic wave transducers (in green colour) are employed and aligned on both sides of the detection area. The transducer A serves as the transmitter/receiver transducer, working with the ultrasonic pulse signal transmitter/receiver (Panametrics 5800PR, highest excitation frequency 35 MHz) in the pulse-echo manner, and the transducer B is the same model but only takes effect in receiving the ultrasonic signals combined with another ultrasonic pulse signal transmitter/receiver (Panametrics 5077PR, highest excitation frequency 10 MHz). The measurement apparatus has a geometrically symmetrical structure and consists of two pieces of plexiglass buffer rods with the thickness 10 mm and measurement cell 6.5 mm, where the buffer rod

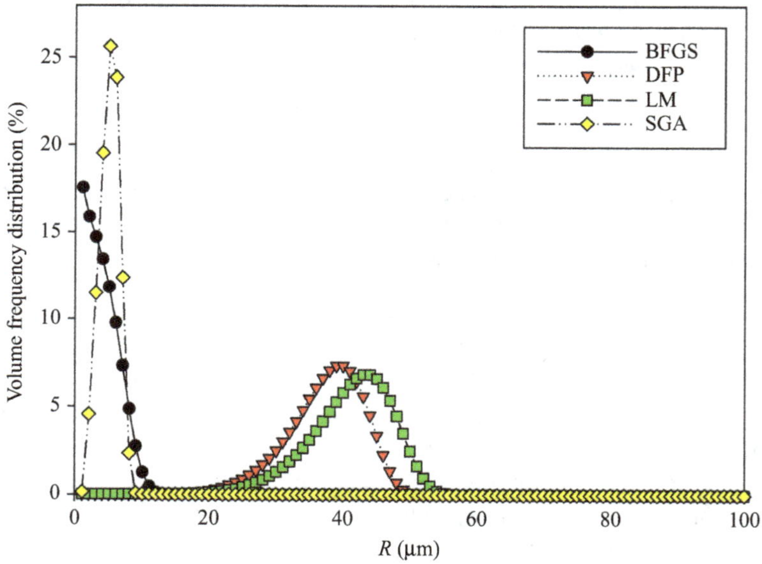

Figure 2.25 Simulated results with four different algorithms for particles ($R = 5$ μm)

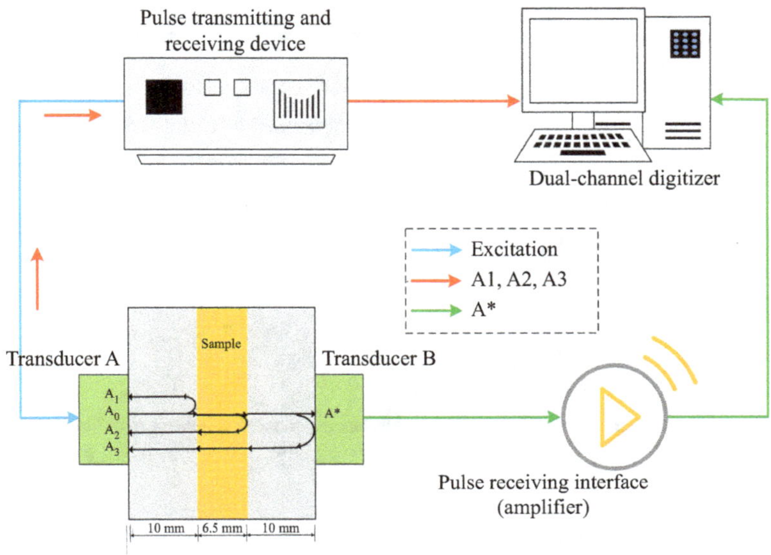

Figure 2.26 Ultrasonic spectrometer used to measure ultrasonic properties of silicon aqueous suspensions

should be optimised carefully in advance to avoid confusion due to echo interference phenomenon.

There are three factors to be considered when selecting an appropriate buffer rod for the optimal experimental results: smoothness, thickness and material. First, the surface of the buffer rod must have a good smoothness to guarantee the ultrasonic transducer can be closely aligned with its surface. Second, the thickness of the buffer rod should separate the reflection signals effectively. Finally, the buffer rod should possess a suitable acoustic impedance to ensure a good SNR. One more concern is that the temperature should be carefully kept around 25 °C by adopting a thermostat bath to avoid the effects of temperature changes on concerning sound propagation. To acquire the data of reflection and transmission signals and analyse the waveform synchronously, a computer with a dual-channel data acquisition card (NI PCI-5114, with a sampling rate of 250 MS/s) were used.

Figure 2.27 shows the measured particle size distribution results for two kinds of micron-sized glass beads-glycerol suspensions (particle density is 4.19 g/cm^3, volume concentration is 1% and sample numbers are SB040103 and SB040107, respectively). The samples were mechanically stirred for 3 min before measurement by ultrasonic system. The comparison results of the particle size distribution for these two kinds of samples were obtained by microscope image analysis (MIA) (PIP8.1, OMEC) and the improved genetic algorithm 3 (IGA3) [70] by ultrasonic attenuation spectroscopy (UAS), respectively. The radiuses obtained by MIA were 9.62 and 27.09 μm. The retrieval results by IGA3 were 9.57 and 27.67 μm. It shows that two kinds of glass beads-glycerol suspension samples could be clearly

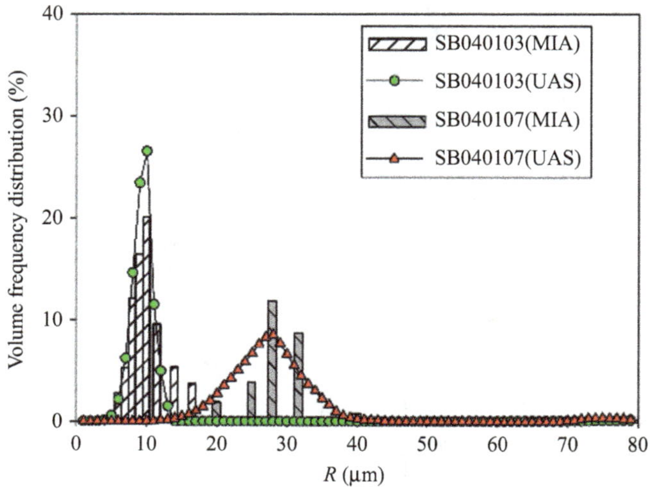

Figure 2.27 Comparison of particle size distribution for two kinds of glycerol-glass beads suspension samples (volume fraction: 1%) measured by MIA and obtained with IGA3 inversion method by UAS

distinguished. The relative error for retrieved particle size distribution between IGA3 and MIA for sample SB040107 was 5.8% larger than that of sample SB040103. Because particle size of sample SB040107 is larger, it resulted in relatively higher speed of particle falling during measurement process.

2.6 Future prospects

Light and ultrasonic scattering have been widely applied for in-line measurement, especially for particles or granular phases in two-phase flow phenomena. There have been rapid progresses in the past decades in developing different kinds of instruments based on these phenomena. It is believed that the most promising of these technologies will soon be found in ultra-fast and in situ characterisation for nano-particles, in-line monitor of ultra-low particle emissions and so on.

References

[1] Tropea C, 'Optical particle characterisation in flows', *Annu. Rev. Fluid Mech.*, 2011, **43**, pp 399–426

[2] Mie G, 'Beiträge zur Optik Trüber Medien, Speziell Kolloidaler Metallösungen', *Ann. Der Physik*, 1908, **25**, pp 377–445

[3] Hergert W, Wriedt T, *The Mie Theory: Basics and Applications*, Springer, 2012

[4] van de Hulst C H, *Light Scattering by Small particles*, Dover, 1981

[5] Shen J, Cai X, 'Algorithm of numerical calculation on Lorentz Mie theory', *PIERS Online 1*, 2005, pp 691–694

[6] Gouesbet G, Maheu B, Gréhan G, 'Light scattering from a sphere arbitrarily located in a Gaussian beam, using a Bromwich formulation', *J. Opt. Soc. Am. A*, 1988, **5** (9), pp 1427–1443

[7] Ren K, Gréhan G, Gouesbet G, 'Localized approximation of generalized Lorenz-Mie theory: Faster algorithm for computations of beam shape coefficients', *Part. Part. Syst. Charact.*, 1992, **9**, pp 144–150

[8] Lock J A, Gouesbet G, 'Rigorous justification of the localized approximation to the beam-shape coefficients in generalized Lorenz–Mie theory. I. On-axis beams', *J. Opt. Soc. Am. A*, 1994, **11** (9), pp 2503–2515

[9] Gouesbet G, Lock J A, 'Rigorous justification of the localized approximation to the beam-shape coefficients in generalized Lorenz–Mie theory, II, Off-axis beams', *J. Opt. Soc. Am. A*, 1994, **11** (9), pp 2516–2525

[10] Lock J A, 'Improved Gaussian beam-scattering algorithm', *Appl. Opt.*, 1995, **34** (3), pp 559–570

[11] Wang J, Gouesbet G, 'Note on the use of localized beam models for light scattering theories in spherical coordinates', *Appl. Opt.*, 2012, **51** (17), pp 3832–3836

[12] Gouesbet G, Lock J A, 'On the description of electromagnetic arbitrary shaped beams: The relationship between beam shape coefficients and plane wave spectra', *J. Quant. Spectrosc. Radiat. Transfer*, 2015, **162**, pp 18–30

[13] Shen J, Jia X, Yu H, 'Compact formulation of the beam shape coefficients for elliptical Gaussian beam based on localized approximation', *J. Opt. Soc. Am. A*, 2016, **33** (11), pp 2256–2263

[14] Barton J P, 'Internal and near-surface electromagnetic fields for a spheroidal particle with arbitrary illumination', *Appl. Opt.*, 1995, **34** (24), pp 5542–5551

[15] Barton J P, 'Internal, near-surface, and scattered electromagnetic fields for a layered spheroid with arbitrary illumination', *Appl. Opt.*, 2001, **40** (21), pp 3598–3607

[16] Han Y, Gréhan G, Gouesbet G, 'Generalized Lorenz-Mie theory for a spheroidal particle with off-axis Gaussian-beam illumination', *Appl. Opt.*, 2003, **42** (33), pp 6621–6629

[17] Xu F, Ren K, Gérard G, Gérard G, Cai X, 'Generalized Lorenz-Mie theory for an arbitrarily oriented, located, and shaped beam scattered by a homogeneous spheroid', *J. Opt. Soc. Am. A*, 2007, **24** (1), pp 119–131

[18] Han Y, Zhang Y, Zhang H, Han G, 'Scattering of typical particles by beam shape in oblique illumination', *J. Quant. Spectrosc. Radiat. Transfer*, 2009, **110**, pp 1375–1381

[19] Lock J A, 'Contribution of high-order rainbows to the scattering of a Gaussian laser beam by a spherical particle', *J. Opt. Soc. Am. A*, 1993, **10** (4), pp 693–706

[20] Lock J A, Jamison J M, Lin C Y, 'Rainbow scattering by a coated sphere', *Appl. Opt.*, 1994, **33** (21), pp 4677–4690

[21] Lock J A, Adler C L, 'Debye-series analysis of the first-order rainbow produced in scattering of a diagonally incident plane wave by a circular cylinder', *J. Opt. Soc. Am. A*, 1997, **14** (6), pp 1316–1328

[22] Gouesbet G, 'Debye series formulation for generalized Lorenz–Mie theory with the Bromwich method', *Part. Part. Syst. Charact.*, 2010, **20** (6), pp 382–386

[23] Xu F, Lock J A, Cameron T, 'Debye series for light scattering by a spheroid', *J. Opt. Soc. Am. A*, 2010, **27** (4), pp 671–686

[24] Xu F, Lock J A, Gouesbet G, 'Debye series for light scattering by a nonspherical particle', *Phys. Rev. A*, 2010, **81**, 043824

[25] Xu F, Lock J A, 'Debye series for light scattering by a coated nonspherical particle', *Phys. Rev. A*, 2010, **81**, 063812

[26] Mishchenko M I, Hovenier J W, Travis L D, *Light Scattering by Nonspherical Particles: Theory, Measurements, and Applications*, Academic Press, 1999

[27] Gouesbet G, Gréhan G, *Generalised Lorenz-Mie Theories*, Springer, 2011

[28] Roth N, Anders K, Frohn A, 'Refractive-index measurements for correction of particle sizing methods', *Appl. Opt.*, 1991, **30** (33), pp 4960–4965.

[29] van Beeck J P, Giannoulis D, Zimmer L, Riethmuller M L, 'Global rainbow thermometry for droplet-temperature measurement', *Opt. Lett.*, 1999, **24** (23), pp 1696–1698

[30] Wang J, Gréhan G, Han Y, Saengkaew S, Gouesbet G, 'Numerical study of global rainbow technique: Sensitivity to non-sphericity of droplets', *Exp. Fluids*, 2011, **51** (1), pp 149–159.

[31] Saengkaew S, Bodoc V, Lavergne G, Grehan G, 'Application of global rainbow technique in sprays with a dependence of the refractive index on droplet size', *Opt. Commun.*, 2013, **286** (1), pp 295–303

[32] Wu X, Li C, Cao K, *et al.*, 'Instrumentation of rainbow refractometry: portable design and performance testing', *Laser Phys.*, 2018, **28**, 085604

[33] Li H, Rosebrock C D, Wriedt T, Mädler L, 'The effect of initial diameter on rainbow positions and temperature distributions of burning single-component n-Alkane droplets', *J. Quant. Spectrosc. Radiat. Transfer*, 2017, **195**, pp 164–175

[34] Ouattara M, Lamadie F, Sentis M, Onofri F, 'Droplet sizing and mixture fraction measurement in liquid-liquid flows with rainbow-angle diffractometry', *Appl. Opt.*, 2017, **56** (29), pp 8109–8120

[35] Wu Y, Li C, Cao J, Wu X, Saengkaew S, Chen L, Gréhan G, Cen K, 'Mixing ratio measurement in multiple sprays with global rainbow refractometry', *Exp. Therm. Fluid Sci.*, 2018, **98**, pp 309–316

[36] Van Beeck J P, Riethmuller M L, 'Rainbow phenomena applied to the measurement of droplet size and velocity and to the detection of nonsphericity', *Appl. Opt.*, 1996, **35** (13), pp 2259–2266

[37] Wu Y, Promvongsa J, Wu X, Cen K, Gréhan G, Saengkaew S, 'One-dimensional rainbow technique using Fourier domain filtering', *Opt. Express*, 2015, **23** (23), pp 30545–30556

[38] Wu Y, Promvongsa J, Saengkaew S, Wu X, Chen J, Gréhan G, 'Phase rainbow refractometry for accurate droplet variation characterisation', *Opt. Lett.*, 2016, **41** (20), pp 4672–4675

[39] Yu H, Hui S, Shen J, Tropea C, 'Measurements of refractive index and size of a spherical drop from Gaussian beam scattering in the primary rainbow region', *J. Quant. Spectrosc. Radiat. Transfer*, 2018, **207**, pp 83–88

[40] Yu H, Xu F, Tropea C, 'Spheroidal droplet measurements based on generalised rainbow patterns', *J. Quant. Spectrosc. Radiat. Transfer*, 2013, **126**, pp 105–112

[41] Yu H, Xu F, Tropea C, 'Optical caustics associated with the primary rainbow of oblate droplets: Simulation and application in non-sphericity measurement', *Opt. Express*, 2013, **21** (22), pp 25761–25771

[42] Marston P, H. Trinh E, 'Hyperbolic umbilic diffraction catastrophe and rainbow scattering from spheroidal drops', *Nature*, 1985, **312**, pp 529–531.

[43] Yu H, Shen J, Tropea C, Xu F, 'Model for computing optical caustic partitions for the primary rainbow from tilted spheriodal drops', *Opt. Lett.*, 2019, **44** (4), pp 823–826

[44] Tscharnuter W, 'Photon correlation spectroscopy in particle sizing', *Encyclopedia of Analytical Chemistry: Applications, Theory and Instrumentation*, 2006, pp 5469–5485

[45] King T A, 'Photon correlation spectroscopy: Technique and scope', in Bevington G A C (ed.), *Comprehensive Polymer Science and Supplements*, Pergamon: Amsterdam, 1989, pp 911–935

[46] Thomas J C, 'Photon correlation spectroscopy: Technique and instrumentation', *Proc. SPIE*, 1991, pp 2–18

[47] Pecora R, 'Quasi-elastic light scattering from macromolecules', *Annu. Rev. Biophys. Bioeng.*, 1972, **1**, pp 257–276

[48] Silbey R, Deutch J M, 'Quasi-elastic light scattering from large macromolecules', *J. Chem. Phys.*, 1972, **57** (11), pp 5010–5011

[49] Liu L, Cai X, Zhang J, Xu C, 'Research on a noval fast imaging dynamic light scattering method for nanoparticle size measurement', *Acta Optica Sinica*, 2015, **35** (5), pp 369–375 (in Chinese)

[50] Xu C, Cai X, Zhang J, Liu L, 'Fast nanoparticle sizing by image dynamic light scattering', *Particuology*, 2015, **19**, pp 82–85

[51] Wang Z, Cai X, Xu C, Liu L, 'Nanoparticle sizing by image processing with dynamic light scattering', *Acta Optica Sinica*, 2014, **34** (1), pp 282–287 (in Chinese)

[52] Zhou W, Zhang J, Liu L, Cai X, 'Ultrafast image-based dynamic light scattering for nanoparticle sizing', *Rev. Sci. Instrum.*, 2015, **86** (11), 115107

[53] Su M, Xu F, Cai X, Ren K, Shen J, 'Optimization of regularization parameter of inversion in particle sizing using light extinction method', *Particuology*, 2007, **5** (4), pp 295–299

[54] Zhang Q, Rubini P, 'Modelling of light extinction by soot particles', *Fire Safety J.*, 2011, **46** (3), pp 96–103

[55] Álvarez D, Castillo M, Payne F A, Cox R B, Xiong Y, 'Application of light extinction to determine stability of beef emulsions', *J. Food Eng.*, 2010, **96** (2), pp 309–315

[56] Kätzel U, Gruy F, Babick F, Klöden W, 'Light extinction at agglomerates of spheres: A practical test on the submicroscale', *J. Colloid Interf. Sci.*, 2005, **289** (1), pp 116–124

[57] Wu X, Qing W, Luo Z, Gao Q, Fang M, Cen K, 'Measurement of particle rotation in CFB riser with high-speed digital video camera technique', *Proc. CSEE*, 2005, **25** (7), pp 72–77 (in Chinese)

[58] Cai X, Su M, Shen J, *Particle Size Measurement Technology and Aplication*, Chemical Industry Press, 2010

[59] Cai X, Wang L, Ouyang X, 'A novel integrated probe system for measuring the two phase wet steam flow in steam turbine', *J. Eng. Themophys.*, 2001, **22** (6), pp 743–746

[60] Yang Y, Li Q, Li W, Chen X, 'Modulation transfer function for colour area CCD based on Bayer filtering', *Optics and Precision Eng.*, 2012, **20** (7), pp 1611–1618 (in Chinese)

[61] Povey M J W, 'Ultrasound particle sizing: A review', *Particuology*, 2013, **11** (2), pp 135–147

[62] Ye Z, Hoskinson E, Dewey R K, Ding L, Farmer D M, 'A method for acoustic scattering by slender bodies . 1. Theory and verification', *J. Acoust. Soc. Am.*, 1997, **102** (4), pp 1964–1976

[63] Alvarez A, Wang C, Ye Z, 'A numerical algorithm, of the multiple scattering from an ensemble of arbitrary scatterers', *J. Comput. Phys.*, 1999, **154** (1), pp 231–236

[64] Dukhin A S, Goetz P J, 'Acoustic and electroacoustic spectroscopy characterizing concentrated dispersions emulsions', *Adv. Colloid Interf.*, 2001, **92** (1–3), pp 73–132

[65] Epstein P S, Carhart R R, 'The absorption of sound in suspensions and emulsions: I. Water fog in air', *J. Acoust. Soc. Am.*, 1953, **25** (3), pp 553–565

[66] Allegra J R, Hawley S A, 'Attenuation of sound in suspensions and emulsions: theory and experiments', *J. Acoust. Soc. Am.*, 1972, **51** (5), pp 1545–1564

[67] Su M, Cai X, 'The numerical study of acoustical attenuation and velocity in the suspension of superfine particles', *Acta Acustica*, 2002, **27** (3), pp 218–222 (in Chinese)

[68] Allen T, *Powder Sampling and Particle Size Determination*, Elsevier, 2003

[69] Schilling R J, Harris S L, *Applied Numerical Methods for Engineers Using MATLAB and C*, Brooks/Cole Publishing Co., 2000

[70] Yang H, Su M, Wang X, Gu J, Cai X, 'Particle sizing with improved genetic algorithm by ultrasound attenuation spectroscopy', *Powder Technol.*, 2016, **304**, pp 20–26

Wu Zhou received her PhD degree in Engineering from Southeast University in 2011 and is now an Associate Professor in the School of Energy and Power Engineering, University of Shanghai for Science and Technology. Her research interests include particles and two-phase flow measurement, combustion measurement, mainly by image-based methods. She has two projects funded by Natural Scientific Foundation of China. She has published more than 20 papers.

Haitao Yu received his Dr.-rer.nat. degree from Technische Universität Darmstadt, Germany in 2013 and joined University of Shanghai for Science and Technology in 2014. His main research interests include theories of light scattering from particle and application technologies of optical particle sizing. He participated in two projects of German Research Foundations and had one project funded by Natural Science Foundation of China. He has published more than 20 papers. He is a youth committee member of Chinese Society of Particuology.

Mingxu Su received his PhD in Thermal Engineerin from the University of Shanghai for Science and Technology (USST) in 2002. In the same year, he got a postdoctoral position at UMR 6614 – CORIA in Rouen, France. He went back to USST being a teacher till now. His main research interests are methods of particle size characterisation based on ultrasonic spectroscopy and light scattering theories, as well as the industrial applications on particulate two-phase flow measurement. He has published more than 100 papers.

Xiaoshu Cai has been a professor in the Scholl of Energy and Power Engineering, University of Shanghai for Science and Technology since 1996. His main research interests include measurement methods and instruments for particle sizing, multi-phase flow, environmental monitoring and combustion, and thermal power engineering and turbomachinery. He has published more than 200 papers and few books. He is the editorial body member of several journals, including *Particuology*, *IMechE Part A: Journal of Power and Energy*, *Frontiers in Energy*, *KONA* and several Chinese journals, guest editor of some journal special issues and Co-joint Professor of Newcastle University of Australia and visiting professors at some other universities.

Chapter 3

Smart visual sensors in robotic systems

Pooja Agrawal[1] and Laxmidher Behera[1]

List of abbreviations

AC	adaptive critic
AGV	autonomous ground robot
CMAC	cerebellar model articulation controller
DHDP	dual Heuristic dynamic programming
FOSMC	fractional order sliding mode controller
HDP	Heuristic dynamic programming
IBVS	image-based visual servoing
IK	inverse kinematic
KSOMSC	KSOM-based sub-clustering
LWPR	locally weighted projection regression
MLP	multiple layer perceptron
PBVS	position-based visual servoing
PI	policy iteration
RBFN	radial basis function network
SMC	sliding mode controller
SOM	self-organising map
VI	value iteration

3.1 Introduction

This chapter presents smart visual sensing in robotic system. Smart robotic systems are based on visual feedback and have many applications in real-time, such as in warehouse, search and rescue operations and health care. The objective of visual sensing is to control a robotic system while an error corrector loop is established in desired visual features. This is called visual servoing or hand-eye coordination.

[1]Department of Electrical Engineering, Indian Institute of Technology, Kanpur, India

Based on the camera and robot placement, visual servoing systems can be categorised in eye-in-hand and eye-to-hand configurations. This can be further classified based on the number of cameras used for visual servoing operations. This chapter provides an overview of camera models and calibration process. Camera calibration is a necessary step to determine the camera intrinsic and extrinsic parameters using data observed by the camera. Several techniques, such as 3D reference object-based calibration, 2D plane-based calibration and self-calibration are reviewed. A review of intelligent hand-eye coordination techniques in the literature is also presented. This chapter includes both position and image-based visual servoing using neural and fuzzy networks, Kohonen self-organising map and other learning models. Comparative studies are provided to highlight the advantages and limitations. This chapter ends with experimental results in some interesting applications, including warehouse automation and implementation issues of visual sensing due to camera, images processing and feature extraction.

Conventional robotic systems highly depend on environment structure. In other words, for successful operations, robot systems require structured environments. For successful operations in dynamic environments, a smart robotic system is required, which can adapt according to the environments.

Smart robotic systems have many applications in warehouse automation, search and rescue operations, health care and entertainment besides various other industrial uses. It is often required that such systems operate autonomously in a mostly unstructured as well as unknown environment. This can be achieved by integrating the robots with vision, which makes them be adaptive and flexible. A smart visual sensing system is capable of real-time processing of visual data for efficient control and decision. For implementation of a smart robot system in real-time to assist human beings, a continuous learning is required to adapt the dynamically changing environments.

To achieve the excellence similar to human, visual feedback-based redundant manipulator is required to work in a dynamically changing and cluttered environment. Vision provides continuous information about dynamic environments. Changes in environment introduce a kinematic redundancy in the manipulator for handling complex tasks. Control of a redundant manipulator using visual feedback is a challenging task. This involves visual servoing and redundant manipulator control. The task of controlling robots using visual feedback is known as visual servoing introduced by Hill and Parker [1] in 1979. Various tutorials on visual servoing approaches are given in [2–5]. In literature, visual servoing term is also referred as hand-eye coordination [6,7], visual-motor coordination [8–11] and visual motor control [12]. Based on the servoing architecture, visual servoing approaches are mainly classified into two categories based on control design scheme. These are dynamic-look-and-move system (indirect visual servo) and direct visual servo system [13].

In general, visual servoing is a process of computing the end-effector velocity to reach the desired position using the image features obtained through the visual feedback [14]. The concept of visual servoing was discussed by Hutchinson *et al.* [2] in 1996. A detailed survey on visual servoing can be found in [3]. Visual feedback gives information on a dynamic environment. It basically assumes that there exists a non-redundant manipulator, which can generate the desired end-effector velocity with its own inverse kinematic algorithm. A non-redundant manipulator can achieve the end-effector velocity estimated from visual servoing with unique joint angle configuration. In a dynamic environment, the available unique joint angle configuration may become infeasible to position the end-effector, due to the presence of obstacles and the physical constraints. This necessitates the use of redundant manipulators for vision-based control in a dynamic environment. Theoretically, many choices of joint angle configuration exist for redundant manipulators to achieve the estimated end-effector velocity. The excess DOF can be effectively used for performing additional constraints introduced by the dynamic environment. An optimal joint angle configuration needs to be selected while these additional constraints are satisfied. This is popularly known as redundancy resolution.

For a manipulator with degree-of-freedom more than six, an efficient solution of inverse kinematic (IK) is needed. As the system is highly nonlinear and there exist multiple solutions, it is difficult to find IK map. There is a possibility to have an infinite number of solutions in case of redundant manipulator, which causes closed form solution impossible. These problems can be solved by classical techniques and learning-based techniques.

Classical visual servoing robot manipulator approaches require an accurate forward kinematic model to resolve the redundancy. Although these approaches work well, their performance deteriorates for inaccurate models. Pseudo-inverse computation is required for visual robot control and the pseudo-inverse is sensitive to parameter variations. Model inaccuracies pose a major challenge for pseudo-inverse computation as it causes parameter variations, which may eventually result in controller instability. In addition, classical approaches need pseudo-inverse of both image and kinematic Jacobian at each instant, which is computationally intensive. The pseudo-inverse computation poses a major challenge in real-time implementation with degree-of-freedom.

On the other hand, learning-based approaches are used to learn either the inverse kinematic map or the forward kinematic map to control the manipulator. Learning-based approaches do not require the complete knowledge of the kinematic model. Moreover, these approaches are computationally efficient as compared to model-based approaches. This chapter discusses the learning-based algorithms to resolve the redundancy while controlling the manipulator directly from the vision space.

In past few decades, learning-based approaches have gained attention due to their excellent performance capability. However, some issues have to be dealt with such as convergence rate, amount of training data needed, accuracy of the solution and smoothness in trajectory. The issues mentioned above are addressed in [11,12,15–17] by new learning-based approaches.

In literature, various learning architectures have been suggested for learning of the manipulator inverse kinematic map. These are multiple layer perceptron (MLP) [18] and radial basis function network (RBFN) [19,20], CMAC networks [21,22], locally weighted projection regression (LWPR) [23], Hopfield Networks [24] and self-organising maps (SOM) [9,25]. The learning-based methods can be divided into three different classes, namely, direct inversion modelling [9,26], distal learning [27] and differential kinematics models [28,29]. Direct inverse modelling learns the inverse kinematics map directly. Due to the topology preserving property, Kohonen's SOM [30] is used extensively for direct modelling for visual motor control. Barreto *et al.* [31] have given a detail survey of SOM-based approaches. A 3D SOM-based neural architecture is used by Martinetz *et al.* for non-redundant [9] and redundant manipulators [32], respectively, which was extended later by Walter *et al.* [10]. To reduce on-line training requirement burden, a hybrid method is proposed by Behera *et al.* [11], where an on-line training network is used for fine-tune of the parameters. The distal learning approach [27] is an indirect method, where the forward kinematic model is trained using a neural network. Differential kinematic represents the forward kinematics in differential form.

3.2 Camera models and calibration

Robot end-effector and target are viewed by cameras. World coordinates of them are not directly available to derive a forward kinematic model. The Cartesian coordinates of a point of interest have to be converted to corresponding image pixel coordinates for the forward model. This shows the requirement of a camera model. The process of finding model from un-calibrated camera is called camera calibration.

3.2.1 Camera models

A camera model describes a mathematical projection of a 3D object space on a 2D image space. This subsection discusses the various camera models [33–37].

3.2.1.1 Perspective camera

Perspective projection is a nonlinear mapping of a point from Cartesian space (3D) to image space (2D). This camera model can be represented by CMOS and CCD cameras. This is most often referred as pinhole camera, which is widely used and simple. Figure 3.1 shows a standard pinhole camera model, where f is the camera focal length and $(\boldsymbol{X}, \boldsymbol{Y}, \boldsymbol{Z})$ is the 3D point projected at $(\boldsymbol{u}, \boldsymbol{v})$ position in 2D image plane.

A line perpendicular to the image plane and passing through the camera centre is known as principal axis or optical axis. Intersection of the image plane with the principal axis is called principal point or optical point. Note that the perspective projection preserves linearity but does not preserve parallelism. The relation between the 2D and 3D point is given as

$$u = f\frac{X}{Z}, \quad v = f\frac{Y}{Z} \tag{3.1}$$

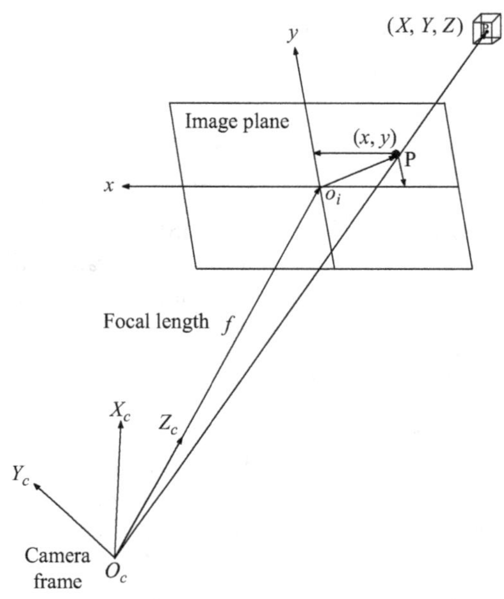

Figure 3.1 Standard pinhole camera model

The above relation shows that in perspective projection nearer object appears bigger than distant object. Since camera sensors are not exactly in square shape, and (u, v) can be written as

$$u = f_u \frac{X}{Z}, \quad v = f_v \frac{Y}{Z} \tag{3.2}$$

where $f_u = \kappa f$ and $f_v = lf$

Terms κ and l are scale parameters in pixels/m. Please note that image centre and principal point c may not always be the same. Considering that the location of the point c in the image plane is at (c_u, c_v), (3.2) can be written as

$$u = f_u \frac{X}{Z} + c_u, \quad v = f_v \frac{Y}{Z} + c_v \tag{3.3}$$

The image plane may not be exactly rectangular. Suppose that θ is a skew angle between the x and y axes. Equation (3.3) can be written as

$$u = f_u \frac{X}{Z} - f_u \cot\theta \frac{X}{Z} + c_u, \quad v = \frac{f_v}{\sin\theta} \frac{Y}{Z} + c_v \tag{3.4}$$

The above equation can be simplified as

$$\tilde{u} = \frac{1}{Z} KX \tag{3.5}$$

where K is intrinsic parameter matrix as

$$K = \begin{bmatrix} f_u & -f_u \cot\theta & c_u \\ 0 & \dfrac{f_v}{\sin\theta} & c_v \\ 0 & 0 & 1 \end{bmatrix} \tag{3.6}$$

and $\tilde{u} = (u \;\; v \;\; 1)$ representing homogeneous coordinates.

The camera frame is not aligned with the world frame. In (3.5), X is in the camera coordinate frame. By rigid transformation, the following equation is obtained:

$$X = RX^W + T \tag{3.7}$$

where W is the world coordinate frame.

The extrinsic camera parameters rotational and translational are denoted by R and T, respectively. By combining intrinsic and extrinsic parameters using (3.5) and (3.7), the following equation is obtained.

$$\tilde{x} = \frac{1}{Z}K(RX^W + T) \tag{3.8}$$

3.2.1.2 Scaled orthographic camera

In many cases, approximation of perspective projection nonlinear mapping can be linearised by orthographic projection. Here, 3D point is assumed to be at infinitive and all the projection lines are parallel to the projective axis. Projection of a 3D point in image plane can be given as

$$u = sX, v = sY \tag{3.9}$$

where s is a fixed scale factor.

Orthographic projection models are only valid for smaller relative depth in the scene compared to the depth between the camera and the scene.

3.2.1.3 Affine camera

Affine camera is nothing more than a special case of scaled orthogonal projection by an arbitrary affine transformation. This camera model is known as a more general non-perspective camera. Primary advantage of the affine camera is that it is a good approximation to the perspective projection.

3.2.2 Camera calibration

Camera model describes a mathematical projection of a 3D object space on a 2D image space. Camera calibration is well studied. There exist many techniques for camera calibration [38–45]. This is a necessary step to determine the camera parameters from images captured by the camera. The camera calibration process is independent of the camera model. Basically, there are two types of parameters. These are intrinsic and extrinsic parameters. The intrinsic parameters

(f, c_u, c_v, f_u, f_v) represent the camera internal geometric and optical characteristics, whereas the extrinsic parameters (R, T) measure the position and orientation of the camera with respect to the world coordinate system. According to the dimension of the object, calibration techniques are broadly divided into three classes.

3.2.2.1 3D reference object-based calibration

For 3D camera calibration, the geometry of the observing calibration object should be known with good precision. The calibration object usually consists of two or three planes. Note that these planes are orthogonal to each other. Sometimes, a plane undergoing a precisely known translation is also used, which equivalently provides 3D reference points.

3.2.2.2 2D plane-based calibration

In the 2D plane-based calibration, observing object is a planar pattern at a few different orientations. 2D plane calibration techniques are different from Tsai's technique [46] as the knowledge of the plane motion is not necessary.

3.2.2.3 Self-calibration

Self-calibration does not require any calibration object. Since only image point correspondence is required. This technique can be considered as 0D approach. Only three images are sufficient to recover the internal and external parameters of the camera if images are taken from the same camera with fixed internal parameters. Although these approaches do not need any object for calibration, a large number of parameters must be estimated. This results in a complex mathematical problem.

Note that calibration performance depends on the user requirement. Therefore, no single calibration technique is the best for all. For a situation that a higher accuracy is demanded, a 3D calibration approach is preferable. However, 2D calibration is a good choice due to its good accuracy and ease of use.

3.3 Visual servoing

Visual feedback provides information on the dynamic environment and the object. Typically, vision-based manipulator control is implemented in open loop, and the manipulator first 'looks' and then 'moves'. Due to the model inaccuracies, this result is poor in positioning accuracy. An alternate approach, which is generally referred as visual servoing, uses a visual control loop. The visual servoing task can be achieved using either single or multiple cameras for giving visual feedback to the manipulator system.

Camera placement in robotic systems plays a vital role. In literature, a camera fixed with an end-effector is called an eye-in-hand configuration as shown in Figure 3.2(a). A camera placed in the environment is called an eye-to-hand configuration. Figure 3.2(b) shows the eye-to-hand configuration, which is also known as stand-alone camera system [3]. The forward differential kinematic relationship

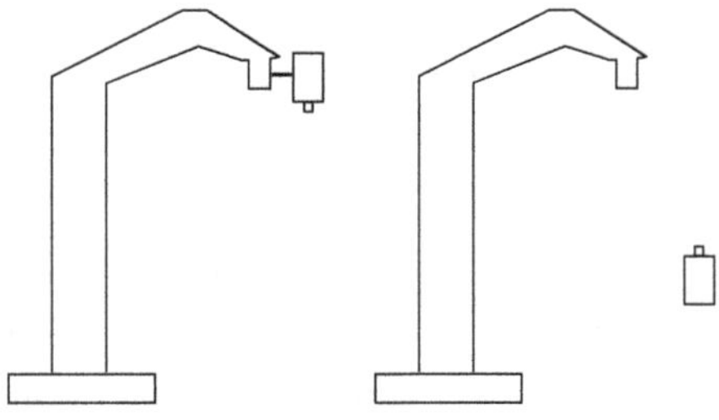

Figure 3.2 (a) Eye-in-hand configuration and (b) eye-to-hand configuration

between the joint angular velocity $\dot{\theta}$ and the Cartesian space velocity $\dot{\boldsymbol{X}}$ can be given as

$$\dot{\boldsymbol{X}} = J\dot{\theta} \tag{3.10}$$

where J is the kinematic Jacobian of the manipulator, which is not a square matrix in case of redundant manipulators.

This shows that there exists infinite joint angular velocity. For the end-effector velocity, a joint angular velocity needs to be computed using an inverse Jacobian, which exits only for square matrices and hence cannot be computed in the case of redundant manipulators. Thus, for redundant manipulator, the pseudo-inverse has to be computed.

For accomplishment of the visual servoing operations, image features are required for estimating the end-effector and the object positions in the vision space. The relation between the image feature motion and the motion of the end-effector current position can be represented using image Jacobian L:

$$\dot{\boldsymbol{u}} = L\dot{\boldsymbol{X}} \tag{3.11}$$

where $\boldsymbol{u}$ and $\boldsymbol{X}$ represent the image features and end-effector current position, $\dot{\boldsymbol{u}} \in R^n$ and L is a $p \times n$, which is known as image Jacobian matrix or interaction matrix [47,48].

Using the forward kinematic relationship given by (3.10), the relationship between $\dot{\boldsymbol{u}}$ and $\dot{\theta}$ may be written as

$$\dot{\boldsymbol{u}} = LJ\dot{\theta} = J_L\dot{\theta} \tag{3.12}$$

where J_L is the $p \times m$ Jacobian matrix.

Based on the visual information, visual servoing systems are mainly classified into three classes as follows.

3.3.1 Position-based visual servoing

In position-based visual servoing (PBVS) approach [49,50], image features $\boldsymbol{u}$ are detected using the camera pose (i.e. position and orientation) in a reference frame.

The approach considers that the camera model is known to estimate the target object pose. Because the PBVS approach requires 3D reconstruction of the object under consideration, it is prone to fail due to calibration errors. Figure 3.3 shows the position-based visual servoing approach.

Note that PBVS-based control schemes are sensitive to disturbance and measurement noise as target pose is estimated from images, which are subject to noises. Although position-based control suffers from image noise, it gives the change to define the task in the standard Cartesian frame.

3.3.2 Image-based visual servoing

Image-based visual servoing (IBVS) control approach uses image features directly to estimate the robot motion. In literature, it is also referred as 2D visual servoing [51,52]. This approach specifies the error signal between the actual and desired image features in a 2D image plane. As the image-based approach is directly based on image features, it may reduce computation delay. Also, it eliminates the necessity of image interpretation and error caused due to sensor modelling and camera calibration. However, controller design is challenging as the system is nonlinear and highly coupled. Image-based visual servoing structure is shown in Figure 3.4.

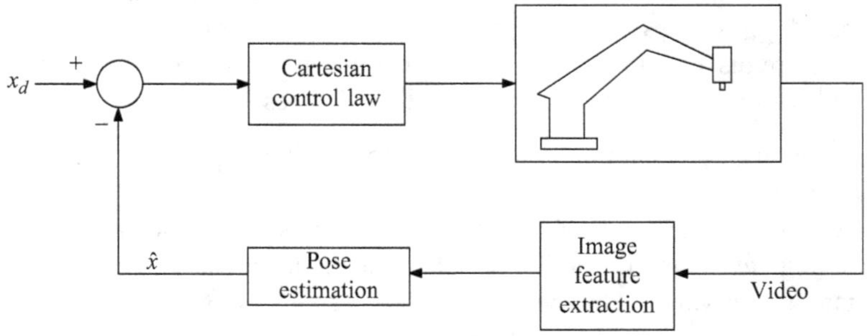

Figure 3.3 Dynamic position-based structure

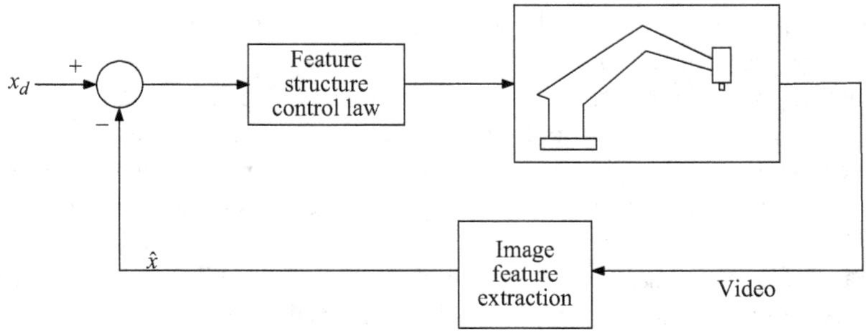

Figure 3.4 Dynamic image-based structure

Because the feedback errors are computed from the image plane, image-based controllers are more robust to disturbances than position-based controllers. However, online computation of image Jacobian is required, which is difficult to evaluate.

3.3.3 2-1/2-D visual servoing

This approach is a combination of both image and position-based approaches to avail the benefit of both the methods [49,53]. The error is defined by combining both the image space and the Cartesian space information to compute the end-effector velocity. It is also known as a hybrid visual servoing approach. Such hybrid schemes decouple the translation and rotation velocity of the end-effector by proper feature selections and resulting in accurate positioning.

3.4 Visual servoing control

3.4.1 Manipulator robot

The objective of the visual servoing control is to compute the desire Cartesian space velocity motion for the end-effector to reach the desired position in vision space asymptotically. Visual servoing approaches are mainly model-based and learning-based. Due to the various advantages of learning-based approaches over model-based approaches, in this chapter, the learning-based visual motor control approaches are discussed. Development of learning-based approaches is motivated due to following drawbacks of model-based approaches.

1. For redundancy resolution, visual servoing requires the exact computation of the image Jacobian L and kinematic Jacobian J. This requires the accurate knowledge of the workspace, which may not be available in the dynamic changing environment, and hence image Jacobian should be computed at each instant in real-time. On the other hand, classical model-based approaches require a pseudo-inverse of both image and kinematic Jacobian at each instant, which is computationally intensive. For higher degree-of-freedom manipulator, the pseudo-inverse computation poses a major challenge in real-time implementation. Thus, the efficient combined estimate of the pseudo-inverse of J and L using learning methodologies is an interesting research problem within learning-based paradigm.
2. Existing redundancy resolution schemes minimise an instantaneous cost function and achieve local optimum in real-time. Minimising an instantaneous cost function leads to instabilities, resulting in higher torque and joint angular velocity, for long trajectories spanning the entire workspace. Optimisation of a global cost function as an integral cost function over the entire trajectory leads to an off-line optimal control problem. Such global optimisation schemes are robust to instabilities, but cannot be implemented in real-time as required in a visual servoing situation. Hence the development of real-time optimal redundancy resolution schemes for global cost function is one of the unsolved

research problem in this area and worth investigating within learning paradigm.
3. Like kinematic visual control, optimality plays an important role in dynamic visual servoing. Optimisation in dynamic visual servoing results in minimisation of torque, effort while avoiding the obstacles and kinematic limits. An optimal control strategy, which is computationally efficient, robust to model inaccuracies and adaptive to environmental changes, is highly desirable. A learning-based paradigm for dynamic visual servoing also poses an interesting research problem.

These challenges necessitate the development of computationally efficient visual control schemes, which are robust to model inaccuracies and could directly affect the error in vision space from joint space. These goals can be achieved through learning-based approaches. Learning-based control schemes can adapt their behaviour according to the feedback received from the environment and are robust to model inaccuracy. An autonomous system, which can improve its performance with time, using the available feedback and cope up with model inaccuracies and changes occurring in a dynamic environment, is highly preferred for real-life implementation of the robotic systems. Robots are active agents and are amenable to learning-based schemes, which do not require the complete knowledge of the camera, manipulator and the environment model.

Learning-based visual servoing control approaches can be solved in open loop and closed loop framework. In open loop as shown in Figure 3.5, the inverse kinematic map using neural network is directly learned from the training data. Although open-loop approaches work well for simple pick-and-place kind applications, they suffer from operations which require more precision. On the other hand, closed-loop approaches can achieve difficult tasks with more precision. The closed-loop framework uses SOM and feed-forward networks for estimating the inverse kinematic relation as shown in Figure 3.6.

This chapter describes the learning-based visual servoing control approaches. The schematic of a typical vision-based manipulator control scheme in learning paradigm is shown in Figure 3.7. It consists of a stereo-vision system and a robot manipulator. Image processing as well as the learning-based controller are executed on a PC. The image processing unit is used to extract 4D image coordinate vectors to represent the current end-effector position $\boldsymbol{u}$, and the desired position $\boldsymbol{u}_d$.

The learning-based controller generates either the joint angle θ or the joint angular velocity $\dot{\theta}$, which is given to the servo unit to drive the robot manipulator so that the end-effector reaches the desired position.

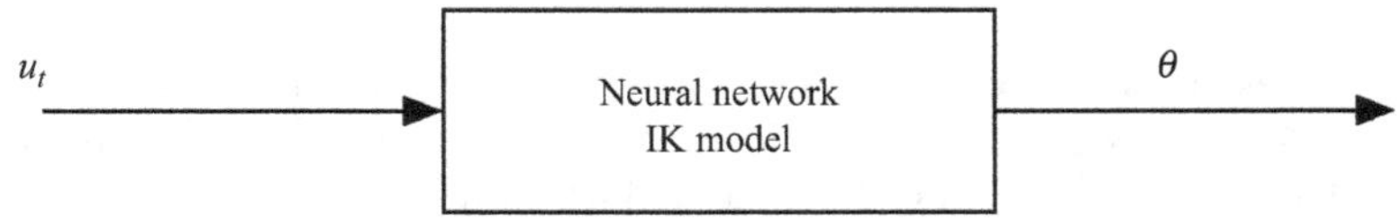

Figure 3.5 Inverse kinematic problem in an open-loop scheme

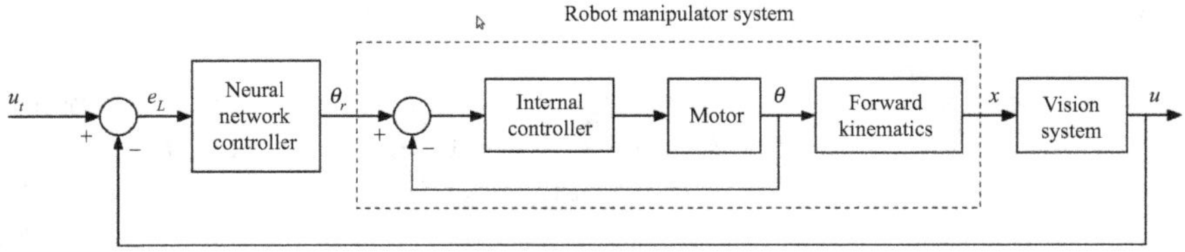

Figure 3.6 Inverse kinematic problem in a closed-loop scheme

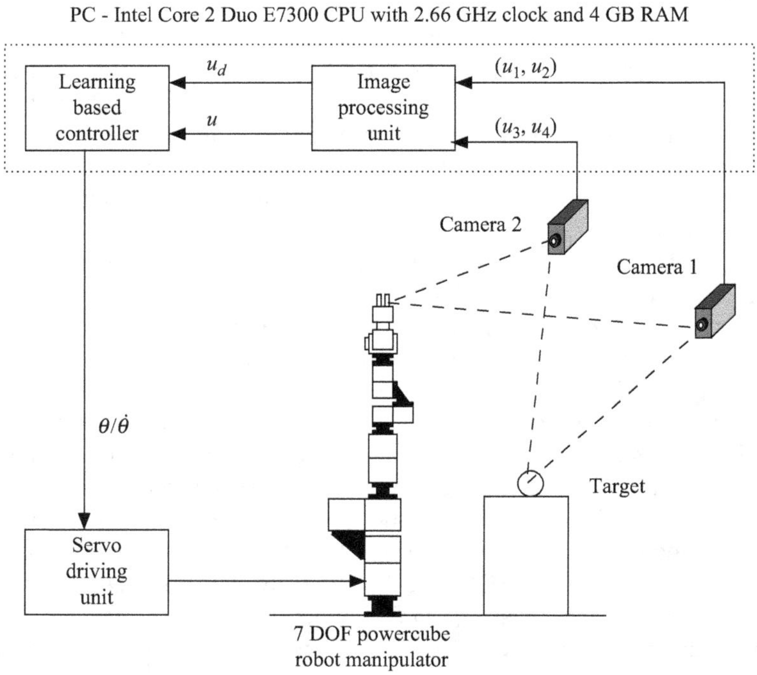

Figure 3.7 Closed-loop learning-based visual motor control system

3.4.1.1 Experimental setup

Figure 3.8 shows the actual setup of a seven DOF PowerCube™ robot manipulator used for experiments. Stereo vision setup fixed on the workspace is used to see the end-effector and target. The two Fire-i™ digital cameras, with a resolution of 320 × 240 pixels, located at the top corner is used for the experiment. The cameras are mounted in such a way that the larger portion of the available workspace can be covered.

A Cartesian workspace, visible by both the cameras, of dimension 600 mm × 500 mm × 500 mm is considered for experiments. A PC with Intel Core 2 Duo E7300 CPU with 2.66 GHz clock and 4 GB RAM is used for image processing and learning operation, where Debian 4.02 operating system is running in multi-user graphics mode with all the services enabled. The manipulator control in

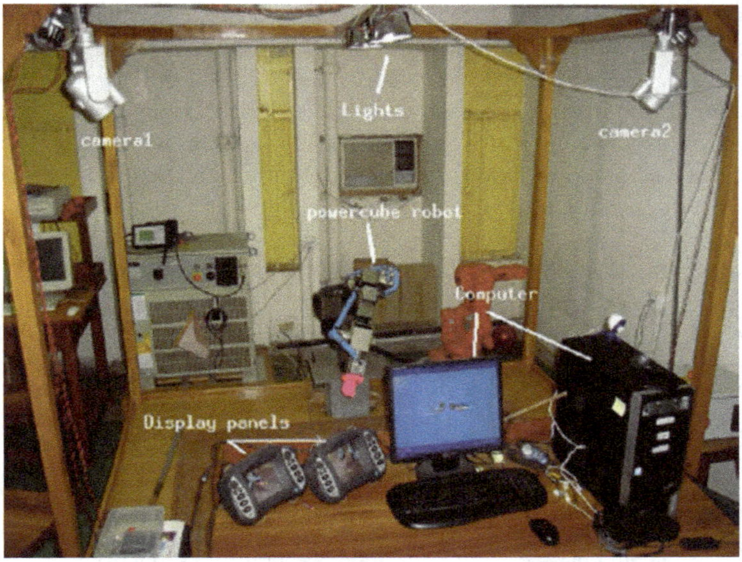

Figure 3.8 Experimental setup for VMC experiment

real-time is written in C programming language and image processing algorithms are developed using OpenCV [54] 4 GB RAM. The camera calibration is done using Wilson's C implementation of Tsai algorithm [55]. Image thresholding and filtering operations are used for extracting region of interest from the images captured by the cameras. For computation of necessary motion of the joint angle, centroid of the region is used. To reach the manipulator end-effector to the desired location, these computed joint angles are applied to the robot.

3.4.2 KSOM-based redundancy preserving network

For applications where high precision accuracy may not be required, SOM-based sub-clustering for learning inverse kinematic solution can be adopted [56]. In SOM-based sub-clustering approach training is done off-line with huge training data set (approximately 5 × 105) with a large size KSOM network (with 10 × 10 × 10 nodes). Due to off-line training, this approach cannot be used for task where high precision is required, such as welding or tool cutting. In addition, the number of sub-clusters have to be given a prior, which may lead to the formation of outliers. Additionally, the inverse kinematic solution is not conservative [57].

In SOM-based approach, inverse kinematic relation is difficult to learn due to one-to-many mapping with the input space mapped to a 3D lattice, where each lattice neuron represents a discrete cell. The linear map structure for the lattice neuron γ is obtained as

$$\theta - \theta_\gamma = A_\gamma (\boldsymbol{u} - \boldsymbol{w}_\gamma); A_\gamma = \frac{\partial r}{\partial \boldsymbol{u}}\bigg|_{(\boldsymbol{w}_\gamma, \theta_\gamma)} \tag{3.13}$$

where w_γ is the centre of the lattice neuron, A_γ is the inverse Jacobian matrix and θ_γ is the joint angle vector associated with γ lattice neuron.

Each lattice neuron γ is represented by a triplet (w_γ, θ_γ, A_γ), where w_γ discretises the 4D input space, θ_γ discretizes the 6D output space and the matrix A_γ of dimension 6×4 represents a linear map from input to output space. Because each weight vector w_γ is associated with only one joint angle vector θ_γ, this method gives a unique inverse kinematic solution even for a redundant manipulator.

Based on the Euclidean distance metric in the image coordinate (as the input) space for a given target position u_t, a winner neuron μ is selected, where the neuron with the weight vector closest to the target is considered a winner as given by

$$\mu = \arg\min_\gamma \|u_t - w_\gamma\| \tag{3.14}$$

A coarse movement θ_0^{out} is obtained as

$$\theta_0^{\text{out}} = s^{-1} \sum_\gamma h_\gamma \left(\theta_\gamma + A_\gamma \left(\boldsymbol{u}_t - \boldsymbol{w}_\gamma\right)\right) \tag{3.15}$$

where $s = \sum_\gamma h_\gamma$ and $h_\gamma = e^{-\left(\frac{\|\mu-\gamma\|^2}{2\sigma^2}\right)}$ are the neighbourhood functions.

Because of this coarse movement, the end-effector reaches a position $\boldsymbol{v}_0$ and a correcting fine movement θ_1^{out} is computed as

$$\theta_1^{\text{out}} = \theta_0^{\text{out}} + s^{-1} \sum_\gamma h_\gamma A_\gamma (\boldsymbol{u}_t - \boldsymbol{v}_0) \tag{3.16}$$

This corrective movement taken by the above equation is a final movement of the end-effector to a position $\boldsymbol{v}_1$. Although several corrective movements can be applied for good tracking accuracy, one corrective movement is enough for good tracking accuracy.

Network parameters can be updated as follows:

$$\Delta \boldsymbol{v} = \boldsymbol{v}_1 - \boldsymbol{v}_0, \Delta\theta^{\text{out}} = \theta_1^{\text{out}} - \theta_0^{\text{out}} \tag{3.17}$$

$$\Delta\theta_\gamma = \frac{h_\gamma}{s} \left[\theta_0^{\text{out}} - s^{-1} \sum_\gamma h_\gamma \left(\theta_\gamma + A_\gamma \left(\boldsymbol{v}_0 - \boldsymbol{w}_\gamma\right)\right)\right] \tag{3.18}$$

$$\Delta A_\gamma = \frac{h_\gamma}{s\|\Delta \boldsymbol{v}\|^2} \left[\Delta\theta^{\text{out}} - s^{-1} \sum_\gamma h_\gamma A_\gamma \Delta \boldsymbol{v}\right] \Delta \boldsymbol{v}^T \tag{3.19}$$

$$\boldsymbol{w}_\gamma = \boldsymbol{w}_\gamma + \eta_w h_\gamma \left(\boldsymbol{u}_t - \boldsymbol{w}_\gamma\right), \boldsymbol{\theta}_\gamma = \boldsymbol{\theta}_\gamma + \eta_t \Delta\theta_\gamma \tag{3.20}$$

$$A_\gamma = A_\gamma + \eta_a \Delta A_\gamma \tag{3.21}$$

where

$$\eta = \eta_{\text{init}} \left(\frac{\eta_{\text{fin}}}{\eta_{\text{init}}}\right)^{(t/t_{\max})} \tag{3.22}$$

In (3.22), $\eta \in (\eta_w, \eta_t, \eta_a, \sigma)$ is a user-defined parameter and varies between initial value η_{init} and η_{fin} with iteration step t from 0 to $t_{\max}$ during the training phase.

The objective of the update law is to reduce the error $(u_t - w_\gamma)$. The desired joint angle θ^* can be computed as

$$\theta^* = \theta_\gamma + A_\gamma(\boldsymbol{u}_t - \boldsymbol{w}_\gamma) \tag{3.23}$$

where A_γ is updated using (3.21) so that the following cost function is minimised:

$$E_1 = \frac{1}{2}\left(\Delta\theta^{\text{out}} - A_\gamma \Delta\boldsymbol{v}\right)^2 \tag{3.24}$$

Now, θ_γ can be updated to minimise the following error function:

$$E_2 = \frac{1}{2}\left(\theta_0^{\text{out}} - \frac{\sum_\gamma h_\gamma(\theta_\gamma + A_\gamma(\boldsymbol{v}_0 - \boldsymbol{w}_\gamma))}{\sum_\gamma h_\gamma}\right)^2 \tag{3.25}$$

It is observed that the SOM-based lattice neurons preserve the topology of the input space, but it fails to preserve the topology of the output space. The standard SOM algorithm gives a unique inverse kinematic solution for any target in the workspace, which may not be required in case of redundant manipulators.

KSOM sub-clustering approaches [56] overcomes the limitations of sub-clustering based SOM approach by deciding the number of clusters during on-line training adaptively and by preserving the smoothness of joint angle trajectories using a neighbourhood concept. Thus, this approach does not require to decide sub-cluster a prior. These modifications result in high precision, reduction in amount of data required for training, preserve conservative property of the inverse kinematic solution and reduce the position accuracy dependency on the initial condition of the network. In addition, the number of steps demanded for obtaining lesser positioning error increases very slowly in case of multi-step movement. The architecture of the sub-clustering in joint space is shown in Figure 3.9, which consists of a 3D SOM lattice. In Figure 3.9, each lattice neuron is connected with weight and joint angle vectors.

Using only three degree of freedom, it may be possible to reach at any point in 3D Cartesian workspace of the manipulator. Higher degree-of-freedom increases the ability of the manipulator to perform more than one task, such as obstacle avoidance meeting joint angle limits or satisfying other motion constraints. Proper utilisation of the extra degree-of-freedom is still an open research problem. Redundancy resolution using KSOM networks has been used by many authors. Martinetz *et al.* [58] applied KSOM algorithm to a five DOF manipulator and claimed that the redundancy is resolved 'naturally' by using the 'lazy arm method' due to neighbourhood function. SOM is also used for obstacle avoidance by Han *et al.* [59] and Zha *et al.* [60]. Note that most of these methods are applied on three, four or five DOF manipulators and learn the network during training phase itself. This means redundancy solution cannot be changed once the network is trained. Redundant manipulator gives many joint angle combinations to reach the same end-effector configuration. The number of sub-clusters to be associated with each

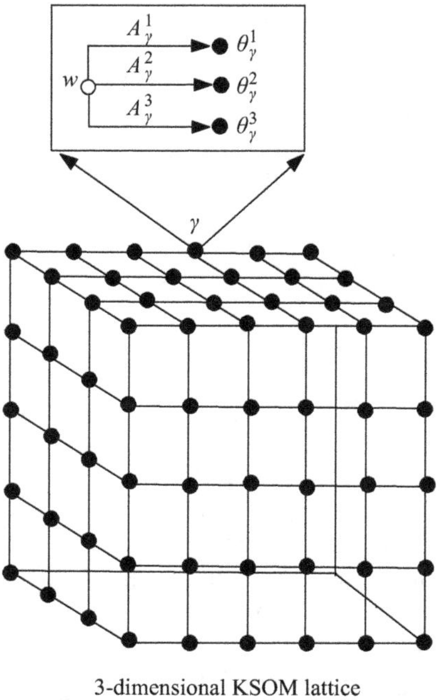

Figure 3.9 Joint angle space sub-clustering

neuron is decided on-line because of variation in redundant solutions across the manipulator workspace.

3.4.2.1 Experimental setup

Figure 3.10 shows some kinematic solutions for same target position. It can be seen that for a given target position, the redundancy preserving network can provide multiple solutions simultaneously. Based on the requirement a particular configuration can be chosen from multiple solutions.

Figure 3.11 shows the experimental result for tracking a circular trajectory, where the desired and actual trajectories are shown as a thin line and in thick points, respectively. Trajectories for the straight line are shown in Figure 3.12 with the desired and actual positions. In both cases, the average positioning error is eight pixels. Accuracy of the end-effector can be improved by multiple-step fine movement steps as shown in Figure 3.13.

3.4.3 Inverse-forward adaptive scheme with a KSOM-based hint generator

In Section 3.4.3, the inverse kinematic is learnt directly using KSOM network. The following will discuss the approach, which computes the inverse kinematic solution by inverting a network (trained a priori to learn). Figure 3.14 shows the approach to solving inverse kinematic problem of a redundant manipulator. In Figure 3.14, the

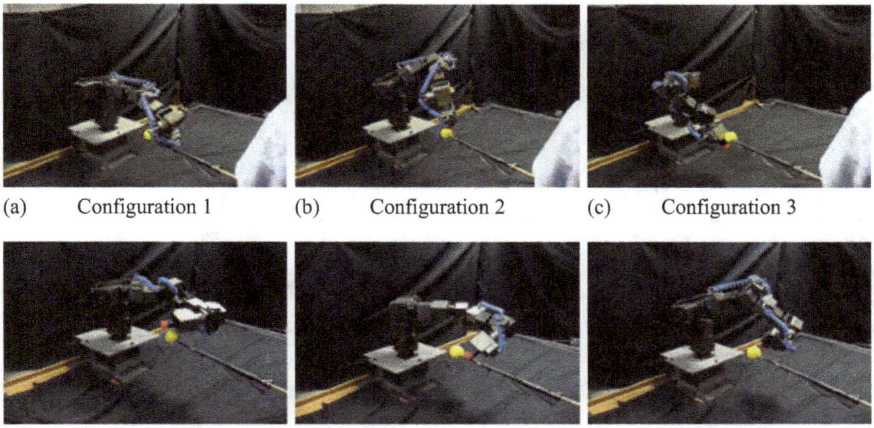

(a) Configuration 1 (b) Configuration 2 (c) Configuration 3

(d) Configuration 4 (e) Configuration 5 (f) Configuration 6

Figure 3.10 Redundant solutions for a given target position

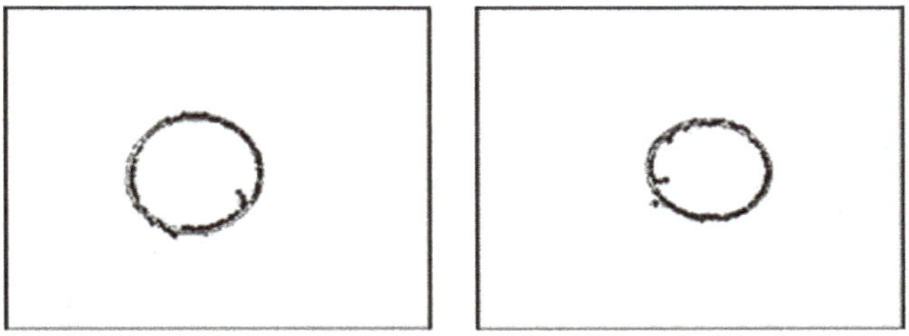

(a) Camera 1 (b) Camera 2

Figure 3.11 Redundant solutions for a given target position as observed in the image plane

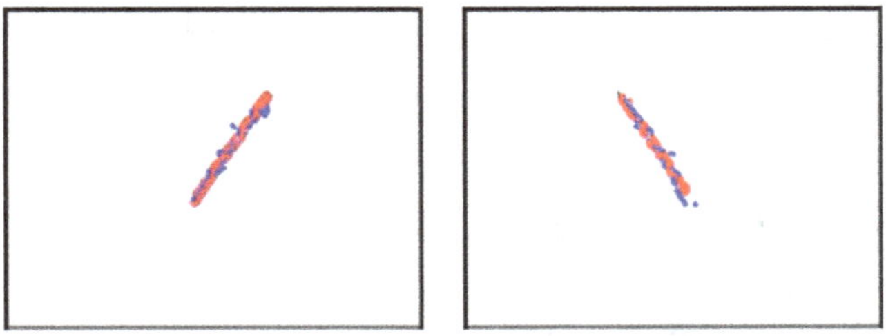

(a) Camera 1 (b) Camera 2

Figure 3.12 Tracking a straight-line trajectory as observed in the image plane

forward training block includes RBFN and robot manipulator. The objective is to compute the suitable joint angle vector $\boldsymbol{\theta}$ lie within joint angle limits to reach the manipulator at a desired position $\boldsymbol{X}_d$. In Figure 3.14, a KSOM-based sub-clustering (KSOMSC) network architecture is used for generating initial suitable guess (hint

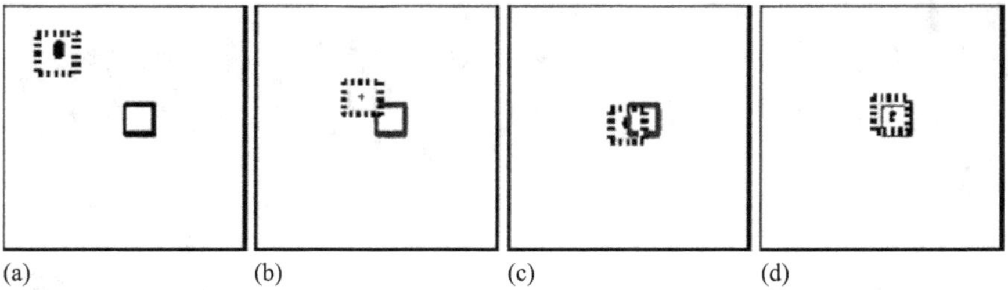

Figure 3.13 Multi-step movement shows (a) the initial state, (b) the state after coarse movement, (c) the end-effector position after five fine movement steps and (d) the final end-effector position obtained after 15 fine movement steps

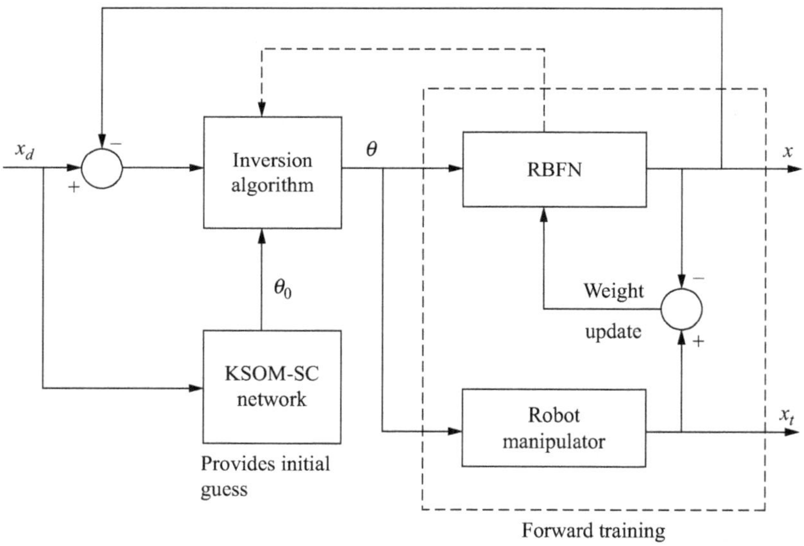

Figure 3.14 Scheme for learning inverse kinematics using RBF network

generator) and RBFN is used for weight update. This approach consists of the following steps.

1. The forward kinematic map is approximated using an RBFN. The forward kinematic map between the input to the RBFN is the joint angle vector $\boldsymbol{\theta}$ and the output of the network is the end-effector position $\boldsymbol{X}$. Here, $\boldsymbol{X_t}$ represents the actual manipulator end-effector position. Training is done using a back-propagation algorithm for minimising the following cost function:

$$\boldsymbol{E}_1 = \frac{\mathbf{1}}{\mathbf{2}}(\boldsymbol{X} - \boldsymbol{X}_t)^2 \tag{3.26}$$

2. An initial hint of the inverse kinematic solution $\boldsymbol{\theta}_0$ for a given target position is generated using KSOMSC network.
3. The input vector of the RBFN is updated to reduce the error function for a given target position $\boldsymbol{X}_d$. The error function is given by

$$E_2 = \frac{1}{2}(X_d - X)^2 = \frac{1}{2}e^T e \tag{3.27}$$

The update law can be written as

$$\theta^{\text{new}} = \theta^{\text{old}} + \Delta\theta \tag{3.28}$$

where $\Delta\theta$ can be computed using gradient descent algorithm as follows:

$$\Delta\theta = -\eta\frac{\partial E_2}{\partial\theta} = \eta(X_d - X)\frac{\partial X}{\partial\theta} = e^T \tag{3.29}$$

where the input Jacobian matrix of the RBFN is represented by $J_i = \partial X/\partial\theta$ and the learning rate for the network inversion is denoted by η.
4. The RBFN is trained again with the new input–output pair $(\boldsymbol{\theta}, \boldsymbol{X}_t)$ if $\|\boldsymbol{X}_d - \boldsymbol{X}\| > \boldsymbol{\varepsilon}$, where $\boldsymbol{\varepsilon} > \boldsymbol{0}$ is user-defined constant based on the positioning accuracy required to be achieved by the end-effector.

Until $\|X_d - X\| < \varepsilon$ is achieved the step (1) to (4) will be repeated. Note that forward training in the training phase is carried out off-line. Whereas in testing phase, as desired position X_d is known and hence trained network undergoes update cycle on-line until the actual manipulator end-effector reaches the desired position. The inverse-forward update cycle is done in two modules. These are inverse module and forward module. In the inverse module, input vector is updated in such a way that the error between network output X and target position X_d reduces. In the forward module, network weights are updated to reduce the error between its output and the actual end-effector position. This forward module improves the accuracy, which can be considered as fine-tuning process.

3.4.3.1 Experimental results

In this experiment, the robot has to reach various points around an obstacle safely. The obstacle is considered to be cuboidal box of dimension 21 cm × 16 cm × 6 cm kept over a wooden pedestal. The obstacle and the pedestal are modelled as spheres of 14 cm and 65 cm, respectively. The pedestal is also considered as an obstacle to avoid the collision with the pedestal. For obstacle avoidance, a ball-covering modelling approach given by Mao *et al.* [26] is used. Figure 3.15 shows the modelling of the obstacle and the final configuration for different points. It also shows the target point and robot final configuration. Figure 3.16 shows some real-time configuration of the robot, for which radius of the spheres has to be selected for avoidance. Note that the radius of the sphere should not be too large or too small to get the right solution without collision.

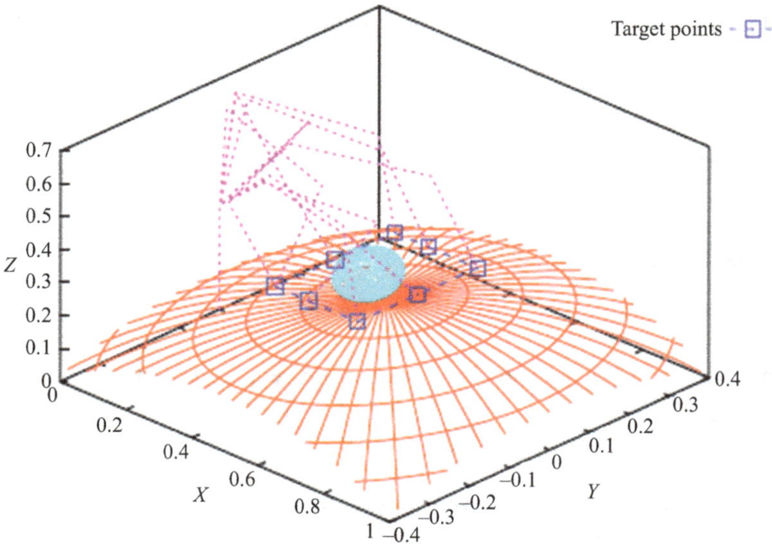

Figure 3.15 Base, over which obstacle is placed is also modelled as an obstacle

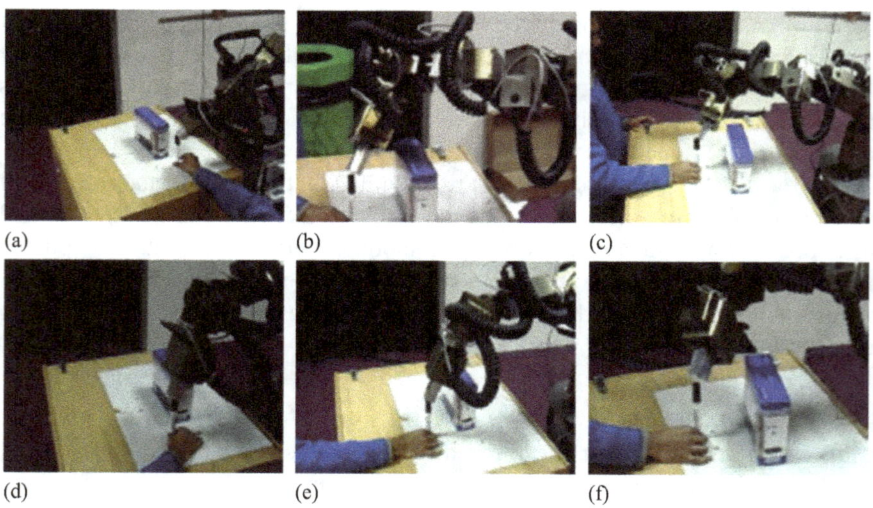

Figure 3.16 Obstacle avoidance experiment. The manipulator reaches various points on a closed trajectory without colliding with the obstacle

3.4.4 Inverse Jacobian matrix estimation using Kohonen's self-organising map

In literature, KSOM-based approach has been widely used for the kinematic control of redundant manipulator [15,16,58,59,61–63]. In [59], KSOM network is used for obstacle avoidance. Kumar *et al.* proposed an inverse-forward adaptive approach using a radial basis function network and a KSOM-based hint generator for redundancy resolution. The approach updates the weights of the learning map online and resulting in an improved positional accuracy. A KSOM network with joint angle space sub-clustering is proposed by Kumar *et al.* in [15]. Although KSOM-based approaches can control the manipulator, there is no discussion provided on the type of solutions and relationship between the solution and the learn KSOM map. In [64], KSOM-based learning for kinematic control algorithm is analysed. A globally asymptotically stable visual servoing approach with learning map is also proposed in [64], which is further generalised in [17].

The objective of the approach [64] is to achieve the global positioning of the end-effector by analysing KSOM-based linear map. The approach analyses the solution learned with KSOM using eigenvalues. Experimental results show that the pseudo-inverse of Jacobian matrix is learnt locally. The accurate positioning of the end-effector is achieved with SOM as

$$\Delta\theta^{\text{out}} = s^{-1} \sum_{\gamma=1}^{N_n} h_\gamma A_\gamma(\Delta \boldsymbol{u}) \tag{3.30}$$

Empirical studies on Cartesian space shows that the linear map of each neuron approximates the inverse Jacobian

$$(LJ)^+ \simeq s^{-1} \sum_{\gamma=1}^{N_n} h_\gamma A_\gamma \tag{3.31}$$

Note that it is easier to develop a globally Lyapunov stable controller with SOM in conjunction with a proportional controller.

3.4.4.1 Experimental results

For experiments kinematic map is used for training. Camera calibration is carried out using Tsai algorithm [46]. Initial and final position of the end-effector are (–0.1, 0.55, 0.15) and (0.1, 0.75, 0.35), respectively. Figure 3.17 shows the manipulator trajectory in real-time. It can be seen that the real-time trajectory has measurement noise.

3.4.5 Near-optimal controllers for input affine nonlinear systems using single network adaptive critic

As discussed in the previous approaches, the redundancy of the solution in vision space can be taken care by minimising the instantaneous cost function. Such approaches do not provide global optimal solution, which can be achieved using integral cost function of end-effector positioning task. Due to the nonlinear behaviour of the redundant manipulator the analytical expression of such cost function may not be possible for real-time implementation. Real-time implementation of such an optimal control problem can be achieved using adaptive critic (AC) learning-based techniques proposed by [65]. Various architectures of AC are defined in [66]. Basically, AC comprises of two networks. These are action network and critic network. The action network shows the relationship between input

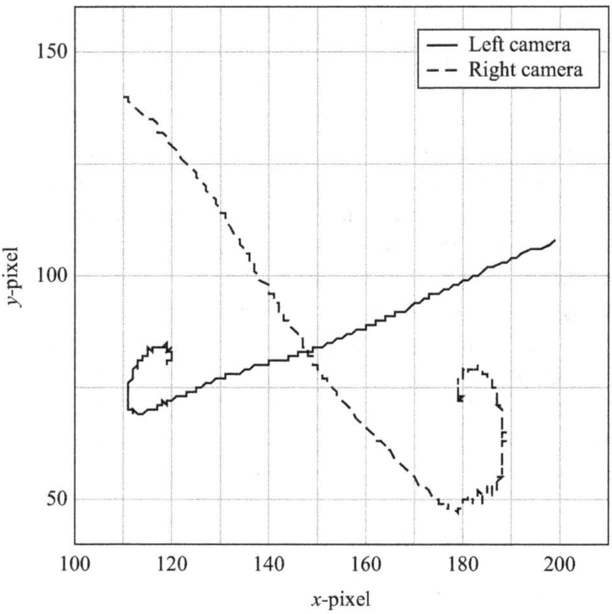

Figure 3.17 End-effector motion in vision space

and state, whereas the critic network shows the relationship between state and co-state. The critic network evaluates the actor performance and provides the feedback to the actor. Based on the feedback, the actor is updated. AC approaches, which approximate the value function, is known as Heuristic dynamic programming (HDP) and co-state vector is known as dual Heuristic dynamic programming (DHDP). Critic network architecture can be action-dependent or action-independent, based on whether the critic network is considered as a function of the system input or not. Classification of the AC network based on the control policy is as policy iteration (PI) and value iteration (VI). In PI-based control policy, critic does not consider the results from the previous iteration and estimated at each iteration. Whereas, in VI-based control policy, critic is updated using previous iteration.

SNAC-based approaches show an advantage over other approaches as critic network only requires to learn co-state vector. In [67], a SNAC-based learning approach for continuous-time systems is proposed. This approach results in under-determined system, which is a major limitation. A SNAC network-based approach using T-S fuzzy model [68] is presented in [64]. This shows the nonlinear controller as a fuzzy cluster, where in small operating zone a nonlinear system behaves as a linear system. In [64], it is claimed that an improved approximation can be achieved.

The architecture of an SNAC with discrete-time system using fuzzy clustering is shown in Figure 3.18, where $x(k)$, $u(k)$, $\lambda(k)$ and $\hat{\lambda}(k+1)$ represent the current state, control input, co-state vector and estimated co-state vector.

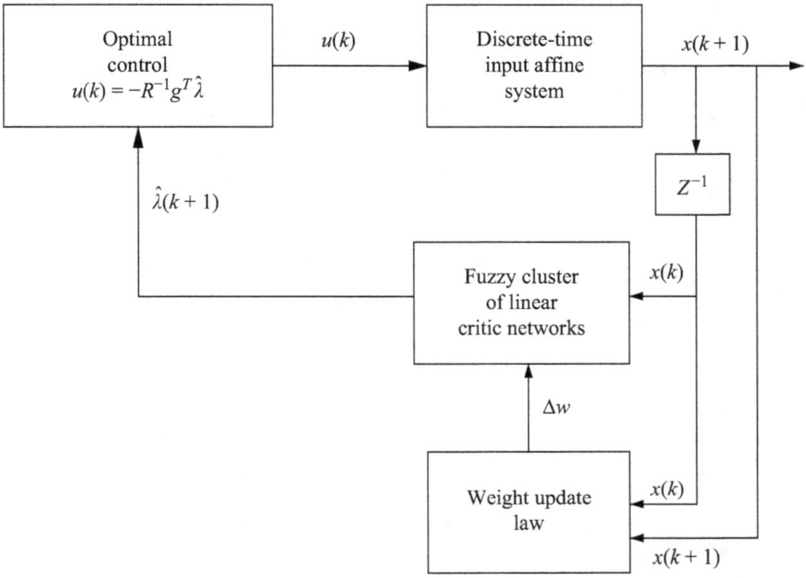

Figure 3.18 Control scheme with discrete-time single network adaptive critic (DT-SNAC)

The expression of the $\hat{\lambda}(k+1)$ can be obtained as

$$\hat{\lambda}(k+1) = \frac{\sum_{i=1}^{N_n} \mu_i W_i x(k)}{\sum_{i=1}^{N_n} \mu_i} \tag{3.32}$$

where $W_i \in \mathbb{R}^{n\times n}$ is the linear map and μ_i is the membership value of the ith fuzzy zone with $i = 1, 2, \ldots, N_n$.

The expression of the membership value can be computed as

$$\mu_i = \prod_{j=1}^{n} \mu_i^j(x_j) \tag{3.33}$$

where $\mu_i^j(x_j) = e^{\left(\frac{-(x_j - c_j^i)^2}{2(\sigma_j^i)^2}\right)}$ with fuzzy centre $c_i = \left[c_1^i, c_2^i, \ldots, c_n^i\right]^T$ and standard deviation $\sigma_i = \left[\sigma_1^i, \sigma_2^i, \ldots, \sigma_n^i\right]^T$.

3.4.6 A single network adaptive critic-based redundancy resolution scheme for redundant manipulators

Figure 3.19 shows the architecture of a single network adaptive critic-based optimal redundancy resolution scheme, where the joint angle is obtained as

$$\Delta\theta(k) = R^{-1} J^T \hat{\lambda}(k+1) \tag{3.34}$$

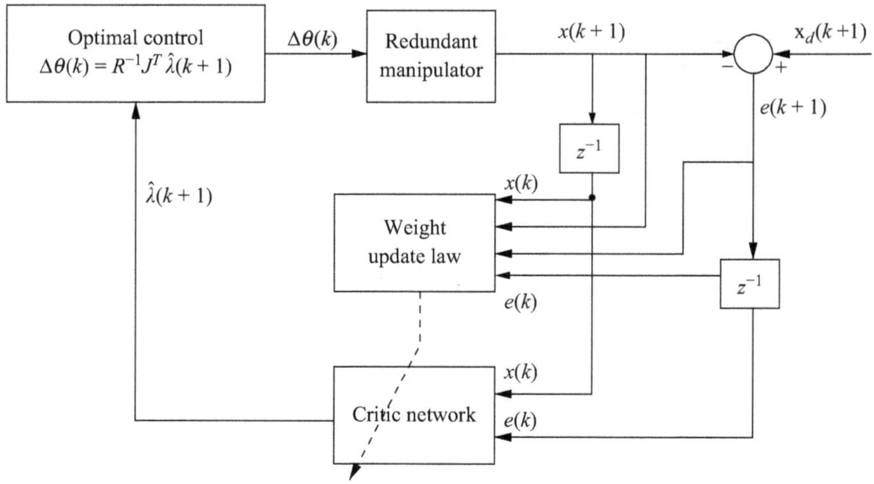

Figure 3.19 A single network adaptive critic based redundancy resolution scheme for redundant manipulators

Such approaches show an advantage in real-time as computation of pseudo inverse is not required for achieving manipulator desired position. For the learning of co-state, this approach also uses T-S fuzzy model with the membership function given as

$$\mu_i(x) = \mu_x^i \mu_y^i \mu_z^i \tag{3.35}$$

where $\mu_x^i(x) = e^{\left(\frac{-(x-c_x^i)^2}{2(\sigma_x^i)^2}\right)}$, $\mu_y^i(y) = e^{\left(\frac{-(y-c_y^i)^2}{2(\sigma_y^i)^2}\right)}$, $\mu_z^i(z) = e^{\left(\frac{-(z-c_z^i)^2}{2(\sigma_z^i)^2}\right)}$

3.4.6.1 Experimental results

In real-time experiments, the PowerCube™ manipulator tracks the elliptical trajectory using visual feedback provided by stereo camera. Based on the computation requirements the sampling interval is taken as 200 ms. Figure 3.20 shows the manipulator trajectories and corresponding error profile with global minimum norm and with avoidance limit.

3.4.7 Reinforcement learning-based optimal redundancy resolution directly from the vision space

The adaptive critic approach discussed in Section 3.4.6 defines the trajectories in the Cartesian space, whereas most of the real-world task requires the trajectories in vision space to improve the robot performance. By defining the trajectories in vision space, the sensor noise can directly be controlled if joint angle space can be estimated from the vision space. Hence, an extension of the SNAC-based approach to the vision space presented is required. Because the approach directly resolves the

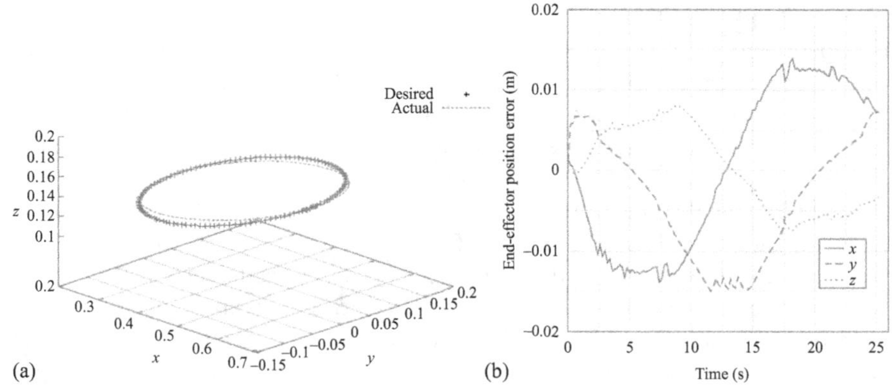

Figure 3.20 End-effector motion: (a) trajectory and (b) position error

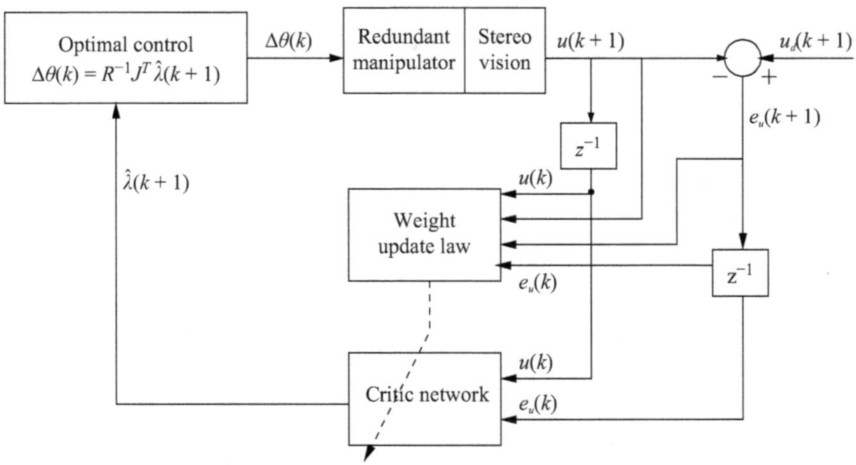

Figure 3.21 Architecture of the SNAC-based optimal redundancy resolution

redundancy from the vision space, it is robust to model inaccuracy. The approach given in [69] also minimises integral cost function as in the case of SNAC discussed in Section 3.4.6.

Note that to resolve the redundancy from the vision space, the correspondence between the join and vision spaces should be known a prior. To learn from the vision space, two neural network-based architectures are presented in [20]. The approach also finds the correspondence between the vision space and joint angle space. These architectures are T-S fuzzy model and KSOM-based critic networks. The approaches are also experimentally tested in real-time using PowerCube™ manipulator integrated with Barrett Hand for grasping the ball. For grasping the ball, the manipulator is visually guided using the adaptive critic based optimal control approach. Figure 3.21 shows the architecture of the SNAC-based reinforcement learning of the optimal redundancy resolution approach.

3.4.7.1 Experimental results

For experiment, an end-effector is wrapped by red tape for identification by stereo camera, where centroid of the wrapped region is used as the end-effector position in the vision space. Experiments are performed in normal lighting conditions. Figure 3.22 shows the tracking of elliptical trajectory by manipulator end-effector using T-S fuzzy model and corresponding error profile.

It can be seen that the maximum tracking error in Cartesian and vision space is of 2.2 mm and five pixels, respectively. Tracking in both the image plane is shown in Figure 3.22, indicating the elliptical trajectory of the manipulator using KSOM and respective error profile. Corresponding trajectories in the image plane for both the camera are shown in Figure 3.23, where elliptical trajectory is tracked with maximum error of 1 mm. Results using both the approaches show that the manipulator can track the trajectory effectively using both the approaches.

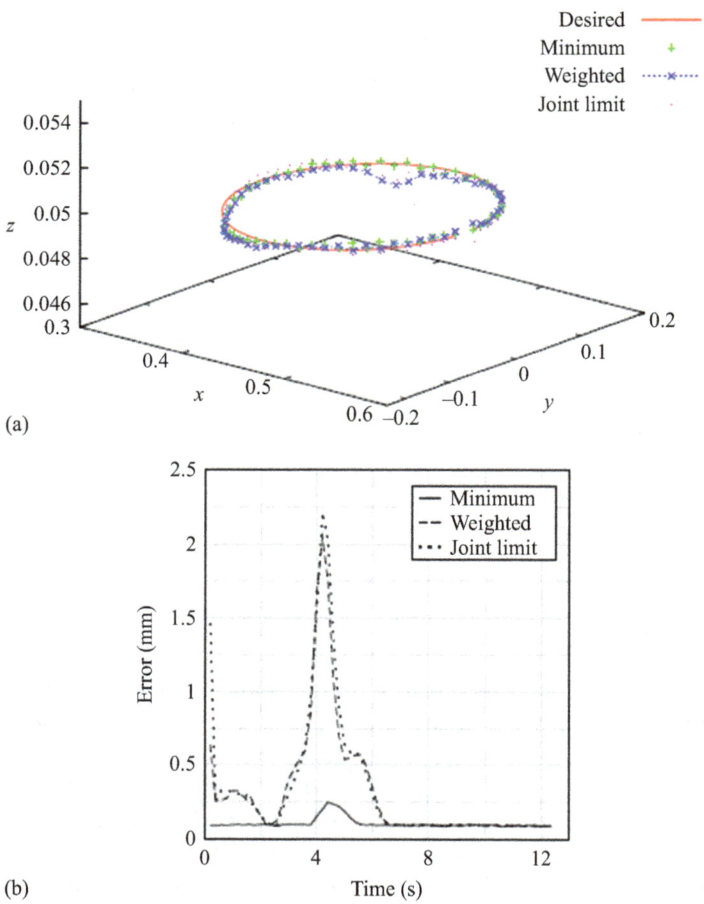

Figure 3.22 Tracking of elliptical trajectory using T-S fuzzy model: (a) trajectory and (b) tracking error

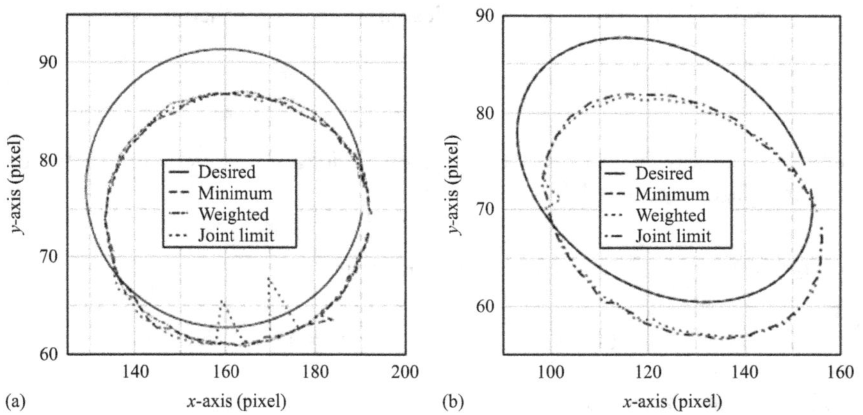

Figure 3.23 Image space trajectory using T-S fuzzy model: (a) left camera and (b) right camera

3.5 Autonomous ground robot

The task of the visual servoing in case of autonomous ground robots (AGVs) is to move the vehicle to a desired position based on visual feedback. The desired position can be static or moving. For a moving target the robot is required to be fixed with a front-facing camera to track the target. Tracking of moving target needs fast computation. Because the image-based visual servoing approaches are independent of camera calibration and do not require target pose as they are preferable for tracking the target. In literature, a sliding mode controller is highly used for such operations due to its inherent robustness capability. Among all the SMC, a FOSMC is preferable due to its control and sliding precision. This makes the FOSMC unique among other controllers. In [70–72], parameters of the sliding surfaces are computed offline, which may not be suitable in varying operating conditions. This problem can be resolved by developing an adaptation scheme for tuning the system parameters. In this regard, Nojavanzadeh *et al.* [73] developed an adaptive fractional order non-singular terminal SMC. Because the robot is controlled through image feature, it is always preferable to integrate the image Jacobian with robot Jacobian in control perspective. The conventional sliding mode control input can be computed by

$$u_c = (LJ)^+(\dot{u}_d - k\ \mathrm{sgn}(\gamma s)) \tag{3.36}$$

where u_d represents the desired feature vector and sliding surface $s = u - u_d$.

Signum function in (3.36) causes chattering, which can be reduced by replacing the signum function by hyperbolic tangent function, which can be written as

$$u_c = (LJ)^+(\dot{u}_d - k\ \tanh(\gamma s)) \tag{3.37}$$

For fractional-order adaptive controller design, a sliding surface is considered as

$$s = \lambda_1 e + \hat{\lambda}_2 \int e\, dt + \hat{\lambda}_3 D^{\alpha-1} e \tag{3.38}$$

Grunwald–Letnikov method is used for the fractional-order derivative computation. The control input is given as

$$u_c = -\frac{1}{\lambda_1 LJ}\left(\hat{\lambda}_3 D^{\alpha} e + \hat{\lambda}_2 e - \lambda_1 \dot{u}_d + k_1 \tanh(\gamma s)\right) \tag{3.39}$$

where $K = \text{diag}(k_1, k_2, \ldots, k_{2n})$, $s = [s_1, s_2, \ldots, s_{2n}]^T$, and $\tanh(\gamma s) = [\tan(\gamma s_1), \tan(\gamma s_2), \ldots, \tan(\gamma s_{2n})]^T$.

Figure 3.24 shows the architecture of the image-based visual servoing of mobile robot using a fractional-order sliding surface controller with adaptation scheme.

3.5.1 Experimental setup and results

For experiments, moving and stationary targets are considered. Targets are represented in squares for visual feedback using a camera fixed on a follower robot. Experiments are carried out in indoor environment using the image-based visual servoing approach with an adaptive fractional order controller. The detail of the hardware is given in Table 3.1. The follower robot is equipped with a laptop and an RGB-D sensor, whereas the target robot has an XU3 system with a WiFi module 3. These robots are differential driven non-holonomic robots. Figure 3.25 shows the experimental setup of the follower and target robots. The sequences of frames in case of moving targets are shown in Figures 3.26 and 3.27, respectively.

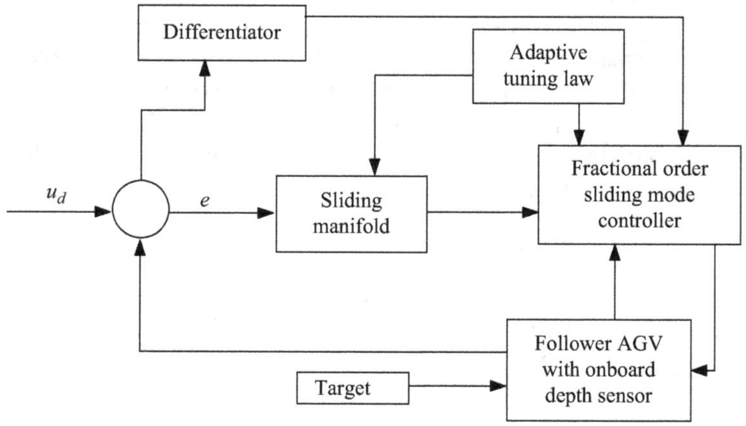

Figure 3.24 Schematic diagram of our proposed scheme

Table 3.1 Description of hardware platform

Component	Description
AGVs	Pioneer P3-DX, which is shown in Figure 3.26
On-board computer	Odroid XU3, laptop
Vision sensor	Kinect V1
WiFi module	Odroid WiFi Module 3
WiFi router	Netgear
Battery	LiPo, 5200 mAh

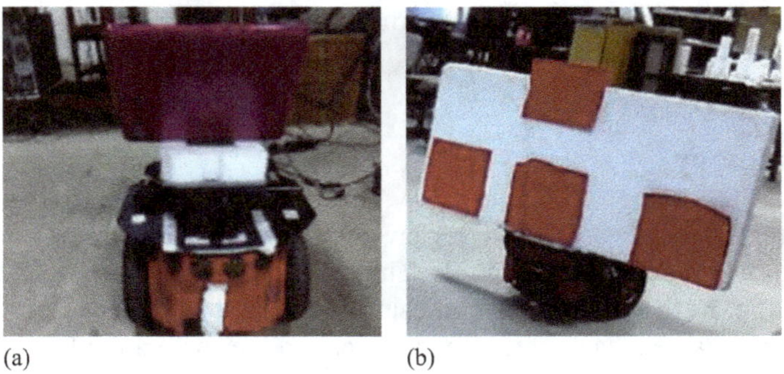

Figure 3.25 Experimental setup: (a) follower robot and (b) target robot

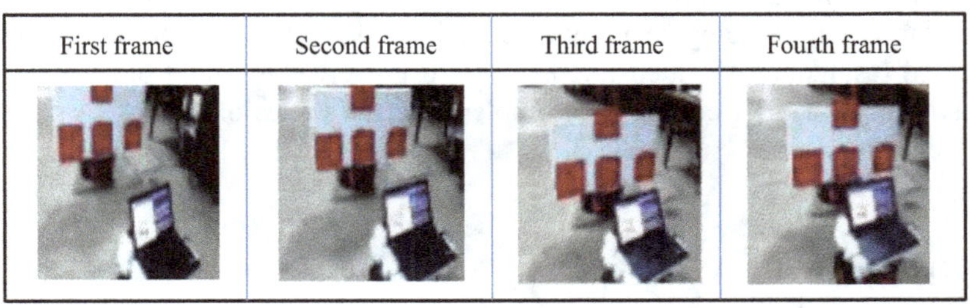

Figure 3.26 Sequence of frames in case of static target

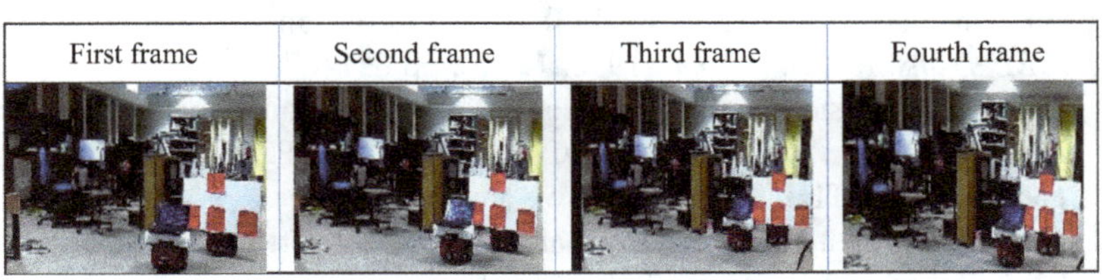

Figure 3.27 Sequence of frames in case of moving target

3.6 Implementation issues

Real-time implementation of the robot visual servoing operations involves various issues, including camera location, camera model, image processing, image feature selection and robot communication. Camera position plays an important role for visual servoing-based real-time operation as in case of eye-to-hand configuration, the camera may be covered by the gripper, whereas in case of eye-in-hand configuration the target may not be covered by the camera field-of-view. Image processing-based tasks are computationally intensive and visual servoing of robot obtains the visual feedback from the images. Target pose is detected using image features. There are various image features that can be selected for target pose estimation. These features can be corner points, straight lines, ellipses and contours.

3.7 Summary

This chapter presented a detailed review of the visual servoing approaches for manipulator and mobile robot controls. The aim of the visual sensing is to control the robot to reach the desired position using visual feedback. Because a camera is required for visual feedback, this chapter described the camera models and camera calibration techniques. Comparative studies of model and model-free visual servoing-based control techniques are given. Detailed discussions and comparisons of existing model-free visual servoing techniques for eye-to-hand configuration using stereo camera are provided. An adaptive fractional-order sliding mode controller for visual servoing of mobile robot is explained. Real-time experimental setup and results are provided for seven DOF PowerCube™ and pioneer P3-DX robots. Finally, a brief comment on implementation issues of the visual seroving approaches is given.

References

[1] Hill J, Park W, 'Real time control of a robot with mobile camera', *in 9th International Symposium on Industrial Robots*, Washington DC, USA, 13–15 March 1979, pp 233–246

[2] Hutchinson S, Hager G D, Corke P I, 'A tutorial on visual servo control', *IEEE Trans. Robot. Autom.*, 1996, **12** (5), pp 651–670

[3] Kragic D, Christensen H I, 'Survey on visual servoing for manipulation', *Technical Report*, ISRN KTH/NA/P-02/01-SE, CVAP259, Royal Institute of Technology (KTH), Stockholm, Sweden, 2002

[4] Chaumette F, Hutchinson S, 'Visual servo control part I: Basic approaches', *IEEE Robot. Autom. Mag.*, 2006, **13** (4), pp 82–90

[5] Chaumette F, Hutchinson S, 'Visual servo control part II: Basic approaches', *IEEE Robot. Autom. Mag.*, 2007, **14** (1), pp 109–118

[6] Kuperstein M, 'Neural model of adaptive hand-eye coordination for single postures', *Science*, 1988, **239**, pp 1308–1311

[7] Jagersand M, Nelson R C, 'Adaptive differential visual feedback for uncalibrated hand-eye coordination and motor controls', *Technical Report*, University of Rochester, USA, 1994

[8] Kuperstein M, 'Adaptive visual-motor coordination in multi-joint robots using parallel architecture', in *Proceedings of IEEE International Conference on Robotics and Automation*, Vol. 4, Raleigh, NC, USA, 31 March–3 April 1987, pp 1595–1602

[9] Martinetz T M, Ritter H J, Schulten K J, 'Three-dimensional neural net for learning visual motor coordination of a robot arm', *IEEE Trans. Neural Networks*, 1990, **1** (1), pp 131–136

[10] Walter J A, Schulten K J, 'Implementation of self-organizing neural networks for visual-motor control of an industrial robot', *IEEE Trans. Neural Networks*, 1993, **4** (1), pp 86–95

[11] Behera L, Kirubanandan N, 'A hybrid; neural control scheme for visual motor coordination', *IEEE Control Syst. Mag.*, 1999, **19** (4), pp 34–41

[12] Kumar S, Patel N, Behera L, 'Visual motor control of a 7 DOF robot manipulator using function decomposition and sub-clustering in configuration', *Neural Process. Lett.*, 2008, **28** (1), pp 17–33

[13] Sanderson A C, Weiss L E, 'Image based visual servo control using relational graph error signals', in *Proceedings of International Conference on Cybernetics and Society*, Cambridge, UK, 8–10 October 1980, pp 1074–1077

[14] Hashimoto K, Ritter T, Kimura H, 'Visual servoing with hand-eye manipulator optimal control approach', *IEEE Trans. Robot. Autom.*, 1996, **12** (5), pp 766–774

[15] Kumar S, Patchaikani P, Dutta A, Behera L, 'Visual motor control of a 7 DOF edundant manipulator using redundancy preserving learning network', *Robotica*, 2010, **28** (6), pp 795–810

[16] Kumar S, Behera L, McGinnity T M, 'Kinematic control of a redundant manipulator using an inverse-forward adaptive scheme with a KSOM based hint generator', *Robot. Autonom. Syst.*, 2010, **58**, pp 622–633

[17] Patchaikani P K, Behera L, 'Visual servoing of redundant manipulator with Jacobian estimation using self-organizing map', *Robot. Autonomous Syst.*, 2010, **58** (8), pp 978–990

[18] Pourboghrat F, 'Neural networks for learning inverse kinematics of redundant manipulator', in *32nd IEEE Midwest Symposium on Circuits and Systems*, Champaign, IL, USA, 14–16 August 1989, pp 760–762

[19] Sun G, Scassellati B, 'A fast and efficient model for learning to reach', *Int. J. Humanoid Robot.*, 2005, **2** (4), pp 391–413

[20] Mayorga R V, Sanongboon P, 'A radial basis function network approach for inverse kinematics and singularities prevention of redundant manipulators', in *IEEE International Conference on Robotics and Automation*, Washington, D.C., USA, 11–15 May 2002, pp 1955–1960

[21] Miller W T, 'Real-time application of neural networks for sensor-based control of robots with vision', *IEEE Trans. Syst. Man Cybernet.*, 1989, **19** (4), pp 825–831

[22] Li Y, Leong S H, 'Kinematics control of redundant manipulators using a CMAC neural network combined with a genetic algorithm', *Robotica*, 2004, **22**, pp 611–621

[23] Vijayakumar S, D'Souza A, Shibata T, Conradt J, Schaal S, 'Statistical learning for humanoid robots', *Autonomous Robots*, 2002, **12**, pp 55–69

[24] Lendaris G, Mathia K, Saeks R, 'Linear hopfield networks and constrained optimization', *IEEE Trans. Syst. Man Cybernet. Part B: Cybernet.*, 1999, **29** (1), pp 114–118

[25] Ritter H J, Martinetz T K, Schulten K J, 'Topology-conserving maps for learning visuo-motor-coordination', *Neural Network*, 1989, **2**, pp 159–168

[26] Mao Z, Hsia T C, 'Obstacle avoidance inverse kinematics solution of redundant robots by neural networks', *Robotica*, 1997, **15**, pp 3–10

[27] Jordan M I, Rumelhart D E, 'Forward models: Supervised learning with a distal teacher', *Cognit. Sci.*, 1992, **16**, pp 307–354

[28] Wang J, Hu Q, Jiang D, 'A Lagrangian network for kinematic control of redundant robot manipulators', *IEEE Trans. Neural Networks*, 1999, **10** (5), pp 1123–1132

[29] Tang W S, Wang J, 'A recurrent neural network for minimum infinity-norm kinematic control of redundant manipulators with an improved problem formulation and reduced architecture complexity', *IEEE Trans. Syst. Man Cybernet. Part B: Cybernet.*, 2001, **31** (1), pp 98–105

[30] Kohonen T, *Self-Organization and Associative Memory*, Springer-Verlag, 1984

[31] Barreto G A, Araujo A F R, Ritter H J, 'Self-organizing feature maps for modelling and control of robotic manipulators', *J. Intell. Robot. Syst.*, 2003, **36**, pp 407–450

[32] Martinetz T, Ritter H, Schulten K, 'Learning of visuomotor coordination of a robot arm with redundant degrees of freedom', in *International Conference on Parallel Processing in Neural Systems and Computers*, New York, USA, 19–20 March 1990, pp 431–434

[33] Hartley R, Zisserman A, *Multiple View Geometry*, Cambridge University Press, 2003

[34] Faugeras O, *Three-Dimensional Computer Vision*, MIT Press, 1993

[35] Trucco E, Verri A, *Introductory Techniques for 3-D Computer Vision*, Prentice Hall, 1998

[36] Gernsheim H, *The Origins of Photography*, Thames and Hudson, 1982

[37] Shashua A, 'Geometry and photometry in 3D visual recognition', *PhD thesis*, MIT, 1992

[38] Brown D C, 'Close-range camera calibration', *Photogram. Eng.*, 1971, **37** (8), pp 855–866

[39] Faig W, 'Calibration of close-range photogrammetry systems: Mathematical formulation', *Photogram. Eng. Remote Sens.*, 1975, **41** (12), pp 1479–1486

[40] Faugeras O, Toscani G, 'The calibration problem for stereoscopic vision', in Casals A. (ed.), *Sensor Devices and Systems for Robotics*, NATO ASI Series (Series F: Computer and Systems Sciences), **52**, Springer, Berlin, Heidelberg, 1989.

[41] Ganapathy S, 'Decomposition of transformation matrices for robot vision', *Pattern Recogn. Lett.*, 1984, **2**, pp 401–412

[42] Gennery D B, 'Stereo-camera calibration', *Proceedings of Image Understanding Workshop*, University of Southern California, USA, 7 November 1979, pp 101–108

[43] Wei G Q, Ma S D, 'A complete two-plane camera calibration method and experimental comparisons', in *4th International Conference on Computer Vision*, Berlin, Germany, 11–14 May 1993, pp 439–446

[44] Maybank S J, Faugeras O D, 'A theory of self-calibration of a moving camera', *Int. J. Comput. Vision*, 1992, **8** (2), pp 123–152

[45] Weng J, Cohen P, Herniou M, 'Camera calibration with distortion models and accuracy evaluation', *IEEE Trans. Pattern Anal. Mach. Intell.*, 1992, **14** (10), pp 965–980

[46] Tsai R Y, 'A versatile camera calibration technique for high-accuracy 3D machine vision metrology using off-the-shelf TV cameras and lenses', *IEEE J. Robot. Autom.*, 1987, **3** (4), pp 323–344

[47] Jgersand M, Fuentes O, Nelson R, 'Experimental evaluation of uncalibrated visual servoing for precision manipulation', in *Proceedings of International Conference on Robotics and Automation*, Albuquerque, NM, USA, 25 April 1997, pp 2874–2880

[48] Mezouar Y, Chaumette F, 'Path planning in image space for robust visual servoing', in *Proceedings of International Conference on Robotics and Automation*, San Francisco, CA, USA, 24–28 April 2000, pp 2759–2763

[49] Malis E, Chaumette F, Boudet S, '1-1/2-D visual servoing', *IEEE Trans. Robot. Autom.*, 1999, **15** (2), pp 238–250

[50] Wilson W J, Hulls C C W, Bell G S, 'Relative end-effector control using Cartesian position-based visual servoing', *IEEE Trans. Robot. Autom.*, 1996, **12** (5), pp 684–696

[51] Chaumette F, Rives P, Espiau B, 'Positioning a robot with respect to an object, tracking it and estimating its velocity by visual servoing', in *Proceedings of IEEE International Conference on Robotics and Automation*, Sacramento, CA, USA, 9–11 April 1991, pp 2248–2253

[52] Espiau B, Chaumette F, Rives P, 'A new approach to visual servoing in robotics', *IEEE Trans. Robot. Autom.*, 1992, **8** (3), pp 313–326

[53] Fang Y, Behal A, Dixon W E, Dawson D M, 'Adaptive 2.5d visual servoing of kinematically redundant robot manipulators', in *41st Conference on Decision and Control*, Las Vegas, NV, USA, 10–13 December 2002, pp 2860–2865

[54] Open Source Computer Vision Library, [online], Available from: http://www.intel.com/technology/computing/opencv/

[55] Wilson R, 'Tsai camera calibration software', [online], Available from: http://www.cs.cmu.edu/rgw/TsaiCode.html
[56] Kumar S, 'Kinematic control of redundant manipulators using neural network', *PhD thesis*, Indian Institute of Technology Kanpur, India, 2009
[57] Tevatia G, Schaal S, 'Inverse kinematics of humanoid robots', in *IEEE International Conference on Robotics and Automation*, San Francisco, CA, USA, 24–28 April 2000, pp 294–299
[58] Martinetz T, Ritter H, Schulten K, 'Learning of visuomotor-coordination of a robot arm with redundant degrees of freedom', in *International Conference on Parallel Processing in Neural Systems and Computers*, North-Holland, 1990, pp 431–434
[59] Han M, Okada N, Kondo E, 'Coordination of an uncalibrated 3-d visuomotor system based on multiple self-organizing maps', *JSME Int. J. Ser. C*, 2006, **49** (1), pp 230–239
[60] Zha H, Onitsuka T, Nagata T, 'A self-organization learning algorithm for visuomotor coordination in unstructured environment', *Artific. Life Robot.*, 1997, **1** (3), pp 131–136
[61] Asuni G, Teti G, Laschi C, Guglielmelli E, Dario P, 'A bio-inspired sensory-motor neural model for a neuro-robotic manipulation platform', in *12th International Conference on Advanced Robotics*, Seattle, WA, USA, 18–20 July 2005, pp 607–612
[62] Barreto G A, Araujo A F R, 'Self-organizing feature maps for modelling and control of robotic manipulators', *J. Intell. Robotic Syst.*, 2003, **36**, pp 407–450
[63] Angulo V R, Torras C, 'Speeding up the learning of robot kinematics through function decomposition', *IEEE Trans. Neural Networks*, 2005, **16** (6), pp 1504–1512
[64] Patchaikani P K, 'Learning based near-optimal redundancy resolution schemes for visually controlled robot manipulators', *PhD thesis*, Indian Institute of Technology Kanpur, India, 2012
[65] Werbos P J, *Approximate Dynamic Programming for Real-Time Control and Neural Modelling*, Multiscience Press, 1992
[66] Seong C Y, Widrow B, 'Neural dynamic optimization for control systems - Part II: Theory', *IEEE Trans. Syst. Man Cybernet., Part B: Cybernet.*, 2001, **31** (4), pp 490–501
[67] Kumar S, Padhi R, Behera L, ' Direct adaptive control using single network adaptive critic', in *IEEE International Conference on Systems of Systems Engineering*, San Antonio, TX, USA, 16–19 April 2007, pp 1–6
[68] Takagi T, Sugeno M, 'Fuzzy identification of systems and its application to modelling and control', *IEEE Trans. Syst. Man. Cybernet.*, 1985, **15**, pp 116–132
[69] Kim S W, Park K B, Lee J J, 'Redundancy resolution of robot manipulators using optimal kinematic control', in *IEEE International Conference on Robotics and Automation*, San Diego, CA, USA, 8–13 May 1994, pp 683–688

[70] Cao D, Fei J, 'Adaptive fractional fuzzy sliding mode control for three-phase active power filter', *IEEE Access*, 2016, **4**, pp 6645–6651

[71] Delghavi M B, Shoja-Majidabad S, Yazdani A, 'Fractional order sliding-mode control of islanded distributed energy resource systems', *IEEE Trans. Sustain. Energy*, 2016, **7** (4), pp 1482–1491

[72] Wang Y, Gu L, Xu Y, Cao X, 'Practical tracking control of robot manipulators with continuous fractional-order nonsingular terminal sliding mode', *IEEE Trans. Industr. Electron.*, 2016, **63** (10), pp 6194–6204

[73] Nojavanzadeh D, Badamchizadeh M, 'Adaptive fractional-order non-singular fast terminal sliding mode control for robot manipulators', *IET Control Theory Appl.*, 2016, **10** (13), pp 1565–1572

Pooja Agrawal received her PhD degree in Aerospace Engineeing from Indian Institute of Science, Bangalore, India, in 2017. From 2016 to 2018, she was a Postdoctoral Fellow in the Department of Electrical Engineering, Indian Institute of Technology Kanpur, India. Her research interests include guidance and control, motion planning and visual servoing of autonomous vehicles.

Laxmidhar Behera is the Poonam and Prabhu Goel Chair Professor at IIT Kanpur with more than 24 years research and teaching experience. He received his BSc and MSc degrees in Engineeing from NIT Rourkela in 1988 and 1990, respectively, and his PhD from IIT Delhi in 1996. He pursued postdoctoral studies in the German National Research Center for Information Technology, GMD, Sank Augustin, Germany, 2000–2001. Previously, he was an Assistant Professor at BITS Pilani, 1995–1999 and a Reader at Intelligent Systems Research Center, University of Ulster, UK, 2007–2009. He was a visiting researcher/professor at FHG, Germany, and ETH, Zurich, Switzerland. His research lies in convergence of machine learning, control theory, robotic vision and heterogeneous robotic platforms. His other research interests include intelligent control, semantic signal/music processing, neural networks, control of cyber-physical systems and cognitive modelling. He received more than INR 170 million research grants to support his research activities. He has established industrial collaboration with TCS, Renault Nissan, BEL, Bangalore and ADNOC, Abu Dhabi, and made significant contribution to robotics-based warehouse automation, vision and drone guided driver assistance system, and drone guided pipeline inspection systems. He has published more than 240 papers in journals and conference proceedings. He has supervised 16 PhD theses to completion. He is a Fellow of INAE and Senior Member of IEEE.

Chapter 4

CCD and CMOS sensors and their applications in flame measurement

Gang Lu[1]

List of abbreviations

ADC	analogue-to-digital converter
AR	anti-reflective coating
CCD	charge-coupled device
CCIR	consultative committee for international radio
CCTV	closed circuit TV
CMOS	complementary metal-oxide semiconductor
ICCD	intensified CCD
EMCCD	electron multiplying CCD
dB	decibel
DR	dynamic range
DSP	digital signal processor
FET	field-effect transistor
FFT	Fast Fourier transform
FOV	field of view
fps	frame per second
FWC	full well capacity
IP	Internet protocol
IR	infrared radiation
LIF	laser induced fluorescence
LMBC	Levenberg–Marquardt with boundary constraint
LRS	laser-sheet-illuminated Rayleigh scattering
MOS	metal-oxide semiconductor
NIR	near IR
NNLS	non-negative least squares
NTSC	National Television System Committee

[1]School of Engineering and Digital Arts, University of Kent, Canterbury, UK

PAL	phase alternating line
PLIF	planar Laser-induced fluorescence
RGB	red, green and blue
QE	quantum efficiency
SNR	signal-to-noise ratio
UV	ultraviolet
UVAR	UV anti-reflective coating

4.1 Introduction

Flame is a complicated object relating to thermal excitation, chemical reaction, fluid flow, heat conduction and molecular transport. The measurement and characterisation of flame are required in not only fundamental research but also many industrial applications. The study of flame has a long history, with early work aiming at investigation into chemical mechanism and the structure of flame. With the development of the imaging sensors, computers and image processing, the measurement and characterisation of flame have been developed rapidly in the past two decades.

This chapter covers digital image sensing technologies and their applications in flame measurement. The commonly used imaging sensors, that are charge-coupled devices (CCD) and complementary metal-oxide semiconductor (CMOS) sensors, are introduced in detail. The characteristics of CCD and CMOS sensors and the main difference between the two sensing technologies are discussed. It also introduces special cameras, camera interfaces and the general guidelines for the selection of cameras.

Vision-based technologies for flame measurement and characterisation are introduced, covering typical 2D and 3D imaging techniques for measuring the characteristic parameters of flame, such as size, shape, brightness, temperature and chemiluminescence. The basic principle, advantages and disadvantages of each technique are described.

4.2 CCD and CMOS imaging sensors

An image is a 2D projection of a 3D scene in the real world. A digital image is a numerical representation of an image and has a finite set of digital values, called pixels (i.e. elements in a picture). A digital image contains a large number of rows and columns of pixels, which are determined by the imaging sensor. The imaging sensor is responsible for transforming the light of object being sensed into electrical signals and then an image. Figure 4.1 shows the generation of a digital image.

Popular imaging sensors can be classified into two different categories based on the technologies used:

- *CCD sensors*, which was developed specifically for the camera industry in 1969 by Willard S Boyle and George E Smith, who were awarded the Nobel Prize for Physics in 2009 for their invention of the CCD technology [1].
- *CMOS sensors*, based on the standard technology that is extensively used in computer systems, for example, memory chips.

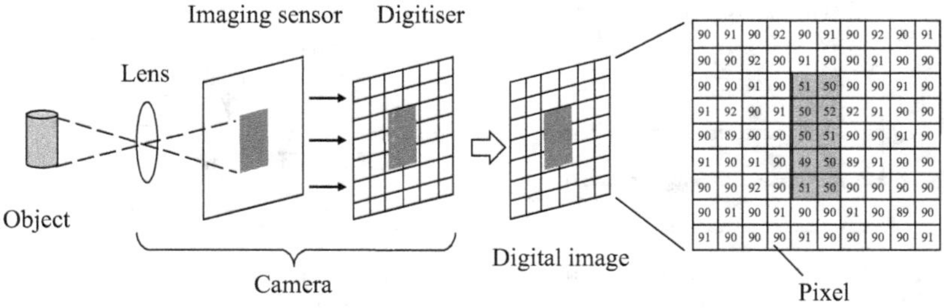

Figure 4.1 Generation of a digital image

CCD and CMOS sensors have their own strengths and weaknesses, depending on a specific application, and hence one type is not categorically superior to another. A digital camera consists of an imaging sensor (either CCD or CMOS) and associated optics to form images, electronics to control sensing elements, readout, amplify and digitise image signals, and interface to transmit image signals to a peripheral device, such as a computer.

4.2.1 CCD sensor

4.2.1.1 Architecture of CCD sensor

A. Physical structure and potential well

A CCD sensor is fundamentally an integrated circuit etched onto a silicon surface forming light-sensitive elements (i.e. pixels). Each element is a metal-oxide semiconductor (MOS) capacitor. The MOS capacitors are not arranged to form individual pixels. In fact, the pixels are defined by the position of electrodes and channel stops above the silicon surface, as shown in Figures 4.2(a) and 4.2(b).

If a positive voltage is applied to an electrode, then the positive potential attracts all negatively charged electrons close to the area under the electrode. In addition, any positively charged holes are repulsed from the area around the electrode. Consequently, a potential well forms, in which all electrons produced by incoming photons are stored, as shown in Figure 4.2(c).

The capacity of potential well to hold the largest number of charges before saturation occurs is known as the full-well capacity of pixel. The physical size of a pixel affects primarily the full-well capacity (FWC). For instance, a great FWC is generally found in a CCD with large pixels. When a pixel exceeds its FWC, the charge starts to fill adjacent pixels, resulting in Blooming problem (refer to Section 4.2.1.2 G).

B. Charge collection

In an MOS, atoms in the silicon crystal have electrons arranged in discrete energy bands, that is the valence band (as the lower energy band) and the conduction band (as the upper energy band). Most electrons occupy the valence band but can be excited into the conduction band by absorption of a photon through the photoelectric effect. The photons entering the MOS create electron-hole pairs. The electrons are then attracted towards the most positive potential well, where they create charge packets. Each packet corresponds to one pixel, as shown in Figure 4.3.

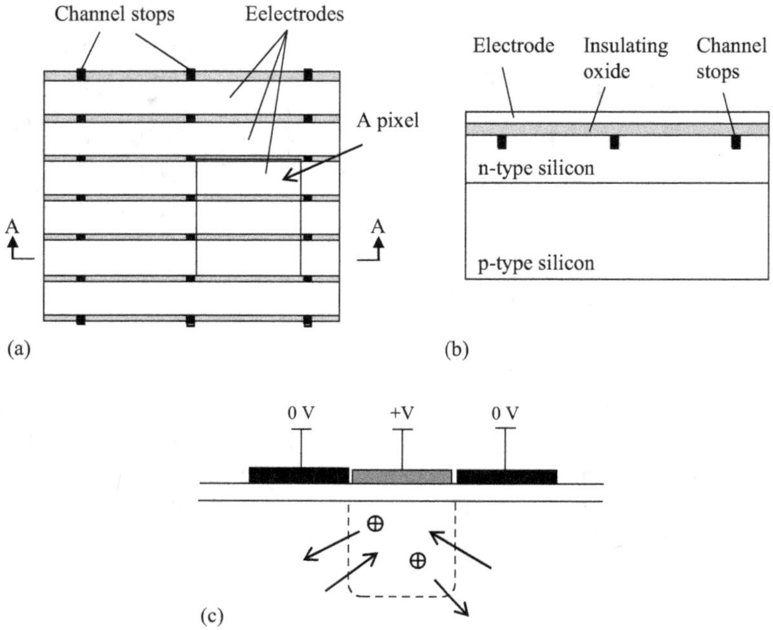

Figure 4.2 Architecture of CCD sensor: (a) top view, (b) A-A view and (c) schematic of potential well

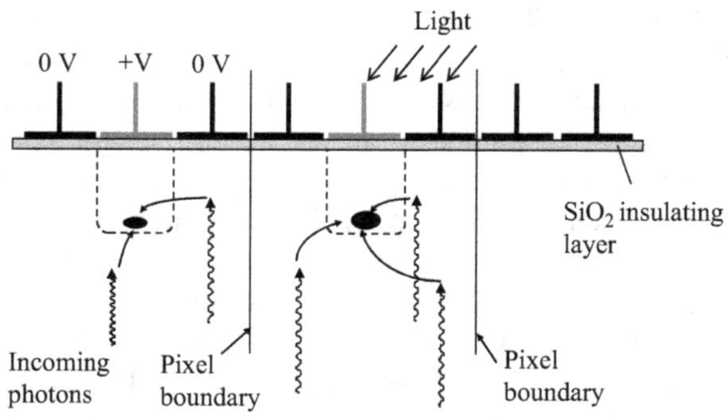

Figure 4.3 Charge collection of CCD sensor

However, not every photon falling onto the CCD produces an electric charge. The percentage of photons that are actually detected is known as the quantum efficiency (QE), which is defined as

$$\mathrm{QE} = \frac{\text{eletrons/second}}{\text{photons/second}} \times 100\% \tag{4.1}$$

QE is the accurate measurement of the electrical sensitivity of CCD to light, which varies with the structure of CCD and the wavelength of light (refer to the

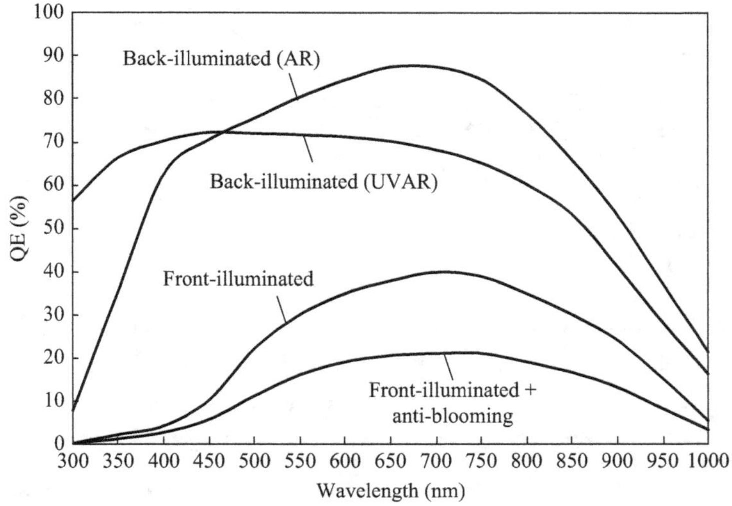

Figure 4.4 Typical QE curves of CCD sensors (AR: anti-reflective coating; UVAR: UV anti-reflective coating)

spectral response). Figure 4.4 shows typical QE curves for front- and back-illuminated CCDs.

C. Charge transfer

Once charges are collected and charge packets are formed, a transmission process takes place to move the charge packets out from the sensing array. Figure 4.5 shows the detailed process how the charge packets are transferred through a row of the sensing array. Each pixel has three electrodes $\Phi 1$, $\Phi 2$ and $\Phi 3$, where only one of these electrodes (i.e. $\Phi 3$) is required to create the potential well for charge collection, the other two (i.e. $\Phi 1$ and $\Phi 2$) are used to transfer the charge packets out of the sensing array. The transmission process is as follows.

(a) The charges being collected are under one of electrodes ($\Phi 3$ is high, $\Phi 1$ and $\Phi 2$ are low), as seen in Figure 4.5(a).
(b) A new potential well is created by holding $\Phi 3$ and $\Phi 2$ high. The charges are now shared between $\Phi 2$ and $\Phi 3$, as shown in Figure 4.5(b).
(c) $\Phi 3$ is then kept low. The charges are fully transferred under $\Phi 2$, as shown in Figure 4.5(c).
(d) To continue clocking out the CCD, taking $\Phi 1$ high and then taking $\Phi 3$ low to ensure that the charges drift across under $\Phi 1$, as shown in Figure 4.5(d).
(e) If $\Phi 2$ is now kept low, the charge is fully transferred under $\Phi 1$, as shown in Figure 4.5(e).
(f) As this process is continued, the charge packets progress either down the column, or across the row, depending upon the orientation of electrodes, until the charge packets of every column (or row) are transferred out the sensing array (see Figure 4.5(d)).

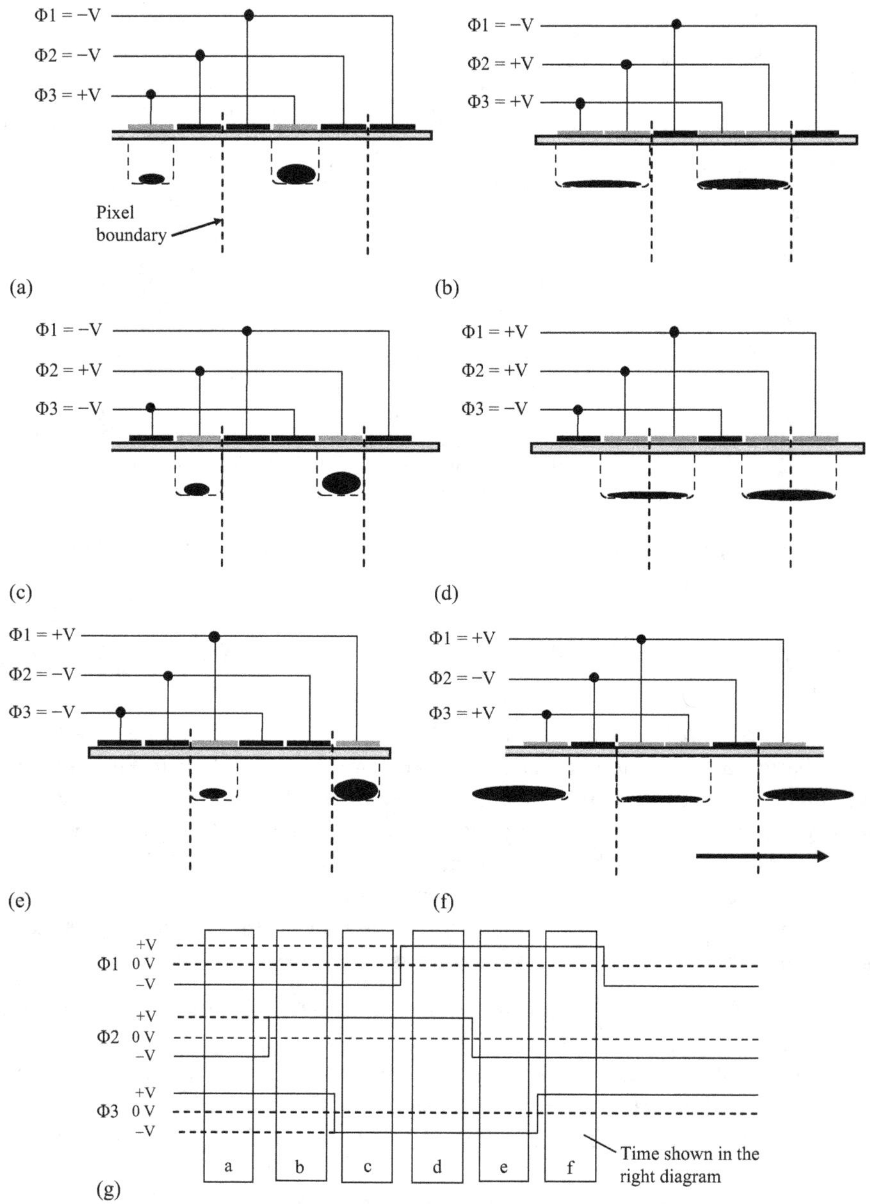

Figure 4.5 Charge transfer of CCD sensor

D. Frame readout

Three techniques are commonly used to transfer a frame out from the CCD sensing array:

1. *Full-frame* (see Figure 4.6(a)). In a full-frame CCD sensor, incoming photons fall on the full sensing array. To readout the signal, accumulated charges must

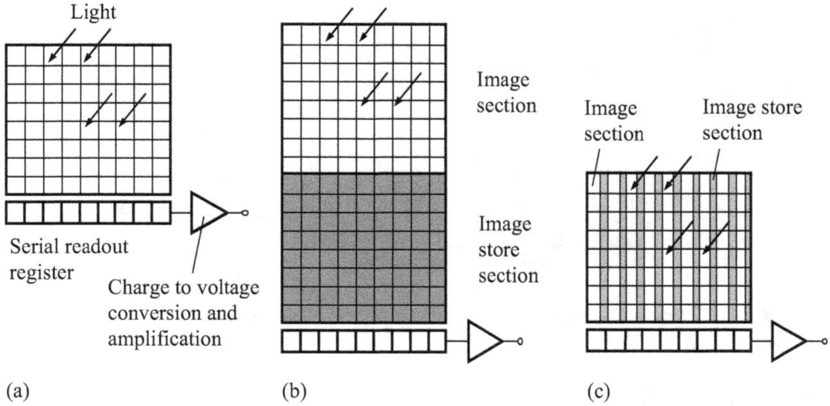

Figure 4.6 Types of frame readout of CCD sensor: (a) full-frame, (b) frame-transfer and (c) interline-transfer

be shifted row by row into a serial (readout) register, and for each row, the serial register must be shifted horizontally to readout each individual pixel (also known as progressive scan readout). The full-frame design features a high fill factor (almost 100%). A disadvantage of such a design is charge smearing caused by the light falling on the sensor whilst the accumulated charge signal is being transferred to the serial register.

2. *Frame-transfer* (see Figure 4.6(b)). A frame-transfer CCD array has two identical regions. One part is used to sense the light and another as a storage region and is protected from light by a light-tight mask. Incoming photons fall on the uncovered light-sensitive portion of the array. The accumulated charges are then rapidly shifted into the masked storage region. While the signal is integrated on the light-sensitive portion, the stored charges are read out to the serial register. The frame-transfer CCD sensor has typically a faster frame rate than the full-frame CCDs.
3. *Interline-transfer* (see Figure 4.6(c)). An interline-transfer CCD consists of charge transfer channels (also called interline masks), which are adjacent to each sensing region. The accumulated charges are shifted into the charge transfer channels right after charge collection. The rapid charge transfer can eliminate possible image smearing effectively. However, the interline masks reduce the light-sensitive area of the sensor.

E. Charge-to-voltage convention

Charge packages are finally transferred into the serial register, where the value of each charge package is converted into a voltage through charge amplifier. In the charge amplifier, as shown in Figure 4.7, charges are passed onto a very low capacitance node (shown as an 'n^+' in the figure) and a voltage is developed. This voltage is sensed with a field-effect transistor (FET), whose output is then sent to a camera circuit. A typical conversion factor is around 5–10 μV per electron with typical full well values being about 100,000 electrons.

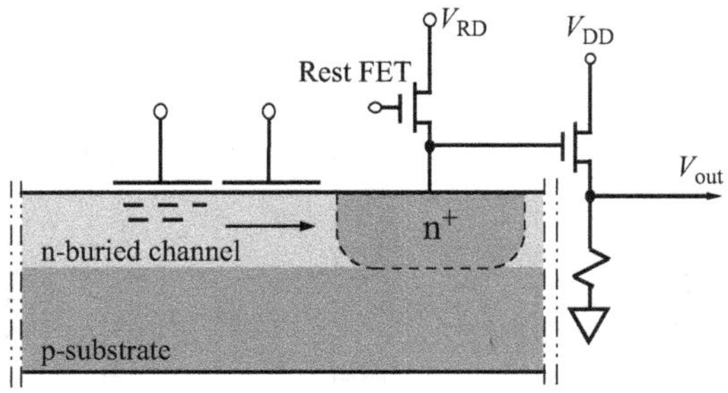

Figure 4.7 Charge-to-voltage convention and charge amplifier of CCD sensor

F. Signal digitisation

A CCD sensor is principally an analogue device producing a series of analogue electronic signals although a CCD camera can be analogue or digital. In an analogue CCD camera, the image information is converted to an analogue signal in a standard video format (e.g. CCIR, RS 170, PAL or NTSC), which can be transferred to a monitor, such as a CCTV display or TV. In a digital camera, the analogue image signal is digitised. The resolution of the digitisation depends on the bits of analogue-to-digital converter (ADC) of the camera, which defines the number of grey-levels, for example 8-bit for 256, 10-bit for 1,024 and 12-bit for 4,096.

G. Readout rate and frame rate

The readout rate of a CCD is regarded as the inverse of the serial conversion time, that is the time required to read out a single pixel. It is often stated as a frequency (Hz), more commonly referred as a pixel clock rate. A typical readout rate is in a range from 20 kHz to 1 MHz. The frame rate of a CCD is defined as the inverse of the time required to acquire an image and to read the data of the complete image to the amplifier, expressed in frames per second (fps). It can approximately be obtained by taking the inverse of the sum of the total pixel readout, digitisation and the exposure (i.e. integration) time of the camera.

4.2.1.2 Characteristics of CCD

A. Sensor size, pixel size, and pixel number

The size of a CCD is measured diagonally in inch, ranging from 1/6″ to 4/3″ for instance. The pixel size affects the resolution of the image. For a given focal length of the camera lens, small pixels result in a high image resolution. However, a small pixel has generally a small full-well capacity, and thus a reduced dynamic range (refer to Section 4.2.1.2 C). On the other hand, a large pixel has a greater full-well capacity and thus a wider dynamic range, resulting in a reduced image resolution.

B. Resolution

The image resolution of a CCD is the measure of its ability to reproduce the detail of an object, which is a function of the number of pixels and their size related to the projected image. The resolution is expressed in terms of the number of line-pairs per millimetre (lp/mm), that is the Nyquist limit, which is defined as a half of the sampling frequency, that is the number of pixels per mm. For instance, if a CCD sensor has an active area of 768 horizontal pixels in 4.8 mm. This represents a horizontal sampling frequency of 160 pixels/mm (= 768 pixels/4.8 mm). The Nyquist limit (resolution) of the CCD is thus 80 lp/mm.

C. Dynamic range

The dynamic range (DR) of a CCD is related to its ability to record simultaneously very low light signals alongside very bright signals. It is defined as the ratio of the FWC and readout noise (refer to Section 4.2.1.2 H), often expressed in decibel (dB), that is

$$\text{DR (dB)} = 20\log(\text{FWC}/\text{readout noise}) \tag{4.2}$$

A wide DR can be achieved through a greater FWC and lower readout noise. As an example, Figure 4.8 gives the comparison of gradation ranges that can be sensed by cameras with different DRs with reference to the human eye.

D. Linearity

One of the important characteristics of a CCD is its ability to respond linearly to any image it views. This means that if it detects 100 photons, it can convert these to 100 electrons, and if it detects 10,000 photons, it can convert these to 10,000 electrons. Good linearity is essential as there is no need for any additional processing on the image to determine the true light intensity of different objects in an image.

E. Spectral response

The spectral response of a CCD covers the complete range of visible spectra (400 nm–700 nm). However, a CCD imaging sensor often has a spectral response in

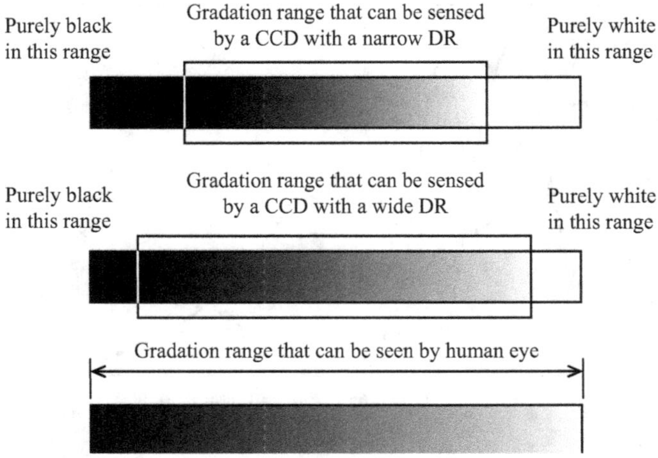

Figure 4.8 Comparison of gradation ranges that can be sensed by cameras with different DRs

both ultraviolet (UV) and near infra-red (NIR) spectral ranges. Figure 4.9 shows the typical spectral responses of mono-chromatic and RGB CCD imaging sensors.

A wide spectral response makes a CCD sensor suitable for infra-red (IR) detection. However, as a CCD responds to both IR and visible lights, the quality of images in a scene can be degraded where a significant IR radiation exists along with the visible light, as many optical lenses have different focal lengths at different wavelengths. Therefore, a focus setting that is correct for the visible light can result in a defocused IR image (often referred to as a ghost image) over the image frame. The problem can be more serious in a B&W CCD than that in a colour CCD. Such a problem can be eliminated by placing an IR cut-off optical filter in front of the CCD sensor.

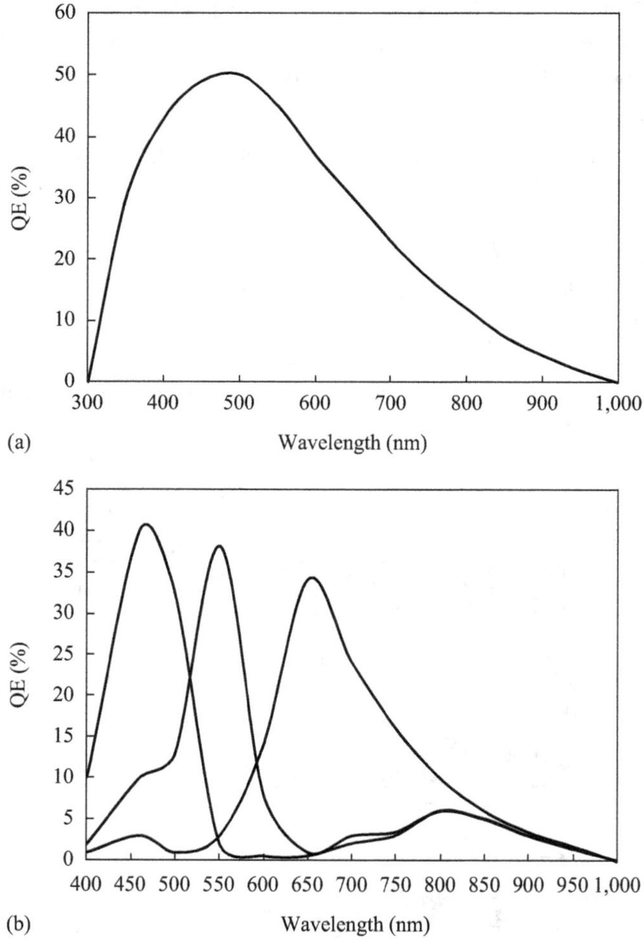

Figure 4.9 Typical spectral responses of CCD sensors: (a) mono-chromatic CCD and (b) RGB CCD

F. Gain

In a CCD imaging sensor, the gain refers to the magnitude of amplification that the CCD sensor produces, and can be set in the electronic circuit of the CCD. The gain is expressed in units of electrons per count (digital unit). For example, a gain of 2 e-/count means that the camera produces 1 count for every 2 recorded electrons.

Note that the use of term 'gain' in the CCD is in the opposite sense to conventional electronics, where the gain refers to an increase in the number of output units in comparison to the number of input units. In addition, the count unit does not provide a direct comparison without knowing the actual gain, because one CCD may use a different gain. Let's consider two cameras, each records 24 electrons in a pixel. If the gain of the first camera is 4.0 and the gain of the second camera is 6.0, the same pixels would produce six counts in the image from the first camera and four counts in the image from the second camera. Thus, without knowing the gain, comparing six counts against four counts is meaningless.

G. Blooming and smearing

Each pixel of a CCD stores electrons depends on its FWC. If a pixel is full of electrons (i.e. saturated), the excess electrons flow to neighbour pixels, and thus blooming occurs. The blooming can cause saturated streaks (i.e. bright strips) over the image, destroying pixel information. The blooming can be prevented by adding a mechanism to CCD to drain off excess charges, called anti-blooming. Smear occurs when a CCD sensor is illuminated (to external light) during readout and electrons leak into the CCD. The smear can be eliminated by closing a mechanical shutter during the readout.

H. Noise sources

Noise is composed of undesirable signal components that arise from various sources in CCD electronics. There are three major noise sources, which affect the performance of CCD, that is dark current noise, photon noise and readout noise.

1. *Dark current noise*. Even in the dark, the silicon lattice of a CCD emits electrons randomly, forming a signal, so called dark current. Those electrons are captured by the CCD's potential wells and counted as a signal, known as dark current noise. The dark current noise can be calculated as the square root of the number of thermal electrons generated within a given exposure time, that is

$$\text{Dark current noise} = \sqrt{\text{Dark current} \times \text{exposure time}} \tag{4.3}$$

The dark current is very temperature dependent. A cooled CCD produces a lower dark current. Typically, the dark current can be halved for every 6–8 °C drop in temperature. Figure 4.10(a) shows the relationship between the dark current noise and its working temperature. Note that individual pixels may have, called a different dark current generation rate across the CCD array. Those pixels may have a higher dark current than the average, known as 'hot' pixels.

2. *Photon noise (photon-shot noise)*. Photons reached on CCD are converted to photo-electrons within the silicon layer. These photo-electrons not only constitute the signal but also carry statistical variation in the photon arrival rate at

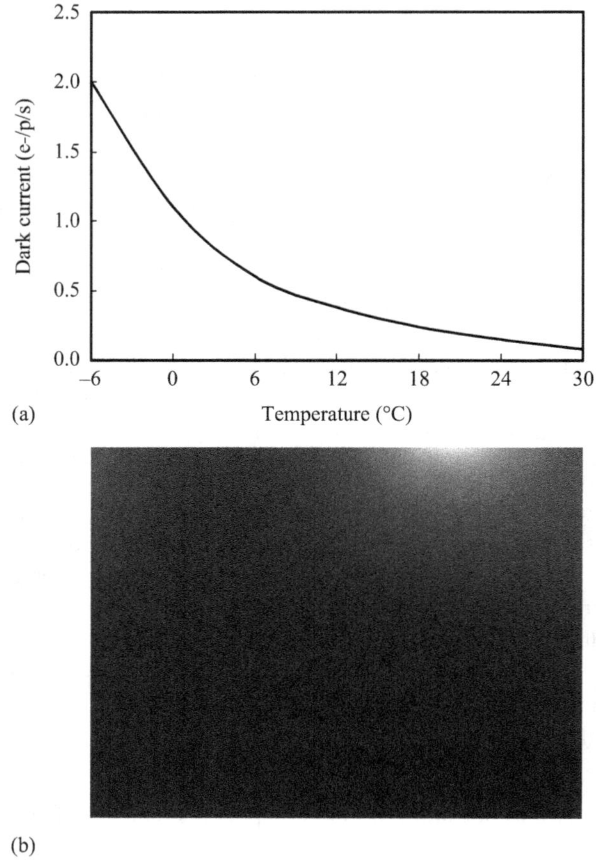

Figure 4.10 Dark current noise: (a) dark current versus temperature and (b) dark current image with 'hot' pixels

a given point, that is photon noise. It refers to the inherent natural variation of the incident photon flux (i.e. light). Photon noise is estimated by the square root of the number of electrons that accumulate in a potential well (in dB), that is

$$\text{Photon noise} = \sqrt{\text{number of electrons}} \tag{4.4}$$

Photon noise usually becomes the dominant noise under high exposure and/or high dark current conditions, and cannot be reduced via CCD design.

3. *Readout noise.* The readout noise is a combination of system noise components inherent to the process of converting charge carriers into a voltage signal for quantification, amplification and A/D conversion. The readout noise is a random noise and usually the dominate noise under short-exposure and/or low-light conditions, where the dark current and photon noises are small.

I. Signal-to-noise ratio

Signal-to-noise ratio (SNR) describes the quality of a measurement. A high SNR is particularly important for precise light measurements. In CCD imaging, SNR refers to the ratio of the measured signal to the overall measured noises (frame-to-frame) at a pixel, that is

$$\mathrm{SNR} = \frac{PQ_e t}{\sqrt{(P+B)Q_e t + Dt + N_r^2}}, \tag{4.5}$$

where P is the photon flux incident on the CCD (photons/pixel/second), B is the background photon flux incident on the CCD (photons/pixel/second), Q_e is the quantum efficiency of the CCD, t is the integration time (seconds), D is the dark current noise (electrons/pixel/second) and N_r is the readout noise (electrons rms/pixel).

4.2.1.3 Colour creation

A CCD imaging sensor is inherently a monochrome device, which does not create colour. To make a colour image with a monochrome sensor requires various filtering techniques so as to assign the image with a particular colour in post processing. The most common filtering techniques include Bayer filtration, three CCD (3-CCD) and Foveon X3® direct image sensors.

A. Bayer filtration

Bayer filtration uses a colour filter array passing red (R), green (G) and blue (B) lights to selected pixels, forming interlaced grids sensitive to R, G and B, respectively, as shown in Figure 4.11. An interpolation process (called demosaicing algorithm) is performed to generate missing colour samples over the sensor array. It is cheap but the image resolution is low. For instance, a colour array has 25% of the full resolution for B and R, and 50% resolution for G.

B. 3-CCD sensor

A 3-CCD sensor has three separate CCD sensing panels, each taking a separate measurement of R, G or B light. This is achieved by placing a trichroic prism assembly in front of the CCD, as shown in Figure 4.12. Light coming into the camera lens is split into three beams, each being directed to the appropriate wavelength

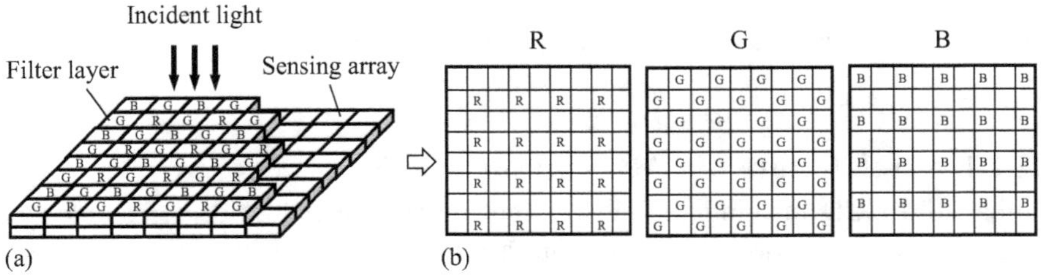

Figure 4.11 Colour generation through Bayer filtration: (a) sensor array and (b) resulted colour patterns

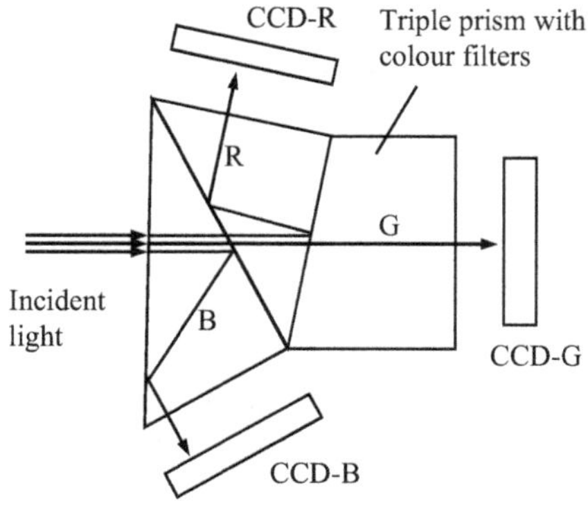

Figure 4.12 Schematic of 3-CCD sensor

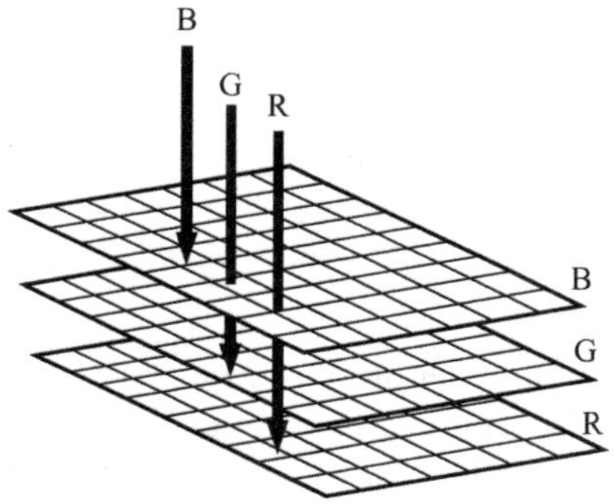

Figure 4.13 Colour generation of Foveon X3 sensor

range of the light (R, G or B) to its respective CCD. The 3-CCD sensor provides a superior image resolution over a Bayer filtration CCD sensor.

C. Foveon X3 sensor

A Foveon X3 sensor [2] has three sensing layers, each recording a different colour (see Figure 4.13). The layers are embedded in the silicon so that R, G and B lights can penetrate the silicon to different depths. As a result, the sensor captures R, G and B lights at every pixel. A Foveon CCD sensor has a good resolution, and improved image sharpness as each pixel can detect more photons.

4.2.2 CMOS imaging sensors

4.2.2.1 Architecture of CMOS sensor

Similar to a CCD imaging sensor, a complementary metal-oxide semiconductor (CMOS) imaging sensor is formed from a grid of light sensitive elements, each being capable of producing an electrical signal/charge proportional to the incident light. However, the process of achieving this is very different from the CCD technology in terms of both the circuit design and the process of implementing that design. A CMOS sensor contains an array of photo-diodes (i.e. pixels) that convert a light to electrons (i.e. photo-electric effect). Unlike a CCD where charges are transferred across the chip and read by a single readout register, a CMOS sensor converts charges to a voltage at individual pixels. At each CMOS' pixel, several transistors amplify and move the charges using more traditional circuits. Table 4.1 gives the comparison between CMOS and CCD architectures.

As shown in Figure 4.14, each CMOS pixel consists of a photo-diode, a capacitor and several transistors. Prior to the start of the integration period, the capacitor is charged to some known voltage. When the integration period begins, the charge on the capacitor is allowed to drain slowly away from the photo-diode, and the rate of drain is directly proportional to the level of incident light reached to that pixel. At the end of the integration period, the charge remaining is read out and digitised. Other circuit arrangements are possible, for example, the capacitor is charged during the integration period, as opposed to being discharged.

As the charge accumulated in the pixel is too small to transfer through column and row signal wires (b, e), it is first converted to a voltage and amplified at the

Table 4.1 Comparison between CMOS and CCD architectures

Imaging sensor	**Light collection and charge accumulation**	**Charge conversion**	**Charge transmission**
CCD	• MOS capacitors • Potential wells	Vertical and horizontal CCDs	Amplifier behind horizontal CCD
CMOS	Photo-diodes	Amplifier within pixel	Signal wire

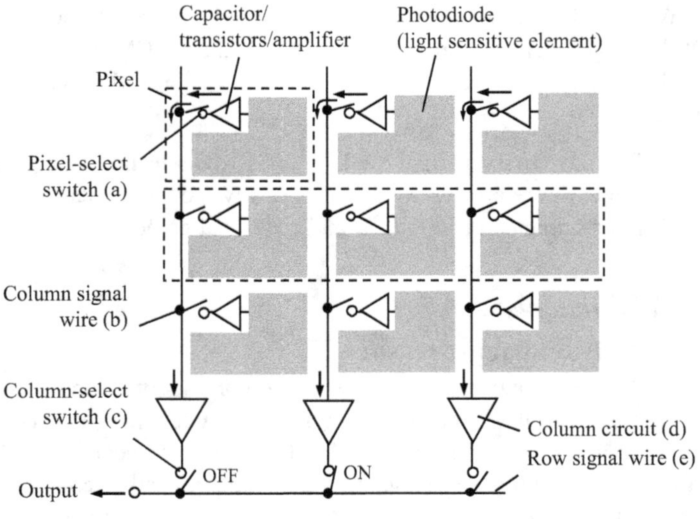

Figure 4.14 Architecture of CMOS sensor

pixel. Any individual signal wire can carry a voltage from only one pixel at a time, as controlled by the pixel-select switch (a). The pixel-select switch, column-select switch (c) and the column circuit (d) are used to control the output of amplified voltages. All the switches act in a precise sequence. All the pixel-select switches on a given row are turned 'ON'. This action outputs the amplified voltages of pixels to their respective column circuit, where they are processed into signal voltages and temporarily stored. The column-selection switches are then turned 'ON' from left, allowing that each pixel in the same row is read out in order. By repeating this operation for all rows from the top to the bottom, the voltage signals of all pixels can be read out from the top-left corner to the bottom-right corner of the image sensor, forming voltage outputs to the signal processor of the camera.

4.2.2.2 CCD versus CMOS

Nowadays, there is no clear division in terms of performance and applications between CCD and CMOS imaging sensors. As a general guide, a comparison between CMOS and CCD is given as follows:

- CCD is more sensitive as it has generally a high fill factor. As an example, if the fill factor of CCD is 100%, the fill factor of the CMOS is less than 70%.
- CCD is good for low-contrast imaging due to its lower inherent noise (readout noise).
- CCD has a wider dynamic range due to its lower readout noise.
- CMOS is flexible. It can be set to frameless data at a higher frame rate. For instance, a CMOS sensor with a resolution of 2,048 × 2,048 and a frame rate of 25 fps can achieve a frame rate of nearly 200 fps by only reading out a 340 × 340 portion of the image.
- CMOS has a much lower power consumption, making it ideal for mobile devices/applications. For instance, whilst a typical CCD consumes a power of 2 to 5 W, a CMOS sensor may only consume only 20 mW to 50 mW.
- CMOS has a good anti-blooming capability, because a CMOS has generally a natural blooming immunity whilst a CCD requires specific circuit design to achieve this.

With the fast development of the CCD and CMOS technologies, the above factors may change. There are CCDs in low-cost, low-power cameras, and CMOS sensors in high-performance professional and industrial cameras.

4.2.3 Special imaging sensors

4.2.3.1 Intensified imaging sensor

An intensified CCD (ICCD) camera, as its name, consists of an image intensifier and a CCD camera. An ICCD camera allows image acquisition at very low light levels (such as at night, the fluorescence of material, or even near-infrared or short wave infrared) due to the high signal gain of the image intensifier coupled optically to the camera lens. Figure 4.15 shows the layout and working principle of a typical image intensifier.

(a)

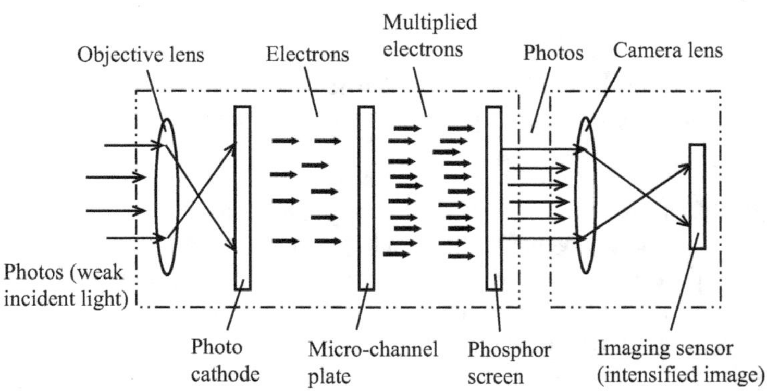

(b)

Figure 4.15 (a) Layout of an image intensifier and (b) working principle of the image intensifier

It is basically a vacuum tube device. The photons of a weak incident light enter the objective lens and strike the photo-cathode. The negatively biased photo-cathode releases electrons which are accelerated to the micro-channel plate with a higher-voltage. Each electron causes multiple electrons to be released from the micro-channel plate. The electrons are then drawn to the higher-voltage phosphor screen. As electrons strike the phosphor screen, the phosphor produces the photons of light which are then acquired by the CCD imaging sensor. An ICCD camera allows long exposure times, high-speed shutters and hence provides very good SNR. A high-end ICCD camera can discriminate single photons from CCD noise.

4.2.3.2 Electron multiplying CCD

Electron multiplying CCD (EMCCD), known as 'on-chip multiplication', is a quantitative digital camera technology that can detect single-photon events without an image intensifier. This is achieved by adding an electron multiplication (EM) register to the end of the serial readout register of a CCD chip (typically a frame-transfer CCD), as shown in Figure 4.16(a). The EM register consists of an additional row of serial readout elements, which are operated at a high clock voltage.

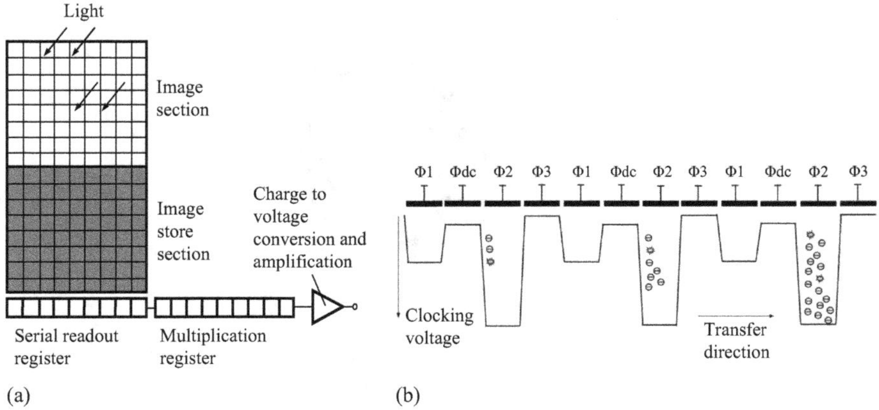

Figure 4.16 Principle of EMCCD imaging sensor: (a) architecture of EMCCD and (b) operation of multiplication register [3]

As the electrons are passed from one element to the next, a small number of additional electrons are generated at each transfer through a process known as impact ionisation. Although the statistical gain is small, typically in the region of 1.01 per element, the EM register has n elements (typically 512 or 1,024), making a gain of 1.01^n. Figure 4.16(b) shows the mechanism of the multiplication gain. The electrons are moved from one pixel to the next by the clocking pulses, **Φ**1, **Φ**2 and **Φ**3, which are applied to the surface electrodes. The EM effect occurs as a result of the enhanced **Φ**2 which is driven by an extra fixed-potential electrode **Φ**dc with a much higher voltage than usual (typically 35–50 V). The EM register allows weak signals to be multiplied before any readout noise is added by the output amplifier, and hence the readout noise is minimised.

4.2.3.3 High-speed camera

A conventional camera has a standard frame rate of 25 fps. High-speed cameras are required for studying transient phenomena, such as flow and combustion processes. Traditionally, high-speed imaging (up to 250,000 fps) can be achieved using a film camera associated with a rotating prism or mirror instead of a mechanical shutter. Specialised CCD sensors can achieve speeds up to or in excess of the tens of millions fps. Some less expensive CCD and CMOS cameras can operate at a high frame rate with a reduced image resolution, that is only reading out a part of the image frame. For example, a Phantom® v2512 camera can record a video at a frame rate over 25,700 fps with a full resolution of 1,280 × 800, and a video at a frame rate up to 1,000,000 fps at a reduced resolution of 256 × 32 [4].

4.2.3.4 Light-field camera

Unlike a conventional camera, which records only light intensity from a scene, light-field (also known as plenoptic) camera can sample a 4D light field on its imaging sensor (CCD or CMOS) in a single photographic exposure, so that not only

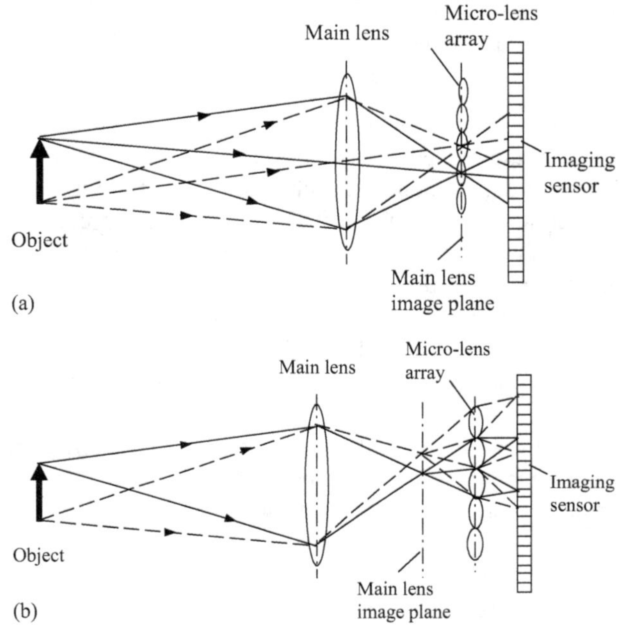

Figure 4.17 Light transmission of light-field camera: (a) standard and (b) focused

the light intensity but also the direction that the light rays are travelling in space can be simultaneously recorded. This is achieved by positioning an internal micro-lens array between the main lens and the imaging sensor. There are two main types of light-field cameras:

1. *Standard light-field camera* has a micro-lens array, which is placed one focal length away from the image plane of a sensor as shown in Figure 4.17(a). Rays from a single point on the object are brought to a single convergence point on the imaging plane of the main lens. The micro-lens separates these rays based on their directions, generating a focused image on the array of pixels. The array of pixels is regarded as a sub-image or a macro-pixel. The direction information of the ray is then determined by both the main lens plane and the micro-lens array of the camera.
2. *Focused light-field camera* has a micro-lens array positioned behind or before the focal plane of the main lens as shown in Figure 4.17(b). In comparison to the standard light-field camera, the focused light-field can produce images with a much higher spatial resolution through a post focusing processing. However, a lower angular resolution can introduce some unwanted aliasing artefacts.

4.2.4 Camera interfaces

The type of output signal of a CCD or CMOS camera is determined by how the data are read out from a camera. If the image data are read out to a video monitor then an analogue output, either in CCIR for monochrome, or in phase alternating line (PAL) for colour, should be used. However, if the data need to be transferred to a computer system, a camera interface should be one of the follows:

- *Camera Link*. It is a digital interface standard which requires a frame grabber to be installed in the computer system. Camera Link is widely used in many high-performance digital cameras, such as EMCCD and high-speed cameras.
- *USB*. USB interface is compatible with USB 2.0 and 3.0 ports on the computer. It is convenient, fast (up to 625 megabits/second, USB 3.0), low cost but has a limited data transmission distance (up to 5 m).
- *Gig-E (Gigabit Ethernet)*. It gains its popularity because it allows huge amounts of data to be transferred quickly and over a long distance.
- *IP (Internet protocol)*. It sends/receives data via a computer network and the Internet.
- *FireWire (IEEE1394)*. It is an old camera interface and offers a 'plug and play' solution if the computer system equipped with a *FireWire* port (it can be added to the computer via a plug-in interface card).

4.2.5 Sizes of imaging sensor and lens

Imaging sensors are available in different sizes, such as 2/3″, 1/2″, 1/3″ and 1/4″, and camera lenses are manufactured to match these sizes. It is important to select a lens suitable for an imaging sensor. The general rule is that the size of the lens must be equal to or greater than the size of the sensor. For instance, a camera lens made for a 1/2″ imaging sensor should work with a 1/2″, 1/3″ or 1/4″ sensor, but not with a 2/3″ sensor. As shown in Figure 4.18, if a lens is smaller than one actually fitted inside the imaging sensor, the four corners of the image will appear to be black (see Figure 4.18(b)). On the other hand, if a lens is larger than the one actually fitted inside the imaging sensor, some part of the image will be outside of the imaging sensor (see Figure 4.18(c)).

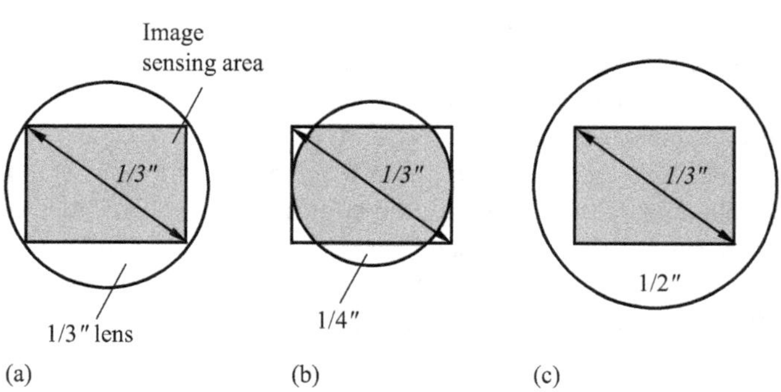

Figure 4.18 Sizes of imaging sensor and lens (using a 1/3″ imaging sensor as example): (a) good, (b) inappropriate and (c) acceptable

4.2.6 Shutters

Whilst a traditional film camera uses a mechanical or a rotary shutter, a digital camera has a shutter controlled by the imaging sensor itself. There are two different shutters which are commonly used in the CCD and CMOS sensors, i.e.:

- *Global shutter.* It allows an entire frame to be captured at the exact same time. The best way to envision how the global shutter works is by imagining an imaging sensor as a slide of film. When light passes through a slide of film, it travels through the entire frame at once. The global shutter produces good quality images in general, but it may make images appear to be blurred as the operator performs a 'swish-pan'.
- *Rolling shutter.* It exposes a frame from top to bottom in a time sequence (defined by exposure time). The imaging sensor captures an incident light by capturing pixels raw-by-raw from top to bottom, creating a rolling effect. The rolling shutter may produce images appearing to be stretched or 'skewed' along the horizontal direction when the object being sensed is moving. With large and very fast change in illumination, it can make frames look split in half, with one-half appearing to be very bright and another very dark.

4.3 Flame imaging and measurement

Flame is the primary reaction zone of a combustion process. As a flame is basically a 3D flow field, the characteristics of the flame are very difficult to conceptualise and comprehend from a direct 2D graphical presentation. With the rapid development of image sensing and computing techniques, visualisation techniques have been enhanced considerably and formed a powerful tool for flame measurement characterisation in both fundamental studies and industrial applications. Once an image is represented in a digital format, appropriate numeral methods can be applied to identify the characteristics of flame qualitatively and quantitatively. This section introduces the typical applications of CCD/CMOS imaging sensors in 2D and 3D measurement of burner flames in fossil fuel combustion systems.

4.3.1 Nature of flame

Flame is an exothermic chain reaction of fuel that propagates through space. The nature of flame is mainly influenced by three factors, that is the initial state of fuel, initial degree of mixing and flow characteristics of the reactants. The initial physical state of the reactant, that is the fuel, is an obvious factor controlling flame properties. Fuel can be gaseous, liquid or solid. Figure 4.19 shows example images of natural gas-, oil- and coal-fired flames, taken from different industry-scale furnaces.

The degree of mixing refers to whether the fuel is premixed with the oxidiser before entering the reaction zone. A premixed flame arises from the combustion of the reactants that are perfectly mixed prior to the combustion. A diffusion flame refers to the combustion of separate fuel and oxidiser streams, which burn as they mix.

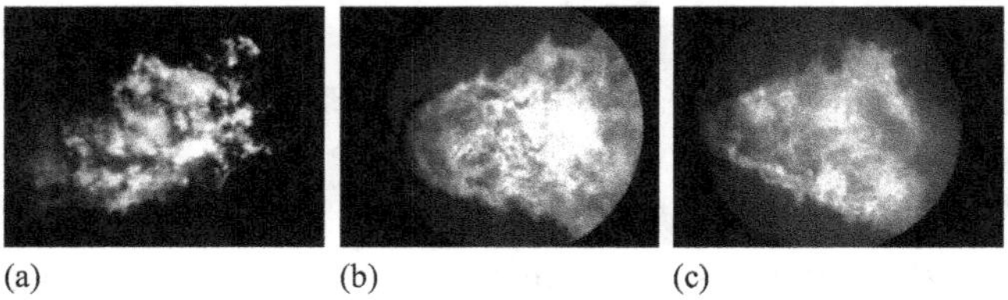

Figure 4.19 Typical images of flame with different fuels in industry-scale furnaces: (a) gas-fired flame, (b) oil-fired flame and (c) coal-fired flame

Flame can also occur in both laminar and turbulent flows. If the flow is laminar, it is called a laminar flame. Otherwise, it is called a turbulent flame. Most practical applications require the use of turbulent flames to produce the volumetric rates of energy production needed for efficiency and compactness. However, concepts of laminar flame propagation are an integral part of studying the turbulent flames.

The radiative spectrum of the flame is one of important characteristics relating closely to the type of fuel and the nature of air-to-fuel mixture. The flame spectrum is highly irregular over a wide range of wavelengths from the UV to IR through visible bands. The spectral emissions of flame consist essentially of continuous and discontinuous (i.e. banded) spectra. The continuous spectra, as shown in Figure 4.20(a), are generally observed in the luminous region of the flame, where radiative energy is emitted by incandescent carbon particles (such as soot and other solid particles). They are distributed over a broad spectral range and complied with the Planck radiation theory. The discontinuous spectra, as shown in Figure 4.20(b), are caused by electronic transition and simultaneous changes in the internal vibrational and rotational energy of gas molecules (such as CO_2 and H_2O) and intermediate species.

4.3.2 Flame visualisation and measurement

Flame visualisation and measurement use an optical or electric/electronic means to visualise the flame zone so as to obtain 2D or 3D information of the flame and are mainly classified into a number of groups according to the nature of each technique, as shown in Figure 4.21.

Flame visualisation and measurement face a number of technical challenges due to the complex nature of the flame and severe operational conditions. The major technical difficulties are as follows:

- *Multiple parameter measurement.* Flame is basically a complex flow field. The most important physical properties of the flame include the geometrical and luminous profiles and temperature. A number of parameters should therefore be defined to describe these characteristics of flame.
- *Dynamic nature.* Variable operation conditions and large disturbances from the environment make that the characteristics of flame are highly nonlinear in

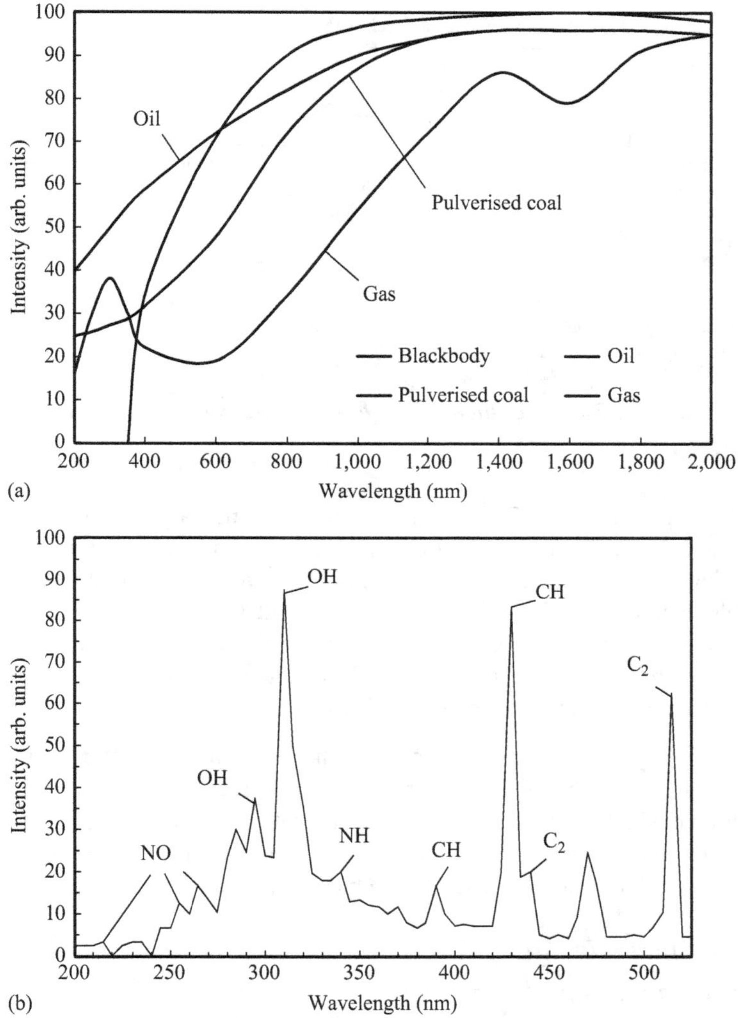

Figure 4.20 Typical radiative spectra of flame: (a) continuous spectra of gas-, oil- and coal-fired flames compared to the blackbody radiation at 1,650 °C [5] and (b) discontinuous emission of a rich premixed butane-air flame [6]

nature. These factors result in difficulties to obtain a high certainty and repeatability of the measurement.

- *2D image of a 3D object*. A 2D image is the projection of a 3D flame, which comprehends characteristic information of the flame along the line-of-sight at each pixel. This makes the interpretation of data very difficult. Some of the information cannot even be spatially resolved.

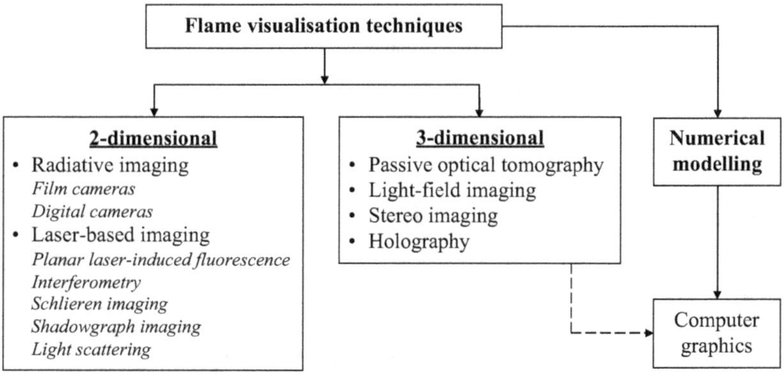

Figure 4.21 Classification of flame visualisation techniques

- *Wide measuring field.* The size and shape of flame vary depending upon the type and scale of the combustion system. To visualise the entire flame, an imaging system with a wide field of view is required, which would cause problems in the optical design as well as the optical distortion of flame images.
- *Hostile environment.* The electronic and optical devices of the system have to be protected from excessive heat radiating from a furnace. The optical lens should be kept dust free during its operation to obtain clear images under dusty conditions.

As a combustion process is a highly nonlinear system, it is difficult to measure flames and interpret the flame data through conventional methods. Appropriate computational algorithms including artificial intelligence are required to acquire and interpret the flame information so as to explore the inherent relationship between flame characteristics and the combustion process.

4.3.3 2D flame imaging

4.3.3.1 Flame radiative imaging

In radiative imaging techniques, the radiative information of flame is directly imaged by an imaging system. The technique offers a number of advantages, such as a simple system set-up (e.g. no additional light source is necessary), a wide field of view and completely remote from the flame, forming one of the most attractive approaches for flame measurement and characterisation.

A. Flame geometry and luminosity

In a digital image of flame, the grey-level of each pixel represents the luminous intensity (i.e. brightness) of the corresponding area of the flame. A range of characteristic parameters of the flame is determined by processing the flame images using appropriate digital image processing techniques. The geometrical and luminous parameters of flame can vary, depending upon the nature of the flame and the accessibility of the camera system to the flame. As a general guide, the following parameters can be defined as shown in Figure 4.22.

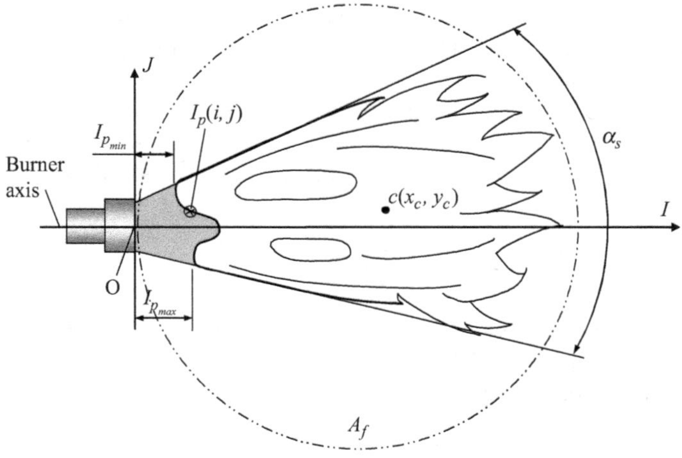

Figure 4.22 Definition of geometrical parameters of flame

- *Luminous region* (R_f) is the cross-sectional area occupied by the flame, donated as $G(i, j)$. The luminous region is commonly normalised to the entire cross-sectional area of the field of view (A_f in Figure 4.22), ranging between 0 and 100% where 0% indicates the absence of flame whilst 100% means the flame occupies the entire field of view:

$$R_f = \frac{1}{A_f}\sum_{i\in R_f}\sum_{j\in R_f}\begin{cases} 1, & \text{if } G(i, j) \geq \delta \\ 0, & \text{else} \end{cases} \times 100\% \tag{4.6}$$

where, δ is a pre-set threshold of grey-level.

- *Centroid* ($C(x_c, y_c)$) is the location of the 'centre-of-mass' of the luminous region, which is a useful parameter indicating the location of the flame:

$$x_c = \left(\sum_{i\in R_f}\sum_{j\in R_f} j \cdot G(i, j)\right) \Big/ \left(\sum_{i\in R_f}\sum_{j\in R_f} G(i, j)\right) \tag{4.7}$$

$$y_c = \left(\sum_{i\in R_f}\sum_{j\in R_f} i \cdot G(i, j)\right) \Big/ \left(\sum_{i\in R_f}\sum_{j\in R_f} G(i, j)\right) \tag{4.8}$$

- *Ignition-point* (I_p) is the point where the fuel is ignited. A stable flame requires the steady ignition point, at which the velocities of combustion flow and flame propagation are well balanced at the ignition temperature of the fuel. The ignition-point is measured as the distance from the burner outlet to the nearest point of the luminous region:

$$I_p(i, j) = \begin{cases} \text{true, if } \dfrac{G(i+1, j)}{G(i-1, j)} \geq \delta_{ip} \\ \text{false, else} \end{cases} \tag{4.9}$$

where δ_{ip} is a pre-set threshold of the ignition-point.

- *Spreading-angle* (α_s) is the angle formed between the two straight lines scribing from the burner outlet to the outer edges of the flame, which is a good indicator of flame shape.
- *Brightness* (B_f) is the average grey-level of luminous region (R_f), and generally normalised to the maximum grey-level (255 for an 8-bit image), ranging between 0 and 100%. A greater the value of B_f means a brighter flame:

$$B_f = \frac{1}{255}\sum\nolimits_{i \in R_f}\sum\nolimits_{j \in R_f}\frac{1}{|2R_f|}G(i, j) \times 100\% \tag{4.10}$$

- *Uniformity* (U_f) represents the relative deviation of the luminous intensities of flame, varying between 0% and 100%. The larger the magnitude of U_f, the more uniform the flame:

$$NU_f = \frac{1}{NB_f}\sum\nolimits_{k=\delta_f}^{255} n(k)\left|B(k) - B_f\right| \times 100\% \tag{4.11}$$

  where $n(k)$ is the number of pixels at the grey-level $B(k)$ (0–255); N is the total number of pixels within the luminous region defined by threshold δ_f.

Note that the brightness and uniformity of flame are relative values, depending not only upon not only the radiation intensities of the flame but also the camera settings, such as the aperture and exposure time being used.

Figure 4.23 shows example results of flame geometry and luminosity measurements, which were conducted using a multi-functional flame monitoring system on a coal-fired combustion test facility (CTF).

B. Two-colour flame pyrometry

Flame pyrometry is referred to as methods for measuring the flame temperature based on the Planck radiation theory. A monochromatic pyrometer measures the temperature by detecting the radiation of the object at a single wavelength if the emissivity of the detected object is known. However, the emissivity of flame is normally either unknown or very difficult to determine, particularly in practical combustion systems. Two-colour pyrometry is therefore commonly used for the flame temperature measurement.

Based on the Planck radiation law, the two-colour method calculates the temperature of flame from the ratio of radiation intensities emitted from the incandescent carbon particles at two different wavelengths for the purpose of eliminating the unknown variable, that is flame emissivity. Thus, the temperature obtained from the two-colour method is actually the one of soot (or other solid) particles inside the flame rather than the one of gas molecules.

Planck's radiation law indicates that the spectral radiation of a blackbody is a function of temperature for a given wavelength, that is

$$M_\lambda = \frac{C_1}{\lambda^5}\left(e^{\frac{C_2}{\lambda T}} - 1\right)^{-1} \tag{4.12}$$

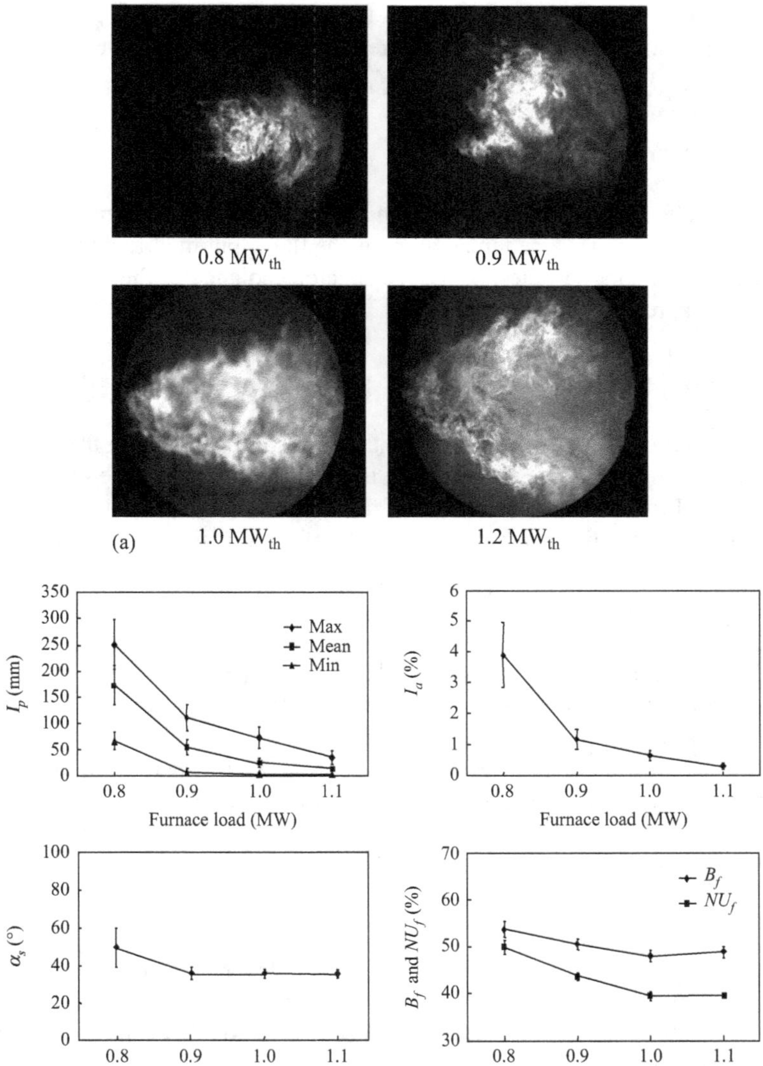

Figure 4.23 Example results of flame geometry and luminosity measurements [7]: (a) flame images, and (b) measured geometric and luminous parameters under different furnace loads

where M is the monochromatic radiance, namely radiant power per unit area of the emitting surface ($W{\cdot}m^{-2}{\cdot}\mu m^{-1}{\cdot}Sr^{-1}$), λ is the wavelength of the radiation (μm), T is the absolute temperature of the blackbody (K), $C_1 = 1.1911 \times 10^8$ ($W{\cdot}\mu m^4{\cdot}m^{-2}{\cdot}Sr^{-1}$) and $C_2 = 1.4388 \times 10^4$ (μm·K) are the first and second Planck's constants.

For short wavelengths and throughout the visible region, (4.12) can be replaced by Wien's radiation law if the temperature considered is up to about 3,500 K (i.e. $e^{C_2/\lambda T} \gg 1$):

$$M_\lambda = \frac{C_1}{\lambda^5} e^{-\frac{C_2}{\lambda T}} \tag{4.13}$$

The radiation of soot/solid particles in a flame is related to the radiation from a blackbody source at the same temperature as the flame but modified by the spectral emissivity of the soot particles. Thus, the spectral radiance of the soot/solid particles can be expressed by

$$M_\lambda = \varepsilon_\lambda \frac{C_1}{\lambda^5} e^{-\frac{C_2}{\lambda T}} \tag{4.14}$$

where ε_λ is monochromatic emissivity of the soot/solid particles in the flame.

It can be proven that the output of the imaging sensor, that is the grey-level G_λ, is proportional to the spectral radiance of the measured object (flame, in this case) and dependent upon the spectral sensitivity S_λ of the imaging system, that is

$$G_\lambda = R_0 S_\lambda S M_\lambda = R_0 S_\lambda \varepsilon_\lambda \frac{C_1}{\lambda^5} e^{-\frac{C_2}{\lambda T}} \tag{4.15}$$

where R_o is the instrument constant, which is independent of wavelength and reflects various factors such as the radiation attenuation due to the configuration of the imaging system.

If the bandwidth of two wavelengths λ_1 and λ_2 is small enough, the ratio between the grey-levels at λ_1 and λ_2 can be expressed as

$$\frac{G_{\lambda_1}}{G_{\lambda_2}} = \frac{S_{\lambda_1} \varepsilon_{\lambda_1} \lambda_2^5 e^{-\frac{C_2}{\lambda_1 T}}}{S_{\lambda_2} \varepsilon_{\lambda_2} \lambda_1^5 e^{-\frac{C_2}{\lambda_2 T}}} \tag{4.16}$$

The ratio between the spectral emissivities, $\frac{\varepsilon_{\lambda_1}}{\varepsilon_{\lambda_2}}$, depends on the properties of the soot/solid particles in the flame. When the two wavelengths are carefully chosen, the difference between the corresponding spectral emissivities can be ignored, that is $\frac{\varepsilon_{\lambda_1}}{\varepsilon_{\lambda_2}} = 1$, and therefore

$$\frac{G_{\lambda_1}}{G_{\lambda_2}} = \frac{S_{\lambda_1} \lambda_2^5 e^{-\frac{C_2}{\lambda_1 T}}}{S_{\lambda_2} \lambda_1^5 e^{-\frac{C_2}{\lambda_2 T}}} \tag{4.17}$$

Rearranging (4.11) yields

$$T = \frac{C_2\left(\frac{1}{\lambda_2} - \frac{1}{\lambda_1}\right)}{\ln\left(\frac{G_{\lambda_1}}{G_{\lambda_2}}\right) + \ln\left(\frac{S_{\lambda_2}}{S_{\lambda_1}}\right) + \ln\left(\frac{\lambda_1}{\lambda_2}\right)^5} \tag{4.18}$$

where G_{λ_1} and G_{λ_2} can be measured by the imaging system, that is flame image intensities at λ_1 and λ_2, which are known from the design of the system. The ratio between the spectral sensitivities, $\frac{S_{\lambda_1}}{S_{\lambda_2}}$, is called instrument factor, which can be determined through a system calibration using a standard temperature source (normally a pre-calibrated blackbody furnace).

Equation (4.18) indicates that the flame temperature can be measured based on the ratios of image grey-level intensities at two different wavelengths without the prior knowledge of the flame emissivity, assuming that the emissivities at the chosen wavelengths, ε_{λ_1} and ε_{λ_2}, are equal. The assumption would be true both for optically thick dust flames and for most optically thin flames if the dust particles are larger than the wavelengths.

Acquisition of two images at two chosen wavelengths simultaneously is a prerequisite for implementing the two-colour method due to the dynamic nature of the flame. This can be achieved using a single monochromic camera associated with a purpose-designed beam splitting facility. Figure 4.24 shows the simple schematic of sensing arrangement of a two-colour flame pyrometric system. Passing through separate narrow band-pass filters located in the two optical paths, two images of flame are projected concurrently onto the imaging sensor (noted that in practical implementation, more prisms may be needed to ensure two optical paths are identical). Figure 4.25 shows the example results of flame temperature measurement through the two-colour pyrometry on a coal-fired CTF.

The two-colour method can also be implemented by using an RGB colour camera, which has three separate channels corresponding to primary colours red (R), green (G) and blue (B). Each image frame produced can be disintegrated into three primary images R, G and B. A combination of the colour-banded images can thus be used to determine the flame temperature based on (4.18). Figure 4.26 shows the R and G images of a coal-fired flame and resulted temperature distribution through the two-colour flame pyrometry.

The use of primary colour images for the temperature measurement offers advantages in simplicity in system design, calibration and operation. However, it would result in some measurement errors because each primary colour is not actually narrowly banded, that is it covers a certain range of wavelengths corresponding to that colour.

For implementation of the two-colour flame pyrometry, several factors must be considered: (1) the two chosen wavelengths should be kept away from the fluorescence emissions from gas molecules and intermediate radicals in the flame (as discussed in Section 4.3.1) as they do not comply with the Planck's radiation law.

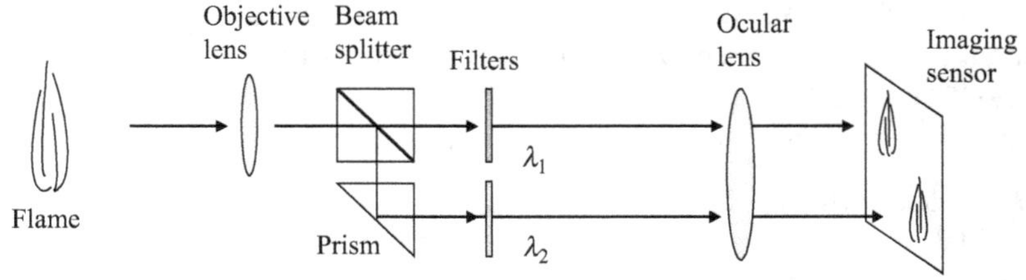

Figure 4.24 Sensing arrangement of a two-colour flame pyrometric system

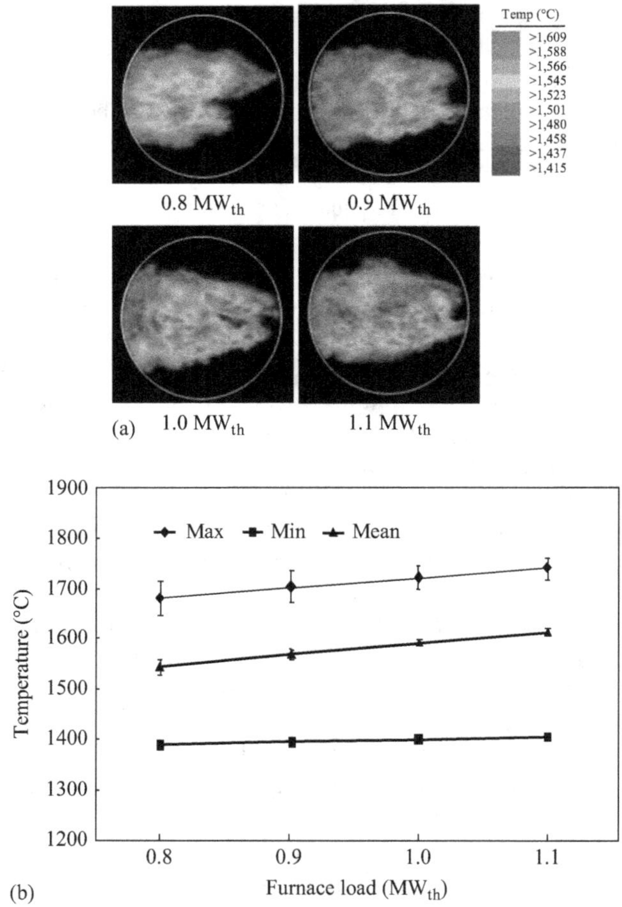

Figure 4.25 Example results of flame temperature measurement through the two-colour pyrometry [7]: (a) temperature distributions and (b) temperature variations under different furnace loads

(2) The two wavelengths are expected to be close to each other to reduce the error caused by the difference in spectral response, especially considering that grey-body behaviour of a flame is assumed. (3) The bandwidths of the two filters should be the narrowest bandwidth, at which satisfactory flame images can be captured for the measured temperature range. When an RGB camera is used, it should make sure that the CCD (or CMOS) sensor of camera gives a higher spectral sensitivity for the chosen channels, so as to achieve a good SNR.

C. Flame oscillation frequency

The oscillatory (i.e. flickering) characteristic of a flame is reflected by the dynamic alternating components (AC) superimposed on the steady state (DC) or 'brightness' level. The oscillation signal of flame can be reconstructed from the grey-level

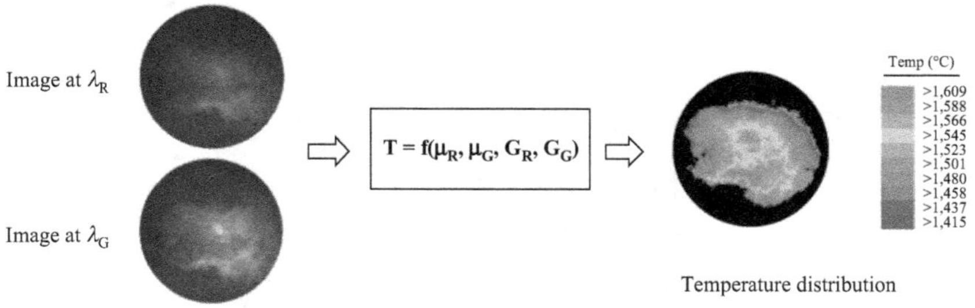

Figure 4.26 RGB based two-colour flame pyrometry

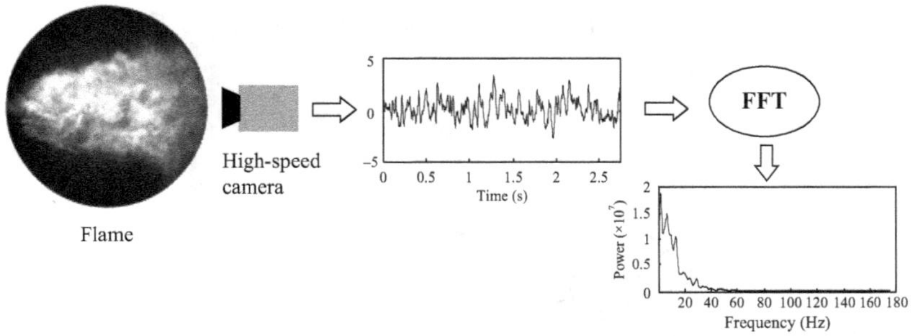

Figure 4.27 Flame oscillation frequency measurement through high-speed imaging [8]

intensity of pixels within flame images captured using a high-speed camera, as shown in Figure 4.27. The quantified frequency is then defined as the weighted average of the power spectra over the measuring range, that is

$$F = \frac{\sum_{n=0}^{N-1} [P_{xx}(f_n) \cdot f_n]}{\sum_{n=0}^{N-1} P_{xx}(f_n)} \tag{4.19}$$

where F is the flicker of a flame, f_n is the nth frequency, P_{xx} is the power density of the nth frequency component and N is the total number of frequency components.

4.3.3.2 Laser-based flame imaging

A. Planar Laser-induced fluorescence imaging

Planar Laser-induced fluorescence (PLIF) is an optical diagnostic technique for the fluorescence detection of radicals (i.e. species), such as OH, NO and O_2 in a flame. Figure 4.28 shows a typical PLIF system for flame measurement. The PLIF can also be used for measuring the flame temperature through the detection of fluorescence and scattering generated by atomic or molecular species stimulated by a Laser beam at a particular wavelength. There is a specific relationship between the temperature and decay lifetime of the fluorescence and the intensity of the scattering of molecular gas. These methods are non-intrusive and have a fast response

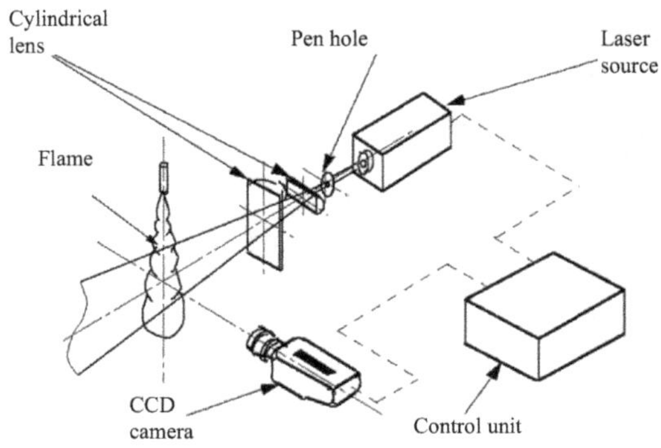

Figure 4.28 A PLIF system for flame imaging [9]

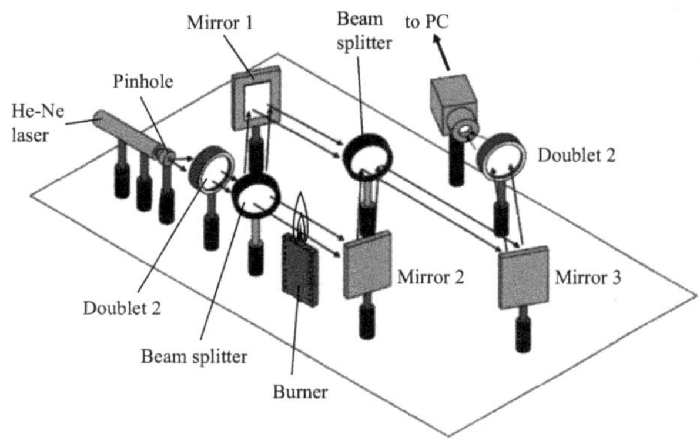

Figure 4.29 A Mech-Zehnder interferometric system for flame measurement [10]

and can be applied to the planar measurement of the flame, particularly the micro-structures and the velocity profile of the flame. In a PLIF system, however, the light intensity can be reduced considerably by the flame species, which would have a serious impact on the measurement accuracy.

B. Interferometry, schlieren and shadowgraph imaging

Interferometry, schlieren and shadowgraph imaging are commonly used to visualise the variations in the indices of refraction induced by changes in density, pressure and temperature of flames. The three techniques work in similar principles but are divided in terms of the interpretation of the observed pattern of light.

Interferometry uses the interference occurring between a coherent beam passing through the region of flame and a comparison beam. When they are properly combined on a screen (i.e. image sensing plane), the two waves interfere, producing an image outlining regions of differing the optical path length. There are a number of optical arrangements for forming paired coherent beams. Figure 4.29 shows a typical Mech-Zehnder interferometric system for flame measurement.

Schlieren and shadowgraph imaging rely upon precise lenses, mirrors and imaging sensors to image refractive disturbances in transparent media. They have been broadly applied for the flame visualisation and measurements including the location of flame front, inhomogeneities of gas temperature and soot concentration along the light path. In schlieren imaging, as shown in Figure 4.30(a), a beam of parallel light from a slit or point source passes through a flame zone and focuses on a stop (i.e. knife-edge). When the beam of light passes through the flame zone, which contains varying refractive indices, it is deflected around the stop and brought to the image sensing plane, producing an image connecting the regions of the same refractive gradient. The first lens is to collimate rays from the light source, while the second lens is to focus the rays on the plane of the stop. The stop blocks off the rays of light that have not been deviated, that is the rays, which are parallel

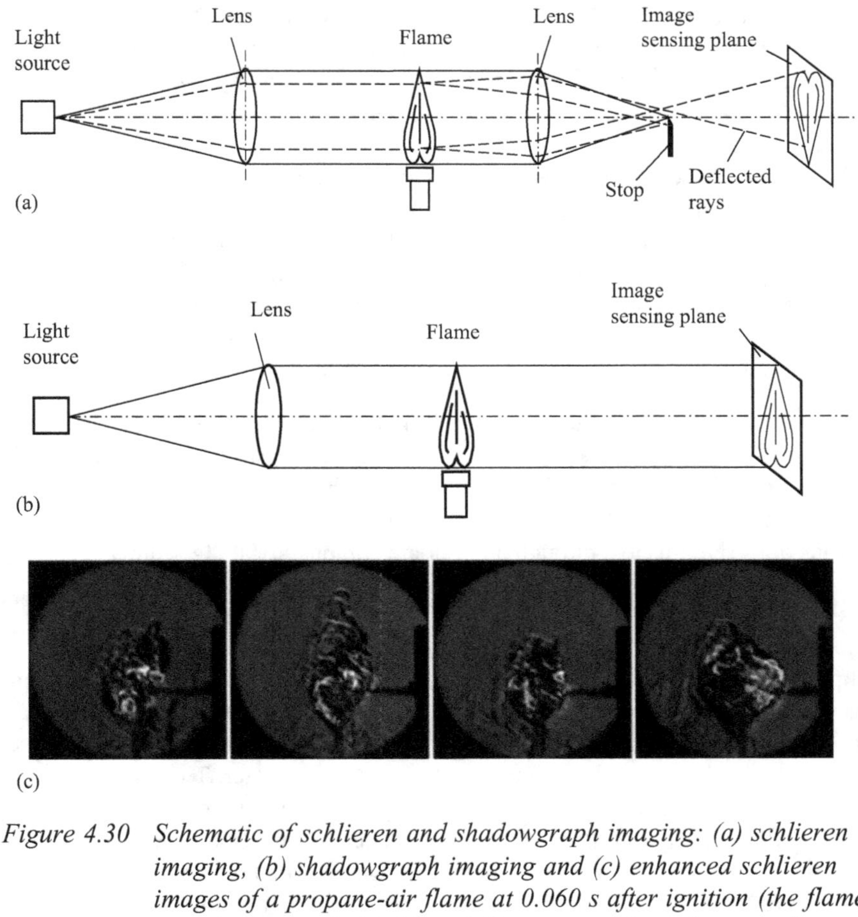

Figure 4.30 Schematic of schlieren and shadowgraph imaging: (a) schlieren imaging, (b) shadowgraph imaging and (c) enhanced schlieren images of a propane-air flame at 0.060 s after ignition (the flames were ignited at 0, 10, 20 and 30 mm away from the centreline, as shown from the left to the right) [11]

to the optical path. In this way, only the rays that were deflected due to the inhomogeneity of the flame zone examined reach onto the image sensing plane.

In shadowgraph imaging, as shown in Figure 4.30(b), the shadow of a flame is projected by a uniform beam of light passing through a region of interest of a flame. Deflection in the shadowgraph is proportional to the steep gradient of the flame density.

C. Light scattering

Light scattering relies on the interaction between radiation and molecules or seed particles in a flame zone. A sheet of light, normally derived from a Laser source, illuminates a section of the flame. The scattered radiation is then collected and imaged onto an imaging sensor. Two techniques regarding light scattering are usually used in the flame visualisation, that is particulate scattering and Rayleigh scattering. In the particulate scattering, the light scattered from particles in a flame can be used to define a flame region and outline its geometry. The particles can either be presented in the flame or be introduced as a tracer. This technique is a powerful diagnostic tool for visualising the particle velocity profiles of the flame. Laser-sheet-illuminated Rayleigh scattering (LRS) imaging systems [12] are also used to study the microscopic structures and temperature distribution of turbulent diffusion flames.

Whilst the Laser-based imaging techniques have been well developed and widely used for the 2D and 3D measurements of the flame, they have several limitations. They are normally only suitable for studying small objects, such as laboratory-scale flames or flow fields (e.g. engine sprays) due to the restriction of light sources and the complicated configurations of optical and electronic systems.

4.3.4 3D flame imaging

A flame is inherently a 3D fluid dynamic flow field. To fully reveal the characteristics of the flame, 3D visualisation and measurement are desirable. As the flame is a luminous and in most cases translucent object, it can be imaged directly using one or more imaging sensors. A 3D model of flame can then be reconstructed based on the 2D projections (i.e. images) through appropriate computational algorithms.

4.3.4.1 3D flame geometric reconstruction

One of simple solutions for 3D flame visualisation and reconstruction is to employ multiple cameras around the burner so that flame information in the three dimensions can be acquired. Figure 4.31 shows a three-camera system for the 3D visualisation and characterisation of gaseous flame. The system consists of three monochromatic CCD cameras, a frame grabber and a computer. The three cameras, placed equidistant and equiangular from each other around the burner, capture 2D images of flame concurrently from the three different directions. Dedicated computing algorithms are used to reconstruct the 3D geometric model of the flame from the acquired 2D images. A set of geometric parameters, including volume, surface area, orientation, length and circularity, are then defined and used to characterise the flame from the model generated.

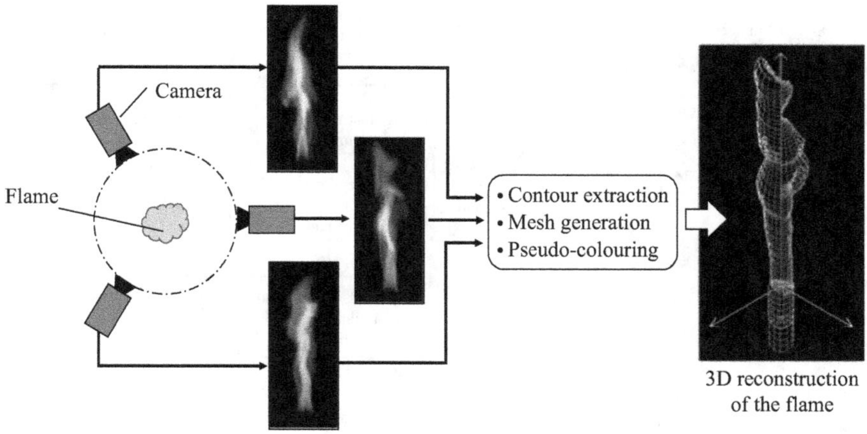

Figure 4.31 A multi-camera system for the 3D geometric reconstruction of flame [13]

4.3.4.2 Passive optical flame tomography

Figure 4.32 shows the schematic diagram and physical implementation of a tomography system for 3D flame reconstruction [14]. It has eight image fibre bundles with associated coupling lenses, two RGB cameras, two frame grabbers and a computer system with application software. Four flexible image fibre bundles are joined onto a single eyepiece, which is attached to the camera through a coupling lens, forming four flame images on the same imaging sensor. The eight fibre bundles are arranged on one side of the burner as a semi-circle and are fixed in positions with an equal separation between two adjacent fibre bundles. With such an arrangement, the imaging system can produce a total of eight flame images concurrently from eight different directions around the flame.

Figure 4.33(a) shows a set of 2D flame images, captured by the imaging system as shown in Figure 4.28. Figure 4.33(b) shows the reconstructed luminosity distribution of flame cross- and longitudinal-sections through the tomography algorithm known as the logical filtered back-projection and simultaneous algebraic reconstruction technique (LFBP–SART) [14]. Combining the optical tomography and two-colour pyrometry, the 3D reconstruction of flame temperature can also be achieved [15].

More 2D projections can be achieved through a multi-lens approach. An example of such a system is shown in Figure 4.34(a). The system is equipped with 40 special lenses arranged around on one side of the burner. A high-speed camera (either analogue or digital) is rotated along the circumference behind the lenses and acquires images in a sequence. The images are then transferred and processed in a computer system. Figure 4.34(b) shows an example of 3D reconstruction of flame using such an approach.

The passive optical tomography can acquire multiple flame projections around the burner and hence have a good spatial resolution. The larger number of projections yields a better reconstruction result. However, the increased number of

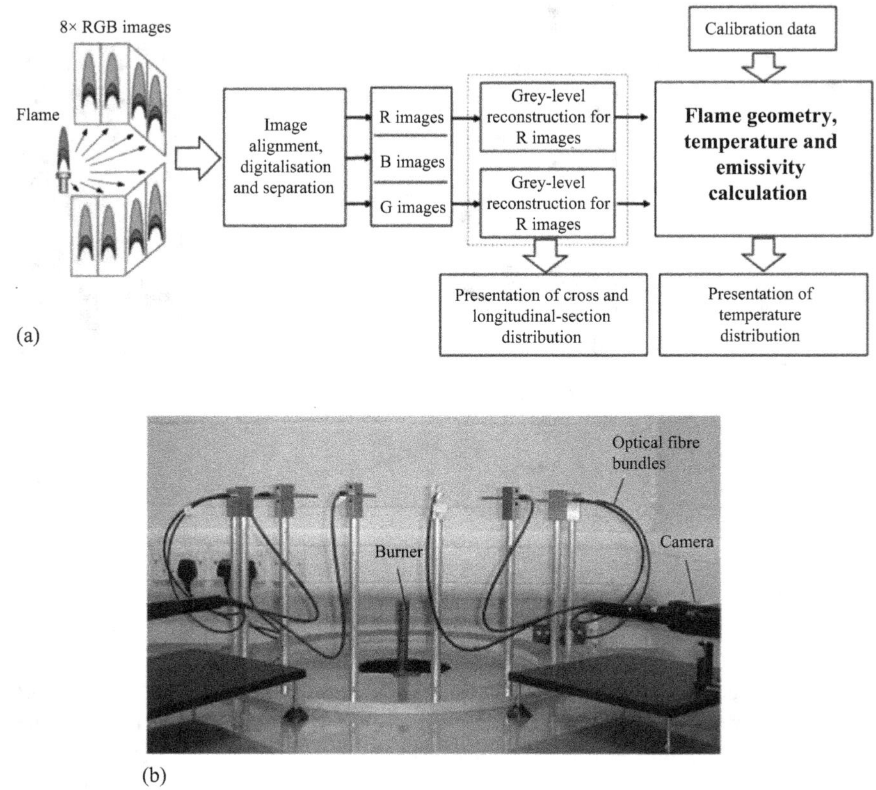

Figure 4.32 Schematic and physical implementation of an optical tomography system for 3D flame reconstruction: (a) schematic and (b) physical implementation [14]

cameras entails high system cost and complexity in system setup. Therefore, they are generally restricted to laboratory uses.

The camera-based tomographic reconstruction of the flame is normally based on an orthographic optical model, that is all pixel's rays are assumed to be parallel to the optical path of a particular view. This is not always true in a practical setup. The inconsistency between the optical model and the system setup would introduce errors in 3D reconstruction results. The problem can be more serious in the region away from the optical path because the impact of the rays passing at an angle through that region cannot be neglected. It would be less problematic if the flame studied is small in comparison to the objective distance. To fully solve the problem, full-fledged 3D reconstruction algorithms should be implemented.

4.3.4.3 Flame light-field imaging

A light-field camera can be used for performing 3D reconstruction of flame. Figure 4.35 shows the experimental setup for the measurement of flame temperature

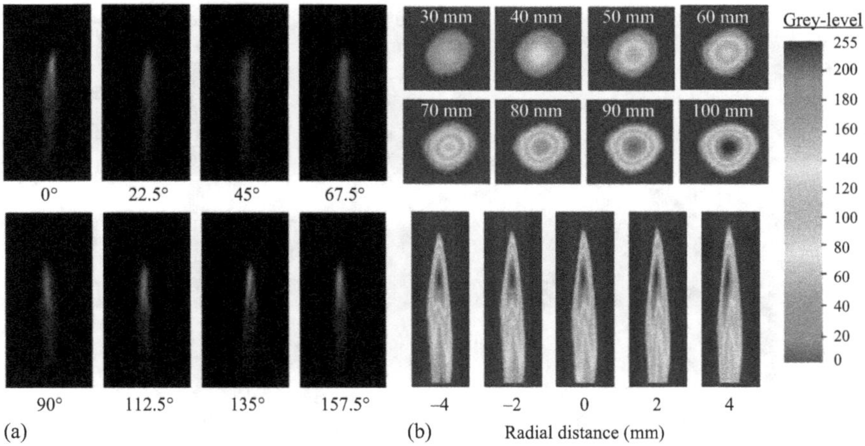

Figure 4.33 2D images and grey-level reconstruction of a gas burner flame using the 3D flame imaging system shown in Figure 4.32: (a) eight 2D flame images and (b) grey-level reconstruction of flame cross-and longitudinal-sections using (a) [14]

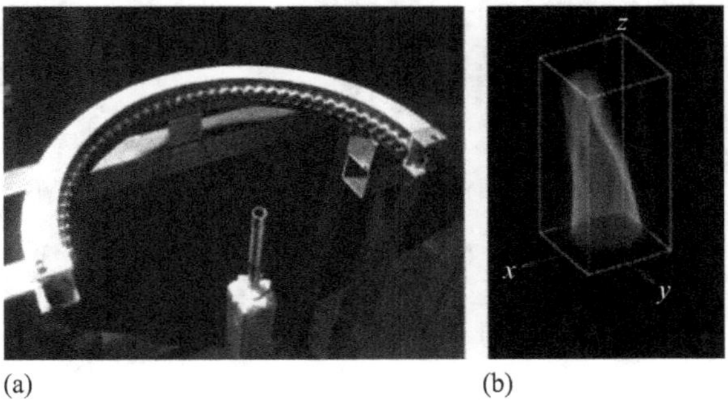

Figure 4.34 (a) Forty-lens-camera tomography system and (b) reconstructed flame image [15]

and absorption coefficient based on the light-field imaging technique. Combining the Levenberg–Marquardt with boundary constraint (LMBC) and non-negative least squares (NNLS) [16], the flame temperature and absorption coefficient are reconstructed simultaneously by sampling the multi-wavelength flame radiation with a coloured light-field camera. Figure 4.36 shows examples of the raw image of flame captured by the light-field camera, the reconstructed image and the reconstructed temperature and absorption coefficient based on the LMBC- NNLS algorithms.

4.3.4.4 Flame stereoscopic imaging

The foundation of stereoscopic imaging is very similar to the 3D perception in human vision, which requires only two different views. The technique can give

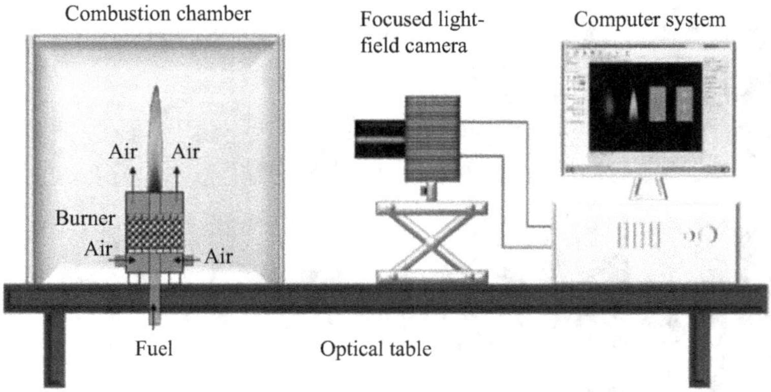

Figure 4.35 Experimental setup for flame temperature and absorption coefficient measurement through light-field imaging [16]

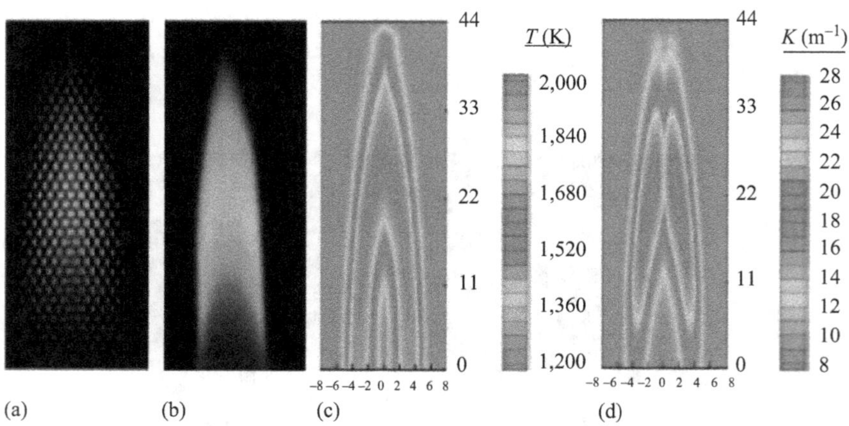

Figure 4.36 Example results [16]: (a) raw image, (b) refocused image, (c) reconstructed temperature distribution and (d) reconstructed absorption coefficient

high spatial resolution with a simple sensing set-up. It has capability of providing a full-field-of-view 3D measurement in an unstructured and dynamic environment.

A. Principle

Binocular stereoscopic imaging uses two off-set imaging sensors to generate pairs of stereo images, that is left and right images of an object, from which the depth of the object in relation to the imaging sensor can be determined and consequently the full-field-of-view measurement can be achieved. Figure 4.37 shows the physical layout and the working principle of the binocular stereo camera. Two imaging sensors observe simultaneously an object (fire flame, in the case) from two slightly differed points of view. In Figure 4.37, two parallel optical axes represent the optical paths of the sensors. The focal length of the camera lens is f. b is the distance

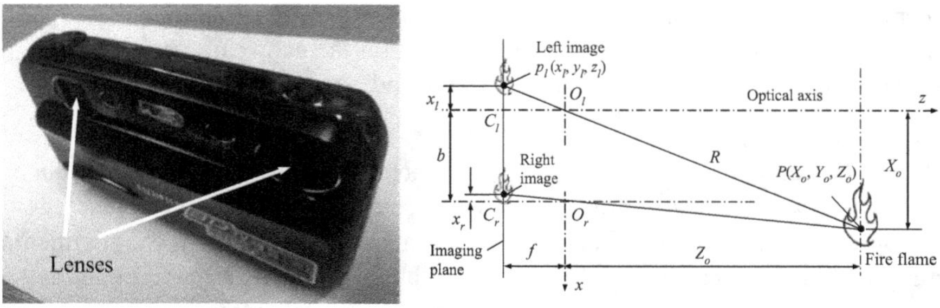

(a) (b)

Figure 4.37 (a) Binocular stereo camera and (b) working principle of (a) [17]

between the two optical axes. C_l and C_r represent the optical centres of the left and right sensors, respectively. O_l and O_r are the principal points of the left and right optical planes. $P(X_o, Y_o, Z_o)$ is a point on the object. $p_l(x_l, y_l)$ and $p_r(x_r, y_r)$ are the projections of the point on the left and right imaging planes (so called corresponding points).

According to the principle of similar triangles

$$x_l = -x_l \frac{f}{Z_o} \tag{4.20}$$

$$x_r = (b - X_o)\frac{f}{Z_0} \tag{4.21}$$

$$y_l = y_r = Y_o \frac{f}{Z_o} \tag{4.22}$$

Rearranging (4.13) and (4.14) yields

$$Z_o = \frac{bf}{x_r - x_l} \tag{4.23}$$

or

$$Z_o = \frac{bf}{d} \tag{4.24}$$

where $d = x_r - x_l$ is called disparity.

Equation (4.24) indicates that when b and f are fixed, coordinate Z_o is inversely proportional to the disparity d. The distance from the origin (O_l) to the point P, R, is then given by

$$R = \frac{b\sqrt{f^2 + x_{lr}^2 + x_l^2}}{d} \tag{4.25}$$

Equation (4.25) is called the true range equation and can be worked out from the coordinates of the corresponding pixels in the left and right images. Disparity d is generally very small compared to R, and thus a small error in computing the disparity can result in a significant inaccuracy in reconstruction. An extra care is taken in camera calibration to ensure a correct image acquisition and consequently data processing.

B. Implementation

A binocular camera can be used to perform binocular stereoscopic imaging. Such a camera has two identical lenses and imaging sensors, which ensure a pair of stereoscopic images are acquired concurrently. Figure 4.38(a) shows the experimental setup of a binocular camera for fire flame detection, where a gas camping burner is used to generate a fire flame in a large engineering laboratory with a complex background. Figure 4.38(b) shows a pair of stereoscopic images of the fire flame.

The two views can be projected through a single imaging sensor if a stereo adapter is used. Figure 4.39 shows the physical layout of such a stereoscopic imaging setup and the schematic of its light transmission. Two images of the object are captured by the imaging sensor at different view angles through a set of mirrors inside the stereo adapter. Such as optical arrangement can be used for reconstructing the surface area of a flame. As example, Figure 4.40 shows a pair of stereo

(a) (b)

Figure 4.38 (a) Experimental setup and (b) left- and right-side images [17]

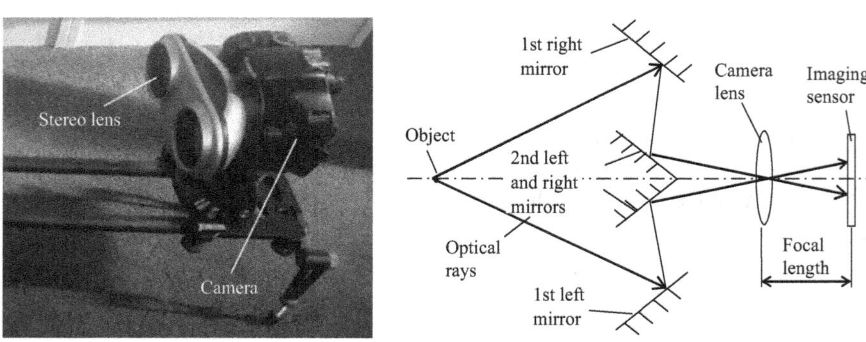

(a) (b)

Figure 4.39 (a) Camera with a stereo adapter and (b) schematic of (a)

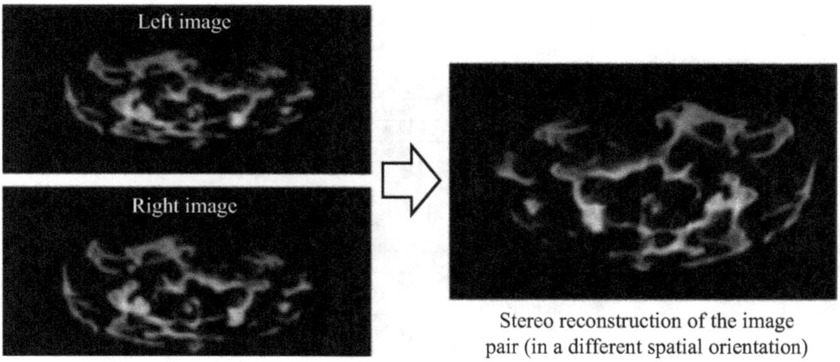

Figure 4.40 A pair of stereo images of an impinging flame and its reconstructed image [18]

images and the reconstructed surface area of a turbulent impinging diffusion flame using a similar imaging setup.

4.3.5 Flame chemiluminescence imaging

Investigation into the physical and chemical nature of the light emitted by a flame, particularly intermediates (or free radicals) of the flame, is an important area in fuel and combustion research. This is commonly regarded as the flame chemiluminescence detection. The magnitude of a chemiluminescence signal is related to the concentration of the corresponding excited species. Due to this property, chemiluminescence imaging has been widely used to determine the location of flame reaction zone and to infer the heat release pattern and rate.

Flame chemiluminescence imaging captures radiative emission from electronically excited species formed in thermal excitation and chemical reactions, such as OH^* (308 nm), CH^* (432 nm), CN^* (387 nm) and C_2^* (514 nm). Current development in optical sensing, image intensifying and image processing has made it possible to obtain high-resolution chemiluminescence profiles of flame radicals.

Figure 4.41 shows the sensing arrangement of a chemiluminescence imaging system, which can detect flame radicals OH^*, CH^*, CN^* and C_2^* concurrently. The system includes bandpass filters at the selected wavelength bands that correspond to the radicals of interest, an EMCCD camera, a frame grabber and a computer with associated application software. A purpose-designed optical beam splitting unit is used to divide the light of flame into four identical beams, each beam passing through the corresponding band-pass filter before being captured by the camera. Note that in practical implementation, more prisms may be needed to ensure four optical paths are identical. Figure 4.42 shows the unfiltered image of a premixed air–propane flame and the images of four radicals captured by the system. The information obtained can be very useful for assessing combustion conditions, such as air-to-fuel ratio [19] and emission formations [20]. By combining the chemiluminescence imaging and computed tomography, 3D reconstruction of flame chemiluminescence can be achieved [21].

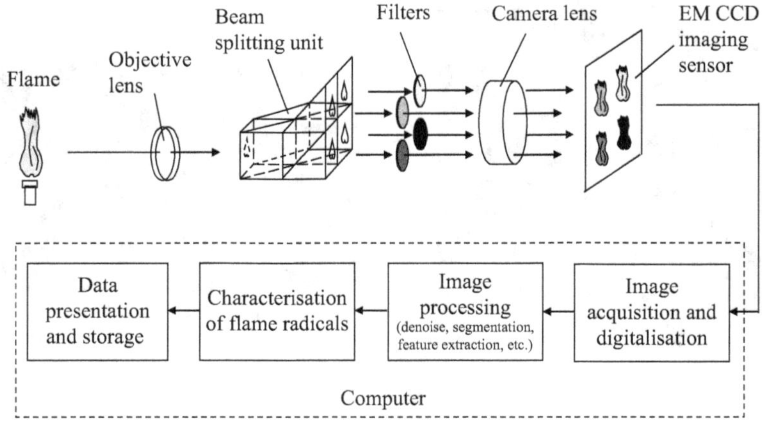

Figure 4.41 Sensing arrangement of a flame chemiluminescence imaging system [19]

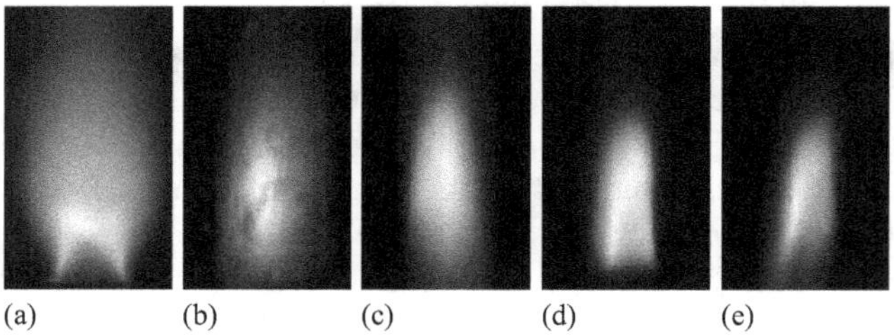

Figure 4.42 Radical images of a premixed air–propane flame (equivalence ratio: 1, exposure time: 1 s) [19]: (a) unfiltered, (b) OH (308 nm), (c) CN (387 nm), (d) CH (432 nm) and (e) C_2 (514 nm)

4.4 Summary

This chapter introduced the detailed architectures, characteristics, as well as advantages and disadvantages of two imaging sensors, that is CCD and CMOS sensors, and CCD/CMOS-based cameras. It also covers visualisation techniques for advanced flame measurement and characterisation. These techniques can be categorised according to the principle employed. The 2D measurement of flame geometry and luminosity can be achieved using relatively simple imaging techniques. Two-colour pyrometry incorporating with digital imaging and image processing has been an effective approach for 2D measurement of flame temperature. A variety of laser-based imaging techniques have been developed and widely used for the 2D and 3D measurement of laboratory-scale flames, particularly the microstructures and the particle velocity profile of the flame. Optical tomography and stereoscopic imaging have been proven to be powerful tools for 3D reconstruction and measurement of the flame. EMCCD imaging has become a practical approach for profiling the chemiluminescence of free radicals in the flame. Combining with the fast-developing computing technology and computational algorithms, 2D and 3D flame imaging technologies have helped combustion researchers and engineers not only for the in-depth understanding of flames but also the control and optimisation of combustion processes.

Whilst significant progress has been made in 2D and 3D flame visualisation and characterisation, substantial work is still needed in developing practically useful flame imaging techniques, both in hardware platforms and in reconstruction and characterisation algorithms. Some attempts have been made to use soft computing techniques, such as neural network, fuzzy logic, support vector machine and deep learning to characterise the flame, and thus the combustion process. These new techniques can be expected to lead to more advanced technologies for flame monitoring and optimisation. Such improvement in flame measurement would result in the co-ordination of automatic control and optimisation, leading to safer, greener and more economical operation of combustion systems.

References

[1] Smith G, 'Willard Boyle (1924–2011): Physicist who helped invent the "eye of the digital camera" ', *Nature*, 2011, **474** (7349), pp 37

[2] Rush A, Hubel P, 'X3 sensor characteristics', *J. Soc. Photogr. Sci. Technol. Jpn.*, 2003, **66** (1), pp 57–60

[3] McWhirter I, 'Electron multiplying CCDs - New technology for low light level imaging', in *Proceedings of 33rd Annual European Meeting on Atmospheric Studies by Optical Methods, IRF Sci. Rep.*, 2008, **292**, pp 61–66

[4] Vision Research Inc., Phantom ultrahigh-speed cameras manual, https://www.phantomhighspeed.com/products/cameras/ultrahighspeed/v2512 (last access on 24th August 2019)

[5] Willson P M, Chappell T E, 'Pulverised fuel flame monitoring in utility boiler', *Meas. Control*, 1985, **18** (2), pp 66–72

[6] Leipertz A, Obertacke R, Wintrich F, 'Industrial combustion control using UV emission tomography', *Symp. (Int.) Combust.*, 1996, **26** (2), pp 2869–2875

[7] Lu G, Yan Y, Colechin M, 'A digital imaging based multi-functional flame monitoring system', *IEEE Trans. Instrum. Meas.*, 2004, **53** (4), pp 1152–1158

[8] Lu G, Yan Y, Colechin M, Hill R, 'Monitoring of oscillatory characteristics of pulverized coal flames through image processing and spectral analysis', *IEEE Trans. Instrum. Meas.*, 2006, **55** (1), pp 1152–1158

[9] Kychakoff G, Howe R D, Hanson R K, McDaniel J C, 'Quantitative visualisation of combustion spices in a plane', *Appl. Opt.*, 1984, **23** (5), pp 704–712

[10] Ashra Z N, Ashjaee M, 'Experimental study of a laminar premixed lfg/air flame in a slot burner using mach-zehnder interferometry', *Therm. Sci.*, 2016, **20** (5), pp 1649–1660

[11] Wang Y, Zhang Y, 'Simultaneous high speed schlieren and direct imaging of ignition process with digitally enhanced visualisation', *Energy Proced.*, 2015, **66**, pp 241–244

[12] Ida T, Ohtake K, 'Microscpic structures in turbulent diffusion flames', *JSME Int. J., Ser. B*, 1994, **37** (3), pp 1629–1635

[13] Bheemul H C, Lu G, Yan Y, 'Three-dimensional visualization and quantitative characterization of gaseous flames', *Meas. Sci. Technol.*, 2002, **13** (10), pp 1643–1650

[14] Hossain M M, Lu G, Yan Y, 'Optical fiber imaging based tomographic reconstruction of burner flames', *IEEE Trans. Instrum. Meas.*, 2012, **61** (5), pp 1417–1424

[15] Ishino Y, Hirano T, Hirano M, Ohiwa N, 'Non-scanning 3D-CT visualizations of premixed flames with a 40-lens camera', *Proceedings of 6th Pacific Symposium on Flow Visualization and Image Processing*, 16–19 May 2007, Hawaii, USA, A2-2, pp 1–6

[16] Li J, Hossain M M, Sun J, *et al.*, 'Simultaneous measurement of flame temperature and absorption coefficient through LMBC-NNLS and plenoptic imaging techniques', *Appl. Therm. Eng.*, 2019, **154**, pp 711–725

[17] Li G, Lu G, Yan Y, 'Fire Detection using stereoscopic imaging and image processing techniques', *Proceedings of IEEE International Conference on Imaging Systems and Techniques* 14–17 October 2014, Santorini, Greece, pp 28–32

[18] Ng W B, Zhang Y, 'Zhang Stereoscopic imaging and reconstruction of the 3D geometry of flame surfaces', *Exp. Fluids*, 2003, **34**, pp 484–493

[19] Krabicka J, Lu G, Yan Y, 'Profiling and characterisation of flame radicals by combining spectroscopic imaging and neural network techniques', *IEEE Trans. Inst. Meas.*, 2011, **60** (5), pp 1854–1860

[20] Li N, Lu G, Li X, Yan Y, 'Prediction of NOx emissions from a biomass fired combustion process based on flame radical imaging and deep learning techniques', *Combust. Sci. Technol.*, 2016, **188** (2), pp 233–246

[21] Floyd J and Kempf A M, 'Computed tomography of chemiluminescence (CTC): High resolution and instantaneous 3D measurement of a matrix burner', *Proc. Combust. Inst.*, 2011, **33** (1), pp 751–758

Gang Lu is a Reader in Electronic Instrumentation in the School of Engineering and Digital Arts, University of Kent, UK. He received his BEng degree from Central South University, China, and PhD degree from the University of Greenwich, UK. He started his career as a mechanical engineer, serving for the iron- and steel-making industry in China, and was a Postdoctoral Research Fellow at the University of Greenwich and the University of Kent. His research interests include sensors, instrumentation, measurement, condition monitoring, digital signal processing, digital image processing and applications of artificial intelligence. He has been involved in a range of projects on advanced monitoring and characterisation of flames in both laboratory- and industrial-scale fossil-fuel combustion systems. Dr Gang Lu is a Chartered Engineer, a Senior Member of the IEEE, and a member of the Energy Institute (UK).

Chapter 5

Tunable diode Laser absorption spectroscopy (TDLAS) sensors

Zhang Cao[1] and Lijun Xu[1]

List of abbreviations

ADC	analogue-to-digital converter
ART	algebraic reconstruction technique
BOS	background-oriented schlieren
CARS	coherent anti-stokes Raman scattering
CPCI	compact peripheral component interconnect
CT	computerised tomography
DAC	digital-to-analogue converter
DAS	direct absorption spectroscopy
DFB	distributed feedback
DSP	digital signal processor
FPGA	field-programmable gate array
INA	instrumentation amplifier
Laser	light amplification by stimulated emission of radiation
LPA	linear power amplifier
LS	least square
MSPS	million samples per second
OPA	operational amplifier
PGA	programmable-gain amplifier
PLIF	planar Laser-induced fluorescence
ppb	part per billion
ppm	part per million
PXI	PCI extensions for instrumentation
PWM	pulse width modulation

[1]School of Instrumentation and Opto-electronic Engineering, Beihang University, Beijing, China

ROI	region of interest
SART	simultaneous algebraic reconstruction technique
SDRAM	synchronous dynamic random-access memory
SNR	signal-to-noise ratio
TDLAS	tunable diode Laser absorption spectroscopy
TEC	thermo-electric cooler
TSVD	truncated singular value decomposition
WMS	wavelength modulation spectroscopy

5.1 Introduction

Tunable diode Laser absorption spectroscopy (TDLAS) sensors are used for non-invasive temperature detection in a wide temperature range. TDLAS can be used in many industrial cases that include absorbable gas species, such as oxygen and water. A tomographic TDLAS sensor enables extraction of temperature distribution. In this chapter, TDLAS sensors are introduced, including principle, hardware design, performance evaluation, image reconstruction as well as two case studies for combustion diagnostics.

In a flame, the applied heat causes fuel molecules to vaporise and they react with oxygen in air. The reaction produces heat in the following exothermic reaction to vaporise more fuel, thus sustaining a consistent flame. High temperature of flame causes the vaporised fuel molecules to decompose, forming various incomplete combustion products and free radicals. These products then react with each other and with the oxidiser involved in the reaction. To achieve sufficient reactions in flame, it poses an urgent need in temperature measurement, and intermediates, such as the methylidyne radical and diatomic carbon. The final products, such as water and carbon dioxide, can be picked up as target species for temperature measurement [1]. Planar Laser-induced fluorescence (PLIF) uses high energy Laser to induce fluorescence on the target radicals and measure the temperature values from the induced intensities. Also, PLIF setups are facilitated for velocity, concentration and pressure measurement [2,3]. However, the Laser source used for PLIF is very expensive to achieve high signal-to-noise ratio (SNR). Typically, two Laser sources at different wavelengths are included in a PLIF system for qualitative measurement of temperature [4]. A PLIF system for quantitative temperature measurement requires two or more wavelengths. It is only suitable for steady flame or requires expensive hardware. More Lasers and cameras are needed in dynamic cases. Coherent anti-stokes Raman scattering (CARS) enables quantitative measurement of temperature at single points in a high temporal resolution up to several kHz, but does not work for high-speed detection of a 2D distribution [5,6]. Background-oriented schlieren (BOS) enables semi-quantitative temperature measurement but requires complicated image processing and pre-calibration for non-disturbed region of interest (ROI) [7–9].

In recent years, TDLAS sensors have attracted strong interest because of their simple structure, fast response, high precision and low cost [10,11]. Tunable diode Lasers are employed to cover the absorption spectroscopy for intermediates, such as NO, CO, CO_2, H_2O, O_2 and CH_4 in the near-infrared region [12–14]. The emitted light of distributed feedback (DFB) Laser has a line width of less than 10 MHz, which is far narrower than the width of gas absorption profile. In this case, when a kind of gas is scanned by tunable DFB Laser, the interference of the absorption lines for other gases can be effectively excluded. Also, the TDLAS technique uses the spectral absorption rate of a specific gas, and so it is affected neither by fluctuation of intensity of the Laser emitted by the source nor the interference of particles, such as soot in flame. If more absorption spectral lines can be covered, intrinsic parameters, such as temperature, concentration, velocity and pressure, for more gases can be evaluated, in the same optical alignment [11,15,16]. The temporal resolution of TDLAS sensors can reach up to more than 10 kHz, and the lower limit of measured gas concentration can be down to one part per million (ppm) or even part per billion (ppb). Because of their simple structure, TDLAS sensors can be easily installed in industrial sites with high reliability and adaptability [17].

A single-path TDLAS sensor measures the average values of the temperature and gas concentration in a Laser path [18]. However, a TDLAS sensor along one Laser path cannot achieve spatial distribution. To measure the temperature and gas concentration distributions in two dimensions, projected data of multiple paths are necessary to retrieve the non-uniform distribution in a tomographic way [19–21]. There are two typical ways of absorption spectrum tomography, that is single-spectral tomography and hyperspectral tomography. In a single-spectral tomography system, projected data of multiple angles are used for absorption tomography. As a result, the detected flame is scanned by rotating a Laser source, and the flame is sometimes required to be axisymmetric, requiring a complicated measurement system but with low temporal resolution [22]. Existing reconstruction methods for single-spectral tomography include the truncated singular value decomposition (TSVD), Landweber algorithm, Tikhonov regularisation and the algebraic reconstruction technique (ART). As for the hyperspectral tomography, projections of fewer angles, for example two angles can be acceptable for rough tomographic imaging in cases of simple profiles, but multiple spectral data are necessary in the tomographic reconstruction [21]. Usually, fewer projection angles mean higher measurement speed and relative low spatial resolution. Hyperspectral tomography of about 200 spatial grid points reached a temporal response up to 50 kHz [23]. The inverse problem is casted into a nonlinear optimisation problem for hyperspectral tomography. Some techniques are applied to address the ill-posed problem and convergence issue, for example regularisation and stochastic minimisation [16].

5.2 TDLAS sensor

A typical TDLAS sensor consists of four parts: (1) a Laser control and emitting unit, (2) a detector, (3) a data acquisition unit and (4) post-processing software.

A DFB Laser is used to scan a narrow spectrum ranging about 1 to 2 nanometers. Gas along the Laser path interacts with the Laser and modulates the shape of the spectral profile. An absorbed intensity profile is captured by a detector located at the end of the path. Intensity values collected by the data acquisition unit are then processed by the post-processing software to recover the absorption spectrum. Typical line-shape model for gas-Laser interaction, for example the Voigt line-shape is applied to extract the intrinsic parameters, such as temperature, pressure, velocity and mole concentration of the gas molecules. If one light path crosses the gas, the average value of each of the gas parameters along the path can be achieved, as shown in Figure 5.1.

More paths from different crossing angles through the gas are needed for scanning and grasping the parametric values at each point in a tomographic way. Similar to medical computerised tomography (CT), an increased number of projections can improve the precision and spatial resolution of a tomographic image. Unlike the medical CT that scans the object using a mechanical rotation, TDLAS tomography for industrial application only uses several fixed projections. Compromise should be made between the spatial resolution and the system complexity. A balance can be achieved if the Laser injects from at least five different locations along the peripheral of ROI [24]. A five-angle TDLAS imaging system is shown in Figure 5.2 as an example. It consists of four parts: a Laser control and emitting unit, an optical alignment and photoelectric detection unit, a multi-channel data acquisition unit and a data processing and imaging unit. Five Laser beams are collimated at five different points, pass Powell prisms and are broadened into fan beams, respectively. Each beam crosses the gas species and arrives at the detector arrays on the opposite side. There are 12 equally spaced photo-detectors in each array. Therefore, 60 independent data acquisition channels are required for all the five arrays. If a Laser beam has a coverage beyond the array in the opposite side, the detectors in more arrays in neighbouring sides can be used to receive the light. For example, besides the 12 photo-detectors in the opposite array, if each fan beam

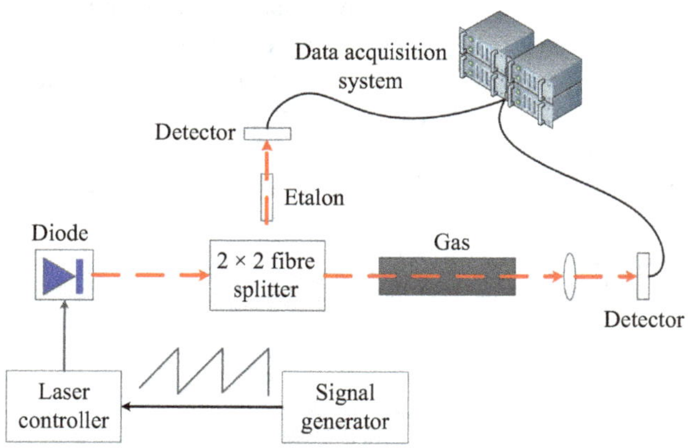

Figure 5.1 Setup of a typical TDLAS measurement system

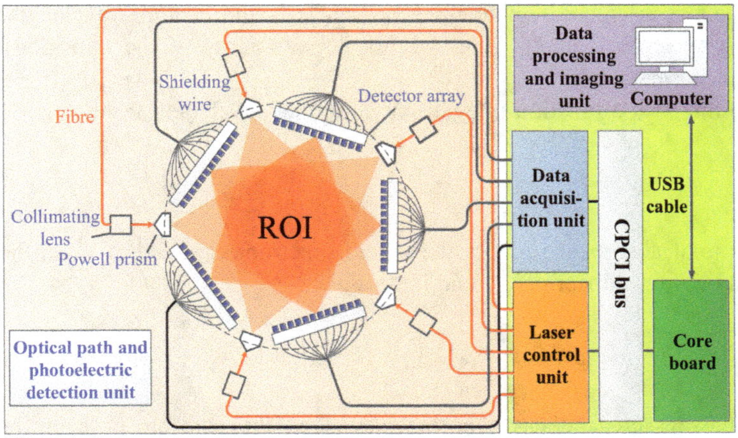

Figure 5.2 Five-angle TDLAS imaging system

can also be received by 12 extra photo-detectors in the neighbouring detector arrays, then 24 detectors are used for each beam and 120 intensity signals in total can be captured in the five-angle tomographic sensor. Photo-electric signals generated in the detector arrays are transmitted through the data acquisition unit to the host computer for post-processing.

5.2.1 Principles for TDLAS sensors

If a collimated Laser at a wavelength of ν [cm^{-1}] passes a gas specie with a thickness of L [cm], it follows the Beer–Lambert law and its intensity absorbance $\alpha(\nu)$ after passing the gas can be expressed as

$$\alpha(\nu) = \int_0^L P(x)X(x)S[T(x), \nu_0]\phi(\nu)\mathrm{d}l \tag{5.1}$$

where $P(x)$ [atm], $X(x)$ and $T(x)$ [K] are the local total pressure, the local molar fraction of the target species noted as gas concentration and the local temperature at x, respectively. $\phi(\nu)$ [cm] is the line-shape function at ν and usually the Voigt line-shape is used for TDLAS sensors.

The line strength of the transition function, $S[T(x), \nu_0]$ [$\mathrm{cm}^{-2}\ \mathrm{atm}^{-1}$], can be written as

$$S(T, \nu_0) = S(T_0, \nu_0)\frac{Q(T_0)}{Q(T)}\left(\frac{T_0}{T}\right)\exp\left(-\frac{hcE''_{\nu_0}}{k}\left(\frac{1}{T} - \frac{1}{T_0}\right)\right) \times \frac{1 - \exp(-hc\nu_0/kT)}{1 - \exp(-hc\nu_0/kT_0)} \tag{5.2}$$

where h [J · s] is Planck constant, c [$\mathrm{cm} \cdot \mathrm{s}^{-1}$] is the light speed in vacuum, k [$\mathrm{J} \cdot \mathrm{K}^{-1}$] is the Boltzmann constant, $Q(T)$ is the partition function of the absorbing molecule,

T_0 [K] is the reference temperature, ν_0 [cm^{-1}] is the central wavelength of the absorbing spectrum and E''_{ν_0} [cm^{-1}] is the lower energy of the transition for the spectrum centred at ν_0.

The line-shape function integration along the wavelength ν satisfies

$$\int_{-\infty}^{+\infty} \phi(\nu)\mathrm{d}\nu = 1 \tag{5.3}$$

The total absorbance across the whole spectrum cantered at ν_0 in the Laser path can be inferred from (5.1) and (5.3) as

$$A_{\nu_0} = \int_{-\infty}^{+\infty} \alpha(\nu)\mathrm{d}\nu = \int_0^L P(x)X_{\text{abs}}(x)S[T(x), \nu_0]\mathrm{d}l \tag{5.4}$$

If the pressure and gas concentration along the path are both uniform and termed as P and X, the total absorbance at two different spectrums centred at ν_1 and ν_2 can be simplified as

$$\begin{cases} A_{\nu_1} = PXS(T, \nu_1)L \\ A_{\nu_2} = PXS(T, \nu_2)L \end{cases} \tag{5.5}$$

The ratio of the total absorbance at two different wavelengths can be derived as

$$\begin{aligned} \frac{A_{\nu_1}}{A_{\nu_2}} &= \frac{S(T, \nu_1)}{S(T, \nu_2)} \\ &= \frac{S(T_0, \nu_1)\frac{1-\exp(-hc\nu_1/kT)}{1-\exp(-hc\nu_1/kT_0)}}{S(T_0, \nu_2)\frac{1-\exp(-hc\nu_2/kT)}{1-\exp(-hc\nu_2/kT_0)}} \\ &\quad \times\exp\left[-\frac{hc\left(E''_{\nu_1} - E''_{\nu_2}\right)}{k}\left(\frac{1}{T} - \frac{1}{T_0}\right)\right] \\ &\approx \frac{S(T_0, \nu_1)}{S(T_0, \nu_2)} \exp\left[-\frac{hc\left(E''_{\nu_1} - E''_{\nu_2}\right)}{k}\left(\frac{1}{T} - \frac{1}{T_0}\right)\right] \end{aligned} \tag{5.6}$$

As a result, the temperature along the Laser path can be derived as

$$T = \frac{\frac{hc}{k}\left(E''_{\nu_1} - E''_{\nu_2}\right)}{\ln \frac{A_{\nu_1}}{A_{\nu_2}} + \ln \frac{S(T_0, \nu_2)}{S(T_0, \nu_1)} - \frac{hc}{k}\frac{\left(E''_{\nu_1} - E''_{\nu_2}\right)}{T_0}} \tag{5.7}$$

In general, the temperature values along the path are non-uniform and (5.7) can produce a weighted value of the distribution, even if the gas concentration is unevenly distributed. In some applications, the temperature distribution is required to indicate the combustion efficiency or safety evaluation, and more light paths are necessary to achieve more projections for the tomographic images.

When the absorbance along the Laser paths of different views can be measured, the tomographic images of the pressure, temperature and gas concentration are deduced. During image reconstruction, a circular ROI is typically selected and discretised into N grids, as shown in Figure 5.3. These grids are of the same size, and in each of them the flame parameters, such as the pressure, temperature and gas concentration, are treated as constants. Five fan-beam Lasers penetrate the ROI and are picked up by five arrays of 12 photo-diodes with equal interval denoted as d, respectively. The numbers of both discretised grids and projections are selected by balancing tomographic image quality and the system complexity.

According to (5.4), the total absorbance $A_{\nu,i}$ obtained from ith Laser path can be formulated as

$$A_{\nu,i} = \sum_{j=1}^{N} \alpha_{\nu,j} L_{ij} = \sum_{j=1}^{N} [PXS(T,\nu)]_{\nu,j} L_{ij} \tag{5.8}$$

where i and j are the indices of the Laser path and a grid along the path, $\alpha_{\nu,j}$ is the averaged density of the integrated absorbance crossing the jth grid, L_{ij} is the absorption path length of ith Laser path in jth grid.

In general, (5.8) can be rewritten in the matrix form, that is

$$\mathbf{L}\boldsymbol{\alpha}_\nu = \mathbf{A}_\nu \tag{5.9}$$

where the $M \times N$ matrix $\mathbf{L}$ is

$$\mathbf{L} = \begin{bmatrix} L_{11} & L_{12} & \cdots & L_{1N} \\ L_{21} & L_{22} & \cdots & L_{2N} \\ \vdots & \vdots & \ddots & \vdots \\ L_{M1} & L_{M2} & \cdots & L_{MN} \end{bmatrix} \tag{5.10}$$

where, $\boldsymbol{\alpha_\nu} = \{\alpha_{\nu,1}, \alpha_{\nu,2}, \alpha_{\nu,3}, \ldots, \alpha_{\nu,N}\}^T$, $\boldsymbol{A_\nu} = \{A_{\nu,1}, A_{\nu,2}, A_{\nu,3}, \ldots, A_{\nu,M}\}^T$ and $\mathbf{L}$ can be pre-determined from the geometrical alignment of the Laser beams and photodiode arrays, just like the sensitivity matrix calculated in the medical CT.

The values of $\boldsymbol{\alpha_\nu}$ are retrieved using image reconstruction algorithms from a limited number of ray measurements, such as Tikhonov regularisation, SART or

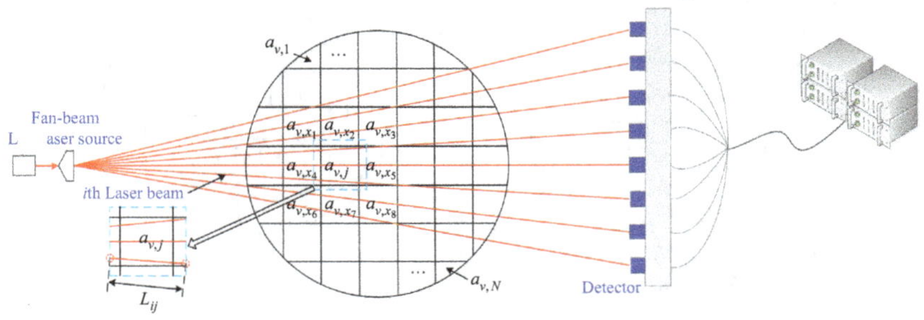

Figure 5.3 Fan-beam configuration with a single view and discretised grids in ROI

Landweber algorithm. Tomographic images of the absorbance at two pre-selected transitions at the wavelengths of ν_1 and ν_2 are reconstructed, respectively. Then the ratio of the reconstructed $\alpha_{\nu_1,j}$ and $\alpha_{\nu_2,j}$ in the *j*th grid is calculated and the temperature T_j in the same grid can be extracted from (5.7). As a result, the temperature distribution in ROI is reconstructed.

5.2.2 Laser control and emitting unit

DFB Lasers are used for TDLAS for their precise and stable wavelength that is critical in spectrum measurement. In a DFB Laser, a diffraction grating acting like an optical filter is etched close to a *p–n* junction of the diode, and a single wavelength is fed back to the gain region for Laser stabilisation. The DFB Laser has a stabilised wavelength that is set during manufacturing by the pitch of the grating and can only be tuned slightly with temperature. A typical DFB Laser has a line width of 10–15 MHz, and the output line width is narrow enough relative to the width of the absorption line. It is of small size and can be flexibly controlled to tune the output wavelength quickly. The output wavelength of the DFB Laser is mainly adjusted by controlling the temperature and finely tuned by the drive current. The output power and wavelength of the DFB Laser vary linearly with the drive current. To provide stable temperature and drive current control signals to the DFB Laser, the control and emitting unit includes dual closed control loops, that is a temperature control loop and a current control loop for a thermo-electric cooler (TEC). In the temperature control loop, a constant current generated from the controller flows through the built-in thermistor R_T of the Laser to monitor the temperature, as shown in Figure 5.4. The voltage across the thermistor R_T acts on an integrator composed of an operational amplifier (OPA). When the voltage deviates from the set value, the integrator works in the forward or reverse direction, and a pulse width

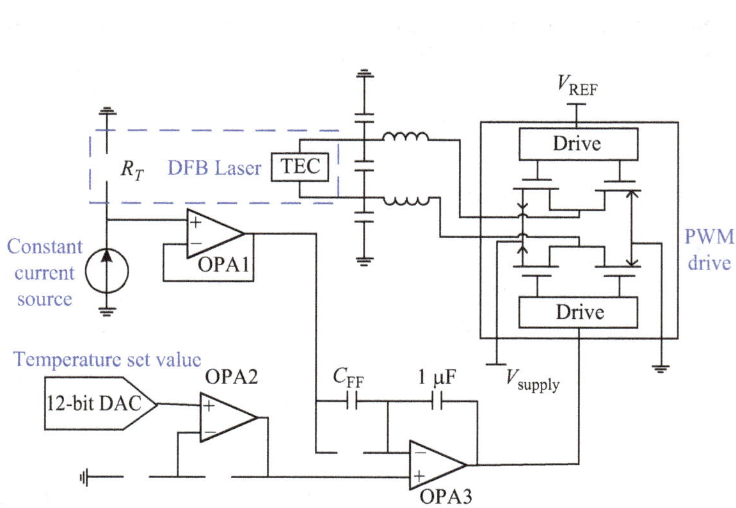

Figure 5.4 TEC temperature control loop for DFB Laser

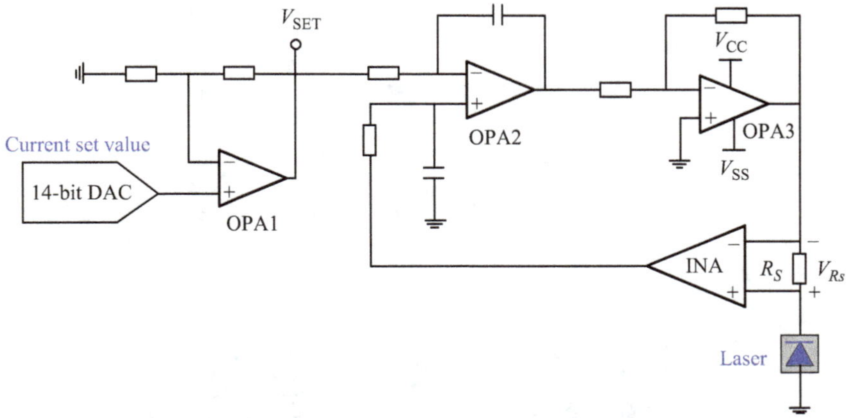

Figure 5.5 Driving current control loop for DFB Laser

modulation (PWM) driver provides a driving current to TEC to realise heating or cooling. Figure 5.5 shows the current control loop for the DFB Laser.

A voltage signal is generated by a 14-bit digital-to-analogue converter (DAC) to set the current value. At the same time, the drive current flowing through the Laser passes through a high-precision ultra-low temperature drift resistor R_S. The voltage V_{Rs} across the resistor R_S is used to detect the current flowing through the Laser and is fed back through an instrumentation amplifier (INA) to the integrator composed by the amplifier, that is OPA1. When the feedback voltage deviates from the set value, OPA1 will drive an LPA and regulate the current flowing through the resistor R_S and the Laser.

5.2.3 Data acquisition unit

To sample and record signals in all 60 channels simultaneously at up to 5 MSPS, the parallel data acquisition unit possessed four 128 MB SDRAMs on board, as shown in Figure 5.6. The TDLAS tomographic system can acquire data at 600 MB/s, which is beyond the data rates of the compact peripheral component interconnect (CPCI) bus, that is 284 MB/s and the USB 2.0 full-speed bus of 60 MB/s [25].

In some cases of steady combustion flame, tomographic images at low speed can meet the measuring requirement and online line-shape fitting can be applied to reduce the data streaming rate. If a high frame rate is required to capture the dynamics of instable flame, there could be too much data to be processed by a field-programmable gate array (FPGA) in real time. For example, if the temporal resolution is 5 kHz, at least 570 k points of the Voigt profile fitting operations are needed to process within 1 s. Data fitting is nonlinear and multiplication and addition operations of each point cost about 1,000 clock periods. The iteration is operated in a cascade way, and the basic frequency of the soft-core must be beyond 500 MHz, which is higher than the maximum soft-core frequency of commercially available FPGA, for example about 384 MHz at present. In this case, SDRAMs can be used as a temporary storage for the intensity data and uploading of raw data

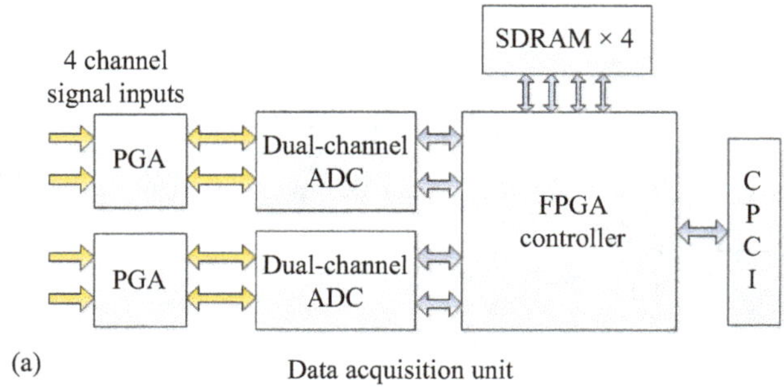

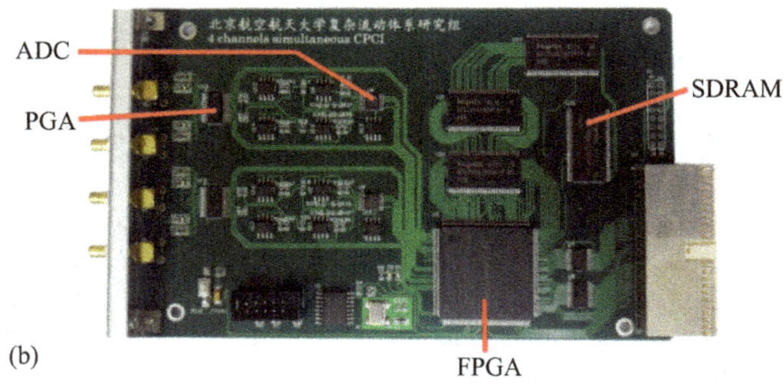

Figure 5.6 Parallel data acquisition unit for four channels: (a) schematic diagram and (b) photo

makes it possible for the post-processing for more details in the target combustion process, usually lasting several seconds. As a result, the data acquisition units digitalise the analogue signals from the photo-electric detectors. The data are saved in SDRAMs temporarily and then transmitted to the host PC after a specific time interval. Each data acquisition unit collects the analogue signals in four channels. As shown in Figure 5.6, the input analogue signals are first conditioned by a dual-channel programmable-gain amplifier (PGA) to make full use of the capacity of the analogue-to-digital converters (ADCs). Then, the FPGA controls the ADCs to collect the output signal of the PGAs and writes the digitalised data to SDRAMs for further data transferring to the host PC.

To evaluate the performance of the unit, a DC signal is imposed on all 60 data acquisition channels. The SNR of each channel is calculated from 200 repetitive measurements. SNR values for the channels are all above 67 dB, and their fluctuation is shown in Figure 5.7. The fluctuation may come from ADC quantisation error, cross-talk of signal lines and the electronic noise in the background.

5.2.4 TDLAS tomographic system

A core control board and 15 parallel data acquisition units, as well as two Laser controller boards, are assembled in a 3U80TE CPCI case with 20 slots as shown in Figure 5.8.

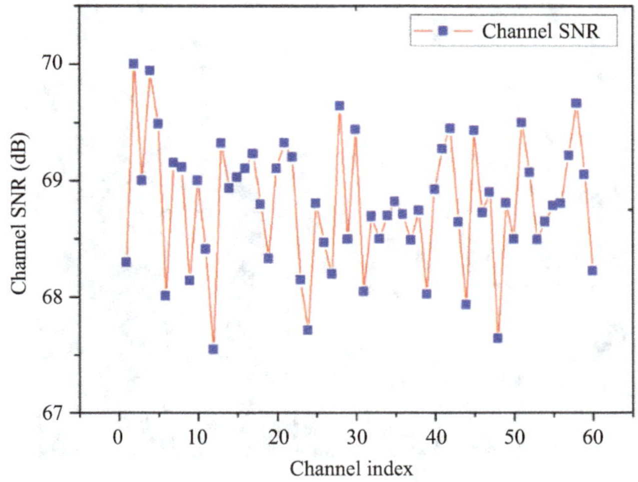

Figure 5.7 SNRs for all 60 data acquisition channels

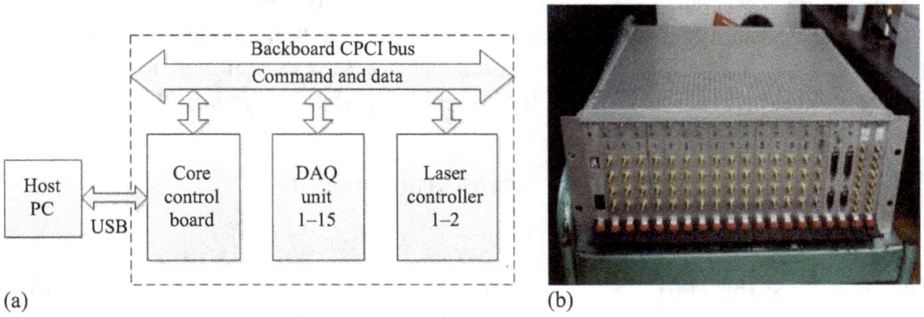

Figure 5.8 Reconfigurable TDLAS tomographic system of 20 slots CPCI case: (a) schematic diagram and (b) photo

The core control board bridges the host PC and the CPCI-case via the USB cable. The sampling rate of each channel in the date acquisition unit can be varied from 1 MSPS to 5 MSPS, by selecting different parameter values in FPGA programing. Total memory size is fixed, and the recording time for one measurement process is decreased from 16 s to 3.2 s, as the sampling rate is changed from 1 MSPS to 5 MSPS. Two DFB Lasers are respectively modulated by two Laser controller boards simultaneously. These boards generated control signals for the temperature and current controllers, and the wavelength scanning speed of the Laser is adjustable up to 10 kHz. However, in normal operation, the Laser modulation frequency is set no more than 5 kHz to make sure every period consisting of at least 1,000 sampling points, and more sampling points improved the precision of the line-shape fitting for the absorption spectrum. The main functions of the software are to configure Laser controller parameters, manage local file storage and

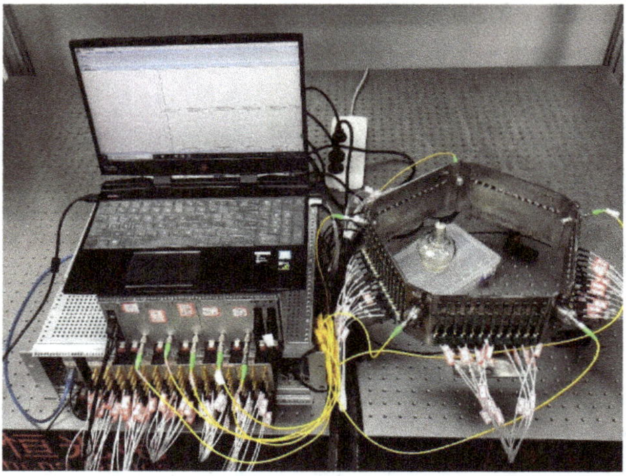

Figure 5.9 Five view TDLAS tomographic system

perform image reconstruction and display. The software uses multi-threads to receive USB data and store the data to hard disk simultaneously. Reconstruction algorithms, such as Landweber iteration and SART are included to reconstruct tomographic images. The TDLAS tomographic system is shown in Figure 5.9.

5.3 Total absorbance extraction and on-chip implementation

The detector array converts the intensity of the transmitted Laser into electrical signals that are collected by ADC and saved for absorbance extraction. Two typical methods, that is direct absorption spectroscopy (DAS) and wavelength modulation spectroscopy (WMS), are used to extract the total absorbance from the light intensity profile. The DAS method aims to recover the absorption spectrum first and then calculate the area covered by the absorption profile. The values of the areas at two different central wavelengths are used to obtain the ratio of the total absorbance and temperature. In the DAS method, the light intensity at each wavelength is collected to achieve the total absorbance, but the data stream is huge for a high frame rate. Also, the DAS method is sensitive to the background disturbance, as each spectral line can be affected by the unavoidable low frequency fluctuation in the background. The WMS method uses a high frequency signal to carry the low frequency absorbance signals and is suitable for harsh environment with mechanical vibration and thermal radiation. The total absorbance is derived from the several frequency components of the detected intensity signal if the mole concentration is assumed to be relatively low. In many applications, this assumption is reasonable, and the WMS method become a highly sensitive and robust approach for the absorbance extraction as well as temperature.

5.3.1 DAS method

In the DAS method, the driving current of the tunable DFB Laser is tuned to make its output wavelength scan across the whole absorption spectrum. A triangular wave excitation signal is imposed on the Laser controller to generate a wavelength-scanning Laser signal. The Laser beam is divided into two paths through a fibre splitter. One is directly transmitted to the etalon and collected by a photo-detector. The etalon signal can be used to convert the Laser light intensity signal into the values of wave numbers. The other Laser beam passes through a collimator to convert the fibre beam into parallel space light. The collimated light penetrates through gas species and is collected by another photo-detector.

Ideally, the spectral light intensity should be first obtained when no absorbing gas exists between the Laser source and the detector, and the intensity profile can be used as the baseline for further absorption spectrum extraction. Then, the gas to be measured is placed in the Laser path, absorption in the intensity profile is generated, depending on the temperature and molar concentration of the gas. The signal obtained in this way can be directly used as a baseline for translation and scaling of the absorption spectrum, which can eliminate the error introduced by the nonlinearity and instability of the Laser output and hence eliminate its influence on the measurement. The intensity profile is transferred to the absorption spectrum using the output of the etalon as a calibration. An additional set of optical paths is required, increasing the complexity and cost of the measurement system.

If the gas to be measured is scanned by two or more different absorption spectra from different projecting views, then the temperature and concentration at each location can be retrieved by a tomographic approach, by solving an optimisation problem. In the actual data processing process, the baseline extraction is different from the ideal case. It is first fitted at the spectrum where no apparent absorption exists. A fitting curve, such as a polynomial, is extended to the range of absorption. Then the absorption spectrum is compared with the fitted baseline, and the integral absorption rate can be achieved in a simple way. In this case, the temperature and concentration of the target gas can be derived from the profile of the absorption rate. The baseline fitting for the DAS method in a TDLAS sensor with a single path is shown in Figure 5.10.

In this way, nonlinear changes introduced in the reference optical path can be eliminated and the light intensity fluctuations caused by the random optical path jitter during the measurement can be suppressed. The DAS method makes hardware implementation relatively simple, and high measurement accuracy can be achieved if the baseline segment without absorption is accurately obtained. However, in a harsh environment, low frequency noise, thermal radiation and mechanical vibration deteriorate the data fitting of the baseline, and the total absorbance data are severely degraded due to the baseline deviation, which may cause a rapid glide in the measurement accuracy. In addition, high pressure can broaden the width of the absorption spectrum, and the DFB Laser will not cover the entire absorption spectrum for an accurate baseline fitting. As a result, the DAS method is usually used in the case of relative low gas pressure and ambient environment.

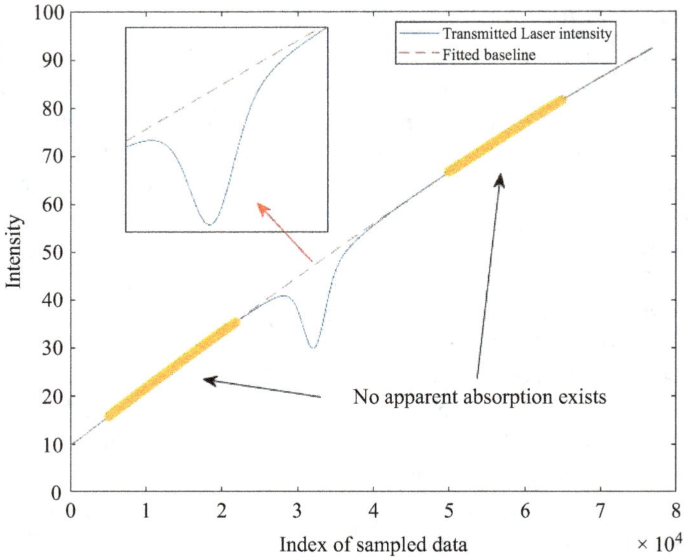

Figure 5.10 Baseline selection and fitting in DAS method

5.3.2 On-chip implementation of DAS method

The online DAS method for TDLAS aims to reduce the data stream rate transmitted to the host PC and improve the real time performance. In this method, raw intensity data are processed in a system-on-chip, to obtain the temperature and gas concentrations online. In the real applications, an FPGA chip is used to implement the online fitting of the absorption profile. In the fitting process, many floating-point operations, such as the multiplications and divisions are required, and an FPGA is used for online implementation. It has many DSP hard cores and logic resources necessary for online fitting operations. First, the pre-processing is performed in the FPGA, the intensity data points near the points for baseline fitting as well as the absorption peak are screened, and the parameters in the Voigt function are also evaluated. The FPGA integrates all the absorption rate under the Voigt profile. The integrated values, that is the total absorbance, are transmitted to the host computer via USB. In the host PC, the temperature of the gas is obtained from the total absorbance of the spectrums centred at two different wavelengths, and the process is shown in Figure 5.11.

Online fitting of the Voigt profile can be realised in the following steps. First, the values of the fitting parameters, such as the half width and the total absorbance, are initially estimated. Initial values of the integrated areas of the line-shape profile, that is the total absorbance are obtained by summing up all absorbance in the original measurements. Initial values of the half widths of the line-shape are obtained from the absorption rates. Then, the calculated absorption rate is iteratively updated using the look-up table, and the deviation between the calculated and

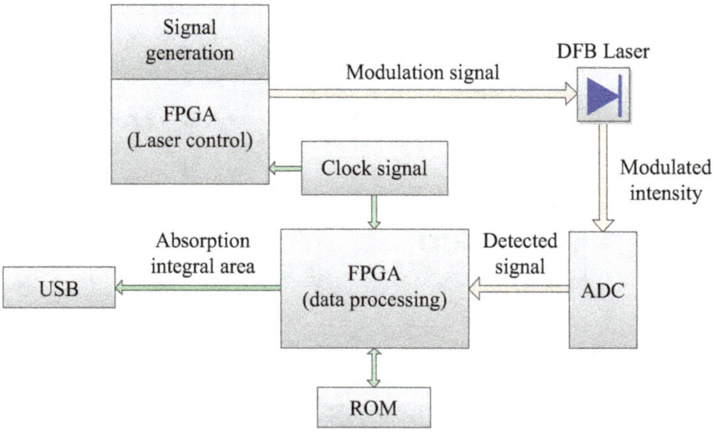

Figure 5.11 On-chip fitting in FPGA for DAS method

measured absorption rates is minimised to approach the optimal values of the fitting parameter. A lookup table for the Voigt function is constructed with a data volume of 1.3 Mbytes, and it consists of 564 × 601 points of the function values. The values beyond the scope of stored points is linearly interpolated, which the Voigt function calculations are enabled to be ready for online iterative updating in the line-shape fitting from the measured data. Because the table is too big to be written directly into the program, a physical memory such as a ROM is used.

In experiments, the rich block RAM resource inside the FPGA is used and changed into a ROM form. The table is stored in the Block Memory IP core and no extra ROM is required. Also, the read or write speed of the lookup table is faster than that of an extra ROM. Light intensity data in two scanning cycles are collected using photo-detectors to evaluate the real time performance of the online fitting algorithm, as shown in Figure 5.12.

After the data of one wavelength scanning period are collected, they are pre-processed and the data points for baseline fitting and line-shape fitting are selected and transmitted to the parameter fitting unit. The centre of the absorption spectrum is estimated using difference operations. Accurate locations of the centre are determined after the scanning saw-tooth is removed from the detected intensities. Data points for the baseline fitting are selected in the vicinity of the centre, once it is determined. In this method, the baseline fitting is achieved by applying a cubic polynomial fitting method. The unabsorbed baseline obtained after baseline fitting is shown in Figure 5.13.

Starred points in Figure 5.14 show the selected data for the Voigt line-shape fitting after extraction of the absorption spectrum. As can be seen, the absorption rate curve is not smooth due to noise in the actual measurements. At the same time, two small protrusions appear on the left side of the absorption profile, due to the presence of other relatively weak absorption peaks around the target absorption peak. In this case, the total absorbance would be severely distorted if it is calculated

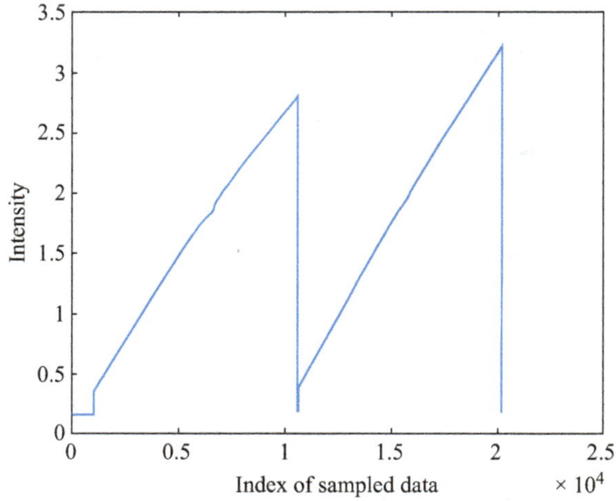

Figure 5.12 Intensity data in two cycles for DAS method

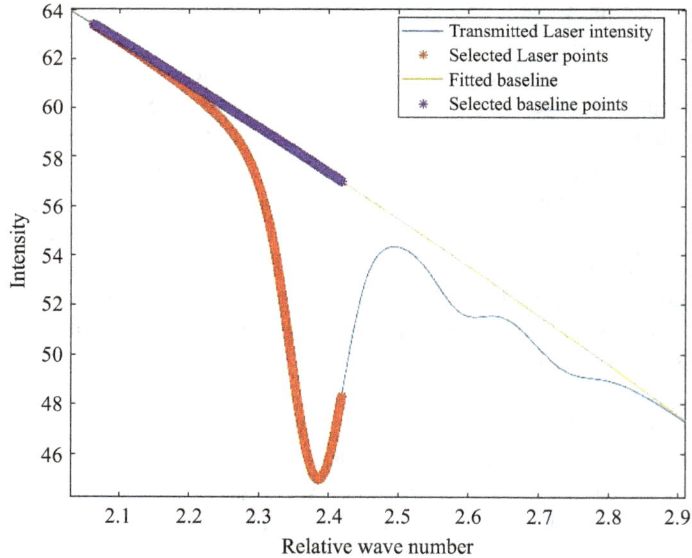

Figure 5.13 Fitted baseline and calculated absorbance rate

as the area covered by the absorption profile. The data to be fitted are carefully selected to avoid these unintended errors. The Gauss–Newton iteration method is used to approach the real values of the total absorbance for the central wavelength set before scanning. The results are shown in Figure 5.15, indicating that the absorbance curve obtained by the Gauss–Newton iterative method is very smooth. In the fitted line-shape, the fluctuation caused by the noise and other weak

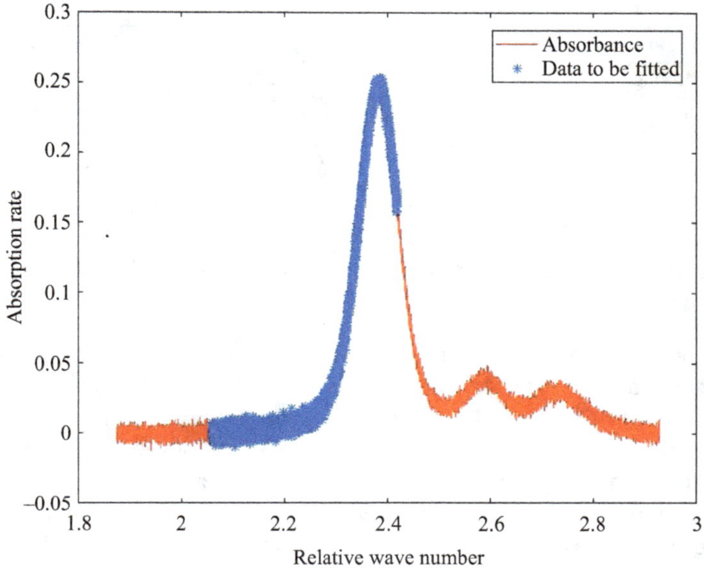

Figure 5.14 Absorption profile and fitted Voigt line-shape

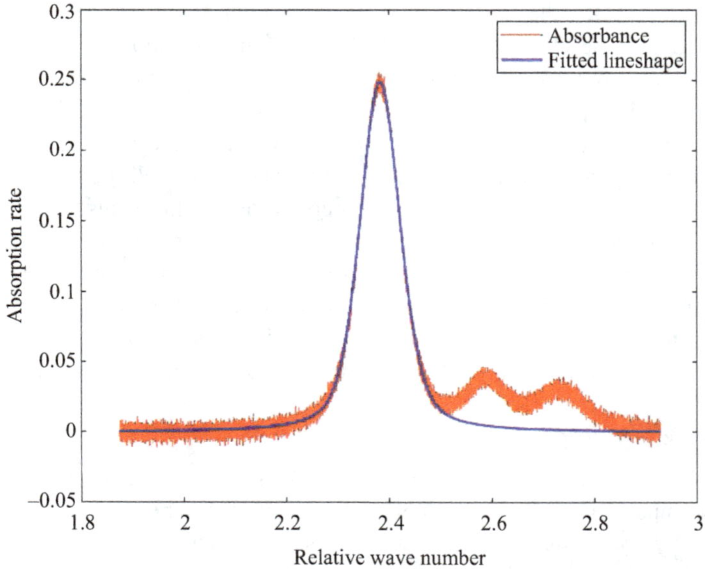

Figure 5.15 Voigt line-shape fitting using Gauss–Newton iterations

absorption peaks are effectively suppressed and the fitted profile agree well with the measured data in the vicinity of the absorption peak. As a result, the total absorbance used for temperature calculation can be reliably ready if the fitting is operated on absorption spectrums centred at two different wavelengths.

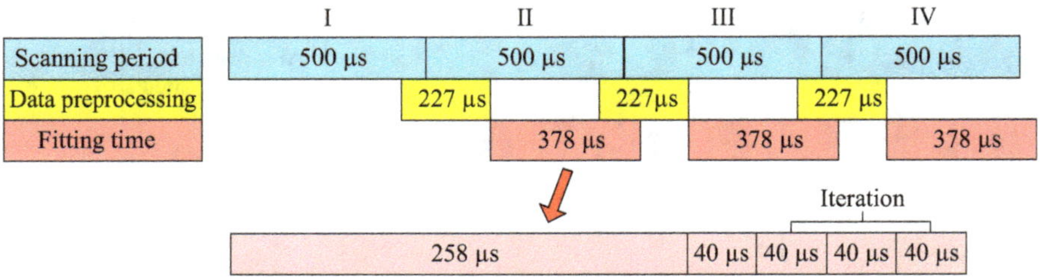

Figure 5.16 Time used for online fitting of line-shape in DAS method

In an online method, the calculation speed is an important indicator to evaluate its pros and cons. The TDLAS sensor uses a 2 kHz saw-tooth wave with an AD sampling rate of 1 MHz, and 500 sampled data are captured in each saw-tooth cycle. The time cost of on-line fitting of each saw-tooth cycle must be within 500 μs. Otherwise, the sampled data of the next cycle cannot be processed in time, and more and more data would be delayed for processing. The online fitting can be completed within a saw-tooth cycle, and a streamline strategy is used to suppress the fitting time. In each fitting, four scanning periods are used and indicated as I, II, III and IV to illustrate the operations of data pre-processing and fitting, as shown in Figure 5.16.

In a 12-bit ADC, the sampling rate is set to be 1 MHz. The floating-point IP core is used and the clock is up to 400 MHz. There are 500 data points in each wavelength scanning period lasting for 500 μs. In each scanning period, it takes 227 μs to complete pre-processing. In the parameter fitting, the maximum number of iterations is six and the relative fitting error is within 0.5% after iterations. In the experiments, three iterations are enough and it takes about 378 μs to obtain a total absorbance value; 258 μs is used on the preparation for the Gauss–Newton iterations and 40 μs per iteration. Even if the iterative number is less than six, it takes 498 μs at most for each parameter fitting, including the preparation time of 258 μs and six iterations of 240 μs, which is within one saw-tooth period. As a result, the parameter fitting for the line-shape can be completed before the next wavelength scanning period starts, and no data congestion occurs for online measurement.

5.3.3 WMS method

In the WMS method, the incident saw-tooth wave intensity is modulated by a high-frequency sinusoidal wave. The modulated intensity wave penetrates the target gas is detected by the diode detector or a detector array. The absorption of the gas is carried in the detected intensity signals and can be demodulated in series of several frequency components. The fundamental frequency of the detected signals is the same as the modulation frequency, denoted as f and two or more components can be used to reduce the distortion caused by the optical alignment. Usually, the ratio of the components of $2f$ to $1f$ is used to derive the total absorbance. This technique is termed as the WMS-$2f/1f$ spectral peak detection method. In the method, the diode Laser's injection current is driven by a ramp signal modulated by a high frequency sinusoidal signal at the frequency f. The tuning of the injection current changes the diode Laser's wavelength and intensity simultaneously. The output wavelength linearly changes with the amplitude of the injection current. The instantaneous wavelength of the Laser along the time can be expressed as

$$v(t) = v_0 + a\cos(2\pi ft) \tag{5.11}$$

where $v(t)$ and v_0 are the instantaneous and averaged wavelengths, and f and a are the modulation frequency and amplitude, respectively.

Because the value of v_0 varies slowly, it can be treated as a constant during each measurement. The related Laser intensity $I_0(t)$ can be expressed as

$$I_0(t) = \bar{I}_0 + i_0\cos(2\pi ft + \psi) \tag{5.12}$$

where $\bar{I}_0$ is the average Laser output intensity, i_0 is the amplitudes of intensity modulation, and ψ is the phase shift between the intensity and the wavelength signals.

After the transmitted Laser is penetrated into the absorbing gas, its intensity $I_t(t)$ can be described by the Beer–Lambert law:

$$I_t(t) = e^{-\alpha(v)L} \cdot I_0(t) = e^{-S(T)\cdot P\cdot X_i\cdot\phi(v)\cdot L} \cdot I_0(t) \tag{5.13}$$

where $\alpha(v)$ is the absorption coefficient and L is the optical absorbing path length, P is the total pressure, T is the temperature, X_i is the local mole fraction of the absorbing species, $\phi(v)$ is the line-shape function and $S(T)$ is the line strength of the transition, which is a function of temperature.

The Laser intensity $I_t(t)$ is multiplied by a sinusoidal signal and the signal with $\pi/2$ phase-shift in a lock-in amplifier to generate the X-component of the WMS-1f signal and the WMS-2f signal (X_{1f}, X_{2f}) and the Y-component of the WMS-1f signal and the WMS-2f signal (Y_{1f}, Y_{2f}) at the frequencies of f and $2f$ [26]. Then, the WMS-2f/1f signal $S_{2f/1f}$ is calculated as

$$S_{2f/1f} = \sqrt{X_{2f}^2 + Y_{2f}^2}\Big/\sqrt{X_{1f}^2 + Y_{1f}^2} \tag{5.14}$$

The peak value of the measured WMS-2f/1f signal, that is $S_{2f/1f}$, is detected to determine the parameters of the target gas, for example CO_2.

Figure 5.17 shows the whole peak detection process for CO_2 transition at 6,357.3 cm^{-1} (1% CO_2, $a = 0.088$ cm^{-1}). It is categorised into three steps: (a) signal modulation, (b) each $S_{2f/1f}$ is the demodulated using a reference signal and (c) the peak value, that is $P_{2f/1f}$ of the measured WMS-2f/1f signal $S_{2f/1f}$ is obtained.

In hardware implementation, the light intensity signal after Laser is penetrated into gas, it is converted into a digital signal by a detector, and then phase-locked calculation is performed by applying the quadrature demodulation in FPGA. The process of the entire system implementation is shown in Figure 5.18.

The harmonics of the intensity profile is demodulated from the acquired light intensity signal. Temperature and concentration along the path can be obtained by comparing the measured demodulated profile with an iteratively updated

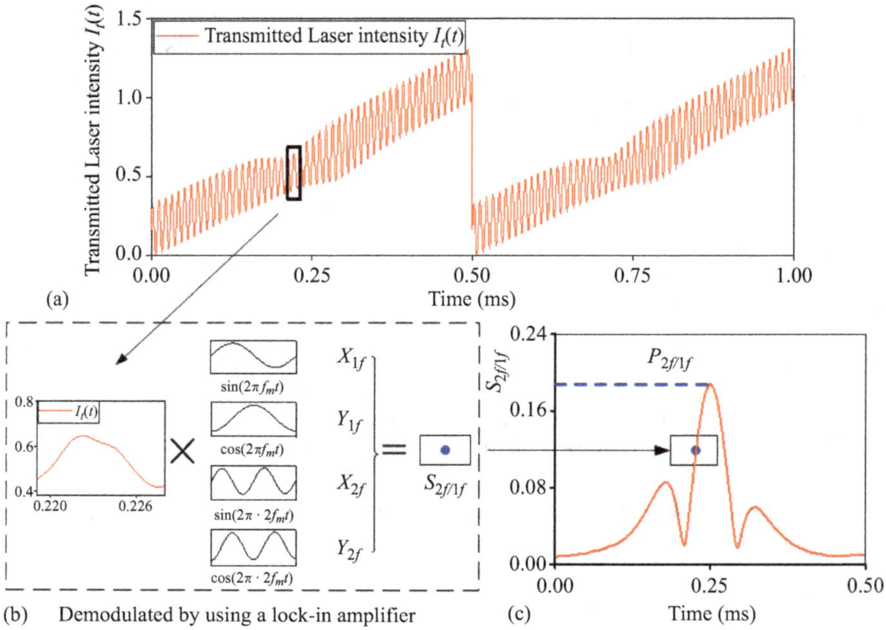

Figure 5.17 Three steps of peak detection process: (a) transmitted Laser intensity, (b) demodulation process and (c) related WMS-2f/1f signal for CO_2 transition at 6,357.3 cm^{-1} (3% CO_2, $a = 0.088$ cm^{-1})

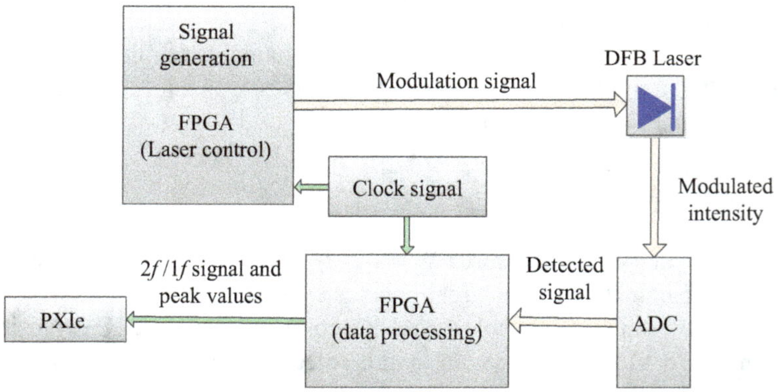

Figure 5.18 Schematic of WMS method in hardware implementation

theoretical values. During online fitting in the wavelength modulation method, a polynomial is used to fit the monotonic portion of the $2f/1f$ signal derivative, that is in the vicinity of the main peak of the harmonic signal. An iterative least squares method is used to calculate the peak value in parallel with data acquisition. Also, the distortion caused by random noise is effectively suppressed by a low-pass filter in the least square method. In the on-chip implementation of the WMS-2*f*/1*f* method, the peak values for the harmonics of 1*f* and 2*f* is calculated in parallel to realise online temperature measurement.

5.3.4 On-chip implementation of WMS method

In the main peak fitting, the demodulation of the harmonics 2*f* and 1*f* is iteratively implemented to locate the peak value of the ratio in a synchronised way, and no extra calculation time is required. The total consumed time for the parameter fitting in the WMS method is mainly determined by the peak detection process. To verify the proposed method, the iterations of the least squares peak detection method are implemented in FPGA. The hardware resources required by using the traditional least square method and the iterative method are compared and listed in Table 5.1.

The slices, LUTs and LUT-FF pairs used in the iterative least square method are about half of those used in the conventional least square method. Only the hard core of DSP48s required by the iterative LS method is doubled. Fewer regular hardware resources, such as Slices, LUTs mean that FPGA has more spare room to accomplish other tasks, and the iterative LS method makes more complicated tasks in FPGA possible. Also, more iterations enable a smaller relative error of solution. Here, 150 iterations can achieve a relative error below 0.2%. In this case, 150 operations of fixed-point multiplications and additions are required. It takes 150 clock cycles to determine the coefficient of the polynomial used for peak fitting. Together with other read and write operations, a total of 195 clock cycles are enough for the WMS-2*f*/1*f* main peak fitting. If the external clock period is 2 μs, the main peak-to-peak value is obtained within 0.39 ms after the main peak region is estimated in one wavelength scanning period. However, the traditional LS method takes 2.08 ms to achieve the results in the same precision. The clock cycles used are about four sinusoidal wave periods, which is within the scanning period lasting about 50 periods in actual measurement. As a result, each online fitting is completed within one wavelength scanning, and continuous online fitting is realised for the total absorbance as well as the temperature measurement.

To evaluate the performance of the online WMS TDLAS sensor, the temperature of air in a copper tube was measured and a thermocouple was installed for comparisons, as shown in Figure 5.19. By online fitting of wavelength modulation spectroscopy in FPGA, the WMS-2*f*/1*f* main peaks and peaks with wavelengths of

Table 5.1 Hardware consumptions of traditional and iterative least square methods in FPGA

Resource type	Iterative LS	Traditional LS	Logic unit number
Slices	2,258 (15.3%)	4,736 (32.2%)	14,720
LUTs	3,975 (6.8%)	7,511 (12.8%)	58,880
LUT-FF pairs	7,521 (12.8%)	14,935 (25.4%)	58,880
DSP48s	20 (3.1%)	10 (1.6%)	640

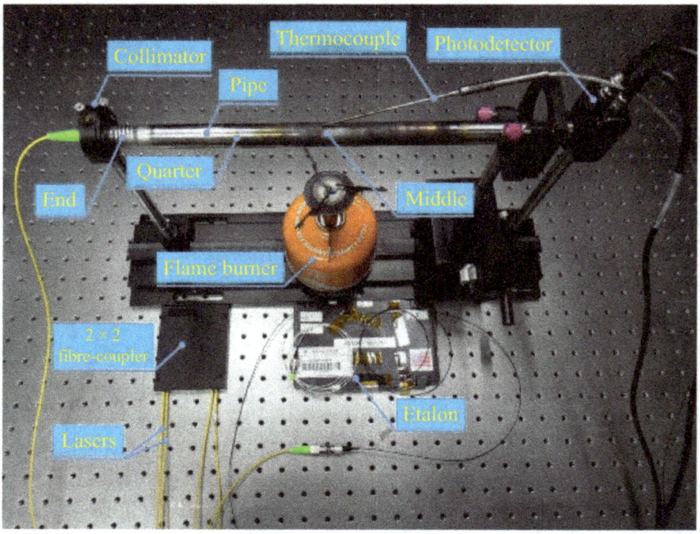

Figure 5.19 Experiment setup for WMS TDLAS sensor

6,799.83 cm^{-1} and 7,185.6 cm^{-1} can be obtained. In the experiment, a 100 Hz saw-tooth signal was used for wavelength scanning and modulated using a 10 kHz sinusoidal signal. The modulated light intensity from the DFB Laser is split into two parts through a fibre coupler, one passing through the gas to be measured and the other to the etalon for wavelength calibrations. The intensity signals are received by photo-detectors and sampled by 14-bit ADCs at a sampling rate of 500 kSPS. A flame burner is used to heat the air in the coper tube and the air flow convection in the region of interest is confined by the tube. Temperature values at three different locations, that is the end, quarter and intermediate position of the copper tube are measured using a thermocouple for comparison.

Before air is heated, air in the tube is measured by a WMS TDLAS sensor and the thermocouple simultaneously, and the results of both methods agree well with each other, as shown in Figure 5.20.

The fluctuation in the TDLAS readings indicates that it is more sensitive and the response is faster to the environment temperature than the thermocouple. The copper tube is heated by an alcohol lamp and the temperature increased from room temperature to about 620 K, then the lamp is removed to allow the copper tube cooled down freely, and the temperature evolution in the copper tube is monitored for 90 s. During the natural cooling of the heated copper tube, thermocouples are also output temperature readings at three different locations of the copper tube. The thermocouples are placed at three locations near the Laser path, the end, quarter, and middle of the copper tube, and the readings of the TDLAS sensor and the thermocouple are shown in Figure 5.21. The temperature obtained by the WMS TDLAS sensor indicates the evolution trend of the temperature inside the copper tube clearly, as it is cooling freely in the environment of room temperature. The

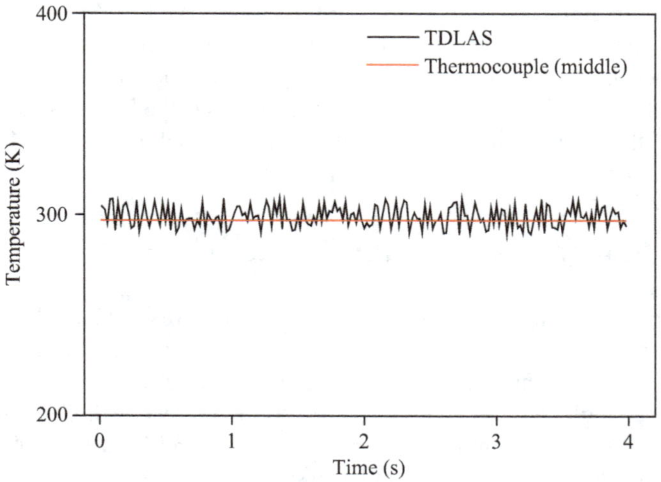

Figure 5.20 Variation of temperature values measured by TDLAS sensor and thermocouple at room temperature with time

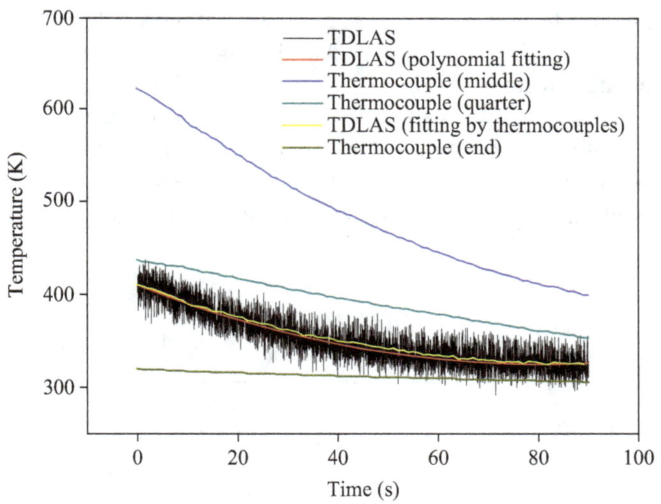

Figure 5.21 Reading variation of TDLAS sensor and thermocouple at different locations in freely cooling tube with time

fitting results of the WMS measurement system are compared with the thermocouple measurements. The results are consistent with each other in the actual situation, and the maximum deviation is less than 3 K, as shown in Figure 5.21.

The temperature measured by the thermocouple in the middle is greater than that measured by the WMS TDLAS sensor. The main reason is that the heated part of the tube is in the central area of the copper tube, and the temperature on both ends is lower than the central area. At the same time, the TDLAS sensor measured the average temperature, which is lower than the temperature in the centre. In addition, thermal radiation to the surrounding low-temperature environment may also cause the deviation. In temperature measurement, the TDLAS sensor reveals the trend of the temperature evolution. Its temporal resolution of 50 Hz enables a quick tracking of the temperature fluctuation [27].

5.4 Calibration of single path TDLAS sensor

The performance of the single path TDLAS sensor was evaluated for the temperature range from –50 °C to 900 °C. Because a single standard device for temperature calibration cannot cover the entire temperature range, two experimental setups for the single-path temperature verification are designed. The two different setups are facilitated to cover the ranges of –50~50 °C and 50~900 °C, respectively. The schematics of both setups are similar in structure and are shown in Figure 5.22. The Lasers used for these two temperature ranges scan different wavelength ranges. The pipe furnace for the low-temperature range is cooled from 50 °C to –50 °C using alcohol as the heat exchange medium, while the other furnace uses electrical heating to vary the temperature in the Laser path from 50 °C to 900 °C. In the experiments, the optical paths for Laser penetration and detection are aligned in the calibration tube. The glass tube is placed in the pipe furnace and then heated. The air temperature inside the tube starts to rise and stabilised at the set value. It usually takes about 2 h for temperature to be stabilised from one to another. At the stabilised states of the temperature in the laser path, both the TDLAS sensor and the thermocouple are used to output the temperature readings. These operations continue until all the pre-set temperature values are measured and compared.

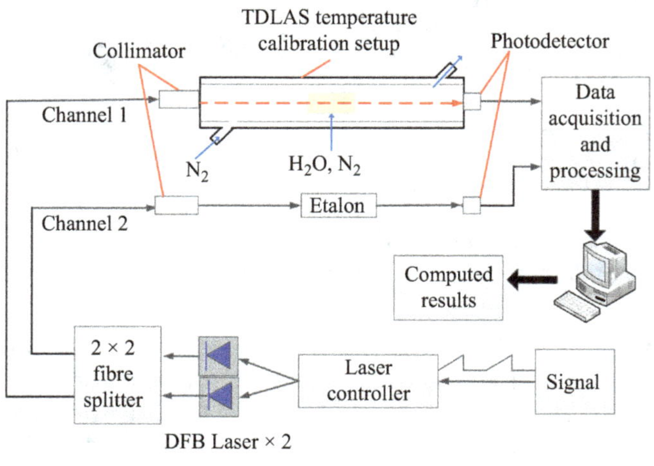

Figure 5.22 Performance evaluation setups for TDLAS sensor

In optical alignment, it is not easy to adjust the positions of the Laser beam, especially as it is an infrared light and a glass cabinet in the tube distorts the Laser positions. A fibre splitter is used to introduce a red light in the same Laser path, and the alignment for the visible red light is easier for manual operations. After the alignment is completed, the red light is turned off and the infrared light used for TDLAS starts to work, followed by the temperature tuning in the furnace and the temperature measurement. The target species used for Laser absorption measurement is water molecule. Two different spectral lines centred at 7,444 cm^{-1} and 7,185 cm^{-1} respectively are selected to extract the total absorbance and hence the temperature along the path. A tube in blue is used to confine the water molecule in the Laser path, as shown in Figure 5.23.

On the Laser path, the temperature distribution is not uniform from the incident light to the detector. The total length of tube is nearly 120 cm and the values decline as the distance from its centre is increasing. In the setup, only the temperature in the vicinity of the tube middle, for example the region 2 of about 10 cm long can be regarded as a region of uniform temperatures. The thermocouple is placed at this region, just on the peripheral of the quartz tube in the furnace. Water molecules are confined in a glass cabinet in the Laser path of uniform temperature, that is region 2. The other parts, such as regions 1 and 3, along the Laser path are full of flowing nitrogen gas, which introduces no absorption in any cases. Flowing nitrogen also blows water vapour away to ensure an extremely low concentration of the water vapour inside the regions.

In this way, the average temperature measured by the TDLAS sensor is the same as the average value in the glass cabinet, and the readings of the sensor indicate the same temperature as the thermocouple in a stabilised state. As a result, the outputs of the TDLAS sensor and the thermocouple are comparable, and the high precision thermocouple is used as a reference for TDLAS sensor evaluations. The length of the chamber in region 2 is short, and the detected signals are very weak. In this case, water is added to enhance the spectral absorption. At the same time, during the experiment, a pressure gauge is added at the gas outlet to obtain the pressure value of the measured gas.

5.4.1 Performance evaluation

In the temperature range of –50 °C~50 °C, water molecules will be frozen, and hence the oxygen molecules are selected as the target specie. Two absorption

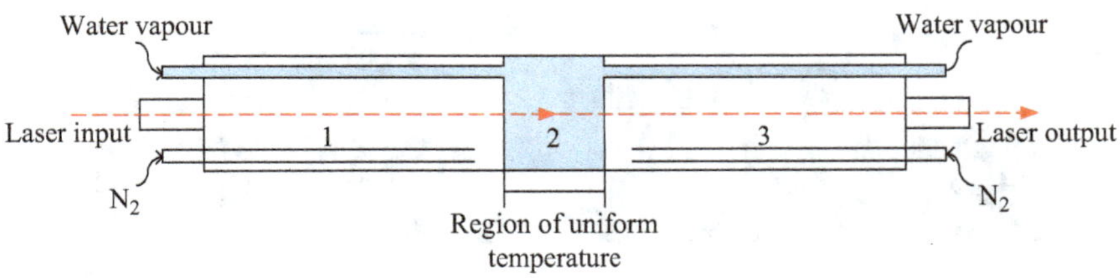

Figure 5.23 Pipe furnace and the tube for gas isolation in Laser path

spectra for the oxygen in the air are selected and central wavelengths of them are 2,005 nm and 2,013 nm, respectively. The experimental setup for the low-temperature evaluation of the TDLAS sensor is shown in Figure 5.24.

Laser signals with two different wavelengths are generated by two DFB Lasers in sequence, and then both connected to a 2 × 2 fibre splitter to generate two Laser paths, in which the intensity values are the same. The light in one of them passes through a Mach-Zehnder fibre interferometer used as an etalon and is collected by a photo-detector (detector model: Thorlabs, PDA10CS-EC, bandwidth 17 MHz, area of the photosensitive surface 0.8 mm^2). The light in the other path passed a collimator and turned to be an approximately parallel spatial light. The light travels through the tube for gas isolation, and the gas chamber of the tube is placed in the uniform temperature region of the pipe furnace. The other parts of the tube except for the chamber are full of the nitrogen gas, and the absorbed intensity signals are collected by a photo-detector. Figure 5.25 shows the raw data of absorbed intensity values of both DAS and WMS methods at the temperature setting of –50.5 °C, which is measured using the thermocouples and the extracted profiles of the absorbed spectral centred at the wavelengths of 2,005 nm and 2,013 nm are shown in Figure 5.26.

Six different temperatures were set in the range of –50 °C and 50 °C, and the readings of the TDLAS sensor are compared with the pre-set values in Table 5.2. All the temperatures were stabilised for about half an hour before the TDLAS sensor started to measure the temperature. The set values were measured and controlled using a platinum resistance temperature sensor. It can be seen from the

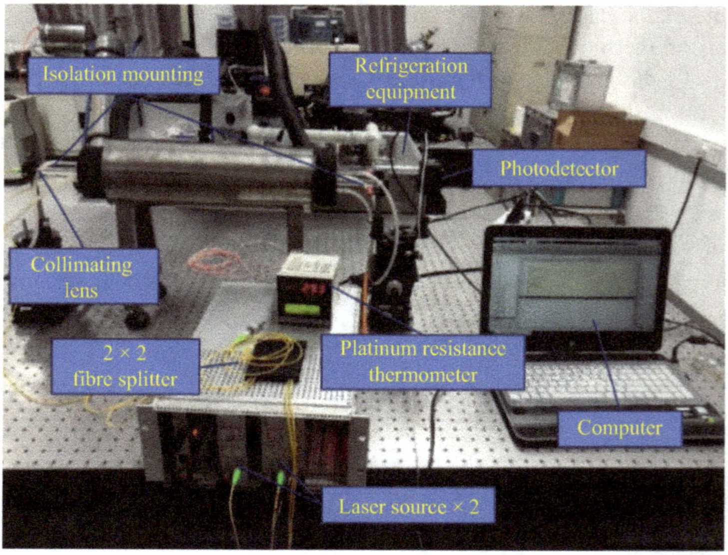

Figure 5.24 Experiment setup for TDLAS sensor evaluation at low temperature between –50 °C and 50 °C

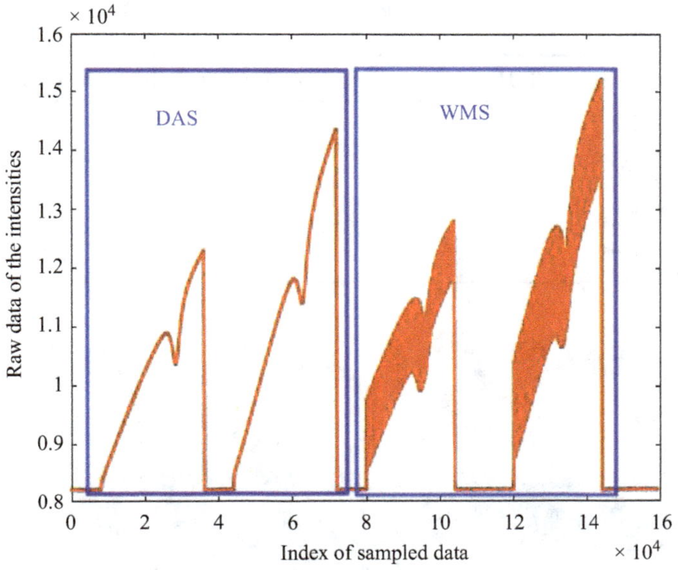

Figure 5.25 Raw data of absorbed intensities for both DAS and WMS methods at –50.5 °C

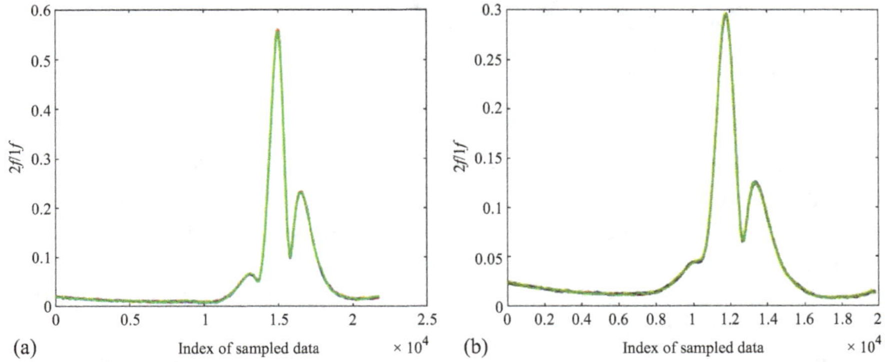

Figure 5.26 Absorption profiles at spectra centred at the wavelengths of (a) 2,005 nm and (b) 2,013 nm

data in Table 5.2 that the maximum deviation of the TDLAS sensor readings from the set values is 2.7 °C, and all the relative errors are within 1.5%.

5.4.2 Performance evaluation at high temperatures

In the temperature range from 50 °C to 900 °C, water molecules are used as the specie for the TDLAS sensor. The experimental system is shown in Figure 5.27.

Table 5.2 TDLAS sensor evaluations in case of low temperatures (°C)

Set value	Readings of TDLAS sensor	Temperature deviations	Relative error (%)
−50.5	−53.2	−2.7	−1.21
−34.6	−33.0	1.6	0.67
−10.1	−7.6	2.5	0.95
−1.0	−0.1	0.9	0.33
28.5	27.7	−0.8	−0.27
38.5	36.7	−1.8	−0.58

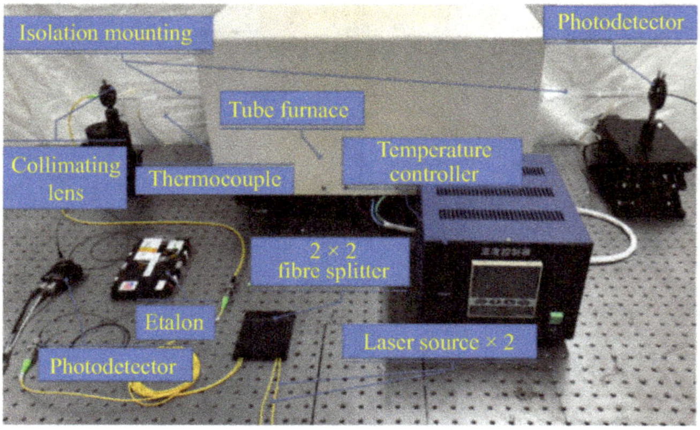

Figure 5.27 Evaluation setup for TDLAS sensor between 50 °C and 900 °C

It consists of DFB Lasers, a fibre splitter, a collimator, a Mach-Zehnder fibre interferometer, a gas isolation tube and a detector. The central wavelengths of the DFB Lasers are 7,185.59 cm^{-1} and 7,444.34 cm^{-1}, respectively. Following almost the same procedure in the case of the low-temperature evaluation, two Lasers are controlled subsequently and combined to be one beam in a 2 × 2 fibre splitter. Two outputs of the splitter are connected into two individual Laser paths. One beam passed the Mach-Zehnder fibre interferometer and is used as an etalon. The other is collimated into a spatial beam and travels through the pipe furnace. The Laser path in the furnace is confined in a quartz tube and nitrogen isolation is applied to make sure that only the part with uniformly distributed temperature contributes to the absorbed profiles. As a result, the temperature readings of the TDLAS sensor are comparable with the values measured by the thermocouple.

Taking the temperature measurement at 400 °C as an example, Figure 5.28 shows the light intensity for two spectral scanning centred at wavelengths of 7,185 cm^{-1} and 7,444 cm^{-1}. Extracted absorption spectra for two central wavelengths are shown in Figure 5.29, indicating that the profiles of the line-shape are smooth around the central wavelengths and high repetitive results are expected.

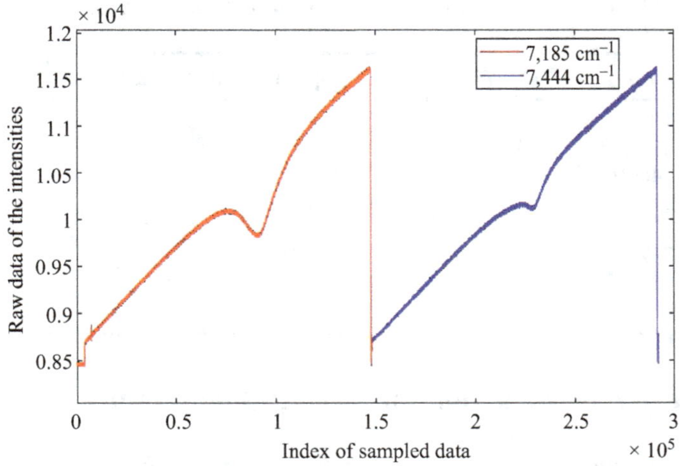

Figure 5.28 Light intensity for spectra near 7,185 cm^{-1} and 7,444 cm^{-1} at 400 °C

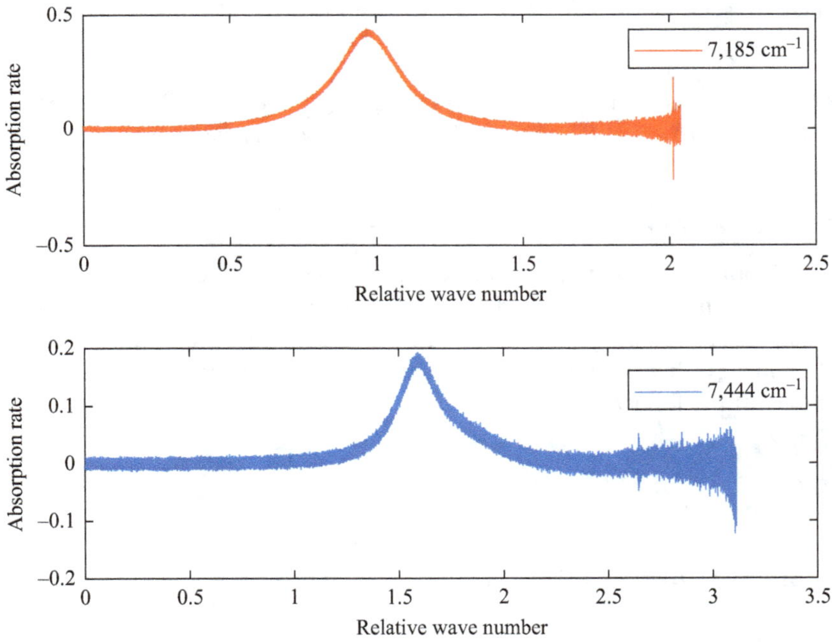

Figure 5.29 Absorption profiles at spectra centred at 7,185 cm^{-1} and 7,444 cm^{-1}

Eight different temperatures are pre-set for performance evaluation in the range from 50 °C to 900 °C, and the comparisons with the set values are listed in Table 5.3. The results indicate that the maximum temperature deviation of the TDLAS gas temperature sensor from the temperature set value is 13.4 °C, and the maximum relative error was 1.3%.

Table 5.3 TDLAS sensor evaluations in case of high temperatures (°C)

Set value	Readings of TDLAS sensor	Temperature deviations	Relative error (%)
102	101.3	−0.7	−0.2
302	303.1	1.1	0.2
350	346.7	3.3	0.9
400	397.7	2.3	0.6
502	492.2	−9.8	−1.3
702	712.9	10.9	1.1
802	804.3	2.3	0.2
902	888.6	−13.4	−1.1

5.5 Image reconstruction for TDLAS tomography

Many reconstruction algorithms can be applied to retrieve the spatial distribution of the temperature and gas concentration. Three typical algorithms: SART, Tikhonov regularisation and Landweber iteration are used to solve the linear equation (5.9) for TDLAS sensors. Equation (5.9) is rewritten in a general form as

$$L\alpha = A \tag{5.15}$$

where $L = \left(l_{i,j}\right)$ is an $M \times N$ matrix, $A = (a_1, a_2, \ldots, a_M)^T \in \boldsymbol{R}^M$ observed data, $\sigma = (\sigma_1, \sigma_2, \ldots, \sigma_N)^T \in \boldsymbol{R}^N$ an underlying image and T is the transpose of a vector or matrix.

5.5.1 SART algorithm

SART algorithm is a basic algebraic reconstruction method designed to solve the linear system in image reconstruction. The sequence generated by SART converges to a minimiser of a weighted least square functional from any initial guess [28]. SART can solve linear unknowns in few iterations for high-resolution tomography situations. If the following is defined

$$L_{i,+} = \sum_{j=1}^{N} l_{i,j} \quad \text{for } i = 1, \ldots, M \tag{5.16}$$

$$L_{+,j} = \sum_{i=1}^{M} l_{i,j} \quad \text{for } j = 1, \ldots, N \tag{5.17}$$

$$\bar{A}(\sigma) = L\sigma \tag{5.18}$$

Then, the general iteration form of SART can be written as

$$\sigma_j^{(k+1)} = \sigma_j^{(k)} + \frac{\omega}{L_{+,j}} \sum_{i=1}^{M} \frac{L_{i,j}}{L_{i,+}} \left(a_i - \bar{A}_i(\sigma^{(k)})\right) \tag{5.19}$$

for $k = 0, 1, \ldots,$ where ω is a relaxation parameter in (0, 2). The value of ω is one for the standard SART implementation.

5.5.2 Tikhonov regularisation method

For the linear inverse problem in (5.15), a general version of Tikhonov regularisation method takes the following form:

$$\min\{\|L\sigma - A\|_2^2 + \lambda\|H\sigma\|_2^2\} \tag{5.20}$$

where λ is a positive constant chosen to control the size of the solution vector and H is a matrix that defines a (semi)norm on the solution through which the 'size' is measured. Often, H represents the first or second derivative operator. If H is the identity matrix, then the Tikhonov problem is said to be in standard form.

The solution σ_λ to (5.20) solves the problem, that is

$$(L^T L + \lambda H^T H)x = L^T A \tag{5.21}$$

As λ increases, the norm $\|Hx\|_2$ of the solution vector decreases monotonically while the residual $\|Ax - b\|_2$ increases monotonically. Tikhonov regularisation has an important equivalent formulation as

$$\min\|Lx - A\|_2 \text{ subject to } \|Hx\|_2 \leq \delta \tag{5.22}$$

where δ is a positive constant.

The problem in (5.22) is an LS problem with a quadratic constraint. Using the Lagrange multiplier formulation

$$\|L\sigma - A\|_2^2 + \lambda(\|H\sigma\|_2^2 - \delta^2) \tag{5.23}$$

If $\delta \leq \|x_{\text{LS}}\|_2$, where σ_{LS} is the least square solution to (5.21), then the solution σ_δ to (5.22) is identical to the solution σ_H to (5.20) for an appropriately chosen λ, and there is a monotonic relation between the parameters δ and λ [29].

5.5.3 Landweber method

If the kth estimate of σ can be denoted by σ_k, then the Landweber method is generally written as

$$\sigma_k = \sigma_{k-1} + \alpha L^T(|A - L\sigma_{k-1}) \tag{5.24}$$

where α is the relaxation factor to control the convergence rate [30].

Usually, the larger the value of α, the faster the convergence of (5.24) can be. Usually, the value of α satisfies

$$\|\alpha L^T L\| < 2 \tag{5.25}$$

where $\|\cdot\|$ is the norm of the matrix.

Equation (5.24) can be rewritten in the form of z-transform as

$$\sigma(z) - \sigma_0 = z^{-1}\sigma(z) + \alpha L^T\left(A\frac{1}{1 - z^{-1}} - Lz^{-1}\sigma(z)\right) \tag{5.26}$$

$$\sigma(z) = \left[I - (I - \alpha L^T L)z^{-1}\right]^{-1}\left(\frac{1}{1 - z^{-1}}\alpha L^T A + \sigma(z)\right) \tag{5.27}$$

When the values of z approaches to 1, the values of $\sigma(z)$ will be close to the exact value of the final iteration. In real applications, z can be selected as a close value to 1 and the part can be calculated in advance to reduce the computing time:

$$[I - (I - \alpha L^T L)z^{-1}]^{-1} \tag{5.28}$$

5.6 Case studies

5.6.1 Case study 1: Acoustic excited flame monitoring

An acoustically excited combustion setup was established, consisting of a Bunsen burner and an acoustic excitation apparatus, as shown in Figure 5.30(a). The fuel gas CH_4 flows from a gas tank and goes into a T-type connector and finally arrives at the Bunsen burner. Oxygen comes from the air outside through the opening window and is fully mixed with CH_4 gas inside the pipeline before the burner. A loudspeaker is used to acoustically modulate the gas flow and generated the periodic flame. A signal generator drives the speaker and the acoustic waves tune the injections of the fuel gas. A trumpet shape sealing film is used to fabricate the input port of the T-type connector, and the acoustic energy can be intensified in this way. A tunable sinusoidal waveform from the signal generator is imposed on the speakers and the frequency varies from 80 Hz to 120 Hz as shown in Figure 5.30(a). As a result of the acoustic excitation, the gas flow varies at the same frequency of the driving signals. A TDLAS tomographic sensor is applied to measure the dynamic temperature distribution of the flames as shown in Figure 5.30(b).

In this experiment, H_2O is chosen as the target absorbing species for its strong and independent absorption spectra. Two H_2O absorption transitions are selected and centred at 7,444.36 cm^{-1}, that is 7,444.35 cm^{-1} combined with 7,444.37 cm^{-1} and 7,185.6 cm^{-1}, which can be generated using two DFB Lasers. In the experiment, the light paths are adjusted to ensure an effective detection of the fan-shaped Laser beams. The Lasers are modulated to produce a saw-tooth wave and etalon is used to characterise the wavelengths of the Laser wave. The fuel gas valve and the loudspeaker are turned on to make periodic flames. Meanwhile, all intensity values from all the channels are collected on the photo-electric detectors and stored in SDRAMs. After that, the data acquisition units transmit the stored data to the host PC for extraction of absorption profiles, as shown in Figure 5.31, and the total absorbance can be derived for temperature readings. Landweber algorithm was applied to reconstruct the tomographic images. The frame rate of the TDLAS tomographic sensor was 1,000 frames per second, which is beyond twice of the exciting frequency, that is 120 Hz, and the fluctuation in the frame can be detailed. Sixteen frames of image are selected to show two whole fluctuating cycles, as shown in Figure 5.32.

(a)
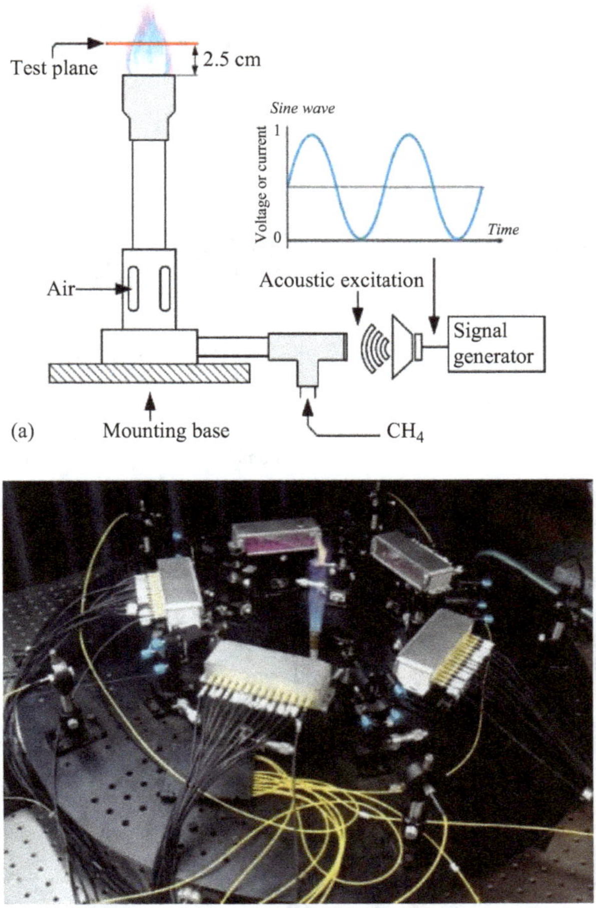

(b)

Figure 5.30 Experiment setup for TDLAS tomography in acoustic excited flame: (a) schematic of acoustically excited Bunsen burner and (b) tomographic setup

The reconstructed images show that an expanding ring shape evolved periodically in the high-temperature region. The results validate that the acoustic excitation flame is a diffusion flame and the cross-section distributions agree well with the tendency in the longitudinal section. The frequency of the periodically varied flames can be calculated from the number of frames in one cycle and the pre-set temporal resolution in the data acquisition system, and it is equal to the setting of Laser modulation frequency. As the uncertainty of gas flow in the Bunsen burner pipeline, the number of frames in one cycle has small fluctuation. Therefore, the average frame number of ten cycles is used as the reference. The frequency of acoustic excitation is linearly related to flame impulses, as shown in Figure 5.33.

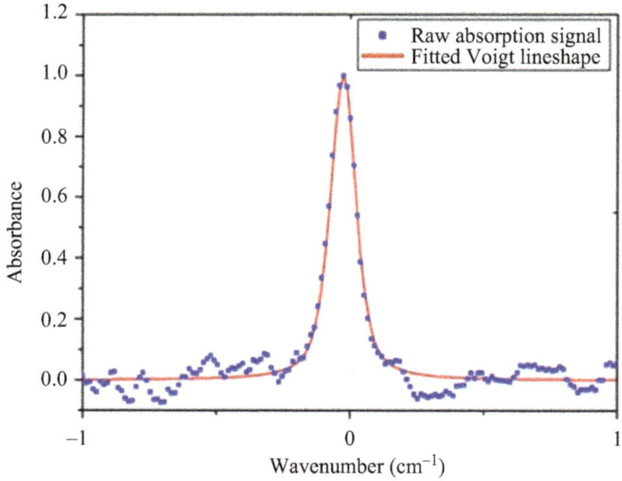

Figure 5.31 Absorption line-shape extraction near central wavelength of 7,185 cm^{-1}

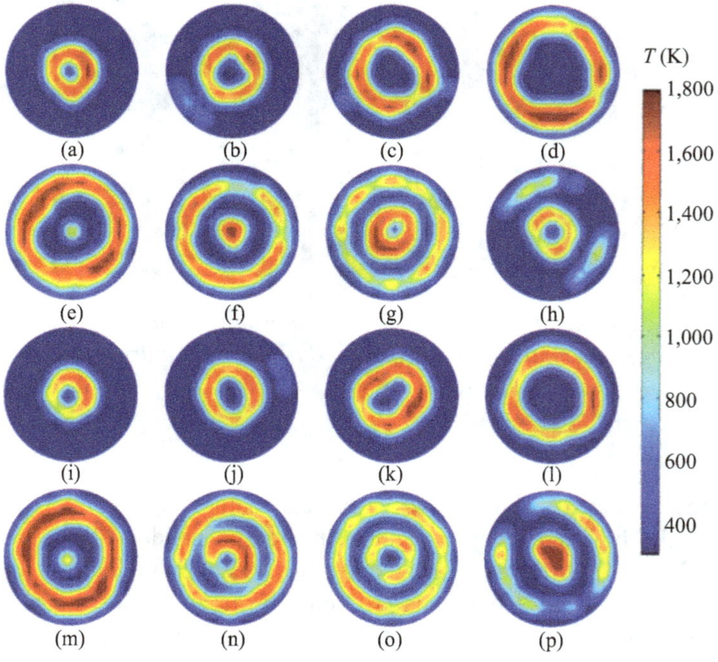

Figure 5.32 Reconstruction results of 16 temperature distributions in two 120 Hz flame impulse periods captured at 1,000 frames per second. The distributions in (a)–(h) are for the first period and those in (i)–(p) are for the second period

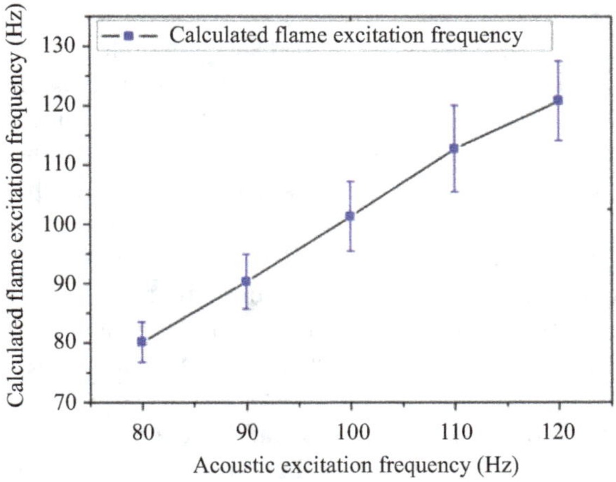

Figure 5.33 Calculated frequency of acoustically excited flames versus acoustic exciting frequency

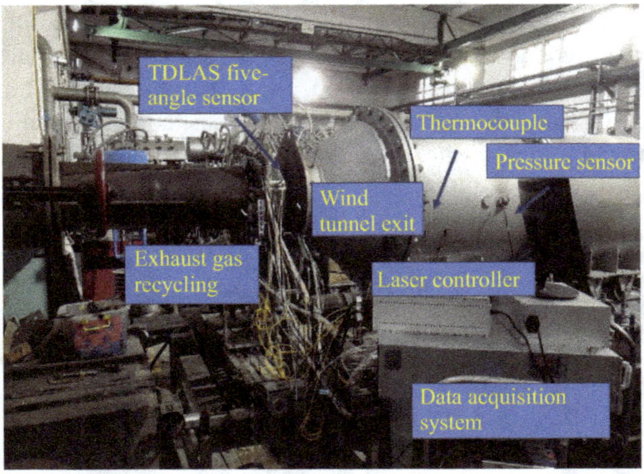

Figure 5.34 Experiment setup for TDLAS tomography in high-temperature wind tunnel

5.6.2 Case study 2: Process monitoring for high-temperature wind tunnel

The diameter of the wind tunnel used for the experiment is about 600 mm. The TDLAS tomographic sensor stands on the ground, just in the test section about 20 cm behind the wind tunnel exit. The outlet airflow is heated to vary between 600 K and 1,200 K and the passes through the TDLAS tomographic sensor at the speed of up to 0.3 Mach. The diagrams of the experimental setup in the hot temperature wind tunnel and the sensor installation are shown in Figure 5.34.

In the experiments, Lasers scanning across the central wavelength of 7,185 cm^{-1} are used to reconstruct the absorbance distributions in the test regions. The TDLAS sensor is first installed on steel framework standing on the ground. All the Lasers are checked before the wind tunnel starts to blow and ignite. After the ignition, no visible flame can be seen at the testing section, but the TDLAS sensor enables the flame visible in terms of the spectral absorbance. In the range from 600 K to 1,300 K, the absorbance of water over the spectral centring at 7,185 cm^{-1} rises along with an increasing temperature. The distributions of water absorbance at different stages of the wind tunnel operations are reconstructed as shown in Figure 5.35. The distributions show that the absorbance in the centre is significantly larger than that in the other parts, and six typical stages for the temperature rising are listed.

The WMS method is used to suppress the fluctuation in the wind turbulence and the normalised harmonics of the method are used to show the three heating processes. During dynamic process of temperature heating, the average values of the water absorbance in each frame are extracted to illustrate the heating dynamics, as shown in Figure 5.36.

The ignition happened around 8 min after the tunnel operation. The average absorbance increased rapidly after ignition, then slowly decreased until it reached a stabilised working condition, that is condition 1. The working condition 2 came following another heating after about 60 min. The real temperature values are provided by the installed thermocouples as references, and the dynamics in general agree well with the actual heating operation of the wind tunnel. It can be concluded

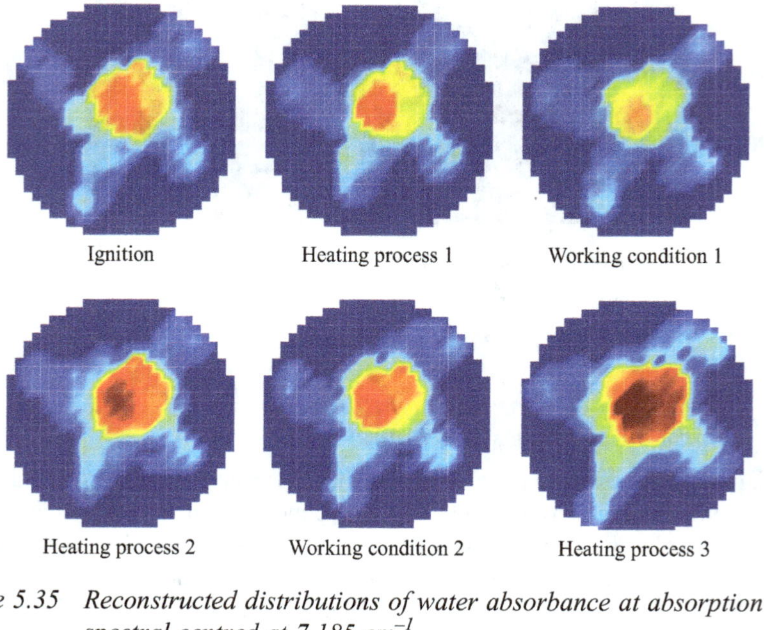

Figure 5.35 Reconstructed distributions of water absorbance at absorption spectral centred at 7,185 cm^{-1}

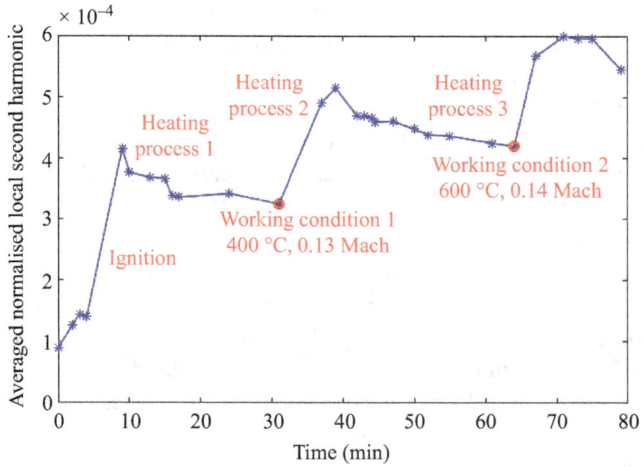

Figure 5.36 Average total absorbance versus time in three heating processes of wind tunnel

that the TDLAS sensor details the fluctuation in the water distribution in the wind tunnel. As a result, the temperature distribution can be expected if more spectral are available.

5.7 Prospects and future work

At present, TDLAS tomographic sensors can be used to capture the distribution of gas temperature and concentration. More parameters, such as the velocity of the gas molecule cluster and the local pressure in a combustion region, can be extracted in future work for a more comprehensive understanding of combustion processes. Simulated results can be found in some recently published literatures and more experimental results are expected in the future. Moreover, deeper fusion of the TDLAS technique with other Laser diagnostic methods, such as PLIF and CARS, may provide more insights how the radicals and products in the flame evolve and react and show combustion flames in more details.

References

[1] Deguchi Y, *Industrial Applications of Laser Diagnostics*, Boca Raton, CRC Press, 2012

[2] Maes N, Meijer M, Dam N, *et al.*, 'Characterization of spray: A flame structure for parametric variations in ECN constant-volume vessels using chemiluminescence and laser-induced fluorescence', *Combust. Flame*, 2016, **174**, pp 138–151

[3] Richardson D R, Roy S, Gord J R, 'Femtosecond, two-photon, planar laserinduced fluorescence of carbon monoxide in flames', *Opt. Lett.*, 2017, **42** (4), pp 875–878

[4] Miller J D, Peltier S J, Slipchenko M N, *et al.*, 'Investigation of transient ignition processes in a model scramjet pilot cavity using simultaneous 100 kHz formaldehyde planar laser-induced fluorescence and CH * chemiluminescence imaging', *Proc. Combust. Inst.*, 2016, **36** (2), pp 2865–2872

[5] Dennis C N, Slabaugh C D, Boxx I G, Meier W, Lucht R P, '5 kHz thermometry in a swirl-stabilized gas turbine model combustor using chirped probe pulse femtosecond CARS, Part 1: Temporally resolved swirl-flame thermometry', *Combust. Flame*, 2016, **173**, pp 441–453

[6] Slabaugh C D, Dennis C N, Boxx I, Meier W, Lucht R P, '5 kHz thermometry in a swirl-stabilized gas turbine model combustor using chirped probe pulse femtosecond CARS, Part 2: Analysis of swirl flame dynamics', *Combust. Flame*, 2016, **173**, pp 454–467

[7] Nicolas F, Todoroff V, Plyer A, *et al.*, 'A direct approach for instantaneous 3D density field reconstruction from background-oriented schlieren (BOS) measurements', *Exp. Fluids*, 2016, **57**, p 13

[8] Raffel M, 'Background-oriented schlieren (BOS) techniques', *Exp. Fluids*, 2015, **56**, p 60

[9] Grauer S J, Unterberger A, Rittler A, Daun K J, Kempf A M, Mohri K, 'Instantaneous 3D flame imaging by background-oriented schlieren tomography', *Combust. Flame*, 2018, **196**, pp 284–299

[10] Cai W, Kaminski C F, 'Multiplexed absorption tomography with calibration free wavelength modulation spectroscopy', *Appl. Phys. Lett.*, 2014, **104**, p 154106

[11] Goldenstein C S, Strand C L, Schultz I A, Kai S, Jeffries J B, Hanson R K, 'Fitting of calibration-free scanned-wavelength-modulation spectroscopy spectra for determination of gas properties and absorption lineshapes', *Appl. Opt.*, 2014, **50** (3), pp 356–367

[12] Hartmann A, Strzoda R, Schrobenhauser R, Weigel R, 'CO2 sensor for mainstream capnography based on TDLAS', *Appl. Phys. B*, 2014, **116** (4), pp 1023–1026

[13] Ren W, Farooq A, Davidson D F, Hanson R K, 'CO concentration and temperature sensor for combustion gases using quantum-cascade laser absorption near 4.7 mm', *Appl. Phys. B*, 2012, **107** (3), pp 849–860

[14] Tang Y, Liu W, Kan R, *et al.*, 'Measurements of NO and CO in Shanghai urban atmosphere by using quantum cascade lasers', *Opt. Exp.*, 2011, **19** (21), pp 20224–20232

[15] Goldenstein C S, Mitchell S R, Hanson R K, 'Fiber-coupled diode laser sensors for calibration-free stand-off measurements of gas temperature, pressure, and composition', *Appl. Opt.*, 2016, **55** (3), pp 479–484

[16] Qu Q, Cao Z, Xu L, Liu C, Chang L, McCann H, 'Reconstruction of two dimensional velocity distribution in scramjet by laser absorption spectroscopy tomography', *Appl. Opt.*, 2019, **58** (1), pp 205–212

[17] Buchholz B, Afchine A, Ebert V, 'Rapid, optical measurement of the atmospheric pressure on a fast research aircraft using open-path TDLAS', *Atmos. Meas. Tech.*, 2014, **7** (11), pp 3653–3666

[18] Carter C D, Gruber M R, Rieker G B, Jeffries J B, Hanson R K, Mathur T, 'Diode laser-based detection of combustor instabilities with application to a scramjet engine', *Proc. Combustion Institute*, 2009, **32** (1), pp 831–838

[19] Liu C, Xu L, Chen J, Cao Z, Lin Y, Cai W, 'Development of a fan beam TDLAS-based tomographic sensor for rapid imaging of temperature and gas concentration', *Opt. Exp.*, 2015, **23** (17), pp 22494–22511

[20] Cai W, Kaminski C F, 'A numerical investigation of high-resolution multi-spectral absorption tomography for flow thermometry', *Appl. Phys. B*, 2015, **119** (1), pp 29–35

[21] Ma L, Cai W, Caswell A W, *et al.*, 'Tomographic imaging of temperature and chemical species based on hyperspectral absorption spectroscopy', *Opt. Exp.*, 2009, **17** (10), pp 8602–8613

[22] Wang F, Cen K F, Li N, *et al.*, 'Two-dimensional tomography for gas concentration and temperature distributions based on tunable diode laser absorption spectroscopy', *Meas. Sci. Technol.*, 2010, **21** (4), p 45301

[23] Ma L, Li X, Sanders S T, *et al.*, '50-kHz-rate 2D imaging of temperature and H2O concentration at the exhaust plane of a J85 engine using hyperspectral tomography', *Opt. Exp.*, 2013, **21** (1), pp 1152–1162

[24] Liu C, Tsekenis S A, Polydorides N, Mccann H, 'Toward customized spatial resolution in TDLAS tomography', *IEEE Sens. J.*, 2019, **19** (5), pp 1748–1755

[25] Jing W, Cao Z, Zhang H, Qu Q, Xu L, 'A reconfigurable parallel data acquisition system for tunable diode laser absorption spectroscopy tomography', *IEEE Sens. J.*, 2017, **17** (24), pp 8215–8223

[26] Sun K, Chao X, Sur R, Jeffries J B, Hanson R K, 'Wavelength modulation diode laser absorption spectroscopy for high-pressure gas sensing', *Appl. Phys. B*, 2013, **110** (4), pp 497–508

[27] Xu L, Xue R, Li Y, Zhang H, Zhang C, 'FPGA-based real-time implementation of temperature measurement via tunable diode laser absorption spectroscopy', *IEEE Sens. J.*, 2018, **18** (7), pp 2751–2758

[28] Jiang M, Wang G, 'Convergence of the simultaneous algebraic reconstruction technique (SART)', *IEEE Trans. Image Process.*, 2003, **12** (8), pp 957–961

[29] Golub G H, Hansen P C, O'Leary D P, 'Tikhonov regularization and total least squares', *SIAM J. Matrix Anal. Appl.*, 1999, **21** (1), pp 185–194

[30] Yang W Q, Spink D M, York T A, McCann H, 'An image-reconstruction algorithm based on Landweber's iteration method for electrical-capacitance tomography', *Meas. Sci. Technol.*, 1999, **10** (11), pp 1065–1069

Zhang Cao received his BSc degree (with distinction) in Automation and MEng and PhD degrees (with distinction) in Measurement Technology and Automatic Devices from Tianjin University, China, in 2003, 2005 and 2008. From 2008, he works with Beihang University, China. Currently, he is a Professor in the School of Instrumentation and Opto-Electronic Engineering. His research interests include process tomography, multiphase flow measurement and inverse problems.

Lijun Xu is a Senior Member of the IEEE. He received his PhD from Tianjin University, China in 1996. From 1997 to 2001, he was an Associate Professor in the School of Electrical Engineering and Automation, Tianjin University. From 2002 to 2006, he was a Research Fellow at the University of Greenwich, and the University of Kent, and a Higher Scientific Officer at the Institute of Cancer Research, University of London, UK. Currently, he is a Professor and the Dean of the School of Instrument Science and Opto-Electronic Engineering, Beihang University, China. His research interests include tomographic imaging, digital imaging and dynamic process monitoring. He has authored and co-authored more than 200 publications. He was awarded the National Science Fund for Distinguished Young Scholars of China in 2012, elected into the Ministry of Science and Technology Innovation Talents Promotion Plan in 2013, chosen as a Chang-Jiang Scholars Program Professor by the Ministry of Education, China in 2014 and elected into a high-level talents special support plan: '10,000 Plan' in 2016.

Chapter 6
Light detection and ranging (LiDAR) sensors

Xiaolu Li[1] *and Lijun Xu*[1]

List of abbreviations

ADC	analogue-to-digital converter
AGC	auto gain control
ALS	airborne Laser systems
APD	avalanche photo-diodes
ASKQ	adaptive-scale key-point quality
CFD	constant fraction discriminator
DC	direct current
DSP	digital signal processor
ECL	emitter-coupled logic
FOV	field of view
FPGA	field-programmable gate array
FWHM	full width at half maximum
GaAs	gallium arsenide
GPS	global positioning system
IMU	inertial measurement unit
IR	infrared radiation
KQ	key-point quality
Laser	light amplification by stimulated emission of radiation
LED	leading-edge discriminator
LiDAR	Light Detection and Ranging
MEMS	micro-electro-mechanical systems
MLS	mobile Laser systems
PIN	p-type, intrinsic, n-type
PM	polygonal mirror

[1]School of Instrumentation and Optoelectronic Engineering, Beihang University, Beijing, China

RGB	red-green-blue
RMS	root-mean-square
TDC	time-to-digital converter
TLS	terrestrial Laser scanners
TOF	time-of-flight

6.1 Introduction

LiDAR is an acronym for light detection and ranging, which is a remote sensing technique for measuring distance to a target by illuminating the target with Laser pulses and then measuring the reflected pulses using optical sensors. According to the sensor platform, LiDAR can be divided into airborne Laser systems (ALS), mobile Laser systems (MLS) and terrestrial Laser scanners (TLS), which has superior point positioning accuracy compared to ALS or MLS systems.

ALS typically includes three primary components: (1) a Laser scanner instrument, (2) a global positioning system (GPS) and (3) an inertial measurement unit (IMU). It combines LiDAR and an aircraft platform, to generate high point density with high accuracy. A LiDAR instrument captures the ranging information that can be combined with IMU and GPS trajectory data to form a geo-referenced point cloud. The quality of the point cloud data produced by an ALS depends on several factors: GPS and IMU accuracy, LiDAR ranging and angular accuracy, system lever arm precision, extended GPS baselines and bore sight calibration. ALS has been commercially used since the mid-1990s, and it has become mature and developed. Today ALS is one of the most effective and reliable means of terrain data collection.

MLS acquires scan data from a moving platform, such as a boat, train and automobile. A Laser scanner is supplemented by position and attitude sensors, like GPS and IMU. Its principle of operation is the same as ALS. MLS data processing workflows are almost identical to ALS. Since the accuracy of a Laser scanner instrument is in the order of few centimetres, the overall accuracy of an MLS is often determined by the accuracy of the navigation solution. MLS can be used to accurately survey challenging environments, reliably capture railway infrastructure, topography and vegetation and for road surface inspection. Moreover, because MLS data can be collected and processed in a short time period, frequent and repetitive urban surveying, mapping and database updating are both possible and affordable.

TLS can rapidly and efficiently provide detailed and highly accurate time-of-flight (TOF) three-dimensional (3D) point clouds of their targets. TLS systems use the same principle as ALS, but they are ground-based. They are a composite of rapid pulse Lasers, precisely calibrated receivers and ranging systems, high-speed controlled motors, accurate mirrors and advanced computing capabilities. The distance between the measured target and a TLS can be obtained by firing Laser pulses at the target and measuring the moment of the pulsed TOF. The quality of the point cloud data produced by a TLS is enabled by the direct geo-referencing of the Laser beam, the measured range and the mounting and scanning geometry. A TLS is more useful for capturing small irregular objects (relative to those viewed from an aircraft), such as buildings, earthworks and landforms like cliff faces, which can be profiled and monitored during mining.

In this chapter, a TLS is introduced as a typical TOF LiDAR sensor, including sections on TLS systems, error analysis and applications.

6.2 TLS system

Theoretically, a typical TLS consists of four sub-systems: (1) a Laser emitting unit, (2) a receiving unit, (3) a ranging unit and (4) a controlling and scanning unit. A Laser emitter is a Laser source with additional components, such as beam-expanding optics. A receiver generally consists of a telescope with a detector placed at its focal plane. The primary optical receiver configuration of LiDAR instruments is coaxial. A Laser beam shares the optical axis of the receiver by implementing central mirrors. The electrical output signal of the detector is fed to the ranging and controlling unit as start/stop signals so that the time between the start and stop signals can be used to calculate the range between a target and a LiDAR system. LiDAR scanners can be used to measure the distance with different angles by rotating parts of the device to achieve a field of view up to 360°. Combining the measured range with the horizontal angle and vertical angle, 3D information can be obtained based on a user coordinate system. The concept of TLS is shown in Figure 6.1 [1].

6.2.1 Laser emitter

A Laser is used to provide a light source for LiDAR, with high power in short pulses, causing an increased return signal to a receiver. A stronger Laser pulse

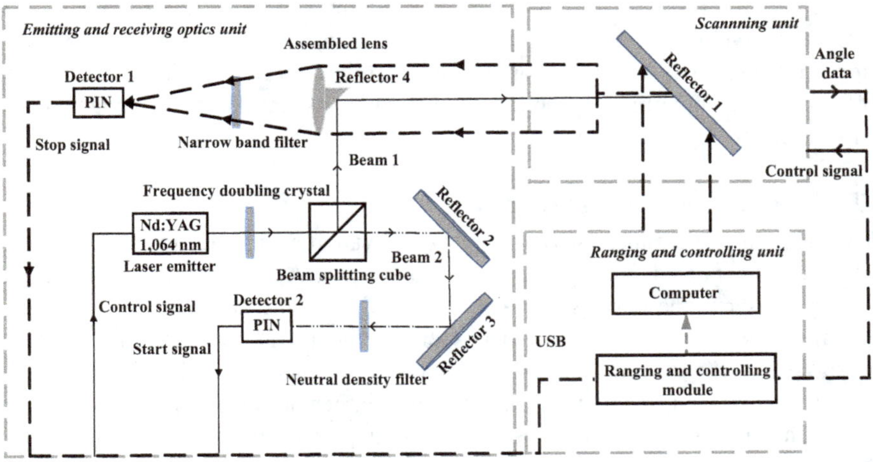

Figure 6.1 Terrestrial Laser scanners system

provides a stronger reflected pulse, which is desirable because the returned pulse becomes very weak at long distance. Mostly based on their availability and size, solid-state Lasers and semiconductor Lasers are the two main types of Laser source, which are generally used.

The solid-state Laser uses crystal as a lasing material and two reflecting mirrors, powered directly from a diode Laser or a flash lamp. Semiconductor Laser is also referred to as diode Laser, commonly using gallium arsenide (GaAs) structure as a gain medium and are powered optically by an electrical current. Semiconductor Laser delivers a Laser pulse quality that is inferior to solid-state Laser and hence requires a lens to function as a Laser beam.

The wavelength of Laser should be selected according to the reflection characteristics of the material. There are three main wavelength regions: around 490–750 nm (visible red), 950–1,064 nm (near-infrared radiation (IR)) and 1,550 nm (IR). The most used wavelength is near-IR for pulsed Lasers. This wavelength is safe because it is not sensitive to human eyes.

To design a TLS optical system, three Laser pulse parameters are considered: pulse width, pulse energy and pulse repetition frequency. In general, a pulse repetition frequency of several to tens of kHz is needed, which is more than enough for a very stable single-point measurement. Pulse width is commonly in nanoseconds for centimetre level accuracy. These three parameters are changeable due to different ranging distance and accuracy requirements. Given these values, a maximum allowed output power can be calculated.

6.2.2 Scanning unit

Optical scanning elements are used to implement TLS, which can sample 3D information about target surfaces. Laser pulse repetition frequency clearly conditions the consecutive positions of Laser footprints on scanning paths. However, the fastest TLS scanners may have small pupils, which reduce sensor dimensions but restrict the levels of received power [2]. Among the variety of Laser scanners that have been developed, several types of the scanner are introduced, including oscillatory mirrors (galvanometer or resonant, micro-electro-mechanical systems (MEMS)), rotating mirrors (monogon, prismatic, pyramidal or irregular), holographic, electro-optical and acousto-optic [2].

6.2.2.1 Galvanometer scanner

The galvanometer scanner uses a one-mirror galvanometer device to collect a Laser pulse that is reflected from the surface of an object. As shown in Figure 6.2, the reflected Laser stripe plane scans across the object's surface, with the rotation of the galvanometer device, making it possible to reconstruct 3D information about the surface of the object. It is highly suitable for industrial applications and less sensitive to measurement position and environmental conditions. Moreover, a galvanometer scanner has an adjustable oscillating frequency, good scan rates and low inertia [3]. However, the galvanometer scanner has a low duty cycle because of mechanical problems at high frequency [3].

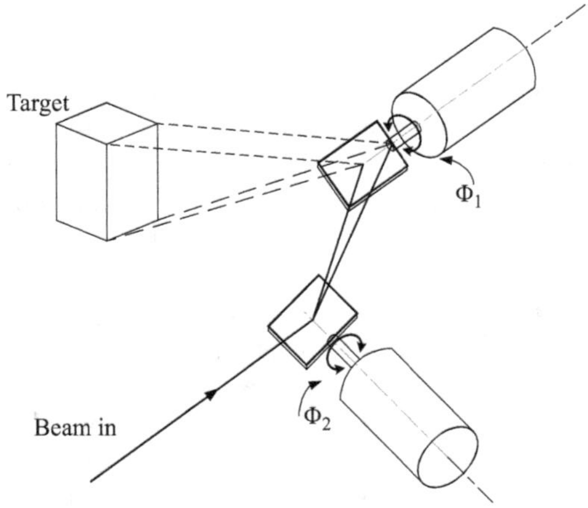

Figure 6.2 Concept image of galvanometer scanner

6.2.2.2 Micro-electro mechanical systems

Micro-electro mechanical systems (MEMS) is made from silicon material, and has good electrical and mechanical properties. MEMS can drive a reflective mirror redirecting an incident beam toward a specific direction, allow miniaturisation, with low power consumption and has superior performance at resonant high frequency actuation [4]. The major disadvantage is its sinusoidal motion and sensitivity to external, as well as self-induced perturbations.

6.2.2.3 Polygonal mirror scanners

A polygonal mirror (PM) scanner incorporates a rotating optical element with three or more reflective facets, which are usually metal mirrors [4]. In most applications, a PM scanner can produce a second axis to beam the steering of pairing with other scanning methods. PM scanners have a major drawback in applications that they require high scan speed, limiting the use of low-speed applications: these polygons typically cost significantly more than regular polygons because their asymmetry precludes cost saving from stacking. Another disadvantage of these scanners is the inherent dynamic imbalance of the polygon during rotation. There are many types of scanning mirrors, but most of them can be categorised in the following.

Prismatic polygonal scanning mirror is the most commonly used. Its mirrored surfaces are parallel to the axis of rotation. Their advantage is lower cost than other scanning mirrors, with less handling, more consistency and less machining time. Their disadvantage is that errors can result from their configurations.

In a pyramidal polygonal scanning mirror, all mirrored surfaces take the same angle with respect to the rotation axis, usually 45°. Pyramidal polygons are primarily used with an input beam parallel to the rotation axis. A given mechanical rotation angle results in an optical scan angle of the prismatic polygon, which is twice of the pyramidal polygon. Therefore, pyramidal polygonal scanners can increase the cloud point density for a given polygon rotation speed. This type of scanner is expensive to manufacture because it cannot stack mirrors together at the same time as with regular prismatic polygons.

An irregular polygonal scanning mirror is defined as one with many plane facets at a variety of angles with respect to the rotation axis. The unique feature of this type of mirror is that it can produce a faster output without a second axis of motion. If the facets are at different angles, the resulting output scans do not superimpose.

6.2.2.4 Monogon scanner

A monogon scanner has only one facet centred on the rotation axis. In a typical TLS system with a monogon, a Laser is directed toward the monogon along the rotation axis and the output sweeps a circle on an internal-drum as the scanner rotates. This type of scan system can produce very accurate spot placement and high resolution and it has many applications in the market. An example of a monogon scanner is shown in Figure 6.3. In our lab, a monogon scanner is used in a LiDAR system [1].

6.2.3 Receiver

An optical system was used as a receiver to collect back-scattered light. LiDAR uses refractive telescopes, or catoptric systems, to collect the reflected light [5]. Depending on a LiDAR application, various receiving optical systems are used. Telescopes should be designed to vary with distance and Laser beam diameter, shape and divergence, the telescope's imaging properties (focal-length-to-diameter ratio), the receiver field of view (FOV) and the location of emitter and receiver optical axes relative to each other. In coaxial systems, a Laser beam is emitted along the optical axis of the receiver telescope. In biaxial systems, the optical axes are spatially separated by at least one radius of the telescope mirror and the Laser beam enters the telescope's FOV from one side.

Although a Laser beam is highly collimated, the telescope's divergence is often reduced further by a beam expansion to values in the order of 100 μrad. FOV

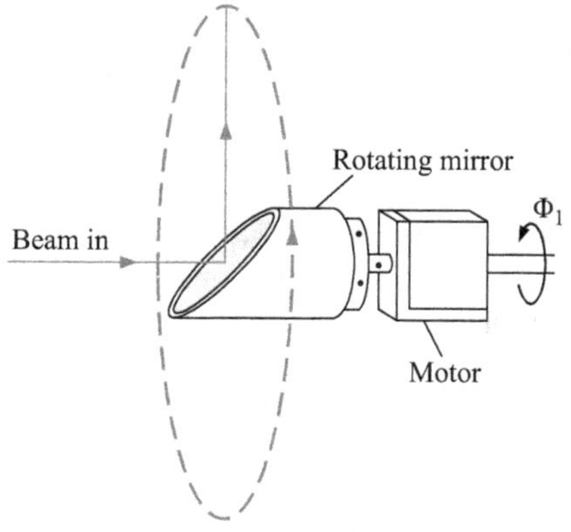

Figure 6.3 Monogon scanner

of the telescope can then be as low as a few hundred μrad. As shown in Figure 6.1, the Laser and telescope are expected to remain in alignment, and that the built-in cooling system would reduce the heat produced by the powerful Laser [1]. The emitted Laser pulses are split into three beams using the beam splitting cube and the beam splitter, which are the emitted pulse toward the target (beam 1), the emitted pulse as start signal (beam 2). The Laser beam reflected from the targets is gathered by the assembled lens with a diameter of 35 mm. The electrical signal produced by the photoelectric detector is fed to the ranging and controlling unit as a stop signal. The electrical signal converted from beam 2 is fed to the ranging and controlling unit as a start signal.

6.2.4 Detector

A detector is used to convert optical energy into an electrical signal. Photon detectors make use of interaction of a quantum of light energy with electrons in the detector material to generate free electrons, which are collected to form a measurable current signal proportional to the intensity of the incoming light pulse. Detectors may be divided into several types. Photoconductive and photovoltaic detectors are commonly used in circuits, in which there is a load resistance in series with the detector. The output is read as a change in the voltage drop across the resistor. Photoconductive detectors contain PIN (p-type, intrinsic, n-type for short) and avalanche photo-diodes (APD) [5].

A PIN photo-diode has a layer of intrinsic material between a thin layer of p-type semiconductor and a thicker layer of an n-type semiconductor, which normally constitutes a photo-diode. APD is a p-n junction photo-detector operated at a high reverse-bias voltage so that charges are rapidly swept from the depletion region. The applied voltage is chosen according to the breakdown voltage of the intrinsic material. It can handle high electric fields generated in the depletion region. It is possible to capture the distance of a far target more effectively than with a PIN photo-diode. A large current may be produced when light strikes the diode.

APD has an internal gain mechanism with a multiplication factor (i.e. gain), which depends on the electric field applied across the avalanche layer, providing a higher internal photo current. Therefore, when the reverse bias voltage is lower than the breakdown voltage of the APD, the gain will increase with the increase of the reverse bias voltage. The multiplication factor can be written as

$$M = \frac{1}{1 - (U_{\mathrm{rev}}/U_{\mathrm{brd}})^n} \tag{6.1}$$

where M is the multiplication factor; U_{rev} and U_{brd} represent actual reverse bias voltage (volt) and breakdown voltage (volt), respectively, and n is a constant related to the material of APD.

Breakdown is an important phenomenon that occurs in high electric fields. Although it is desirable to increase the multiplication factor in an APD, it is restricted by the breakdown voltage. Because shot noise is a main source of noises, particular at a short-range detection, which is amplified by the multiplication factor in an APD. It is essential to determine the multiplication factor for optimal APD gain. An APD detector includes a temperature compensation circuit, a reverse bias voltage circuit and a pre-amplifier circuit.

The main function of a temperature compensation circuit is to maintain APD operation constant with the optimal multiplication factor. Because the APD multiplication factor is temperature-dependent, the multiplication factor at a certain reverse bias voltage becomes smaller with the increase in temperature. The breakdown voltage of an APD is quite sensitive to change in temperature, leading to variation in the multiplication factor. Thus, to obtain a constant signal output, it is necessary either to adjust the reverse bias voltage according to the change in temperature, or to keep the temperature constant.

A reverse bias voltage circuit provides an actual reverse bias voltage, ensuring that the multiplication factor of an APD is high enough to detect a weak Laser echo signal. However, because a small change in the reverse bias voltage can cause fluctuation in the multiplication factor, the reverse bias voltage has a strict requirement of low noise and small voltage ripple.

The purpose of the pre-amplifier circuit is to amplify the APD output. The output of useful signals can easily be contaminated by interference of ambient light and circuit noise. Although the signal of an APD output is amplified using the multiplication factor, the output voltage signal needs further amplification to provide a superior echo signal of ranging.

6.2.5 Ranging system

TOF-based LiDAR measures the distance between the Laser transmitter and the target, by accurately recording duration between start/stop signals of TOF, as shown in Figure 6.4. The time data measured by a ranging system can be processed by a field-programmable gate array (FPGA) and sent to a terminal PC. Combining the range data with the information of Laser beam steering, 3D point cloud can be generated in an arbitrary Cartesian mapping system. The methods of time interval measurement can be divided into two categories: time-to-digital converter

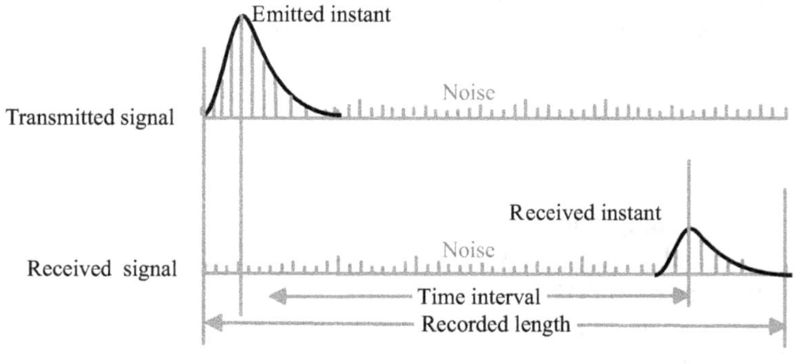

Figure 6.4 Waveform based time determination and distance measurement between start/stop signals

(TDC)-based measurement and analogue-to-digital converter (ADC)-based measurement. With TDC ranging measurement the time between trigger signals is counted. With ADC ranging measurement the time interval between start/stop signals is estimated from the echo waveform using an appropriate algorithm.

6.2.5.1 TDC-based ranging system

A TDC-based ranging system mainly solves problems concerning LiDAR ranging measurement in the analogue domain. It requires a high-performance analogue-discrete return system with pulses of fast rise time and a high-receiving bandwidth. The peak power of the returned pulse can be written in the following form:

$$P_r = \frac{E_T}{p_w} \rho \frac{A_r}{\Omega R^2} \eta_a^2 \eta_o \tag{6.2}$$

where E_T refers to energy per transmitted pulse (J), p_w is the FWHM pulse width (FWHM, full width at half maximum) (ns), ρ is target reflectance, Ω is the scattering angle of the targets (sr), A_r equals the area of the assembled lens (m^2), R represents the target range (m), η_a is one-way atmospheric transmission (also, atmospheric transparency) and η_o is the system transmission.

According to the LiDAR equation (6.2), the pulse amplitude is inversely proportional to the square of the distance and proportional to the target reflectance. The returned pulse amplitude varies widely in a large dynamic range. Due to noise interference and target uncertain reflectivity, the analogue discrete return system cannot meet the requirements for an input signal, resulting in shape variation in the echo waveform, poor ranging accuracy and precision. As the main systematic error, the walk error is caused by variation in the amplitude and shape of input pulse. Timing jitters are the timing deviation caused by statistical fluctuations in the shape of the detected pulses as the largest random error [5,6]. To improve ranging accuracy and precision, the main errors in the analogue discrete return, that is walk errors and timing jitter, have to be eliminated.

Because of its insensitivity to pulse amplitude, a TDC-based circuit, combined with a constant fraction discriminator (CFD), is usually used to reduce the walk error [7]. However, because of the increased input overdrive in the comparator, an extra walk-error can be generated by the comparator of the CFD. In addition, the distorted signal detector saturation also introduces errors at a short detection distance, causing an inaccurate time discriminator. Before the input pulse is sent to the timing discriminators, its amplitude is stabilised to a fixed range by auto gain control (AGC). A schematic diagram of AGC is shown in Figure 6.5 [1].

A peak holding circuit is designed to maintain the pulse peak for a certain period of time, to easily sample the input pulse with an ADC. A holding capacitor and a Schottky diode are used to complete peak holding (highly advantageous in trans-conductance peak holding). A low-performance ADC can be used to detect and sample the output of the peak holding circuit. Selection of PGA depends on the range of returned pulse amplitude, which must be calculated. A timing discriminator is used to convert the analogue electrical pulses to the emitter-coupled logic (ECL) pulses at the right instant, as shown in Figure 6.6. The output of PGA is

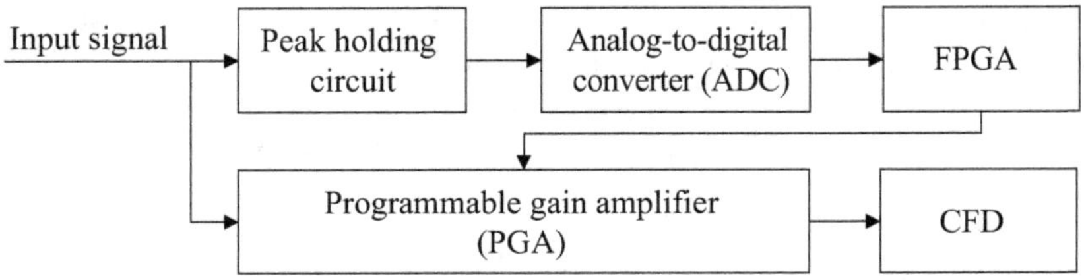

Figure 6.5 Auto gain control (AGC) to stabilise echo amplitude

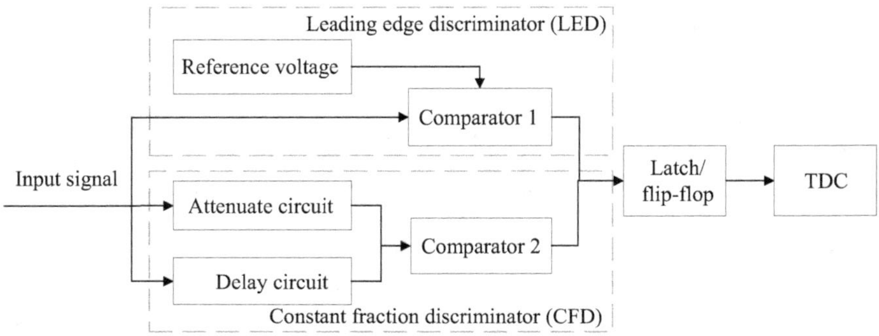

Figure 6.6 Timing discriminators including a constant fraction discriminator

fed into a timing discriminator, which is composed of a CFD and a leading-edge discriminator (LED). The LED is used to avoid false triggers by filtering out noise below the reference voltage. The CFD is used to reduce the walk error caused by the pulse amplitude and the rise time.

A zero-crossing point can be obtained, when the AGC output signal is fed to a delay circuit and an attenuation circuit to produce a logic pulse to trigger the TDC. The zero-crossing points are regarded as the returned moment, and so the measured time interval can be calculated from the difference between the emitted and the returned moments. Theoretically, the zero-crossing point is independent on the pulse amplitude. The delayed time (t_d) and the attenuation fraction (k), which are two most significant parameters of CFD, strongly impact acquisition of the right timing point. The delayed time t_d can be calculated by

$$t_d = t_r(1 - k) \tag{6.3}$$

In literature [1], it is implied that the attenuation fraction (k) can be adjusted in the range from 0.2 to 0.5, mainly depending on the pulse rise time (t_r). Otherwise, it can significantly deteriorate range precision.

6.2.5.2 ADC-based ranging system

Using a waveform-digitising system, a high-speed ADC-based LiDAR has several advantages in performance over those based on a TDC [8]. A signal from the detector is converted to numbers for further analysis. The sampling data are uploaded to a digital signal processor (DSP), or host PC via PXIe Bus or PCIe Bus, which executes appropriate algorithms to analyse the digitised waveform and estimate the location of the target in the return signal. At evenly spaced intervals, which can be determined by a clock, an ADC is used to digitise the voltage at the input and then stored the measured value in a high-speed memory. The shorter the required interval between measurements, the faster the digitising rate and the higher the signal frequency to be resolved. The range measurement accuracy is directly related to the ADC sampling rate and time discriminator algorithms.

The ADC in a digitiser can measure voltage in a certain fixed range by dividing that range into several equally spaced intervals. The width of each interval is the digitiser voltage range divided by the total number of intervals. For example, an N-bit digitiser has a $2^N - 1$ voltage interval. The width of the voltage interval represents the minimum voltage difference that can be resolved. The ADC resolution is defined as the smallest incremental voltage that can be recognized. Week echo signals can be well sampled with high-resolution ADC. Many digitisers have a programmable amplifier in front of the ADC to match the size of the signal to the voltage range of the ADC. Matching the size of the signal to the full ADC range is important in LiDAR systems, to satisfy the dynamic range of the signal.

When the echo pulse waveform is recorded by a digitiser, the time instance must be determined by algorithms. Here, five time-discrimination methods are introduced: (1) a leading-edge discriminator, (2) a peak discriminator, (3) a centre of gravity discriminator, (4) an inflection discriminator (i.e. zero crossing of the second derivative) and (5) a constant fraction discriminator, as shown in Figure 6.7.

To estimate the time instance located on the digitised waveform as introduced in the literature [8], the Gaussian function is used to fit the digitised pulse waveform as

$$f(t; A_p, \tau_p, pw) = A_p \exp\left\{-4 \ln 2\left(\frac{t-\tau_p}{pw}\right)^2\right\} \tag{6.4}$$

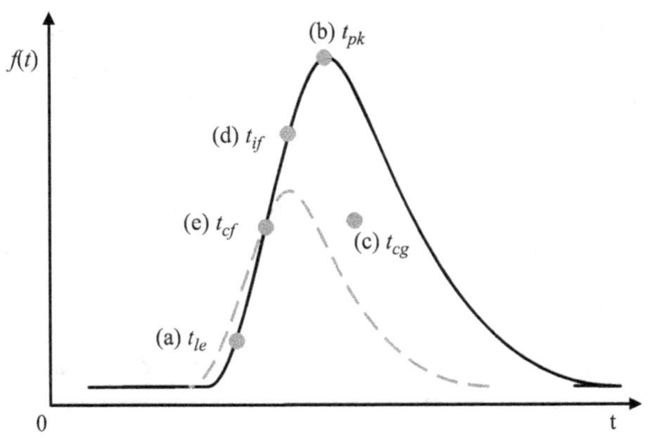

Figure 6.7 Time discrimination methods: (a) leading edge, (b) peak, (c) centre of gravity, (d) inflection and (e) constant fraction

where t is the flying time of the pulse, A_p is the pulse amplitude, τ_p is the sampling position of pulse peak and pw is the FWHM pulse width. The transmitted and received times are determined based on the Gaussian function, using these five methods as follows.

1. *Leading edge discriminator.* A leading-edge discriminator selects the determining time using a leading edge that exceeds a fixed threshold (V_{th}). Its disadvantage is that the determining time accuracy is affected by variations in pulse width and pulse amplitude, resulting in the walk error in range measurement. As shown as point (a) in Figure 6.7, the time instant (t_{le}) of the received pulse is written as

$$t_{le} = \tau_p - \frac{pw}{2\sqrt{\ln 2}}\sqrt{\ln\left(\frac{A_p}{V_{th}}\right)} \tag{6.5}$$

2. *Peak discriminator.* Peak detection is independent on variation in pulse width and amplitude, but it is highly influenced by noise and local spikes of pulse waveform. It estimates the maximum position of the received pulse, as

$$t_{pk} = t_{\max(f(t))} \tag{6.6}$$

As shown as point (b) in Figure 6.7, the time instant (t_{pk}) of Gaussian pulse of the received waveform is written as

$$t_{pk} = \tau_p \tag{6.7}$$

3. *Centre of gravity discriminator.* As shown as point (c) in Figure 6.7, the centre of gravity of the received waveform t_{cg} is used to calculate energy centre of the pulse waveform:

$$t_{cg} = \frac{\int_{t_1}^{t_2} tf(t)dt}{\int_{t_1}^{t_2} f(t)dt} \tag{6.8}$$

where t_1 and t_2 represent integration boundaries on the pulse.

When integration boundaries are symmetrical for Gaussian pulse, the time instant (t_{cg}) can be described as

$$t_{cg} = \tau_p \tag{6.9}$$

4. *Inflection discriminator.* By calculating the zero crossing of the second derivative, the inflection discriminator method determines time instance by

$$t_{if} = t_{\frac{\partial^2 f(t)}{\partial t^2}=0,\text{left}} \tag{6.10}$$

It means the left point of zero crossing of the second derivative. As shown as point (d) in Figure 6.7, the time instant (t_{if}) of the received waveform can be written as

$$t_{if} = \tau_p - \frac{pw}{2\sqrt{2\ln 2}} \tag{6.11}$$

5. *Constant fraction discriminator*. Like the theoretical definition in the analogue ranging system, the method of constant fraction is insensitive to variation in the input amplitude. As shown as point (e) in Figure 6.7, the time instant (t_{cf}) of the received waveform is described by

$$t_{cf} = \tau_p + \frac{t_d}{2} + \frac{pw^2 \ln k}{t_d \cdot 8 \ln 2} \tag{6.12}$$

where k is the attenuation fraction ($0 < k < 1$) and t_d is the delay time. t_{cf} locates on the rising edge of a pulse. More details are presented for the definition of constant fraction in [8,9].

6.2.6 Software

6.2.6.1 FPGA controller system

A LiDAR ranging system can be designed based on AGC and timing discriminators. The system construction is shown in Figure 6.8. In the first part, a controlling unit consists of an FPGA, its peripheral circuits and a motor control board. The second part measures the time interval, which is completed by a TDC. The third and fourth parts, containing AGC and timing discriminators, support the key algorithms running in the time pickoff circuit. The specific design of each part is described as follows [1].

In the controlling unit, the FPGA played an important role in configuring the parameters of the Laser emitter, the motor, the AGC and TDC, and the receiving feedback signals from the AGC. The repetition frequency and duty ratio of the Laser emitted pulse were controlled by the FPGA. In addition, the frequency and speed of the stepper motor (one of the scanning devices) were controlled by the FPGA and a motor-driven device. The time interval is calculated by the TDC. Measurement processes of the TDC are shown in Figure 6.9.

The main methods of time interval measurement using a TDC-based ranging system include electronic counting, analogue interpolation and time-to-amplitude conversion. But all analogue devices can result in inevitable nonlinear errors, and they are integrated into the system with difficulty. The LiDAR system applies a TDC chip to measure the time interval for high integration and high linearity. The

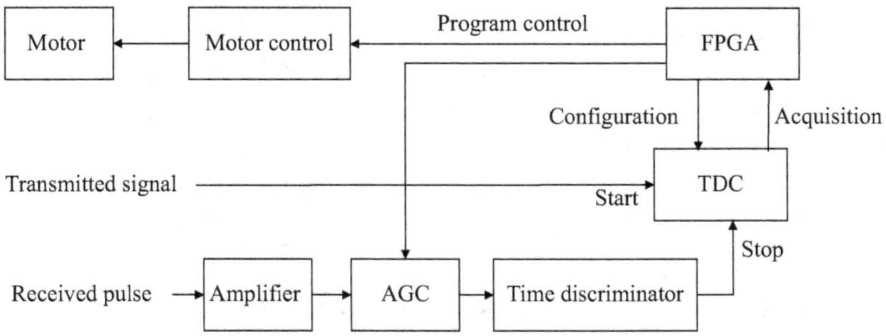

Figure 6.8 Construction of LiDAR ranging system

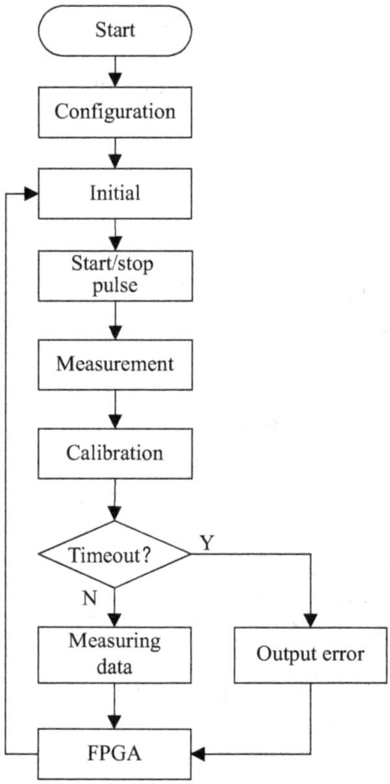

Figure 6.9 Measurement process of TDC ranging system

starting signal is determined by the photodetector in the Laser emitting channel. The stop signal is provided by the returned pulse in the receiving channel. The parameter configuration contains the options of the measurement mode, the reference clock, Laser shot numbers, calibration, data processing, and input sensitivity and interrupt behaviour. After calibration, the results of time measurement were transferred to FPGA.

6.2.6.2 Data structure

As the control centre, the FPGA also gathers, processes and uploads data to a computer via a USB connector. The format of the data package is shown in Figure 6.10. The first part includes the angle and Laser shot number. The second part is time interval data obtained by the TDC. By adopting a ping-pong buffering strategy, the data package is stored in the FIFO of the FPGA, which speeds data transmission and data processing and meets the real-time requirement.

6.2.6.3 Point cloud data processing and displaying

The main functions of the upper computer are to set up the parameters of the scanning system, analyse the data uploaded by the lower computer and calculate

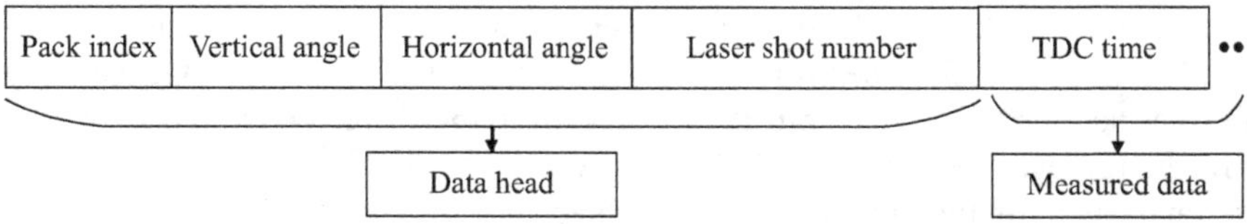

Figure 6.10 LiDAR data structure definition

the 3D coordinates of point cloud of measured targets. A USB communication driver is used to spread data, which are uploaded and stored in a buffer, and finally saved in a file. In the monitoring zone of the software, the operation parameters and measured data are displayed, including the data header information of each cycle. It is a sign of the packet index, horizontal angle and speed, and of vertical angle and velocity, and the TDC time data, which can monitor the running state system in real time. There are two modes to read data in the software area: the software system, which reads data from the USB, or reads offline BIN file data to form the 3D point clouds.

6.3 TLS errors and self-calibration

TLS is a complex tool with many moving parts, which introduces systematic and random errors into the point-clouds of targets. Systematic errors can be repeated in defined conditions, and to some extent calibration can compensate for these errors. Because random errors are not repeatable, they cannot be removed by calibration. Because the systematic errors can be determined directly, some of them can be recovered individually while others are not separable and hence are determined as groups. The systematic errors of LiDAR can be categorised into four groups: (1) ranging errors, (2) scan angle errors, (3) mounting angle errors and (4) INS errors. The final vertical and horizontal accuracy, achieved in the cloud point, is affected by several sources during the process of LiDAR data capture. The main errors are position errors due to GPS, INS and GPS-INS integration. The angle error occurs when the Laser instrument is not perfectly aligned with the platform roll, pitch or yaw axis. The ranging error is due to a time measurement error, wrong atmospheric correction and ambiguities in target surface, resulting in walk errors.

6.3.1 Angle errors model

By combining the measured distance with the 3D scanning mechanism, the point-clouds of targets are obtained. As one of the main systematic errors, the mounting angle error of TLS is estimated by modelling, calibrating and correcting to improve the scanner's manufacturing and positional accuracy of point-clouds. Because it can lower the cost involved in the calibration process, researchers, engineers in industry and academia have explored self-calibration methods for improving the positional accuracy of a TLS without a reference instrument or calibrated artefacts.

TLS self-calibration is divided into two steps: the establishment of error models and parameter estimation of error models [10]. In a scanning unit of the lab-scale TLS, a rotary table and a 45° rotating mirror installed on a mirror motor rotate the incident Laser beam in horizontal and vertical directions, respectively. The rotating mirror is an elliptical mirror with a short diameter of 50 mm. The mirror motor is a direct current (DC) motor with a speed of 720°/second. The main rotational shaft of the rotary table is defined as the vertical axis (Z-axis). The incident Laser beam is along the horizontal direction (i.e. the mirror motor shaft, X-axis). Y-axis is determined according to the right-hand rule. Thus, the angle between the rotating mirror surface and the incident Laser beam should be 45°, which makes the reflective Laser beam rotating in the YOZ plane under ideal conditions. A schematic diagram of this scanning mode is shown in Figure 6.11, which is composed of a rotary table and a 45° rotating mirror installed on a mirror motor. A panoramic scanning can be carried out when the incident Laser beam rotates in horizontal and vertical directions.

As shown in Figure 6.11, the middle point of the rotating mirror is defined as the origin O, that is the intersection of the rotating mirror and the incident Laser in a perfect case. The positive direction of X-axis is coincident with the incident direction of Laser. The positive direction of Z-axis pointed up. Y-axis is determined according to the right-hand rule. A target point P in the polar coordinate is defined as follows. The length of OP is the distance L. The angle between the OP projection line on the XOY plane and X-axis is the horizontal angle ϕ_h, varying from 0° to 360° counter-clockwise from the positive direction of the X-axis. The angle between the projection line and OP is the vertical angle ϕ_v, varying from –90° to 90°.

The established angle measurement error model included the beam tilt error (with parameters β_1 and β_2), transit tilt error (with parameter γ), standing tilt error (with parameters α_1 and α_2), mirror tilt error (with parameter θ) and the vertical index offset error (with parameter δ), deduced from the scanning mechanism of the

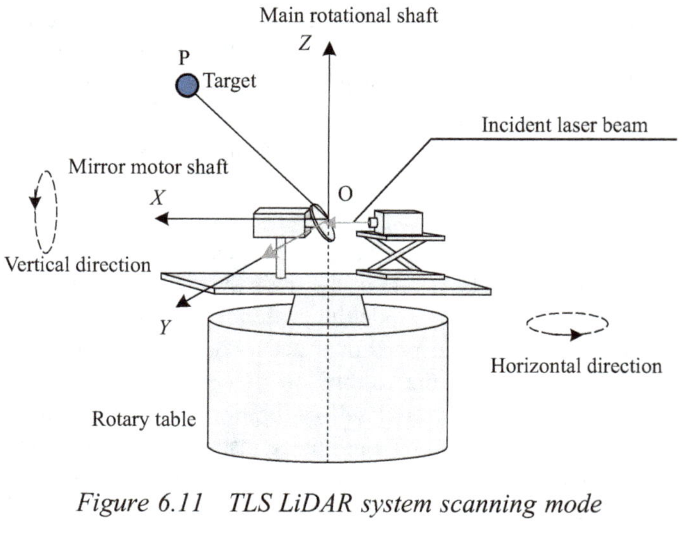

Figure 6.11 TLS LiDAR system scanning mode

Table 6.1 Angle measurement error model parameters of lab-scale TLS

Parameter	Description	Two-face sensitivity
β_1	Beam tilt error along Z	Horizontal angle direction
β_2	Beam tilt error along Y	Vertical angle direction
γ	Transit tilt error	Horizontal angle direction
α_1	Standing tilt error along Y	None
α_2	Standing tilt error along X	None
θ	Mirror tilt error	Horizontal angle direction
δ	Vertical index offset error	Vertical angle direction

lab-scale TLS system [10,11]. A list of parameters in the angle measurement error model is provided in Table 6.1.

In the modelling process, the axis eccentricity caused by non-intersecting between the horizontal axis and the vertical axis is not considered. After integrating five types of mounting angle errors, the angle measurement error model ($\Delta\varphi_h$ for horizontal angle and $\Delta\varphi_v$ for vertical angle) involving seven parameters of the lab-scale TLS system can be expressed as

$$\begin{aligned}\Delta\varphi_h = {} & -\beta_2 + (2\gamma - \beta_1)\tan\varphi_v + \alpha_1\cos\varphi_h\tan\varphi_v \\ & -\alpha_2\sin\varphi_h\tan\varphi_v - \frac{2\theta}{\cos\varphi_v}\end{aligned} \tag{6.13}$$

$$\Delta\varphi_v = (\beta_1 - \gamma)\cos\varphi_v - \beta_2\sin\varphi_v - \alpha_1\sin\varphi_h - \alpha_2\cos\varphi_h + \delta \tag{6.14}$$

6.3.2 Self-calibration methods

Three existing self-calibration methods for parameter estimation are introduced here: (1) network method, (2) length-consistency method and (3) two-face method [10,11]. The network method measures different coordinates for one fixed target at several different physical TLS locations. To estimate the parameters of self-calibration model, the coordinates of the different locations should be transferred to one when the difference between the transferred coordinates is minimised. However, a major disadvantage is that the location transformation can introduce unknown rotation and translation parameters into calculation. Hence, the network algorithm is suitable for estimating a small number of model parameters.

The length-consistency method does not need rotation and translation parameters for the calculation. It derives the similarity criterion of distance between any pair of two targets from the different TLS locations be equal. However, the length-consistency method is not suitable for sparse point-cloud that can reduce range accuracy. To avoid measurement error generated from multiple TLS locations, a two-face method can be used to estimate the parameter with a higher speed. In a full panoramic scan, the same target would be scanned twice, that can be used to determine parameters directly from the difference between front-face and back-face measurements of the targets. The disadvantage of the two-face method is that some of the model parameters cannot be estimated due to lack of two-face sensitivity.

6.3.2.1 Two-face method

Because it has a simple optimisation process with fast speed, a two-face method is chosen to calibrate the lab-scale TLS system. The same target can be scanned twice during the one circle of the rotating mirror on the stationary rotary table, assuming that it is counter-clockwise, around the motor shaft. As the vertical angle ranged from –90° to 90°, the point-cloud data of the targets are obtained in the front-face mode. As the rotating mirror spins past 90° towards 270°, the point-cloud data of the targets are obtained in the back-face mode.

For example, the self-calibration model, where the vertical angle φ_v in the front-face mode described in the angle measurement error model varies from –90° to 90°. Therefore, $\pi - \varphi_v$ and $\pi + \varphi_h$ in the back-face mode replace φ_v and φ_h in the front-face mode. A coefficient k_{tf} is added to the original angle measurement error model. The self-calibration model can be rewritten as

$$\Delta\varphi_h = -\beta_2 + k_{tf}(2\gamma - \beta_1)\tan\varphi_v + \alpha_1 \cos\varphi_h \tan\varphi_v \\ -\alpha_2 \sin\varphi_h \tan\varphi_v - k_{tf}\frac{2\theta}{\cos\varphi_v} \quad (6.15)$$

$$\Delta\varphi_v = (\beta_1 - \gamma)\cos\varphi_v - k_{tf}\beta_2 \sin\varphi_v - \alpha_1 \sin\varphi_h \\ -\alpha_2 \cos\varphi_h + k_{tf}\delta \quad (6.16)$$

Theoretically, the corrected angles obtained from the front- and back-face mode should be equal.

$$\varphi_{fh,i,j} - \varphi_{bh,i,j} = (2\gamma - \beta_1)\left(\tan\varphi_{fv,i,j} + \tan\varphi_{bv,i,j}\right) \\ -2\theta\left(\frac{1}{\cos\varphi_{fv,i,j}} + \frac{1}{\cos\varphi_{bv,i,j}}\right) \quad (6.17)$$

$$\varphi_{fv,i,j} - \varphi_{bv,i,j} = -\beta_2\left(\sin\varphi_{fv,i,j} + \sin\varphi_{bv,i,j}\right) + 2\delta \quad (6.18)$$

where $\varphi_{fh,i,j}$ is the measured horizontal angle of the jth target point in the ith location in front-face mode; $\varphi_{fv,i,j}$ is the vertical angle of the jth target point in the ith location in front-face mode; $\varphi_{bh,i,j}$ is the measured horizontal angle of the jth target point in the ith location in back-face mode; $\varphi_{bv,i,j}$ is the vertical angle of the jth target point in the ith location in back-face mode, where $i = 1, 2, \ldots, m; j = 1, 2, \ldots, n$ for n target points and m TLS locations. Five parameters β_1, β_2, γ, θ, δ can be estimated using $Q = 2 \times m \times n$ equation. The remaining two parameters α_1, α_2 are determined using the other methods.

6.3.2.2 Network method

The network method is used to estimate the model parameters using a data set obtained from n target points at m TLS locations, as shown in Figure 6.12. It can be outlined as follows:

1. TLS calibration is derived from point clouds in the positions of various fixed targets at different coordinates, which are placed in several different physical locations.

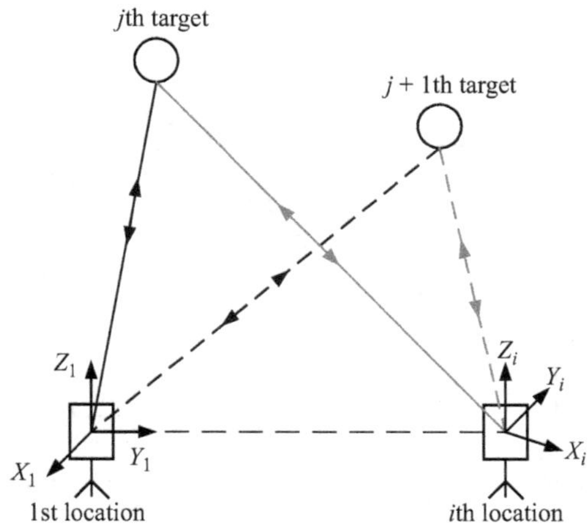

Figure 6.12 Self-calibration model using network method

2. Self-calibration model parameters are estimated by the difference of the overlap coordinates transferred from one location (*i*th location) to a given location (1st location).

   This optimisation problem is described as

$$\begin{pmatrix} x_j^1 \\ y_j^1 \\ z_j^1 \end{pmatrix} = R^i \begin{pmatrix} x_j^i \\ y_j^i \\ z_j^i \end{pmatrix} + T^i \tag{6.19}$$

where x_j^i, y_j^i, z_j^i are the calibrated Cartesian coordinates of the *j*th target points at the *i*th TLS location; x_j^1, y_j^1, z_j^1 are the calibrated Cartesian coordinates of the *j*th target points at the 1st TLS location; R^i and T^i are respectively the rotation matrix and the translation vector from the *i*th location to the 1st location.

According to the experimental condition, they can be described as

$$R^i = \begin{pmatrix} \cos\omega & -\sin\omega & 0 \\ \sin\omega & \cos\omega & 0 \\ 0 & 0 & 1 \end{pmatrix}, \quad T^i = \begin{pmatrix} x_0 \\ y_0 \\ 0 \end{pmatrix} \tag{6.20}$$

where ω is the rotation angle (*i*th coordinate system relative to 1st coordinate system) counter-clockwise around Z_i axis, and $(x_0, y_0, 0)^T$ are the Cartesian coordinates of origin point the *i*th coordinate system (O_1) in the 1st coordinate system ($O_1 - X_1Y_1Z_1$).

6.3.2.3 Length-consistency method

The length-consistency method is applicable to a dense point-cloud, as shown in Figure 6.13. It involves in measurement of a common set of *n* targets from *m* TLS

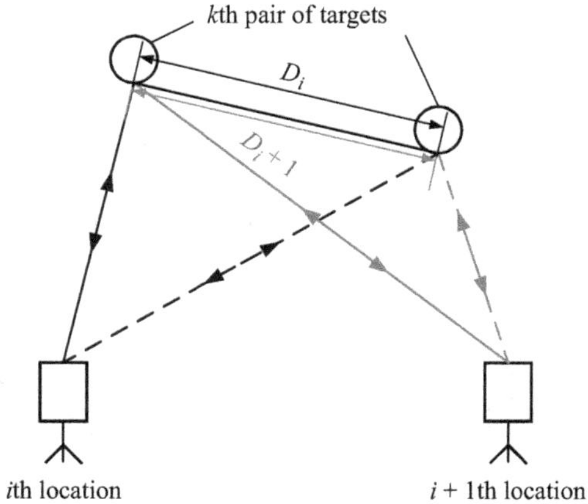

Figure 6.13 Self-calibration model using length-consistency method

locations. The optimising criterion of the length-consistency method is that the distances (i.e. the lengths) between any pair of two targets determined from the different TLS locations should be as close to each other as possible. This optimisation problem is described as

$$\min f(x) = \min \sum_{i=1}^{m} \sum_{k=1}^{n(n-1)/2} \left(D_i^k - \overline{D^k} \right)^2 \tag{6.21}$$

where D_i^k is the distance between the kth pair of targets from the set of n targets determined from the ith TLS location, and $\overline{D^k}$ is the mean of the distances of the same two targets scanned from all different location.

6.3.3 Feature points extraction

To solve self-calibration model parameters, it is necessary to extract some feature points from point-clouds of the designated area as the input from the self-calibration model. Feature points extraction methods are divided into two categories, based on their data source: (1) auxiliary target-based method and (2) background target-based method. The auxiliary target-based method extracts feature points from designated targets and deals with fewer point-clouds, with prior knowledge of the target size and location. The background target-based method obtains feature points from background point-clouds without auxiliary targets or prior knowledge, but the feature points extracted directly from the background point-cloud were refined to ally the two-face feature points as input from the self-calibration model.

Relevant object features (e.g. corner points or edges) are not directly recorded. Instead, they are modelled from the point clouds in a separate process. While it is possible to record the same object several times from different observation points, it is impossible to record the very same points in these repeated surveys. Therefore, deviations are only noticed after objects have been extracted from the point clouds and modelled. Even though the geometric properties of the objects are known, the deviation of single points from the surface of the object can be an indication of accuracy.

To autonomously estimate the parameters of the self-calibration model, several TLS point-clouds obtained from the front-face mode and back-face mode of different locations are determined as the input from the self-calibration model. Two categories of feature point determination methods are applied and compared in the following.

6.3.3.1 Determination of auxiliary target-based feature points

There are two properties of the auxiliary target-based method: first, the feature points are extracted from the geometrical characteristic of several auxiliary targets. Second, the target size and location are prior knowledge for determination of feature points, making the self-calibration process awkward and complicated. As shown in Figure 6.14 the point-cloud data of three planar targets are obtained from the panoramic scanning. The plane boundary is determined based on the margin of the point-cloud. Thus, four corner points of the target are extracted to further estimate the model parameters.

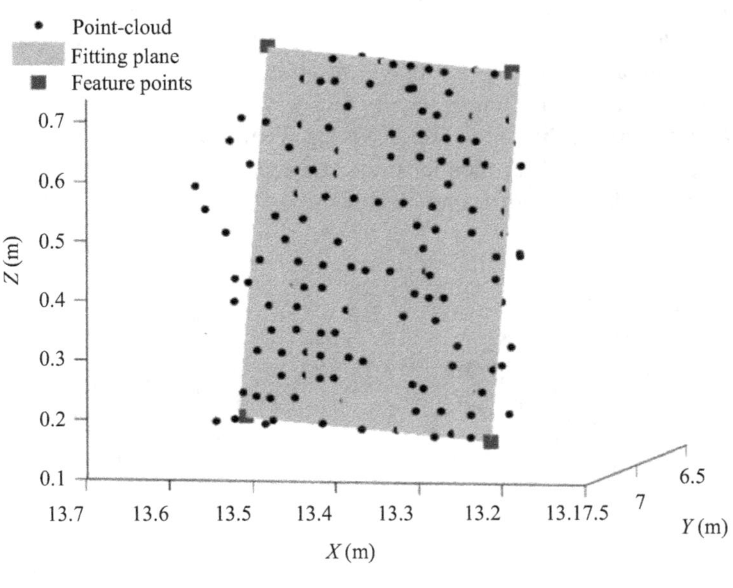

Figure 6.14 Feature points extraction (with corner points of fitting planes) using auxiliary target-based method

6.3.3.2 Determination of background target-based feature points

There are two properties of the background target-based autonomous method for feature points determination: (1) the feature points are extracted from point-clouds of the actual environmental targets, and (2) the entire procedure does not require prior knowledge and extra auxiliary targets, making the process in determination of feature points more autonomous.

The background target-based autonomous method for determination of feature points as introduced in the literature [11] is implemented using a key-point quality measure (i.e. the KQ algorithm) [12]. After the background target-based autonomous method is used, the features of one key-point in the front-face mode are crossed to match features of the key-point in the back-face mode one by one, and feature point pairs are eventually determined. The components of the autonomous self-calibration method include extraction of initial feature points using a KQ algorithm (see Figure 6.15(a)); refining the initial feature points to attain coarse feature points using a K-means algorithm (see Figure 6.15(b)) and optimising coarse feature points into fine feature points based on measurement precision. Finally, using these feature points, two-face and network methods are used to solve the self-calibration model parameters.

6.3.4 *Self-calibration results*

Experiments were carried out to verify the feasibility and availability of the self-calibration model in an empty hall with three targets, sized 0.64 m × 0.64 m (see Figure 6.16). A panoramic view of the point-cloud obtained by the lab-scale TLS for the self-calibration appears in Figure 6.17. Three planar targets were placed around the lab-scale TLS system, from 10 to 20 m.

The experimental procedure was as follows: The point-cloud data of three planar targets were obtained by TLS scanning from multiple TLS locations. The TLS scanned three rounds at each location, and each scan contained the front- and the back-face modes in a horizontal TLS scan.

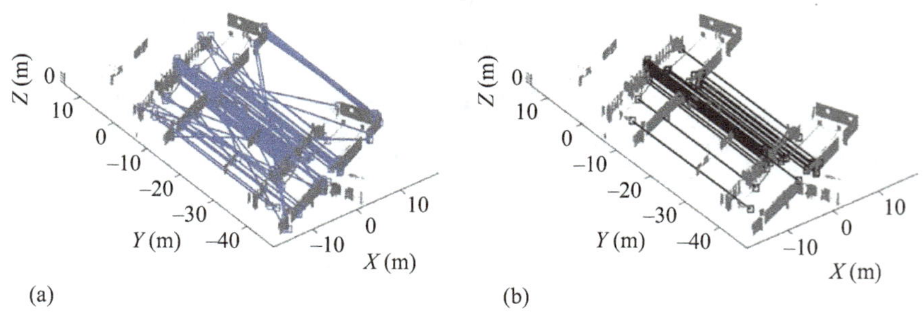

Figure 6.15 Feature points extraction using background target-based method: (a) KQ-based feature point extraction and (b) KQ and K-means-based feature point extraction

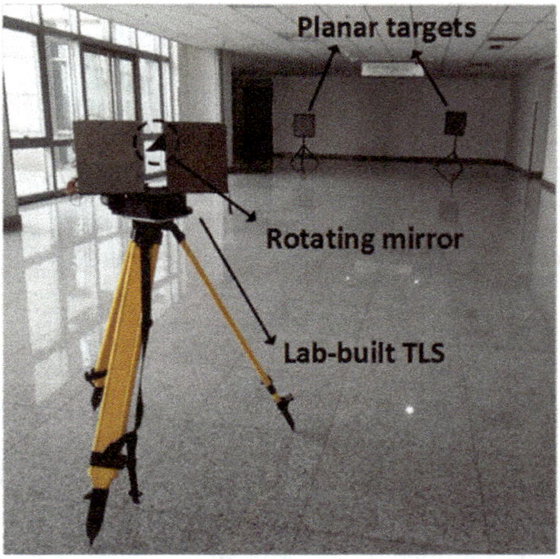

Figure 6.16 Self-calibration experiments using lab-scale TLS

To evaluate the measurement accuracy, two metrics σ_{φ_h} and σ_{φ_v} are used, the standard deviation of the half of difference between two-face measurements for horizontal and vertical angle, separately:

$$\sigma_{\varphi_h} = \frac{1}{mn}\sum_{i=1}^{m}\sum_{j=1}^{n}\left|\frac{\varphi_{fh,i,j} - \varphi_{bh,i,j}}{2}\right| \tag{6.22}$$

$$\sigma_{\varphi_v} = \frac{1}{mn}\sum_{i=1}^{m}\sum_{j=1}^{n}\left|\frac{\varphi_{fv,i,j} - \varphi_{bv,i,j}}{2}\right| \tag{6.23}$$

Take the angle measurement error model and lab-scale TLS system introduced above as an example, the measurement error of horizontal angle and vertical angle (σ_{ϕ_h} and σ_{ϕ_v}) is 0.3629° and 0.0954°, respectively. The measurement error of horizontal angle and vertical angle (σ_{ϕ_h} and σ_{ϕ_v}) after self-calibration is 0.0469° and 0.0258°, respectively. It is clear that the self-calibration model can improve the positional accuracy of the lab-scale TLS. Correct estimation can improve the positional accuracy of the measured point-cloud. There is still a slight difference between two-face point-cloud coordinates in the horizontal direction, indicating that some other systematic errors and random errors exist in the TLS scanning.

6.3.5 TLS ranging errors

6.3.5.1 Definitions of the performance indicators

Four main indicators demonstrate ranging performance: precision, accuracy, bias error and nonlinearity error [9]. Precision represents random errors in the system, as a scatter in the measured data. The accuracy is a systematic error, which can be calibrated by bias error and nonlinearity error by a look-up table. Nonlinearity error

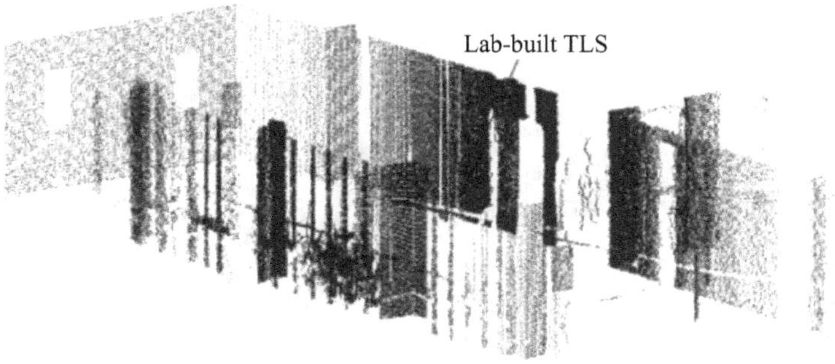

Figure 6.17 Point cloud scene obtained in self-calibration experiment

is a value of deviating from the calibrated linear relationship between the input (e.g. distance) and the output (e.g. accuracy), which is a type of systematic error.

Ranging precision is a quantity of how well a measurement can be made. A single-shot precision describes the uncertainty limited by random errors in the measurement without reference to a true range value. It is defined with statistical variance (σ-value) of the range measurement distribution and calculated from the standard deviation of measured data as

$$E_{\text{pre},j} = \sqrt{\frac{1}{N_j}\sum_{i=1}^{N_j}(R_{i,j} - \overline{R}_j)^2} \tag{6.24}$$

where $E_{\text{pre},j}$ is the ranging precision measured in the jth ranging position, N_j is the sampling number in the jth ranging position, $R_{i,j}$ is the ith measured data and $\overline{R}_j$ is the average value of measured data in the jth ranging position.

Accuracy indicates the largest total error in a certain measurement range and time period. Ranging accuracy is the degree of conformity of a measured quantity with its actual value, which is defined as

$$E_{\text{acu},j} = \frac{1}{N_j}\sum_{i=1}^{N_j}(R_{i,j} - R_{\text{true},j}) \tag{6.25}$$

where $E_{\text{acu},j}$ is the ranging accuracy measured in the jth ranging position, and $R_{\text{true},j}$ is obtained from the reference ranging system as the true value.

Bias error is the difference between the average value, which is equivalent to a correction that negates the systematic error, made by adjusting for the bias. The bias is expressed by

$$E_{\text{bias}} = \frac{1}{M}\sum_{j=1}^{M} E_{\text{acu},j} \tag{6.26}$$

where E_{bias} is the ranging bias, M indicates the total number of ranging positions at different distances and $E_{\text{acu},j}$ represents the ranging accuracy calculated from the jth ranging position.

Nonlinearity error defines the deviation of the measurement results from a fitting line between the smallest and largest measured distances, which is usually given the smallest root-mean-square (RMS) deviation for the individual measurement points, expressed as

$$E_{\text{nonlinearity}} = \sqrt{\frac{1}{M}\sum_{j=1}^{M}(\overline{R}_j - R_{\text{line}})^2} \tag{6.27}$$

where R_{line} is the fitting line drawn in coordinates between the smallest and largest measured distance.

In summary, the ranging precision indicates the measurement system ability. The ranging accuracy indicates how accurate the measurement system can achieve. The bias error can provide a correction to negate the systematic error, further enhancing the ranging accuracy. Nonlinearity error of ranging accuracy indicates complexity of establishing look-up table to correct the ranging errors varying with distance.

6.3.5.2 TDC-based ranging errors

The measured time interval is usually not fixed due to atmospheric absorption and scattering, uncertainty of surface reflectivity, noise interference and other factors, so that LiDAR measurement accuracy and precision are inferred. The main inaccuracy in the time-pick-off circuit includes walk errors, timing jitters, drift and nonlinearity. Walk error is defined as the timing error resulting from variation in the amplitude and shape of the input pulses as the main systematic error in LiDAR. Timing jitter refers to the timing deviation caused by noise as statistical fluctuations in the shape of the detected pulses, which is the largest random error. Drift and nonlinearity can be reduced to a negligible level by periodic calibration. Therefore, to reduce walk error and timing jitter it is important that the time-pick-off circuit is designed to guarantee an accurate time interval.

Because of its insensitivity to pulse amplitude, a CFD in a time-pick-off circuit is theoretically used to eliminate the walk error. However, two imperfections in a CFD must be overcome for actual application. First, the response time of the CFD comparator is limited by the over-drive of the comparator and the slope of the input pulses, which can generate extra walk error. Second, timing jitter in a CFD is influenced by the increase in the pulse amplitudes. Therefore, the walk error and the timing jitter are not sufficiently reduced if only the CFD is used in the time-pick-off circuit, especially when the input pulse amplitude is varied in a wide dynamic range.

Timing jitter is defined as the timing error caused by noise as statistical fluctuations existing in the shape of the detected pulses, as shown in Figure 6.18. Time jitter is used to indicate the high limit of the standard deviation of the range measurement data. It is calculated by the ratio of the RMS of noise to the slope of the returned pulse at the timing moment.

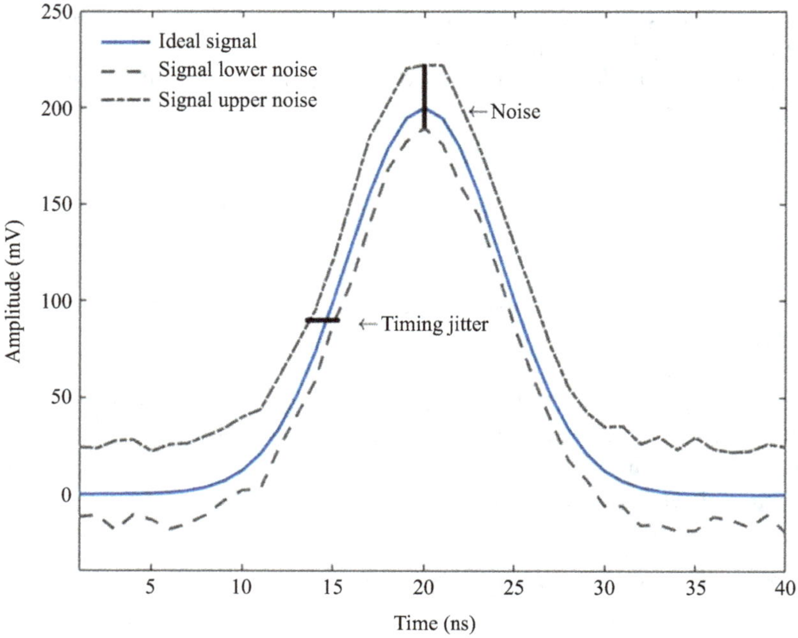

Figure 6.18 Definition of time jitter caused by noise as statistical fluctuations existing in the shape of the detected pulses

6.3.5.3 ADC-based ranging errors

The key analogue component in the front-end receiving path of such a system is the ADC used to digitise the narrow pulses reflected from near and/or distant objects. An ADC requires very fast sampling rates with high dynamic performance, large analogue input bandwidth and low power consumption. For example, a 2-nanosecond (ns) width Laser pulse needs sampling at 1 GS/s to Nyquist. A multiple ADC synchronisation feature may allow the sampling frequency to increase beyond 2 GS/s, if required, to meet the requirements of high accuracy LiDAR systems. The range measurement accuracy that can be achieved is directly related to the ADC sampling frequency. At 1 GS/s, the ADC clock period is 1 ns. In a 1 ns sampling instant, light travels 30 cm. For example, over a distance of 100 m, an initial measurement accuracy of ± 15 cm is achievable at 1 GS/s. With appropriate algorithms, the initial measurement accuracy can be improved. The error will increase as the sampling frequency is reduced. To measure the change of a 2 ns width pulse, the ADC should have a sampling rate in the order of 1 GHz or higher.

The shape of a received pulse also contains information regarding the properties of the target. The shape can only be determined by a significant over-sampling ratio, benefiting the digital domain in terms of processing gain. For most applications, the digital device used to capture and process data is an FPGA. Readily available off-the-shelf FPGAs can handle data rates in the range of 500 Mbit/s to 1 Gbit/s. However, cost increases significantly as the data rate handling capacity increases.

6.4 TLS applications on civil and agriculture

TLS can produce high precision, high density point cloud data for the region of interest. They have been used successfully in a variety of environments to support a wide range of geoscience study, including detailed mapping of fault scarps, geologic outcrops, volcanoes, lava lakes, fissures, glaciers, hill slopes and drainage morphology. Using the 3D point cloud and intensity data obtained, a target can be extracted and classified, and a 3D model can be reconstructed. Today, point cloud processing methods are mainly used for the geometry characteristic-based object filtering and target extraction and filtering of target. However, intensity information which is closely related to object characteristics, has not been used sufficiently. Therefore, it is necessary to combine geometric features and intensity information to the study of point cloud classification and feature extraction of the target.

6.4.1 Tunnel seepage detection

Compared to traditional techniques, TLS produces highly accurate, high-density tunnel point clouds, displaying good performance in tunnel deformation monitoring, 3D modelling and other applications. Besides point cloud data, intensity of Laser shot is also recorded via TLS scanning. The intensity is related to the tunnel surface attributes and is used to check leaking or non-leaking areas in tunnel. Because objects absorb diverse energy from Laser light, with the amount absorbed directly dependent on the proportion of water occupied, intensity information reflected from leaking or non-leaking areas are distinguished. In other words, Laser intensity reflected from the seepage area is less than that from non-seeping area.

To acquire point cloud and intensity data, an experiment was conducted on the tunnel of the Nanhuan Railway Bridge, on Fengke Road, in Fengtai district, Beijing. The intensity image of the tunnel, captured by the RIEGL VZ400i, and the grayscale images converted from red-green-blue (RGB) images recorded by camera are shown in Figure 6.19. Experimental facility is RIEGL VZ400i, which has a repetition rate of 1,200 kHz, a horizontal scanning speed of 3°/second, a vertical scanning speed of 30 circles/second, a minimum scanning distance of 0.5 m and maximum scanning distance of 250 m (@object reflection rate = 90%) [13,14].

The intensity correction was focused mainly on one side of the tunnel with serious seepage, removing the leaking area afterwards. Because of the distance and the incident angle effect, the mid-section of the illustration is brighter, with greater intensity, while the side of illustration is darker, with less intensity. Based on the experiments, it is concluded that TLS point clouds can be used to monitor leaking or non-leaking areas in dark environment with good performance. Its 3D model and intensity information reflected from leaking or non-leaking areas can be applied in detecting bridge structure and tunnelling examination of engineering application.

6.4.2 Individual maize plant extraction

TLS can also be used in agriculture for crop monitoring, 3D structural reconstruction of crop plants, and crop extraction and classification. Using the same

Figure 6.19 Tunnel points cloud scanned by RIEGL VZ400i

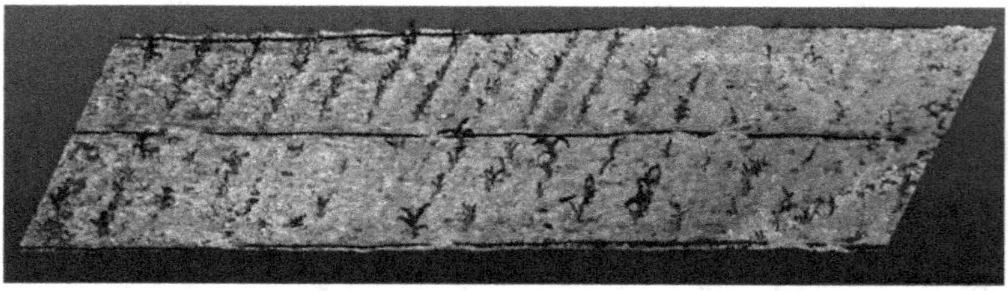

Figure 6.20 Points cloud of crop field collected by RIEGL VZ400i

facilities as in tunnel seepage detection, intensity data of the crop were recorded. Moreover, TLS uses a near-infrared Laser, which is sensitive to crop water content. Thus, it can be used to monitor drought, pests and diseases, and crop extraction and classification. In addition, compared with computer vision methods, TLS is an active detection method, regardless of light conditions. The grayscale points cloud of a crop field, collected by TLS, is shown in Figure 6.20.

Using points cloud (a set of numerical coordinates of targets) alone to extract maize seedlings can lead to considerable error because seedlings are short and corn fields are uneven. A way to perform extraction is by combining corrected intensity data and point cloud, first filtering point cloud data (x, y, z) to eliminate the noise of the crop field, then finding the threshold between soil and seedlings by plane fitting, and last, applying a filter algorithm to pick out the maize seedlings. By correcting the distance effect to intensify the data, and combining the regulated intensity data and point cloud to classify maize seedlings and soil, using the random forest method, extraction can finally be performed by a combination of point cloud

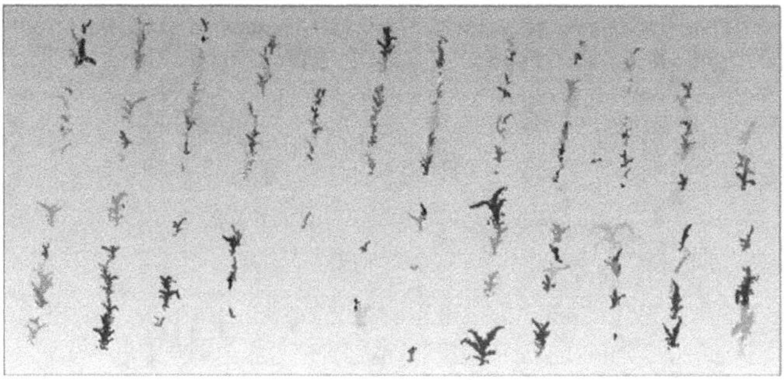

Figure 6.21 Maize seedlings extraction using point cloud

and corrected intensity. The grayscale image of extracted crops is shown in Figure 6.21.

It can be seen from results that TLS intensity data can be used to distinguish the corn seedling and bare soil, and corn seedlings can be effectively extracted from the cropland. The corn seedlings classification accuracy can be improved by using geometrical features and radiometric features (intensity information). RGB values can also be added to the radiometric features to improve the ability of crop plants, and crop extraction and classification. Misclassification mainly resulted from the crop stubbles with lower intensity values, which are sometimes inevitable.

6.4.3 Registration for urban structures

The photo-realistic modelling of large-scale scenes from images and LiDAR is an important research aid in photo-grammetry, for use with urban structures, environmental mapping and cultural heritage applications, which require a combination of range sensing technology and traditional digital photography. In the fusion process of photo-realistic modelling, TLS is responsible for location because of its dependable accuracy. The camera provides information, such as colour and texture. Therefore, the combination of camera and LiDAR has become popular in many disciplines, such as texture mapping, navigation and localisation. In the integration of point clouds obtained from camera and LiDAR, registration is a main procedure of concern.

Registering the complimentary point clouds captured by ALS, MLS/TLS provides an aligned data source for many purposes. For small scanning areas or dense building areas, TLS is used instead of ALS/MLS. A main processing flow for registration has three key steps as shown in Figure 6.22 [15,16]. First, two sets of point clouds were acquired with different scales. These were point clouds reconstructed from a camera and point clouds scanned from LiDAR. Point clouds from a camera are referred to as PCC, and point clouds from LiDAR as PCL. Second, an adaptive-scale key-point quality (ASKQ) algorithm was used to detect and match key-points from a PCC and a PCL as the input of coarse registration [12]. Third, a registration method was used to estimate the transformation parameters to align the

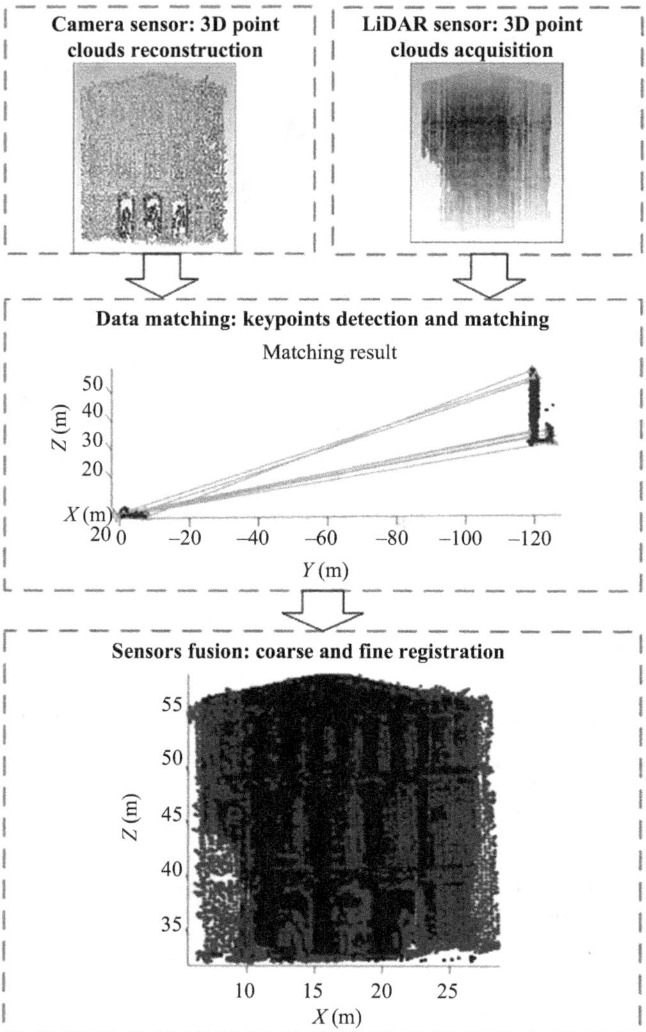

Figure 6.22 Main processing flow for LiDAR and camera sensors data matching

PCC and PCL. Finally, robustness and accuracy were verified using field data for the fine registration.

As shown in Figure 6.22, in the outdoor experiments, PCC were reconstructed from images taken by a camera with pixels of 3,264 × 2,448. PCL were obtained using a Laser scanner RIEGL LMS-Q120i, combined with a pan-tilt, whose highest speed was 1,000 steps/second and position resolution was 0.0125°. The number of PCC was 12,789; and the number of PCL was 32,232. It is apparent that the accuracy of the PCC is less than that of the PCL, and the density of the PCC is sparser than that of the PCL. But there is an area of missing points in the PCL. The

ASKQ algorithm is not suitable for detecting key-points of a plane, because key-points detected by the ASKQ algorithm are based on the local texture features of the object. If other key-point detection methods replace the ASKQ algorithm, the registering process can still be performed.

Registration accuracy and execution time are two crucial issues concerned. Registration accuracy is affected by feature extraction and scale factor determination involved in coarse registration. Algorithm execution time is influenced by the complexity of searching algorithms. When the registration accuracy and executing time satisfy the requirements of user, the point clouds registration can be generalised in the field of 3D modelling.

6.5 Future prospects

TLS, also known as ground-based LiDAR, relies on the following components: equipment for data collection, software for data processing and applications for 3D of various scenes, such as rock or ice outcrops, trees, vegetation and urban objects. TLS instruments mainly provide measurement range, size and portability, scan speed, FOV, multi-return and full waveform data, which support a wide range of applications.

For future applications, TLS measurement results must meet certain specifications to provide the necessary accuracy for particular applications. Therefore, any geometric surveying task includes not only the derivation of the relative positions of points and objects but also an estimation of the accuracy of the results. If the number of observations is not sufficient for adjustment, the accuracy of the results may be estimated by propagating the errors of the observation instruments to the results. The results can be improved by applying calibration values to correct systematic errors.

As with all surveying instruments, errors can occur in the results from TLS for many reasons, such as the instrument itself, environmental factors or the surface permeability of the object surveyed. In these circumstances, it is crucial to identify the accuracy and precision levels of all surveying systems, including TLS, to ensure that the survey returns the best quality data. Therefore, geometric-shaped objects are scanned from different distance using a ranging system. Scanning quality of point cloud can be enhanced by using error self-calibration. Consequently, with the accurate point cloud of TLS, 3D point data can be processed to produce 3D models. These models make it possible to access much of the necessary geometric and visualisation data.

References

[1] Li X, Wang H, Yang B, Huyan J, Xu L, 'Influence of time-pickoff circuit parameters on LiDAR range precision', *Sensors*, 2017, **17** (10), 2369 (20pp)

[2] Chazette P, Totems J, Hespel L, Bailly J P, 'Principle and physics of the LiDAR measurement', in *Optical Remote Sensing of Land Surfaces*, ISTE Press, Elsevier, 2016, pp 201–247

[3] Yu C, Chen X, Xi J, 'Modelling and calibration of a novel one-mirror galvanometric laser scanner', *Sensors*, 2017, **17** (1), 164 (14pp)

[4] Holmstrom S T S, Baran U, Urey H, 'MEMS laser scanners: A review', *J. Microelectromech. Syst.*, 2014, **23** (2), pp 259–275

[5] Marshall G F, Stutz G E, *Handbook of Optical and Laser Scanning*, John Wiley & Sons, 2004

[6] Jaya G L, Chen S, Siek L, 'The design of clocked-comparator-based time-interval measurement circuit for pulse ToF measurement', *IEEE Sens. J.*, 2017, **17** (20), pp 6699–6706

[7] Ruotsalainen T, Palojärvi P, Kostamovaara J, 'A wide dynamic range receiver channel for a pulsed time-of-flight Laser radar', *IEEE J. Solid-State Circuits*, 2001, **36** (8), pp 1228–1238

[8] Li X, Yang B, Xie X, Li D, Xu L, 'Influence of Waveform Characteristics on LiDAR ranging accuracy and precision', *Sensors*, 2018, **18** (4), 1156 (16pp)

[9] Kilpelä A, 'Pulsed time-of-flight Laser range finder techniques for fast high-precision measurement applications', *PhD Thesis*, University of Oulu, Finland, 2004

[10] Li X, Li Y, Xie X, Xu L, 'Lab-built terrestrial Laser scanner self-calibration using mounting angle error correction', *Opt. Exp.*, 2018, **26** (11), pp 14444–14460

[11] Li X, Li Y, Xie X, Xu L, 'Terrestrial laser scanner autonomous self-calibration with no prior knowledge of point-clouds', *IEEE Sensors J.*, 2018, **18** (22), pp 9277–9285

[12] Mian A, Bennamoun M, Owens R, 'On the repeatability and quality of keypoints for local feature-based 3D object retrieval from cluttered scenes', *Int. J. Comput. Vision*, 2010, **89** (2), pp 348–361

[13] Xu T, Xu L, Yang B, Li X, Yao J, 'Terrestrial laser scanning intensity correction by piecewise fitting and overlap-driven adjustment', *Remote Sensing*, 2017, **9** (11), 1090 (16pp)

[14] Xu T, Xu L, Li X, Yao J, 'Detection of water leakage in underground tunnels using corrected intensity data and 3D point cloud of terrestrial laser scanning', *IEEE Access*, 2018, **6**, pp 32471–32480

[15] Xu L, Feng J, Li X, Chen J, 'A LiDAR data-based camera self-calibration method', *Meas. Sci. Technol.*, 2018, **29** (7), 075205 (14pp)

[16] Xu L, Feng J, Li X, Liu C, 'Automatic registration method of TLS LiDAR data and image-based reconstructed data', *IEEE Geosci. Remote Sens. Lett.*, 2019, **16** (3), pp 482–486

Xiaolu Li is a Member of the IEEE. Since 2009, she has been working with Beihang University, China and became an Associate Professor in the School of Instrumentation and Optoelectronic Engineering in 2014. Her main research interests include LiDAR instrumentation and ranging measurement, LiDAR imaging and signal processing. She has published nearly 100 papers and has 24 invention patents granted in China. She has received several awards and honours, including third prize of Technological Invention Award from the Ministry of Industry and Information Technology in 2011, first prize of Technological Invention Award from the Ministry of Education in 2012.

Lijun Xu is a Senior Member of the IEEE. He received his PhD from Tianjin University, China in 1996. From 1997 to 2001, he was an Associate Professor in the School of Electrical Engineering and Automation, Tianjin University. From 2002 to 2006, he was a Research Fellow at the University of Greenwich, and the University of Kent, and a Higher Scientific Officer at the Institute of Cancer Research, University of London, UK. Currently, he is a Professor and the Dean of the School of Instrument Science and Opto-Electronic Engineering, Beihang University, China. His research interests include tomographic imaging, digital imaging and dynamic process monitoring. He has authored and co-authored more than 200 publications. He was awarded the National Science Fund for Distinguished Young Scholars of China in 2012, elected into the Ministry of Science and Technology Innovation Talents Promotion Plan in 2013, chosen as a Chang-Jiang Scholars Program Professor by the Ministry of Education, China in 2014 and elected into a high-level talents special support plan: '10,000 Plan' in 2016.

Chapter 7

Microwave imaging sensors

Alessandro Fedeli[1], *Matteo Pastorino*[1] *and Andrea Randazzo*[1]

List of abbreviations

3D	three-dimensional
AVA	antipodal Vivaldi antenna
DRA	dielectric resonator antenna
EM	electromagnetic
GPR	ground penetrating radar
IoT	Internet-of-things
MST	modulated scattering technique
MWI	microwave imaging
NDT	non-destructive testing
PCB	printed circuit board
SUT	structure under test
TEM	transverse electromagnetic
TM	transverse magnetic
TWI	through-wall imaging
UAV	unmanned aerial vehicle
UWB	ultra wide band
VNA	vector network analyser
VSWR	voltage standing wave ratio

7.1 Introduction

Microwave imaging (MWI) systems are nowadays acquiring an ever-growing importance in numerous application fields, such as non-destructive testing (NDT)

[1]Department of Electrical, Electronic, Telecommunications Engineering, and Naval Architecture, University of Genoa, Genoa, Italy

in civil and industrial applications, subsurface sensing and ground penetrating radar (GPR) prospection, through-wall imaging (TWI) for security and biomedical imaging. MWI systems use an electromagnetic (EM) radiation to illuminate the structure under test (SUT) and exploit the interaction with it to extract information about its physical and geometrical parameters. MWI sensors are constituted by antennas, which are used to generate the incident radiation and collect the field resulting from the EM interaction with the SUT. This chapter, after a brief introduction to the MWI problem and the configuration of the measurement systems, discusses several typologies of MWI antennas and their applications.

In MWI systems, the SUT is assumed to be inside a known investigation domain A and it is illuminated by one or more antennas, producing an incident electric field $\mathbf{E}_{\text{inc}}$. As the name suggests, the working frequencies are comprised in the microwave frequency range, which extends from 300 MHz to 300 GHz and it is nowadays widely used by other apparatuses (e.g. telecommunication ones). By definition, $\mathbf{E}_{\text{inc}}$ is the field generated by the antennas when no targets are present in A, and can be assumed as a known quantity. Such a field is perturbed by the presence of the targets. When SUT is present, the imaging sensors measure a total electric field $\mathbf{E}_{\text{tot}}$, which is in general different from the incident radiation. This difference resides in the scattered electric field, defined as $\mathbf{E}_{\text{scatt}} = \mathbf{E}_{\text{tot}} - \mathbf{E}_{\text{inc}}$, which is a function of the dielectric properties of the SUT.

The mathematical foundation of the MWI problem is already covered by several comprehensive books [1–4]. Therefore, only some basic concepts are briefly recalled here for providing an essential background to the chapter. As an explanatory example, let us consider the case of dielectric targets. In particular, it is assumed that the SUT is located in free space and it is composed of linear, isotropic, spatially non-dispersive and non-magnetic materials, that is with magnetic permeability approximatively equal to the one of vacuum, $\mu_0 \cong 1.26 \times 10^{-6}$ H/m. A time-harmonic radiation with angular frequency ω is considered, omitting the corresponding $e^{j\omega t}$ term. Under such conditions, the total electric field is related to the dielectric properties of A by means of the integral equation [2]

$$\mathbf{E}_{\text{tot}}(\mathbf{r},\omega) = \mathbf{E}_{\text{inc}}(\mathbf{r},\omega) - k_0^2 \int_A \left[\frac{\tilde{\epsilon}(\mathbf{r}',\omega)}{\epsilon_0} - 1\right] \mathbf{E}_{\text{tot}}(\mathbf{r}',\omega) \cdot \overline{\overline{\mathbf{G}}}_0(\mathbf{r},\mathbf{r}',\omega) d\mathbf{r}' \tag{7.1}$$

where $k_0 = \omega\sqrt{\epsilon_0\mu_0}$ is the free-space wavenumber ($\epsilon_0 \cong 8.85 \times 10^{-12}$ F/m being the vacuum dielectric permittivity), and $\tilde{\epsilon}$ is the complex dielectric permittivity, that is

$$\tilde{\epsilon}(\mathbf{r},\omega) = \epsilon(\mathbf{r},\omega) - j\frac{\sigma(\mathbf{r},\omega)}{\omega} \tag{7.2}$$

in which ϵ and σ are the space-dependent dielectric permittivity and electric conductivity in the investigation domain A.

Moreover, $\overline{\overline{\mathbf{G}}}_0$ is the dyadic Green's function for free space, given by [5]

$$\overline{\overline{\mathbf{G}}}_0(\mathbf{r},\mathbf{r}',\omega) = -\left(\overline{\overline{\mathbf{I}}} + \frac{1}{k_0^2}\nabla\nabla\right)\frac{e^{-jk_0|\mathbf{r}-\mathbf{r}'|}}{4\pi|\mathbf{r}-\mathbf{r}'|} \tag{7.3}$$

Relationships similar to (7.1), but with different Green's functions, can be also found when other background configurations, for example the multi-layered structures arising in subsurface imaging and TWI, are considered [6]. By employing the definition of the scattered field and (7.1), a relationship $\mathbf{E}_{\text{scatt}} = \mathscr{F}(\tilde{\epsilon})$ can be written, where $\mathscr{F}$ is an operator that describes the scattering phenomena and provides a mapping between the electric-field samples measured by the antennas and the dielectric properties of the SUT. The scattered electric field is collected in the so-called measurement domain D_{meas}, which is a region outside A. When a tomographic setup is considered, the previous equations can be simplified. In fact, by assuming that the incident field is transverse magnetic (TM) and z-polarised (i.e. $\mathbf{E}_{\text{inc}}(\mathbf{r},\omega) = E^z_{\text{inc}}(x,y,\omega)\hat{\mathbf{z}}$) and that the dielectric properties of the target are independent form the axial coordinate z (i.e. $\epsilon(\mathbf{r}) = \epsilon(x,y)$), the previous scattering equation (7.1) is simplified to

$$E^z_{\text{tot}}(x,y,\omega) = E^z_{\text{inc}}(x,y,\omega) - 0.25jk_0^2 \int_A \left[\frac{\tilde{\epsilon}(x',y',\omega)}{\epsilon_0} - 1 \right] \times E^z_{\text{tot}}(x',y',\omega) H^2_0(k_0\rho) dx'dy' \tag{7.4}$$

where E^z_{tot} is the z-component of the total electric field, H^2_0 is the Hankel function of zeroth order and second kind and $\rho^2 = (x - x')^2 + (y - y')^2$ [2].

Such relationships can be inverted to retrieve information on the dielectric properties of the SUT, that is from the set of measurements of the scattered EM field, the spatial distribution of $\tilde{\epsilon}$ can be estimated. However, the involved inverse scattering problem turns out to be nonlinear, since the total electric field, which in turn depends on the dielectric properties of the targets, is also present inside the integral in (7.1) and (7.4). Furthermore, as it is well known, this problem is severely ill-posed [7]. In the literature, many different approaches have been proposed to address the problem of retrieving information on the targets from the measured scattered-field data [2]. These approaches are often classified into two main groups: qualitative and quantitative methods. In the first class, it is possible to further consider two cases: methods able to provide only some specific information on the target, such as position and dimension, and approaches based on approximate EM models, such as the Born and Rytov models [6].

Radar imaging methods belong to this first class. They often provide maps of the reflectivity profile of the object by using reconstruction procedures based on beamforming concepts, which synthetically focus on a specific zone of the SUT. This approach is often associated to scanning methods that allow to 'see' different zones of the target. Conversely, quantitative methods aim at solving the full nonlinear inverse scattering problem, without approximations different from those introduced by the discretisation needed for implementing the inversion procedures on a computer. In this case, the goal is to reconstruct the full distributions of the dielectric properties of the investigation domain. Approaches belonging to this class can be grouped into two categories: deterministic and stochastic methods [6]. Although being effective from a computational point of view, deterministic inversion strategies may converge to false solutions. Consequently, a special care should be devoted to the selection of a proper starting point, which should be close enough to the solution. Stochastic methods can overcome such a drawback, since they are usually able, at least in principle, to reach the global optimal solution. However, their main drawback is related to the highly demanding computational resources. Qualitative and quantitative inversion strategies are sometimes combined to develop hybrid imaging approaches, capable of exploiting the advantages of both classes [6].

For a detailed description of the inversion procedures applied in MWI, which is beyond the scope of this Chapter, the reader is referred to the specific literature on this topic, for example [2–4,6].

7.2 Microwave imaging systems

As outlined in Section 7.1, MWI systems rely on processing the scattered electric-field data. Such data are collected by a microwave measurement system, which is composed of one or more antennas (the actual MWI sensors) used to illuminate the SUT and to measure the total EM field arising from the interaction between the incident radiation and the target, and by an apparatus for generation and reception of the microwave signals, for example a vector network analyser (VNA). Depending on the application, different configurations, characterised by different numbers of antennas, are adopted. In this respect, three main classes of MWI measurement setups can be identified, as shown in Figure 7.1.

- *Monostatic* – A single antenna is used to illuminate the target and to measure the scattered EM field (see Figure 7.1(a)). Only reflection data are obtained in this case. This approach is often adopted in standard radar systems. However, differently from conventional radar applications, for example for remote sensing, in which far-field conditions usually hold, in MWI applications the target is typically in close proximity to the illuminating/measuring elements.
- *Bistatic* – A pair of antennas scans the investigation domain (see Figure 7.1(b)). The two antennas can move together as in GPR applications or they can be independently positioned in a pre-defined domain, such as a rectangular planar region or all around the scatterer to acquire multi-view data (as in MW tomography). In the latter cases, more information can be collected about the scattered field, but it is necessary to design a proper positioning system to move the antennas and synchronise the movement with the acquisition.
- *Multi-static* – A set of M antennas are located around the investigation domain (see Figure 7.1(c)). Each one is in turn used in transmission mode, whereas the remaining $M-1$ elements collect the scattered EM field. In most cases, a $2 \times M$ microwave switching matrix is required to connect each antenna to the TX/RX ports of the measurement system. This configuration has an advantage of not requiring moving parts.

MWI systems can also be characterised in terms of antenna arrangements. Some standard setups of the measurement domain D_{meas}, shown in Figure 7.2, are the following:

- *Spherical or hemispherical* (see Figure 7.2(a)) – The antennas are located on the surface of a sphere or a portion of it, for example a hemisphere, and are usually polarised along the azimuth and/or elevation directions.

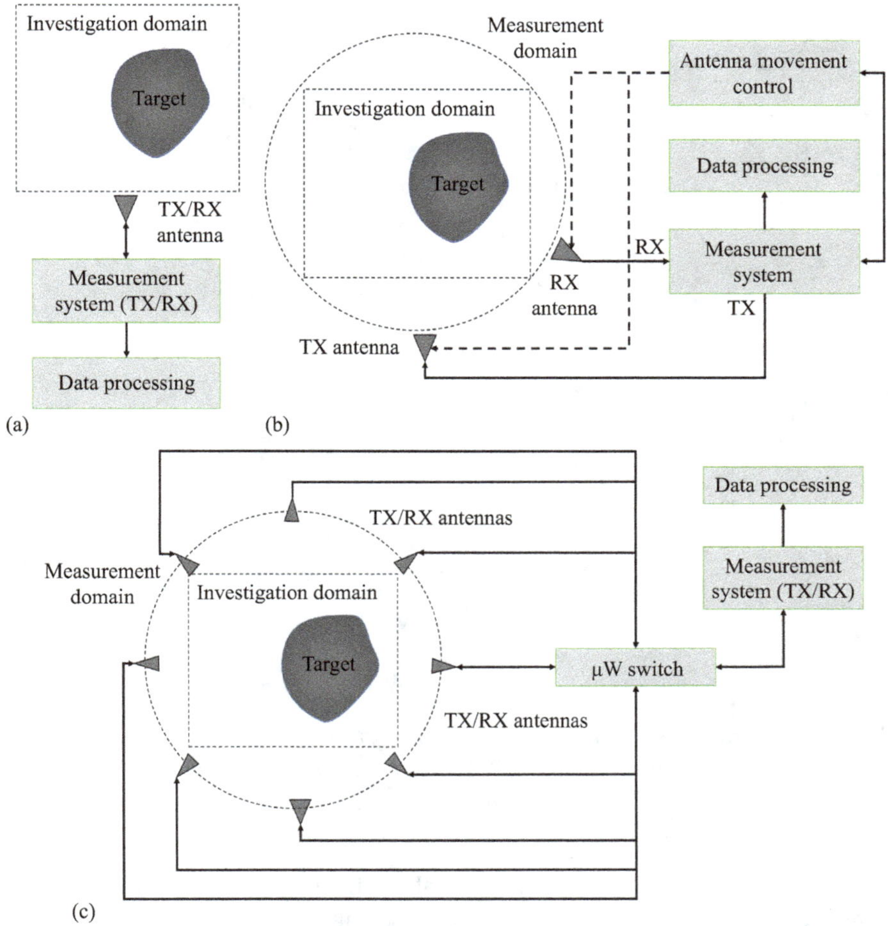

Figure 7.1 Basic typologies of MWI systems: (a) monostatic, (b) bistatic and (c) multi-static configurations

- *Cylindrical* (see Figure 7.2(b)) – The antennas are positioned along circular probing lines located at different heights, that is on a cylindrical surface. To reduce the mutual coupling among adjacent elements, antenna positions can be angularly shifted between adjacent probing lines.
- *Surface scanning* (see Figure 7.2(c)) – The antennas scan a planar surface with equally spaced or random sampling.
- *Tomographic* (see Figure 7.2(d)) – In tomographic systems, a cylindrical target is considered. Consequently, antennas are positioned on a measurement line (e.g. a circular one) around the target, located at a given height corresponding to the considered cross-section to be inspected.

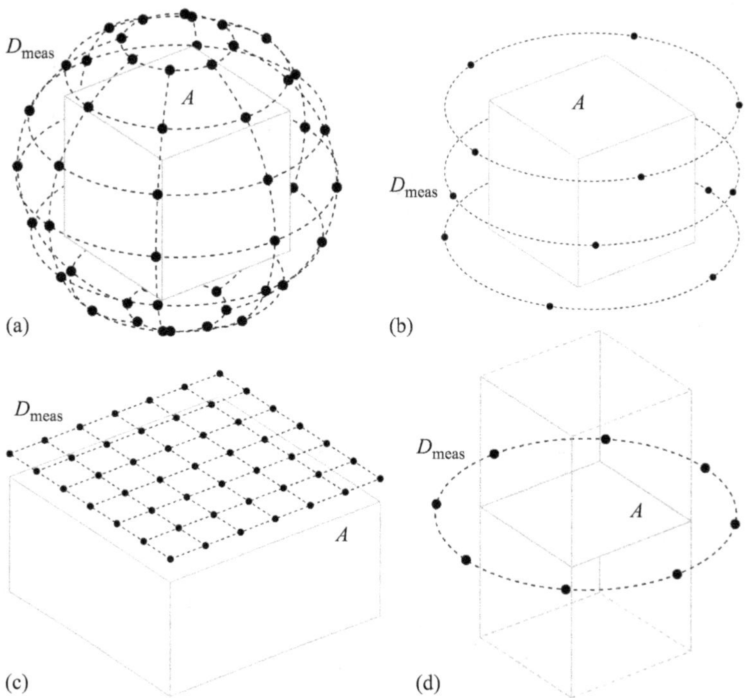

Figure 7.2 Examples of measurement configurations. (a) 3D spherical measurement domain, (b) cylindrical, (c) surface scanning configuration and (d) tomographic setup

Spherical/hemispherical and cylindrical configurations are often adopted in biomedical applications, such as for breast imaging [8] or head imaging [9]. For example, in the breast imaging system developed in [10], 16 antennas are positioned on a circumference and can be moved at different heights in order to synthesise a cylindrical setup. A similar configuration was considered in [11], where a single monostatic antenna is used to collect data along a cylindrical surface. In [12], a hemispherical cup with 60 ultra-wide-band (UWB) antennas is adopted, and another example of hemispherical setup was reported in [13].

Sub-surface imaging in geophysics, civil engineering and NDT is a typical application of surface scanning systems [14–16]. In these cases, the sensors used for scanning the imaging surface are often open waveguides, horn or printed antennas (see Section 7.3).

Tomographic systems are commonly adopted when the SUT has an elongated shape and can be approximated as a cylindrical structure. Such a condition is satisfied in different practical cases, such as detection of buried utilities [14], archaeological prospection [15], inspection of pillars and wood structures [17] and biomedical imaging [18]. In all these situations, the investigation domain is a slice of the target and the antennas are located on the corresponding transverse plane.

An example of multi-static tomographic setup is shown in Figure 7.3, where 16 wideband monopole antennas are hold in contact with the target by means of a custom-made 3D printed structure and are connected to a VNA by means of a two-level microwave switching matrix. In particular, the first level is composed by two 1:4 switches, whereas the second one by four 2:4 switches.

In several applicative scenarios, the dielectric properties of the SUT may assume very high values (e.g. the complex relative dielectric permittivity of muscle is equal to $54.8 - j17.58$ at 1 GHz). This fact may result in different issues affecting imaging systems. First, a high reflection from the external surface of the SUT is expected, which reduces the power of the EM field that effectively enters inside the body and may mask the scattering response from the internal parts of the target. Moreover, the reconstruction of very strong scatterers is generally more challenging for inversion procedures. To overcome such limitations, matching media (e.g. saline solutions and water/glycerin mixtures) are adopted for increasing the EM field penetration and reducing the dielectric contrast through the body under examination [9,19,20]. Additionally, the use of matching media with high values of the dielectric permittivity allows a reduction of the antenna size, since the wavelength inside the matching medium is smaller than the wavelength in air.

Proper containers should be adopted for enclosing the SUT, especially when liquid matching materials are used. In the literature, both plastic and metallic structures have been considered. However, the presence of containers affects the

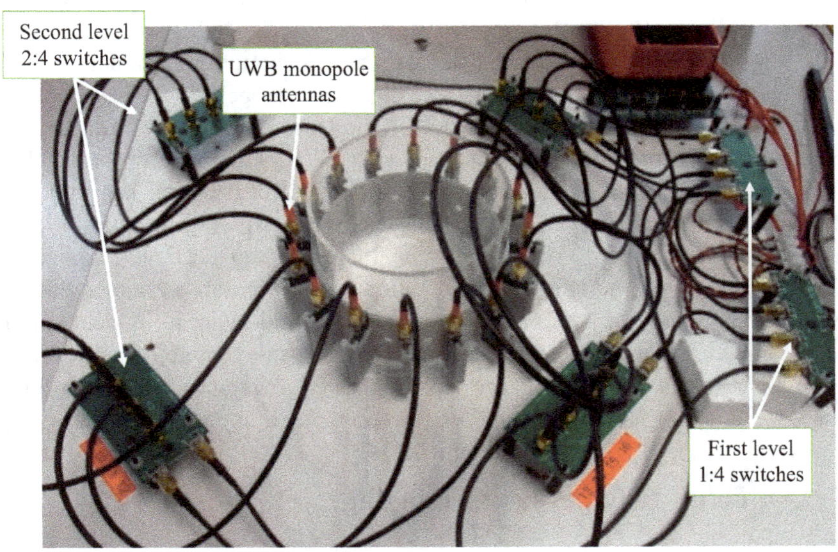

Figure 7.3 Multi-static tomograph prototype for quantitative imaging [18]. [This work is a derivative of Figure 5 in [18], used under Creative Commons Attribution (CC BY 4.0) license (http://creativecommons.org/licenses/by/4.0/CC). The original version can be found at https://doi.org/10.3390/jimaging4120139]

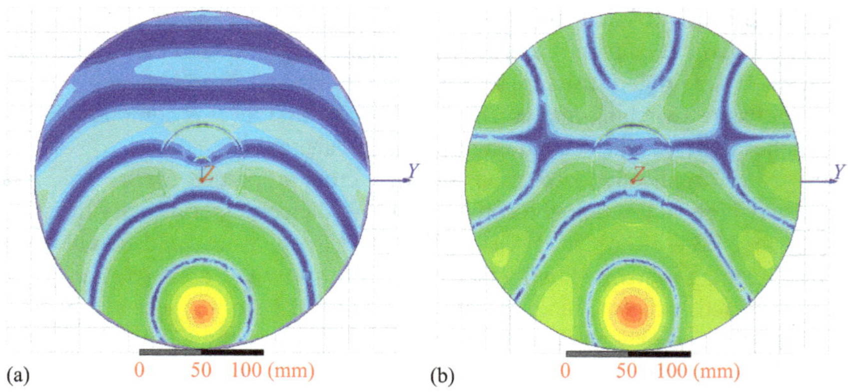

Figure 7.4 Amplitude of the electric field on a horizontal slice of a MWI prototype system at 2 GHz with (a) infinite matching medium, and (b) finite matching medium. © 2015 IEEE. Reprinted, with permission, from [21]

antenna radiation, and need to be properly taken into account in the EM scattering model. The influence of a plastic container on the field radiated by a monopole antenna is shown in Figure 7.4, where the amplitude of the z-component of the electric field on a horizontal slice of a MWI prototype system is simulated in two different conditions [21].

In Figure 7.4(a), the antenna (located at the bottom) is radiating in an unbounded matching medium with the dielectric properties of vegetable oil ($\epsilon = 2.51\epsilon_0, \sigma = 0.03$ S/m). In Figure 7.4(b), the same antenna is radiating inside a plastic cylindrical container of 270 mm diameter, filled with the same vegetable oil and surrounded by air [21]. Clearly, the electric field distribution has significant differences, which should be considered inside the MWI procedure. Metallic chambers have been adopted for enclosing the SUT, too [22,23]. Compared to dielectric enclosures, metallic chambers allow more accurate system modelling, because the boundary of the domain is well defined and easier to model, and provide a complete shielding of the imaging scenario. An example of an MWI aluminium chamber is shown in Figure 7.5. Its faceted structure is shown to further reduce the modelling error and improve the reconstruction quality [24].

7.3 Antennas for microwave imaging

As previously introduced, antennas are the key elements of an MWI system, because they are needed for illuminating the SUT and collecting the resulting scattered fields. In the literature, a wide variety of antennas have been proposed, frequently designed on a case-by-case basis. Several kinds of antennas can be distinguished based on their structure. Some examples of the most common ones

(a) (b)

Figure 7.5 Faceted MWI aluminium chamber: (a) side view and (b) top view. © 2017 IEEE. Reprinted, with permission, from [24]

are schematised in Figure 7.6. In Sections 7.3.1–7.3.9, their main operating principles are recalled, and some recently proposed sensing elements for MWI are discussed.

7.3.1 Monopole and dipole antennas

Simple wire antennas, such as monopoles or dipoles, are common in many experimental imaging systems proposed in the past few decades. These antennas, in their basic form, are constituted by one (in the case of the monopole) or two (in the case of the dipole) straight wires of length l, as shown in Figure 7.6(a) and (b). The radiation pattern of a typical half-wavelength dipole (for which $l = \lambda/4$) is essentially omnidirectional on a plane perpendicular to the antenna axis [25]. These antennas are used around their resonant frequency, which is approximately given by

$$f_{r1} \simeq \frac{1}{4l\sqrt{\epsilon\mu}} \tag{7.5}$$

where ϵ and μ are the dielectric permittivity and the magnetic permeability of the propagation medium.

Monopole antennas behave similarly. Indeed, by exploiting the image theory, it can be proven that a vertical monopole of length l over a ground plane is equivalent to a dipole with total length $2l$, but with a double directivity and a halved impedance [25].

Monopole antennas have been successfully integrated in many MWI systems devoted to breast tumour detection, as well as for the imaging of other parts of the body. An example of monopole antenna for a biomedical imaging system is shown in Figure 7.7 [21]. It has a length $l = 20$ mm and has been designed to work with a centre frequency of about 5 GHz in a vegetable oil solution used as matching

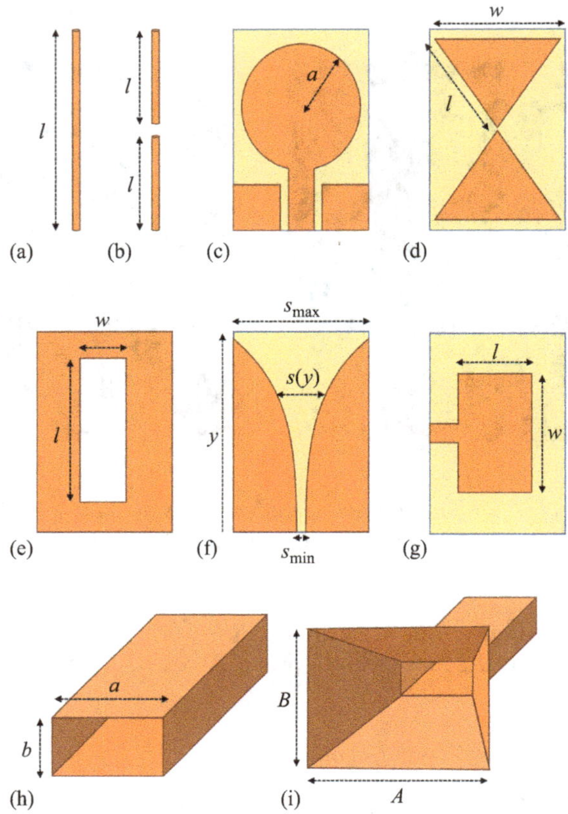

Figure 7.6 Examples of antennas commonly adopted in MWI systems: (a) monopole; (b) dipole; (c) wideband printed monopole; (d) bowtie; (e) slot; (f) tapered slot; (g) patch; (h) open rectangular waveguide and (i) pyramidal horn

Figure 7.7 Monopole antennas used in the biomedical MWI prototype in [21].

medium. Such an antenna is mounted on a moving arm to collect multi-static data in a tomographic configuration [21]. Another setup with the same kind of antenna is presented in [10], where 16 monopoles are used to collect multi-static data at a frequency of 1.3 GHz. In [26], 20 monopole antennas are mounted in a circle over a ground plane and are enclosed inside a square tank made with Perspex sheets. A quantitative MWI apparatus, made of a cylindrical plastic tank equipped with 32 horizontal monopole antennas, has also been proposed in [8]. This antenna arrangement, which adopts the configuration of Figure 7.2(b), allows to simultaneously acquire a full set of measurements for 3D reconstruction of a breast without the need of mechanically moving the sensors on the vertical axis.

The narrow bandwidth of standard dipole and monopole antennas may represent a limiting factor for the MWI applications in which wide-band frequency-stepped signals or ultra wide band (UWB) pulses should be used. To overcome this problem, wideband monopoles, whose basic structure is shown in Figure 7.6(c), may be used. A printed circular disc wideband monopole is taken as an example. This structure is made by etching a metallised dielectric substrate with a circular shape and is fed by a micro-strip or coplanar line. The ground plane does not extend under the disc and is slightly separated from it. The main idea behind this kind of structure is to overlap several resonances to obtain a wide bandwidth. The first (i.e. lowest) resonance is given by the frequency at which the diameter of the disc is approximately equal to $\lambda/4$, that is

$$f_{r1} \simeq \frac{1}{8a\sqrt{\epsilon\mu}} \tag{7.6}$$

where a is the radius of the disc.

The finite length of the ground plane contributes to the insurgence of additional resonances, whose frequencies can be tuned by numerical simulations. If resonances are sufficiently close, a wide operating band is obtained.

The possibility of arbitrarily shaping the printed metallic element and the backing ground has led to many variations, tailored to the different application scenarios [27]. To give some examples, wideband monopoles have been proposed in [28,29] for breast cancer detection and in [30] for radar imaging. In particular, [28] proposed a rectangular printed structure with a semi-circular tapering, which produces a smooth transition between resonant modes. The result is a good matching in the frequency band between 3.4 and 9.9 GHz. In [29], a triangular shape is proposed, and the antenna is designed to work immersed in a Triton X-100 liquid matching medium between 500 MHz and 3 GHz. As regards the shape of the backing ground, an interesting structure is devised in [30], where the bottom conducting layer has been optimised to ensure a good isolation between two adjacent wideband monopole antennas in a bistatic configuration, in a frequency band of 3–18 GHz.

Similar considerations can be applied to dipole antennas, which were one of the first kinds of sensors applied to experimental MWI systems for characterisation of dielectric targets [31]. More recently, two dipole antennas immersed in a plastic tub filled with a matching fluid have been applied to obtain 3D dielectric reconstructions [32]. Several variations based on the classic wire dipole structure have been suggested to face different applications: for instance, printed broadband dipoles have been used for TWI [33] and for testing qualitative breast cancer imaging in [34], where a comparison with a Vivaldi antenna (see Section 7.3.5) has been made.

7.3.2 Folded structures

Although wideband printed structures are attractive because of their bandwidth and their easy fabrication, the increase in the bandwidth (especially when low frequencies are required) is often associated with an increased size of the antenna. To tackle this problem, several folding techniques have been proposed and tested [35]. Some experimental MWI prototypes have antenna elements derived from this folding principle, whose result is a 3D structure sometimes equipped with slots and parasitic elements on the back side [36]. Such antennas have been mainly used in free-space configuration for qualitative imaging purposes, that is without any matching medium.

Other folded structures have been devised to obtain improved wideband coupling with the SUT. For instance, in [17] the problem of coupling the antenna with wood samples has been addressed. The resulting geometry, originated from a wideband monopole, is shown in Figure 7.8. Such an antenna has been found to have a voltage standing wave ratio (VSWR) lower than 2.5 in the frequency band of 0.4–1.1 GHz when it is placed in direct contact with wood materials characterised by a dielectric permittivity $\epsilon = 10\epsilon_0$.

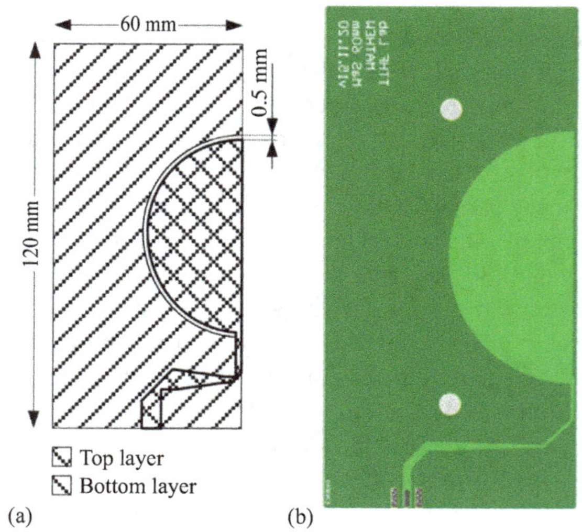

Figure 7.8 Folded quasi self-complementary wideband antennas in an MWI system for wood characterisation: (a) antenna design and (b) top view of the antenna PCB with actual matching feed. © 2018 IEEE. Reprinted, with permission, from [17]

7.3.3 Bowtie antennas

A structure derived from the dipole antenna is the bowtie, as shown in Figure 7.6(d). Bowtie antennas are composed of two triangular elements, in most cases printed on dielectric substrates. It can be shown that a standard printed bowtie antenna has a first resonant frequency (related to the TM_{10} mode) at

$$f_{r1} \simeq \frac{0.288}{lR_t\sqrt{\epsilon_r^{\text{eff}}}} \tag{7.7}$$

where l is the diagonal length of each arm, and R_t is a parameter that depends on the dimensions of the antenna, the substrate height, and its effective relative dielectric permittivity ϵ_r^{eff} [25].

One of the main features of the bowtie antenna versus the conventional wire dipole is the increased bandwidth due to the linear flaring of its arms, while another benefit still remains the rather easy modelling of the radiated EM field, which resembles the one of a dipole antenna. Because of these features, accompanied to a convenient design, bowtie and bowtie-like antennas have been proposed and exploited in several fields.

The most popular imaging apparatuses adopting bowtie antennas are GPRs [37]. In such devices, two printed bowties are paired in a quasi-monostatic configuration and backed by a metallic shield and/or by absorbing materials [38]. Bowtie antennas have been used for breast tumour detection [39], too. For example, in [40], a bowtie switched sensor array for MW holography is presented, whereas the implementation of a cavity bounded by conducting walls equipped with 36 printed bowtie antennas, which operate at 2.75 GHz in a cylindrical configuration, was described in [41]. Brain stroke detection is another scenario where bowtie-like antennas are used: a slotted cavity-backed bowtie structure, as shown in Figure 7.9(a), is employed in the microwave system for

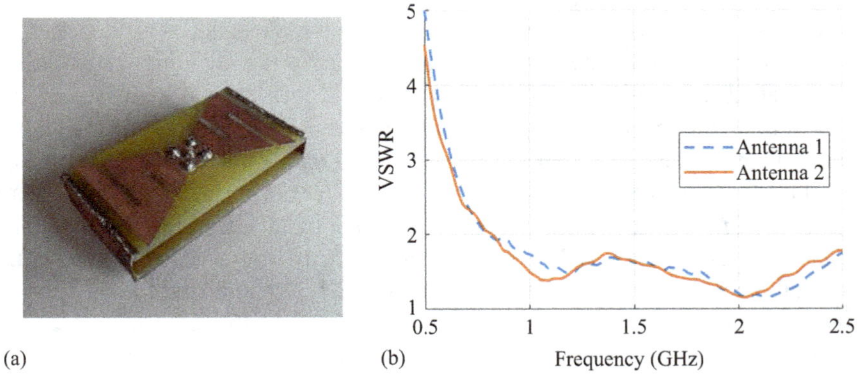

Figure 7.9 Prototype of a slotted cavity-backed bowtie antenna developed for brain stroke microwave imaging [42]: (a) photograph of the sensor and (b) measured VSWR of two antenna prototypes

brain stroke imaging in [42]. In more detail, this antenna has two parallel faces of 45 × 25 mm, connected at a distance of 7 mm with two copper sheets on the short sides. The upper face presents a bowtie shape, in which four slots have been etched to increase the coupling with the SUT and reduce the antenna dimensions. Both faces have been realised on an FR4 dielectric substrate. This antenna has been optimised for operating in contact with a glycerin/water coupling medium enclosed in a polyethylene bag (to be placed between the sensor and the SUT). The obtained VSWR, measured with two antennas placed at the opposite sides of a 180 mm plastic cylinder filled by a mixture of glycerin (70% volumetric) and water, is shown in Figure 7.9(b).

More complex variations on the bowtie structure have been made, such as the double-layer bowtie antenna described in [43], which includes a balun and is optimised for radiating inside a human body. Other variants based on the bowtie are the 'dark eyes' antenna, which presents a surface resistive loading on its arms [44], or slot-loaded structures [45]. To observe cross-polarised field components, two bowtie antennas forming a Maltese cross have been proposed [46], as well as arrays of 'flower' antennas made with two orthogonal bowtie-like elements [47].

7.3.4 Slot antennas

Slot antennas, whose basic geometry is shown in Figure 7.6(e), are composed by an aperture in a conducting plane and are printed on a dielectric substrate in most cases. When one of the sides of the slot is small, these antennas behave similarly to dipoles and can be modelled by means of the equivalence principle (where the electric current distributions of wire antennas are replaced by equivalent magnetic currents) [25]. Several alternative designs have been proposed, including folded and absorber-backed structures [48], and even dual polarised slot elements [49].

Along with the standard narrow slot antennas, wide-slot sensing elements have been developed. Such structures are usually made on dielectric substrates, and they are customarily excited by using micro-strip feeds. Wide-slot sensors and their derivatives have been extensively used for microwave imaging [50,51]. For instance, an absorber-filled cavity-backed wide-slot aperture was described in [51] for MWI of arms. This prototype is shown in Figure 7.10. The short and long edges of the slot have dimensions $w = 28$ mm and $l = 40$ mm, and the aperture is realised on a substrate with dielectric permittivity $\epsilon = 10.2\epsilon_0$. The backing cavity has a depth of 67.5 mm, whose last 57 mm are occupied by an absorbing material. This antenna is excited by a stub-terminated micro-strip line and has an operating bandwidth between 1 and 4 GHz approximately [51]. Rotated and folded versions of the wide-slot have also been proposed for UWB applications [52].

Slot antennas are frequently adopted in MWI as the basic element to build antenna arrays, such as the ones presented in [53] for damage detection in civil structures and in [12,51,54] for biomedical imaging. Elliptical slots, arranged in a square array, were used in an MW camera presented in [55], in conjunction with the modulated scattering technique (MST).

Figure 7.10 Large cavity-backed wide-slot antenna for MWI. © 2017 IEEE. Reprinted, with permission, from [51]

(a) (b)

Figure 7.11 Array of tapered slot antennas used inside an MW camera: (a) front and (b) back. © 2017 IEEE. Reprinted, with permission, from [57]

7.3.5 Tapered slot and Vivaldi antennas

When wideband and UWB antennas have to be employed to illuminate the SUT, a standard choice is to use tapered slot radiators. Essentially, they are realised on a dielectric substrate or shaped from two pieces/sheets of conducting material. Tapered slot antennas can be found in many application areas, ranging from TWI (e.g. a linearly tapered slot array is used in [56]) to microwave cameras [57]. In particular, one section of the 3D camera described in [57], which contains 16 active elements on a printed circuit board (PCB), is shown in Figure 7.11.

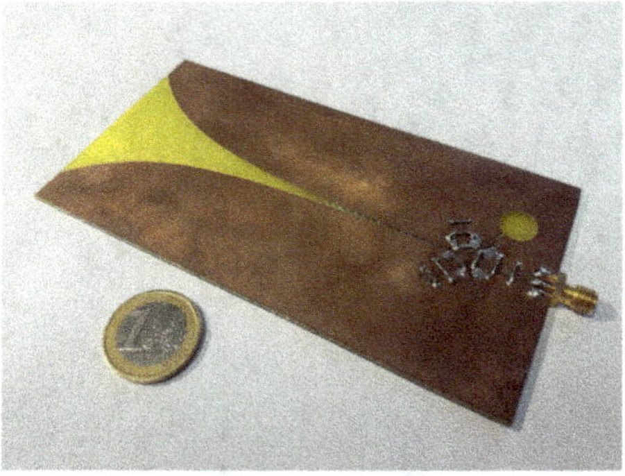

Figure 7.12 Example of Vivaldi antenna for MWI [58]

This array is designed for operating between 20 and 30 GHz, and the length of each tapered slot element is equal to $l = 12.5$ mm. Each slot has a minimum width $s_{min} = 0.4$ mm and maximum width $s_{max} = 6$ mm.

A special case, widely adopted in practice and known as Vivaldi antenna, has a tapered slot described by an exponential function:

$$s(y) = s_{min}e^{py} \tag{7.8}$$

where s is the slot separation, p is the exponential tapering coefficient, y is the position along the antenna and s_{min} is the slot width at the feed gap [25].

The Vivaldi antenna is shown in Figure 7.6(f). It turns out that the antenna length l should be more than one wavelength at the lowest frequency. Moreover, the minimum and the maximum slot widths s_{min} and s_{max} influence the highest and the lowest operating frequencies, respectively [25].

An example of Vivaldi antenna for MWI purposes, whose details are given in [58], is shown in Figure 7.12. This antenna works between 2 and 6 GHz and is fed through a coplanar waveguide and a balun. In more detail, the length of the tapered structure is $l = 102.8$ mm, the minimum and maximum slot widths are $s_{min} = 1.2$ mm and $s_{max} = 51.4$ mm, respectively, and the exponential tapering parameter p is equal to 36.55 m^{-1}. A common variation on the standard Vivaldi structure is represented by the antipodal Vivaldi antenna (AVA), which eliminates the need for a balun on the feed line [59]. AVAs have been used in several forms [60] and even with dielectric directors [61]. They have been employed inside MST measurement systems, combined with dual-polarised probes [62]. Several other variations exist, mainly devoted to further improve the operating bandwidth: from the use of corrugations [63,64] to fractal leaf structures [65].

7.3.6 Open waveguides

Open-ended waveguide antennas are obtained by truncating a metallic waveguide. The typical case of an open-waveguide antenna with rectangular cross-section is represented in Figure 7.6(h). The lowest limit of the operating frequency band is given by the cut-off of the TE_{10} mode [25], that is

$$f_{\min} = \frac{1}{2a\sqrt{\epsilon\mu}} \tag{7.9}$$

where ϵ and μ are the dielectric permittivity and the magnetic permeability of the medium filling the waveguide and a is the length of its largest side.

Ordinarily, open waveguides are used in their single-mode band. Therefore, the highest frequency limit is the cut-off frequency of the higher-order TE_{20} mode if $b < a/2$ or the TE_{01} mode if $b > a/2$, that is

$$f_{\max}^{b<a/2} = \frac{1}{a\sqrt{\epsilon\mu}}, \; f_{\max}^{b>a/2} = \frac{1}{2b\sqrt{\epsilon\mu}} \tag{7.10}$$

where b is the width of the smallest side of the waveguide.

Open waveguide antennas have been extensively employed in NDT (see e.g. [66]), and their characterisation for this kind of application field has been discussed in details in [16]. Open waveguides can be easily integrated in metallic imaging chambers, like in the MWI system for brain stroke detection presented in [9]. In particular, 177 rectangular waveguide antennas are mounted on a stainless-steel chamber with spherical topology. Each waveguide has sides of length $a = 21$ mm and $b = 7$ mm and is internally filled with a ceramic material with dielectric permittivity $\epsilon = 60\epsilon_0$. The resulting cut-off frequency is 922 MHz approximately. In this system, the accurate modelling of the structure allows to detect small changes in the measured scattering parameters due to the presence of stroke-affected tissues [9]. The use of open waveguides inside a chamber filled by a matching medium further reduce their size, allowing a convenient placing of a rather large number of antennas even in portable devices.

7.3.7 *Horn*

Horn antennas are derived from open waveguides, where the last portion of the structure is flared as shown in Figure 7.6(i). Horns have a narrower beam compared to open waveguides and their gain is proportional to the physical area of their aperture, which is described by its dimensions A and B. The working frequency band is, however, the same as the feeding waveguide (see Section 7.3.6). Horn antennas are largely used for MWI, and they have been considered for realisation of the well-known dataset of the Institut Fresnel, which has become a reference for testing MWI algorithms [67]. The wide frequency band of horn antennas, which can be further extended in the ridged versions [68], makes them suitable for radar imaging in different situations [69,70]. Dual-polarised horn antennas have also been proposed, for example for TWI in [71].

Starting from a (linearly or exponentially) flared parallel plate waveguide, transverse electromagnetic (TEM) horn antennas can be constructed. Antennas of this kind have been used for near-field MWI [72], as well as in GPR [73]. An example of TEM horn sensor used for this purpose is shown in Figure 7.13. This antenna operates

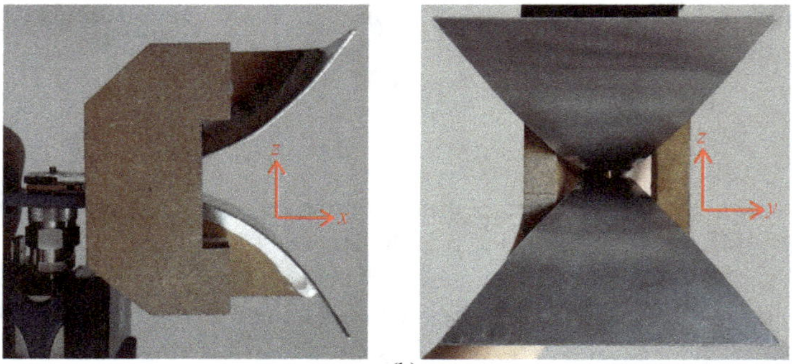

(a) (b)

Figure 7.13 TEM horn antenna used for GPR imaging: (a) side view and (b) front view. © 2017 IEEE. Reprinted, with permission, from [73]

approximately between 1.5 and 3 GHz and has an exponential flaring in the xz plane. The plate separation is 2.6 mm at the connector and 100 mm at the aperture, which is square and 100 mm wide in the y direction [73]. A modification of the TEM horn structure, called dielectric wedge antenna, has been proposed in [74].

7.3.8 Patch

Patch antennas are often used in MWI systems, too. Normally, they are etched from a metallised dielectric substrate, as shown in Figure 7.6(g). Realised in different shapes, they are typically characterised by a small and compact size. Furthermore, the possible use of high-constant dielectric substrates makes it possible to match the properties of the external layers of the SUT or the coupling medium. One of the simplest typologies is the rectangular patch antenna as shown in Figure 7.6(g), which can be modelled by using the resonant cavity theory. In this case, the resonant frequency is given by

$$f_{r1} = \frac{1}{2l_{\text{eff}}\sqrt{\epsilon_r^{\text{eff}}}\sqrt{\epsilon_0\mu}} \tag{7.11}$$

where ϵ_r^{eff} is the effective relative dielectric permittivity and $l_{\text{eff}} = l + 2\Delta l$ is the effective length of the patch, which takes into account a fictitious extension Δl of the physical length l due to fringing field effects [25].

In their original form, patch antennas have a very narrow bandwidth around the main resonant frequency. However, different designs have been suggested to address multi-band operations [75]. For example, an extensive study of several kinds of multi-band patch antennas for breast tumour detection is presented in [76]. To increase the bandwidth and/or the coupling with SUT, it is possible to add slots, modify the antenna shape or create stacked structures [77]. An interesting comparison between stacked patch and wide-slot antennas is reported in [78], where it has been found that the size of a wide-slot antenna can be significantly smaller than a stacked patch designed for the same application.

7.3.9 Other sensor elements

Due to the developments in MWI, the typologies of sensors seen so far are not exhaustive, and sometimes they are not mutually exclusive. However, several other kinds of antennas have been used. Among them, log-periodic antennas play an important role. Thanks to their wide bandwidth, they are not only used for instance for TWI [79] but also are very useful in prototype systems, in which the operating frequency is a parameter to be selected [80]. Moreover, travelling-wave tapered and loaded transmission-line antennas have been used in medical imaging, even in clinical studies [81]. Antennas derived from loop structures have been recently considered, for example inside the metallic MWI chamber described in [24]. In addition, dielectric resonator antennas (DRAs) have been proposed [82], although they have not been extensively applied yet to MWI experimental measurement apparatuses.

7.4 Future trend

The academic research and commercial development of MWI sensors are continuously evolving and intrinsically connected with the expanding number of possible applications. For instance, with the advent of Internet-of-things (IoT), there has been an increasing interest in portable and wearable MWI devices in many areas, ranging from the biomedical field (e.g. portable brain stroke detection systems) to NDT (e.g. GPR apparatuses), TWI and even unmanned aerial vehicle (UAV) applications. Clearly, such new requirements will affect not only the electronic circuits needed to generate the radiation and to receive the scattered signals but also the design of the MWI sensors that can be adopted.

An interesting development is fabrication of sensors on flexible substrates, which allow them to operate in tight contact with the part of the body to be imaged. This can be a significant benefit especially in biomedical applications, for example breast imaging [83] or brain stroke detection [84] because it would allow realisation of wearable imaging devices. However, the properties of the antenna, as well as the radiated EM field, will change with modification in its shape, which may be difficult to predict.

Finally, the exploitation of meta-materials [85] as well as the possibility of realising 3D printed sensing structures [86,87] constitute two research directions that would be interesting to follow even in the MWI community.

References

[1] Pike E R, Sabatier P C (eds.), *Scattering: Scattering and Inverse Scattering in Pure and Applied Science*, Academic Press, 2002

[2] Pastorino M, *Microwave Imaging*, Wiley, 2010

[3] Nikolova N D, *Introduction to Microwave Imaging*, Cambridge University Press, 2017

[4] Chen X, *Computational Methods for Electromagnetic Inverse Scattering*, John Wiley & Sons, 2018

[5] Chew W C, *Waves and Fields in Inhomogeneous Media*, IEEE Press, 1995

[6] Pastorino M, Randazzo A, *Microwave Imaging Methods and Applications*, Artech House, 2018

[7] Colton D, Kress R, *Inverse Acoustic and Electromagnetic Scattering Theory*, Springer New York, 2013

[8] Zhurbenko V, Rubaek T, Krozer V, Meincke P, 'Design and realisation of a microwave three-dimensional imaging system with application to breast-cancer detection', *IET Microw. Antennas Propag.*, 2010, **4** (12), pp 2200–2211

[9] Hopfer M, Planas R, Hamidipour A, Henriksson T, Semenov S, 'Electromagnetic tomography for detection, differentiation, and monitoring of brain stroke: A virtual data and human head phantom study', *IEEE Antennas Propag. Mag.*, 2017, **59** (5), pp 86–97

[10] Grzegorczyk T M, Meaney P M, Kaufman P A, di Florio-Alexander R M, Paulsen K D, 'Fast 3-D tomographic microwave imaging for breast cancer detection', *IEEE Trans. Med. Imaging*, 2012, **31** (8), pp 1584–1592

[11] Fear E C, Bourqui J, Curtis C, Mew D, Docktor B, Romano C, 'Microwave breast imaging with a monostatic radar-based system: A study of application to patients', *IEEE Trans. Microw. Theory Tech.*, 2013, **61** (5), pp 2119–2128

[12] Byrne D, Sarafianou M, Craddock I J, 'Compound radar approach for breast imaging', *IEEE Trans. Biomed. Eng.*, 2017, **64** (1), pp 40–51

[13] Lee S, Kwon S, 'Instantaneous microwave imaging with time-domain measurements for breast cancer detection', *Electron. Lett.*, 2013, **49** (10), pp 639–641

[14] Benedetto A, Pajewski L, *Civil Engineering Applications of Ground Penetrating Radar*, Springer, 2015

[15] Persico R, *Introduction to Ground Penetrating Radar: Inverse Scattering and Data Processing*, Wiley, 2014

[16] Zoughi R, *Microwave Non-Destructive Testing and Evaluation*, Kluwer Academic Publishers, 2000

[17] Boero F, Fedeli A, Lanini M, *et al.*, 'Microwave tomography for the inspection of wood materials: Imaging system and experimental results', *IEEE Trans. Microw. Theory Tech.*, 2018, **66** (7), pp 3497–3510

[18] Fedeli A, Maffongelli M, Monleone R, *et al.*, 'A tomograph prototype for quantitative microwave imaging: Preliminary experimental results', *J. Imaging*, 2018, **4** (12), p 139

[19] Meaney P M, Fox C J, Geimer S D, Paulsen K D, 'Electrical characterization of glycerin: Water mixtures: Implications for use as a coupling medium in microwave tomography', *IEEE Trans. Microw. Theory Tech.*, 2017, **65** (5), pp 1471–1478

[20] Bisio I, Fedeli A, Lavagetto F, *et al.*, 'A numerical study concerning brain stroke detection by microwave imaging systems', *Multimed. Tools Appl.*, 2018, **77** (8), pp 9341–9363

[21] Maffongelli M, Poretti S, Salvadè A, *et al.*, 'Preliminary test of a prototype of microwave axial tomograph for medical applications', *Proceedings of 2015 IEEE International Symposium on Medical Measurements and Applications*, Turin, Italy, 7–9 May 2015, pp 46–51

[22] Tournier P H, Bonazzoli M, Dolean V, *et al.*, 'Numerical modeling and high-speed parallel computing: New perspectives on tomographic microwave imaging for brain stroke detection and monitoring', *IEEE Antennas Propag. Mag.*, 2017, **59** (5), pp 98–110

[23] Asefi M, LoVetri J, 'Use of field-perturbing elements to increase non-redundant data for microwave imaging systems', *IEEE Trans. Microw. Theory Tech.*, 2017, **65** (9), pp 3172–3179

[24] Nemez K, Baran A, Asefi M, LoVetri J, 'Modeling error and calibration techniques for a faceted metallic chamber for magnetic field microwave imaging', *IEEE Trans. Microw. Theory Tech.*, 2017, **65** (11), pp 4347–4356

[25] Balanis C A, *Antenna Theory: Analysis and Design*, John Wiley & Sons, 2015

[26] Zeng X, Fhager A, He Z, Persson M, Linner P, Zirath H, 'Development of a time domain microwave system for medical diagnostics', *IEEE Trans. Instrum. Meas.*, 2014, **63** (12), pp 2931–2939

[27] Ojaroudi N, Ojaroudi M, Ghadimi N, 'UWB omnidirectional square monopole antenna for use in circular cylindrical microwave imaging systems', *IEEE Antennas Wirel. Propag. Lett.*, 2012, **11**, pp 1350–1353

[28] Jafari H M, Deen M J, Hranilovic S, Nikolova N K, 'A study of ultrawideband antennas for near-field imaging', *IEEE Trans. Antennas Propag.*, 2007, **55** (4), pp 1184–1188

[29] Casu M R, Vacca M, Tobon J A, *et al.*, 'A COTS-based microwave imaging system for breast-cancer detection', *IEEE Trans. Biomed. Circuits Syst.*, 2017, **11** (4), pp 804–814

[30] Capobianco A D, Khan M S, Caruso M, Bevilacqua A, '3-18 GHz compact planar antenna for short-range radar imaging', *Electron. Lett.*, 2014, **50** (14), pp 1016–1018

[31] Pichot C, Jofre L, Peronnet G, Bolomey J, 'Active microwave imaging of inhomogeneous bodies', *IEEE Trans. Antennas Propag.*, 1985, **33** (4), pp 416–425

[32] Yu C, Yuan M, Zhang Y, Stang J, George R T, Ybarra G A, Joines W T, Liu Q H, 'Microwave imaging in layered media: 3-D image reconstruction from experimental data', *IEEE Trans. Antennas Propag.*, 2010, **58** (2), pp 440–448

[33] Huang Q, Qu L, Wu B, Fang G, 'UWB through-wall imaging based on compressive sensing', *IEEE Trans. Geosci. Remote Sens.*, 2010, **48** (3), pp 1408–1415

[34] Salvador S M, Vecchi G, 'Experimental tests of microwave breast cancer detection on phantoms', *IEEE Trans. Antennas Propag.*, 2009, **57** (6), pp 1705–1712

[35] Mobashsher A T, Bialkowski K S, Abbosh A M, 'Design of compact cross-fed three-dimensional slot-loaded antenna and its application in wideband head imaging system', *IEEE Antennas Wirel. Propag. Lett.*, 2016, **15**, pp 1856–1860

[36] Zamani A, Abbosh A M, Mobashsher A T, 'Fast frequency-based multistatic microwave imaging algorithm with application to brain injury detection', *IEEE Trans. Microw. Theory Tech.*, 2016, **64** (2), pp 653–662

[37] Ardekani M R M, Lambot S, 'Full-wave calibration of time- and frequency-domain ground-penetrating radar in far-field conditions', *IEEE Trans. Geosci. Remote Sens.*, 2014, **52** (1), pp 664–678

[38] Serhir M, Lesselier D, 'Wideband reflector-backed folded bowtie antenna for ground penetrating radar', *IEEE Trans. Antennas Propag.*, 2018, **66** (3), pp 1056–1063

[39] Fear E C, Li X, Hagness S C, Stuchly M A, 'Confocal microwave imaging for breast cancer detection: Localization of tumors in three dimensions', *IEEE Trans. Biomed. Eng.*, 2002, **49** (8), pp 812–822

[40] Amineh R K, McCombe J, Khalatpour A, Nikolova N K, 'Microwave holography using point-spread functions measured with calibration objects', *IEEE Trans. Instrum. Meas.*, 2015, **64** (2), pp 403–417

[41] Haynes M, Stang J, Moghaddam M, 'Microwave breast imaging system prototype with integrated numerical characterization', *Int. J. Biomed. Imaging*, 2012, **2012** (706365), pp 1–18

[42] Bisio I, Estatico C, Fedeli A, *et al.*, 'Brain stroke microwave imaging by means of a Newton-conjugate-gradient method in Lp Banach spaces', *IEEE Trans. Microw. Theory Tech.*, 2018, **66** (8), pp 3668–3682

[43] Li X, Jalilvand M, Sit Y L, Zwick T, 'A compact double-layer on-body matched bowtie antenna for medical diagnosis', *IEEE Trans. Antennas Propag.*, 2014, **62** (4), pp 1808–1816

[44] Kanj H, Popovic M, 'Miniaturized microstrip-fed "dark eyes" antenna for near-field microwave sensing', *IEEE Antennas Wirel. Propag. Lett.*, 2005, **4**, pp 397–401

[45] Yurduseven O, Smith D, Elsdon M, 'Printed slot loaded bow-tie antenna with super wideband radiation characteristics for imaging applications', *IEEE Trans. Antennas Propag.*, 2013, **61** (12), pp 6206–6210

[46] Yun X, Fear E C, Johnston R H, 'Compact antenna for radar-based breast cancer detection', *IEEE Trans. Antennas Propag.*, 2005, **53** (8), pp 2374–2380

[47] Browne K E, Burkholder R J, Volakis J L, 'Through-wall opportunistic sensing system utilizing a low-cost flat-panel array', *IEEE Trans. Antennas Propag.*, 2011, **59** (3), pp 859–868

[48] Kang W, Lee S, Kim K, 'A ground-folded slot antenna for imaging radar applications', *IEEE Antennas Wirel. Propag. Lett.*, 2011, **10**, pp 155–158

[49] Krishna R V S R, Kumar R, 'A dual-polarized square-ring slot antenna for UWB, imaging, and radar applications', *IEEE Antennas Wirel. Propag. Lett.*, 2016, **15**, pp 195–198

[50] Klemm M, Leendertz J A, Gibbins D, Craddock I J, Preece A, Benjamin R, 'Microwave radar-based differential breast cancer imaging: Imaging in homogeneous breast phantoms and low contrast scenarios', *IEEE Trans. Antennas Propag.*, 2010, **58** (7), pp 2337–2344

[51] Gibbins D, Byrne D, Henriksson T, Monsalve B, Craddock I J, 'Less becomes more for microwave imaging: Design and validation of an ultrawide-band measurement array', *IEEE Antennas Propag. Mag.*, 2017, **59** (5), pp 72–85

[52] Rezaeieh S A, Zamani A, Abbosh A M, '3-D wideband antenna for head-imaging system with performance verification in brain tumor detection', *IEEE Antennas Wirel. Propag. Lett.*, 2015, **14**, pp 910–914

[53] Kim Y J, Jofre L, Flaviis F D, Feng M Q, 'Microwave reflection tomographic array for damage detection of civil structures', *IEEE Trans. Antennas Propag.*, 2003, **51** (11), pp 3022–3032

[54] Sugitani T, Kubota S, Toya A, Xiao X, Kikkawa T, 'A compact 4x4 planar UWB antenna array for 3-D breast cancer detection', *IEEE Antennas Wirel. Propag. Lett.*, 2013, **12**, pp 733–736

[55] Ghasr M T, Abou-Khousa M A, Kharkovsky S, Zoughi R, Pommerenke D, 'Portable real-time microwave camera at 24 GHz', *IEEE Trans. Antennas Propag.*, 2012, **60** (2), pp 1114–1125

[56] Charvat G L, Kempel L C, Rothwell E J, Coleman C M, Mokole E L, 'A Through-dielectric ultrawideband (UWB) switched-antenna-array radar imaging system', *IEEE Trans. Antennas Propag.*, 2012, **60** (11), pp 5495–5500

[57] Ghasr M T, Horst M J, Dvorsky M R, Zoughi R, 'Wideband microwave camera for real-time 3-D imaging', *IEEE Trans. Antennas Propag.*, 2017, **65** (1), pp 258–268

[58] Chiappe M, Gragnani G L, 'Vivaldi antennas for microwave imaging: Theoretical analysis and design considerations', *IEEE Trans. Instrum. Meas.*, 2006, **55** (6), pp 1885–1891

[59] Moosazadeh M, Kharkovsky S, Case J T, Samali B, 'Miniaturized UWB antipodal Vivaldi antenna and its application for detection of void inside concrete specimens', *IEEE Antennas Wirel. Propag. Lett.*, 2017, **16**, pp 1317–1320

[60] Zhuge X, Yarovoy A G, 'A sparse aperture MIMO-SAR-based UWB imaging system for concealed weapon detection', *IEEE Trans. Geosci. Remote Sens.*, 2011, **49** (1), pp 509–518

[61] Bourqui J, Okoniewski M, Fear E C, 'Balanced antipodal Vivaldi antenna with dielectric director for near-field microwave imaging', *IEEE Trans. Antennas Propag.*, 2010, **58** (7), pp 2318–2326

[62] Ostadrahimi M, Zakaria A, LoVetri J, Shafai L, 'A near-field dual polarized (TE-TM) microwave imaging system', *IEEE Trans. Microw. Theory Tech.*, 2013, **61** (3), pp 1376–1384

[63] Mohammed B J, Abbosh A M, Mustafa S, Ireland D, 'Microwave system for head imaging', *IEEE Trans. Instrum. Meas.*, 2014, **63** (1), pp 117–123

[64] Abbak M, Akinci M N, Cayoren M, Akduman I, 'Experimental microwave imaging with a novel corrugated Vivaldi antenna', *IEEE Trans. Antennas Propag.*, 2017, **65** (6), pp 3302–3307

[65] Biswas B, Ghatak R, Poddar D R, 'A fern fractal leaf inspired wideband antipodal Vivaldi antenna for microwave imaging system', *IEEE Trans. Antennas Propag.*, 2017, **65** (11), pp 6126–6129

[66] Kempin M, Ghasr M T, Case J T, Zoughi R, 'Modified waveguide flange for evaluation of stratified composites', *IEEE Trans. Instrum. Meas.*, 2014, **63** (6), pp 1524–1534

[67] Geffrin J M, Sabouroux P, 'Continuing with the Fresnel database: Experimental setup and improvements in 3D scattering measurements', *Inverse Probl.*, 2009, **25** (2), p 024001

[68] Dehmollaian M, Thiel M, Sarabandi K, 'Through-the-wall imaging using differential SAR', *IEEE Trans. Geosci. Remote Sens.*, 2009, **47** (5), pp 1289–1296

[69] Henriksson T, Joachimowicz N, Conessa C, Bolomey J C, 'Quantitative microwave imaging for breast cancer detection using a planar 2.45 GHz system', *IEEE Trans. Instrum. Meas.*, 2010, **59** (10), pp 2691–2699

[70] Laviada J, Arboleya-Arboleya A, Alvarez-Lopez Y, Garcia-Gonzalez C, Las-Heras F, 'Phaseless synthetic aperture radar with efficient sampling for broadband near-field imaging: Theory and validation', *IEEE Trans. Antennas Propag.*, 2015, **63** (2), pp 573–584

[71] Yoon Y S, Amin M G, 'Spatial filtering for wall-clutter mitigation in through-the-wall radar imaging', *IEEE Trans. Geosci. Remote Sens.*, 2009, **47** (9), pp 3192–3208

[72] Amineh R K, Ravan M, Trehan A, Nikolova N K, 'Near-field microwave imaging based on aperture raster scanning with TEM horn antennas', *IEEE Trans. Antennas Propag.*, 2011, **59** (3), pp 928–940

[73] Jezova J, Lambot S, Fedeli A, Randazzo A, 'Ground-penetrating radar for tree trunk investigation', *Proceedings of 9th International Workshop on Advanced Ground Penetrating Radar*, Edinburgh, UK, 28–30 June 2017, pp 1–6

[74] Yarovoy A G, Savelyev T G, Aubry P J, Lys P E, Ligthart L P, 'UWB array-based sensor for near-field imaging', *IEEE Trans. Microw. Theory Tech.*, 2007, **55** (6), pp 1288–1295

[75] Burfeindt M J, Behdad N, van Veen B D, Hagness S C, 'Quantitative microwave imaging of realistic numerical breast phantoms using an enclosed array of multiband, miniaturized patch antennas', *IEEE Antennas Wirel. Propag. Lett.*, 2012, **11**, pp 1626–1629

[76] Aguilar S M, Al-Joumayly M A, Burfeindt M J, Behdad N, Hagness S C, 'Multiband miniaturized patch antennas for a compact, shielded microwave breast imaging array', *IEEE Trans. Antennas Propag.*, 2014, **62** (3), pp 1221–1231

[77] Rokunuzzaman M, Samsuzzaman M, Islam M T, 'Unidirectional wideband 3-D antenna for human head-imaging application', *IEEE Antennas Wirel. Propag. Lett.*, 2017, **16**, pp 169–172

[78] Gibbins D, Klemm M, Craddock I J, Leendertz J A, Preece A, Benjamin R, 'A comparison of a wide-slot and a stacked patch antenna for the purpose of breast cancer detection', *IEEE Trans. Antennas Propag.*, 2010, **58** (3), pp 665–674

[79] Chen P, Shastry M C, Lai C, Narayanan R M, 'A portable real-time digital noise radar system for through-the-wall imaging', *IEEE Trans. Geosci. Remote Sens.*, 2012, **50** (10), pp 4123–4134

[80] Pastorino M, Randazzo A, Fedeli A, *et al.*, 'A microwave tomographic system for wood characterization in the forest products industry', *Wood Mater. Sci. Eng.*, 2015, **10** (1), pp 75–85

[81] Porter E, Coates M, Popović M, 'An early clinical study of time-domain microwave radar for breast health monitoring', *IEEE Trans. Biomed. Eng.*, 2016, **63** (3), pp 530–539

[82] Sabouni A, Kishk A A, 'Dual-polarized, broadside, thin dielectric resonator antenna for microwave imaging', *IEEE Antennas Wirel. Propag. Lett.*, 2013, **12**, pp 380–383

[83] Porter E, Bahrami H, Santorelli A, Gosselin B, Rusch L A, Popović M, 'A wearable microwave antenna array for time-domain breast tumor screening', *IEEE Trans. Med. Imaging*, 2016, **35** (6), pp 1501–1509

[84] Bashri M S R, Arslan T, 'Low-cost and compact RF switching system for wearable microwave head imaging with performance verification on artificial head phantom', *IET Microw. Antennas Propag.*, 2018, **12** (5), pp 706–711

[85] Alqadami A S M, Bialkowski K S, Mobashsher A T, Abbosh A M, 'Wearable electromagnetic head imaging system using flexible wideband antenna array based on polymer technology for brain stroke diagnosis', *IEEE Trans. Biomed. Circuits Syst.*, 2019, **13** (1), pp 124–134

[86] Genovesi S, Costa F, Fanciulli F, Monorchio A, 'Wearable inkjet-printed wideband antenna by using miniaturized AMC for sub-GHz applications', *IEEE Antennas Wirel. Propag. Lett.*, 2016, **15**, pp 1927–1930

[87] Molaei A, Bisulco A, Tirado L, *et al.*, '3-D-printed E-band compressive horn antenna for high-sensing-capacity imaging applications', *IEEE Antennas Wirel. Propag. Lett.*, 2018, **17** (9), pp 1639–1642

Matteo Pastorino is a Full Professor of Electromagnetic Fields at the University of Genoa, Italy, where is the Director of the Department of Electrical, Electronic, Telecommunications Engineering and Naval Architecture. His current research interests include microwave and millimeter wave imaging, direct and inverse scattering problems, industrial and medical applications, smart antennas and analytical and numerical methods in electromagnetism. He has co-authored about 500 papers in scientific journals and conference proceedings.

Andrea Randazzo is a Full Professor of Electromagnetic Fields at the Department of Electrical, Electronic, Telecommunication Engineering and Naval Architecture of the University of Genoa. His primary research interests are in the field of microwave imaging, inverse scattering techniques, numerical methods for electromagnetic scattering and propagation, and smart antennas. He is co-author of about 250 papers published in journals and conference proceedings.

Alessandro Fedeli is an Assistant Professor of Electromagnetic Fields at the University of Genoa, Italy, in the Department of Electrical, Electronic, Telecommunications Engineering and Naval Architecture. He received the MSc and PhD degrees from the University of Genoa in 2013 and 2017, respectively. His research is mainly focused on microwave imaging and computational methods for the solution of forward and inverse electromagnetic scattering problems. He has co-authored more than 90 papers in international scientific journals and conference proceedings.

Chapter 8

Electro-magnetic imaging with ultra-wideband (UWB) sensors

Mohd Zaid Abdullah[1]

List of abbreviations

CDF	central difference
CF	coherence factor
DAS	delay and sum
EDAS	enhanced delay and sum
EM	electro-magnetic
FDTD	finite difference time domain
FQ	focusing quality
GPIB	general purpose interface bus
PML	perfectly matching layers
SMA	sub-miniature version A
SNR	signal-to-noise ratio
TM	transverse magnetic
UWB	ultra-wideband
VNA	vector network analyser

8.1 Introduction

Ultra-wideband (UWB) is an emerging technique, which can provide reliable information on broad range of inhomogeneity. Because it is non-invasive and tissues are not subjected to harmful radiation, UWB has become a potential for medical sensing applications. UWB imaging relies on solving the nonlinear inverse scattering problem to recover the unknown dielectric profile of an imaging domain. However, solving the inverse problem is mathematically complicated, and requires intensive computation to find an optimum solution. This complexity increases

[1]School of Electrical and Electronic Engineering, Universiti Sains Malaysia, USM Engineering Campus, Penang, Malaysia

when dealing with biological materials, because dispersion causes on UWB waves, resulting in different response modes. This effect must be fully accounted for accurate image reconstruction. Recently, the trend is moving towards an iterative approach to obtaining more stable reconstruction and more accurate solution than single-step methods. Such algorithms [1,2] employ an adjoint method to calculate a sensitivity matrix by forward propagation of the wave-field, and backward propagation of the residual waves, resulting in nonlinear optimisation and requiring Lagrange multipliers to impose constraints. In [3,4], both single-step and iterative solutions are computed with this method. A second strategy is to calculate the gradient using the central difference (CDF) approximation. Unlike the adjoint method, where the computational demand is bounded to the speed of the forward and backward solvers, the computational demand of the CDF approximation is a function of the number of unknown variables, and therefore, imposes limits to the choice of domain and pixel sizes. To reduce the computational demand of the CDF calculations, Jacobian updating is achieved by means of Broyden's formula. The relatively small imaging domain size makes the implementation of this technique manageable.

Another much simpler and faster approach is to calculate the response of the scatterers by aligning the arriving signals via a focusing algorithm. This leads to a single-step approach, such as the popular delay and sum (DAS) algorithm and its variations [5,6]. Compared to iterative methods, the single-step methods are computationally less demanding because they do not solve the inverse problem directly. However, the final outcome is limited to qualitative images of varying energy intensity.

Both the multi-step and singe-step solutions to UWB inverse scattering problem are first discussed in this chapter. In the former, the CDF-Broyden update using the refined Polak-Rebiere's non-quadratic approximation will be focused. Meanwhile, the enhanced delay and sum (EDAS) algorithm, which uses the coherence factor and multiply in pair will be the principal topic of the later technique. In both cases the dispersive properties of the media are modelled using Debye first-order system, while the conformal mapping is deployed to accurately map curved geometries. The second part of the chapter focuses on an UWB radar data acquisition system, which is specifically developed to assess the performance of these algorithms experimentally. The hardware consists of custom-designed wideband antennas mounted on arrays and connected to a vector network analyser (VNA) via a 16-channel multiplexer switch unit. This set-up allows measurements be performed in the time domain automatically. Both limited-view and full-view geometries are used for measurements.

8.2 Image reconstruction

Implementation of an UWB imaging system requires a solution to two problems: the forward (or direct) problem, in which electro-magnetic (EM) wave propagation is calculated as a function of the media's EM properties, and the inverse problem, in which scattered signals are used to reconstruct the spatial dielectric field distribution of an imaging domain.

8.2.1 Forward problem

The forward problems deal with the propagation of EM waves, which in linear, isotropic and non-dispersive media is governed by Maxwell's curl equations given by

$$\nabla \times E + \mu \frac{\partial H}{\partial t} = 0 \tag{8.1}$$

$$\nabla \times H - \varepsilon \frac{\partial E}{\partial t} - \sigma E = 0 \tag{8.2}$$

where E and H are the electric and magnetic flux respectively in (x,y,z) dimensions; ε, μ and σ are the electrical permittivity, magnetic permeability and electrical conductivity, respectively.

Given the wide-band excitation, it is more suitable to solve (8.1) and (8.2) in the time domain than the frequency domain. In this case the finite difference time domain (FDTD) method [7] is the most popular because of its simplicity conceptually and in terms of implementation. For TM polarisation, FDTD equations for 2D can be expressed as

$$\begin{aligned} &\frac{E_z^{n+\frac{1}{2}}(i,j) - E_z^{n-\frac{1}{2}}(i,j)}{\Delta t} \\ &= \frac{1}{\varepsilon_0 \varepsilon_r(i,j)} \left(\frac{H_y^n\left(i+\frac{1}{2},j\right) - H_y^n\left(i-\frac{1}{2},j\right)}{\Delta x} - \frac{H_x^n\left(i,j+\frac{1}{2}\right) - H_x^n\left(i,j-\frac{1}{2}\right)}{\Delta y} \right) \\ &- \frac{\sigma(i,j)}{\varepsilon_0 \varepsilon_r(i,j)} \left(E_z^{n+\frac{1}{2}}(i,j) - E_z^{n-\frac{1}{2}}(i,j) \right) \end{aligned} \tag{8.3a}$$

$$\frac{H_x^{n+1}\left(i,j+\frac{1}{2}\right) - H_x^n\left(i,j+\frac{1}{2}\right)}{\Delta t} = -\frac{1}{\mu_o} \frac{E_z^{n+\frac{1}{2}}(i,j+1) - E_z^{n+\frac{1}{2}}(i,j)}{\Delta y} \tag{8.3b}$$

$$\frac{H_y^{n+1}\left(i+\frac{1}{2},j\right) - H_y^n\left(i+\frac{1}{2},j\right)}{\Delta t} = \frac{1}{\mu_o} \frac{E_z^{n+\frac{1}{2}}(i+1,j) - E_z^{n+\frac{1}{2}}(i,j)}{\Delta x} \tag{8.3c}$$

Solving for $E_x^{n+\frac{1}{2}}(i,j), H_x^{n+1}\left(i,j+\frac{1}{2}\right)$ and $H_y^{n+1}\left(i+\frac{1}{2},j\right)$ and assuming a uniform grid ($\Delta x = \Delta y$), the following update equations are obtained:

$$\begin{aligned} E_z^{n+\frac{1}{2}}(i,j) &= E_z^{n-\frac{1}{2}}(i,j) \left[\frac{1 - p(i,j)\sigma(i,j)}{1 + p(i,j)\sigma(i,j)} \right] \\ &+ \left[\frac{p(i,j)/\Delta x}{1 + p(i,j)\sigma(i,j)} \right] \left(H_y^n\left(i+\frac{1}{2},j\right) - H_y^n\left(i-\frac{1}{2},j\right) \right) \\ &- \left[\frac{p(i,j)/\Delta y}{1 + p(i,j)\sigma(i,j)} \right] \left(H_x^n\left(i,j+\frac{1}{2}\right) - H_x^n\left(i,j-\frac{1}{2}\right) \right) \end{aligned} \tag{8.4a}$$

$$H_x^{n+1}\left(i,j+\frac{1}{2}\right) = H_x^n\left(i,j+\frac{1}{2}\right) - \frac{\Delta t}{\mu_o \Delta y}\left(E_z^{n+\frac{1}{2}}(i,j+1) - E_z^{n+\frac{1}{2}}(i,j)\right) \quad (8.4b)$$

$$H_y^{n+1}\left(i+\frac{1}{2},j\right) = H_y^n\left(i+\frac{1}{2},j\right) + \frac{\Delta t}{\mu_o \Delta x}\left(E_z^{n+\frac{1}{2}}(i+1,j) - E_z^{n+\frac{1}{2}}(i,j)\right) \quad (8.4c)$$

where $p(i,j) = \Delta t/\varepsilon_0\varepsilon_r(i,j)$.

In implementing (8.3) and (8.4), let's consider 2D bounded domains with limited-view and full-view geometries and shown in Figure 8.1. These domains can be occupied by a single or multiple dielectrics, described by spatial distribution of their dielectric properties, and terminated with absorbing boundaries. Both the Mur boundary conditions [8] and the perfectly matching layers [9] are considered. Theoretically, such a truncation allows EM waves to be absorbed with a reflection as short and as small as possible. However, numerical experiments showed that some residual reflections occur at the boundary interfaces but the magnitude is too small and hence has little effect on measurement accuracy. Also solving (8.3) and (8.4) requires the determination of time resolution Δt in FDTD modelling. Ideally, Δt is as large as possible so that simulation can run quickly. However, if this parameter is too large, then the EM field cannot be adequately sampled, resulting in the measurement error due to aliasing. For the scheme presented, the maximum time step is determined by the Courant's stability condition [10].

8.2.1.1 Conformal FDTD

The FDTD method divides the computational domain into Cartesian cubical cells in 3D or rectangular cells in 2D. This presents a problem if curved objects are

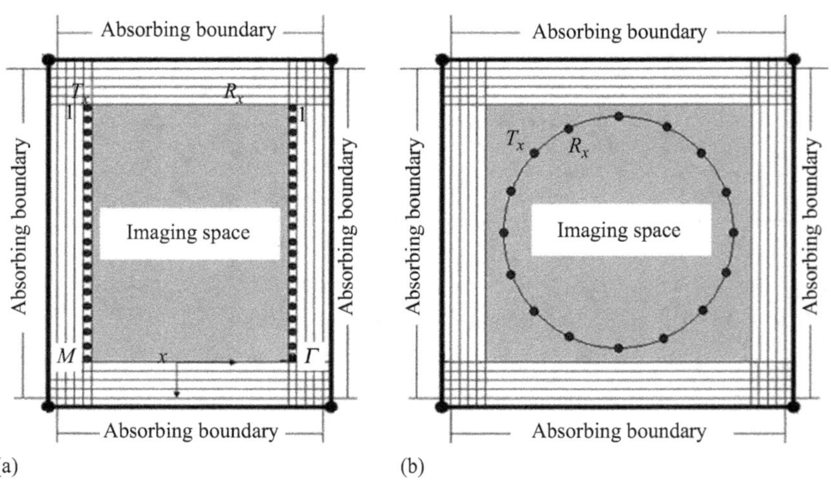

Figure 8.1 Computational domain showing imaging plane and absorbing boundaries: (a) limited-view and (b) full-view geometries. In both cases 'R_x', 'T_x' and 'Γ' denote transmitter, receiver and imaging space, respectively

modelled. To solve this problem, rectangular cells can either be made smaller to increase the resolution of the domain, that is decrease the spatial step to increase the staircase approximation of the curved surface, or use other approximating techniques, such as the conformal mapping. The major drawback of the first approach is the computational cost. The second technique, however, approximates cells partially filled with different dielectric materials by using a weighted area or volume average to calculate its effective dielectric constant. The same technique has been applied to calculate the effective dielectric constant using the individual electric field component along the edge of the cell, as shown in Figure 8.2.

With reference to the curved geometry in Figure 8.2, the updated FDTD equations for deformed cells containing mixed dielectric materials are modified to incorporate the effective dielectric constant defined as the weighted volume average of the two dielectrics (for the 3D case), and expressed as

$$\varepsilon^{\text{eff}}(i,j,k) = (V_2(i,j,k)\varepsilon_2 + (V - V_2(i,j,k))\varepsilon_1)/V(i,j,k) \tag{8.5a}$$

where V is the total cell volume, $V_2(i,j,k)$ is the fractional volume of the (i,j,k)th cell that contains the dielectric material ε_2 and ε_1 is the relative permittivity of the medium in the remaining volume.

Subsequently for the 2D case and using the area instead of volume, the effective dielectric constant is expressed as

$$\varepsilon^{\text{eff}}(i,j) = (A_2(i,j)\varepsilon_2 + (A - A_2(i,j))\varepsilon_2)/A(i,j) \tag{8.5b}$$

where A is the total cell area, $A_2(i,j)$ is the fractional area of the (i,j)th cell that contains the dielectric material ε_2 and ε_1 is the relative permittivity of the medium in the remaining area.

This procedure can be extended to cells containing more than two different dielectric media. Once the effective dielectric constant of the cells has been computed, the conventional FDTD method can be used to update the electric-field

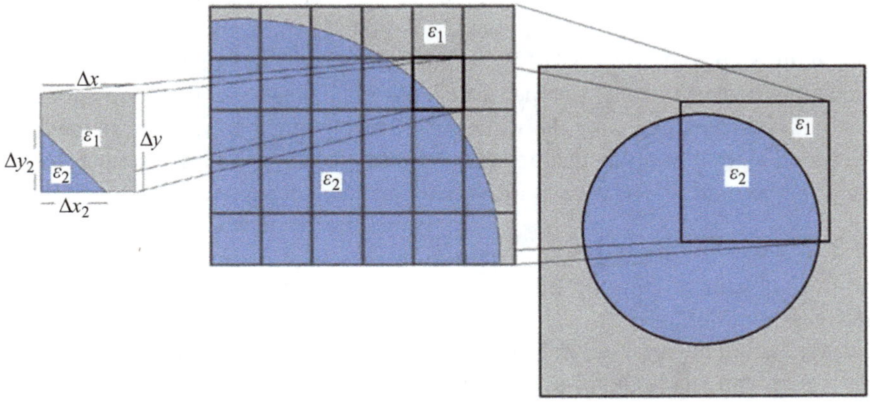

Figure 8.2 FDTD cell containing mixed materials obtained from intersection of FDTD grid and actual curved dielectric

components [11]. With reference to (8.3a–8.3c), the effective dielectric constant can be incorporated in the FDTD calculation by modifying $p(i,j)$ so that

$$p(i,j) = \Delta t / \varepsilon_0 \varepsilon_r^{\text{eff}}(i,j) \tag{8.6}$$

where the suffix $\varepsilon_r^{\text{eff}}$ denotes the effective dielectric constant.

8.2.1.2 Complex permittivity and dispersive model

Biological tissues exhibit a varying dispersive dielectric response, depending on the amount of water content. For higher frequencies in excess of a few hundred MHz, the dipolar orientation of the water molecules is the dominant polarisation mechanism [12]. In general, the complex permittivity is given as

$$\varepsilon = \varepsilon' + j\frac{\sigma}{\omega} \tag{8.7}$$

where ε' is the complex permitivity, σ is the conductivity and ω is the frequency.

The dispersive response of biological tissues can be accurately and effectively incorporated into the FDTD code using the Debye first order model [13,14]. Debye [15] described the frequency dependence of the complex relative permittivity as

$$\varepsilon(\omega) = \varepsilon_\infty + j\frac{\sigma}{\omega\varepsilon_s} + \frac{\varepsilon_s - \varepsilon_\infty}{1 + j\omega\tau} \tag{8.8}$$

where ε_∞ and ε_s are the infinite and static relative permittivities, respectively, and τ is the relaxation time.

Equation (8.8) can be used to model the dispersive behaviour of both fat and skin tissues. Following [16], the Debye's first order media are modelled into the FDTD solver.

8.2.2 Inverse problem

Like other tomography systems, solving the image reconstruction problem in UWB requires solutions to two distinct problems – the forward and inverse problems. Both multi-step and single-step algorithms are considered in this chapter. The distinction between the two types of algorithm is that the first one produces quantitative image, depicting the values of permittivity for each pixel. In contrast, the second algorithm offers qualitative solution, resulting in an image depicting the change in permittivity relative to an initially acquired reference.

8.2.2.1 Adjoint-Broyden

The difficulty in solving the inverse problem arises from two main aspects: ill-posedness and nonlinearity [2,17]. While the nonlinearity of the problem can be addressed using iterative optimisation methods, the use of regularisation and priori information can tackle the ill-posedness of the problem. This priori information is obtained from the known physical characteristics of the media [18].

The aim of optimisation is to solve the material properties $\boldsymbol{x} = (\varepsilon, \sigma)$ of the medium Γ that minimises the cost function F defined as the least square between the calculated field E^{calc} and measured field E^{meas} [2,4]. Assuming a total of M transmitters and N receivers, the residual error of the N receiving locations for the mth projection over the time t in transverse magnetic (TM) mode is defined as

$$r(\boldsymbol{x}) = \frac{1}{2}\sum_{n=1}^{N}\int_{T}\int_{\Gamma}\left(E_z^{\text{calc}} n(\boldsymbol{x}) - E_n^{\text{meas}}\right) d\Gamma dt \tag{8.9}$$

Subsequently, the cost function for the total M projections is

$$F(\boldsymbol{x}) = \sum_{m=1}^{M} r_m^2 \tag{8.10}$$

The Jacobian of the residuals $\boldsymbol{r}$ for a single mth projection is defined as

$$\boldsymbol{J}_m = \partial r_m / \partial \boldsymbol{x} \tag{8.11}$$

For $P = i \times j$ defined as the total number of unknown variables, the $M \times P$ Jacobian of the vector of residuals r can be written as

$$\boldsymbol{J} = \begin{bmatrix} \boldsymbol{J}_1 \\ \boldsymbol{J}_2 \\ \vdots \\ \boldsymbol{J}_M \end{bmatrix} = \begin{bmatrix} \dfrac{\partial r_1}{\partial x_1} & \cdots & \dfrac{\partial r_1}{\partial x_p} \\ \vdots & \ddots & \vdots \\ \dfrac{\partial r_M}{\partial x_1} & \cdots & \dfrac{\partial r_M}{\partial x_p} \end{bmatrix} \tag{8.12}$$

The gradient $\nabla F(\boldsymbol{x})$ can be calculated by

$$\frac{\partial F(x)}{\partial \boldsymbol{x}} = \frac{\partial}{\partial \boldsymbol{x}}\sum_{m} r_m^2 = \sum_{m} 2 r_m \frac{\partial r_m}{\partial \boldsymbol{x}} \tag{8.13}$$

Consequently,

$$\boldsymbol{g} = \nabla F(\boldsymbol{x}) = \frac{\partial F(x)}{\partial \boldsymbol{x}} = 2\boldsymbol{J}^T r(\boldsymbol{x}) \tag{8.14}$$

The evaluation of the gradients using CDF approximation is a computationally intensive process, especially with a large number of pixels for higher resolution. To reduce the computational cost, a good estimate of the gradient is possible using secant methods for multi-variables, from which the Jacobian of the next iterative step is obtained. The best performance of these methods in practice is the one resulting from Broyden formula, where $\boldsymbol{J}^{k+1}$ can be obtained by making the least change to $\boldsymbol{J}^k$ consistent with the generalised form of the multi-variable secant equation [19,20]. The resulting formula is given as

$$\boldsymbol{J}^{k+1} = \boldsymbol{J}^k + \frac{\left(\Delta \boldsymbol{F}^k - \boldsymbol{J}^k \cdot \Delta \boldsymbol{x}^k\right) \otimes \Delta \boldsymbol{x}^k}{\Delta \boldsymbol{x}^k \cdot \Delta \boldsymbol{x}^k} \tag{8.15}$$

Unconstrained gradient-based optimisation solution for multivariable problems range between the steepest descent to Newton methods. The former converges orthogonally as the reduction step is chosen to achieve the maximum amount of decrease in F. Newton methods, on the other hand, converge faster by calculating the Hessian matrix. The solution, however, requires inverting the Hessian and therefore should be positive definite. Although converging in fewer steps, the computational cost of each step is very high, especially if a singular matrix is encountered, where a memory consuming approximate inversion solution may be needed. An intermediate method in the convergence performance is the conjugate directional method, which does not use pre-specified directions, but computes the direction as the algorithm progresses. This method can be applied for non-quadratic problems using the Polak-Rebiere's formula [21] given by

$$\beta_k = \frac{g^{(k+1)T}\left[g^{(k+1)} - g^{(k)}\right]}{g^{(k)T}g^{(k)}} \tag{8.16}$$

Using this approach, the first gradient can be evaluated using the CDF solution, while Broyden approximation together with the CDF may be evaluated interchangeably thereafter. The Jacobian is re-initialised to the CDF value after a few iterations as the Broyden estimates have been observed to deteriorate, resulting in slower convergence or lost the downhill direction. The same trend was reported elsewhere [22].

The ill-posedness of the inverse problem is a major difficulty as it produces ill-conditioned matrices in numerical inversion methods. The minimisation approach uses the gradient only, and therefore the problem of matrix inversion is avoided. Nevertheless, the ill-posedness and local minima characteristics have to be carefully considered by choosing a suitable initial guess with good proximity to the actual solution, and the utilisation of priori information, which can be used to set boundaries on the solution, enforcing it to lie within an appropriate range and limiting outlier solutions [23].

8.2.2.2 Enhanced delay and sum (EDAS)

Although the adjoint-based method discussed above is the most accurate solution to an inverse scattering problem, this algorithm is very computationally intensive because the minimisation involves the calculation of large and nearly singular Jacobian matrix. Most importantly, this method exhibits ill-conditioned property, and hence cannot perform satisfactorily in the presence of large amount of noise. Solving these problems lead to the development of another relatively simple image reconstruction algorithm based on the focusing technique. This method is similar to DAS algorithm which is among the earliest qualitative radar techniques developed for breast cancer detection [24]. The resulting algorithm is referred to as EDAS. The aim is primarily to create a pure response signal of the scatterer from measured data. Principally, the received information includes back-scattered signals from the scatterer and clutter (late arrivals) in addition to the back-scattered from medium-object interface and the input signal (early arrivals). Signal processing is used to remove the early arrivals, which dominate the back-scattered signals, and filtering them leads to improving the signal-to-noise ratio (SNR) of the scatterer response. An averaging procedure is used to suppress the artefact and other unwanted components. The noise is removed using a calibration process that avoids the use of a phantom or *a priori* information. In so doing, a reference signal is first generated by averaging all recorded signals as follows:

$$k(t) = \frac{1}{MN}\sum_{m=1}^{M}\sum_{n=1}^{N} x_{m,n}(t) \tag{8.17}$$

where $x_{m,n}(t)$ is the recorded back-scattered signal from mth transmitter and nth receiver.

Then the reference signal is subtracted from each of the back-scattered signals, resulting in the calibrated signals, $x'_{m,n}(t)$, which contains only the scatterer response and clutters:

$$x'_{m,n}(t) = k(t) - x_{m,n}(t) \tag{8.18}$$

The time delay for all recorded signals is then calculated for all transmitter and receiver locations and for each focal point, r, as follows:

$$\tau_{m,n}(r) = \frac{\|r - r_m\| + \|r - r_n\|}{v\Delta t} \tag{8.19}$$

where the speed of propagation is given by

$$v = \frac{c}{\sqrt{\varepsilon_r}} \tag{8.20}$$

and $\|r_m - r\|$ and $\|r_n - r\|$ are the distances between the focal point r and mth transmitter and nth receiver, Δt is the sampling time, c is the velocity of light in vacuum and ε_r is the relative permittivity of the medium.

Each calibrated signal in (8.18) is time-shifted with an appropriate delay time calculated in (8.19). Following time alignment, a focusing-quality procedure using the coherence factor (CF) is implemented. Traditionally, CF is used for anomaly correction and side lobe suppression in ultrasound imaging applications [25]. This procedure is based on the use of the coherence quality weighting of the back-scattered signals at specific focal points inside the domain. The same method can be applied for UWB imaging, which can significantly enhance the image as results from ultrasonic experiments suggest [26]. In this case, the coherence factor is defined as the ratio between the energy of the coherent sum of the total incoherent energy. Using (8.18) and (8.19), CF(r) is calculated as follows:

$$\mathrm{CF}(r) = \frac{\text{coherent energy}}{\text{total incoherent energy}} = \frac{\left|\sum_{m=1}^{M}\sum_{n=1}^{N} x'_{m,n}\left(t - \tau_{m,n}(r)\right)\right|^2}{\sum_{m=1}^{M}\sum_{n=1}^{N} \left|x'_{m,n}\left(t - \tau_{m,n}(r)\right)\right|^2} \tag{8.21}$$

The focusing quality $FQ_{m,n}(r)$ is then obtained by multiplying the coherence factor with the calibrated signal after time-shifting as

$$FQ_{m,n}(r) = CF(r)x'_{m,n}(t - \tau_{m,n}(r)) \tag{8.22}$$

The high-coherence signals (high coherence factor) represent the main lobe part, which creates an image with good focusing quality, and their amplitude should be maintained. On the other hand, the incoherent signals (low coherence factor) represent the side lobe part, which creates an image with poor focusing quality, and their amplitude should be suppressed. Consequently, the amplitude of each image pixel is adaptively weighted by the corresponding CF so that the unwanted side lobes are effectively reduced.

The enhancement in the detection capability of EDAS algorithm is based on the supplementary weighting process involving pairing multiplication of the back-scattered signals. In this case, each signal pair measured from the same transmitter is multiplied with each other and their products summed. The results from these procedures are squared and then integrated over a window length, w, producing the energy at each focal point:

$$E(r) = \int_0^w \left[\sum_{m=1}^{M} \sum_{n=1}^{N-1} FQ_{m,n}(r) FQ_{m,n+1}(r) \right]^2 dt \tag{8.23}$$

The assembly of $E(r)$ at each focal point produces the desired image, depicting the distributions of energy reflected by the scatterer. The working principle of EDAS is summarised in Figure 8.3.

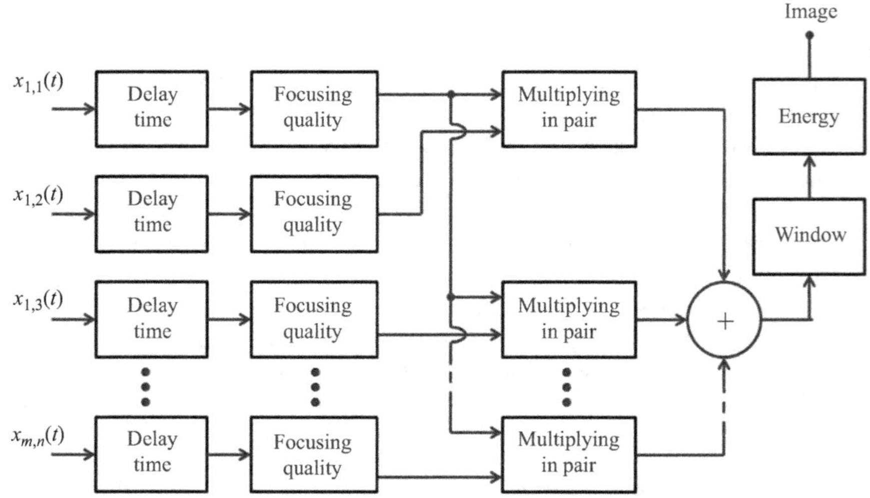

Figure 8.3 Block diagram of EDAS

8.3 Sensor and instrumentation system

To experimentally validate the methods and procedures discussed above, a radar measurement hardware needs to be built. At the forefront of this hardware is the sensor responsible for transmitting UWB waves into the medium and measuring the resulting energy scattered from various angles and viewing points. Another requirement is data automation because the measurements need to be repeated at different transmitter and receiver positions. These requirements are met with the use of 18 GHz multiplexer switch connected to personal computer via the general purpose instrument bus (GPIB). Data automation and signal processing are performed in LabView.

8.3.1 UWB antenna

UWB systems are defined as the systems, which use EM signals having greater than 500 MHz bandwidth or around 20% bandwidth around its centre frequency. There are various types of UWB antennas, which fulfil this requirement, ranging from the conventional omni-directional horn geometry to micro-strip patch type. Geometrically, the UWB antenna for tissue sensing has to be compact while exhibiting important characteristics like low return loss, high gain and cost effective. Two types of UWB antennas are considered: (1) the standard bi-conical antenna and (2) the printed type.

8.3.1.1 Bi-conical antenna

Bi-conical antenna is the first design of UWB sensor. This is a high-performance UWB antenna, which is suitable for microwave imaging. The power transmitted is constant with respect to the azimuth but varies as a function of the elevation, cone heights, cone angles and frequency [27]. The cone height of a bi-conical antenna primarily influences the lower cut-off frequency, while the cone angle influences the input impedance. A bi-conical antenna with cone angle of 45° and cone height of 10 mm is considered in this study. The antenna is driven via a 50 Ω coaxial feeder, which incorporates an in-line balun in the form of ferrite torroid, through which the coaxial cable is threaded [28]. This arrangement ensures a balanced antenna feed and stable frequency response. In the fabrication, the coax centre conductor is connected to the apex of the lower cone, while the coax outer conductor is connected to the upper cone. The height of upper and lower cones, H_U and H_L is 4 mm, respectively. The total height of the whole bi-conical antenna, H_C is 10 mm [29]. A cylindrical globule of epoxy resin (Araldite) at the cone apex provides mechanical strength to the antenna. Figure 8.4(a) shows the schematic of the antenna design while Figure 8.4(b) illustrates the internal connection between the antenna and coaxial cable. Meanwhile Figure 8.4(c) shows an example of a sensor array formed using this antenna design. In this application the coaxial cable with ferrite torroid is used to connect each sensor to the receiving electronics.

Both antenna cones shown in Figure 8.4 are machined from pure copper material, while an RG-58 coaxial is used to drive the antenna, achieving an

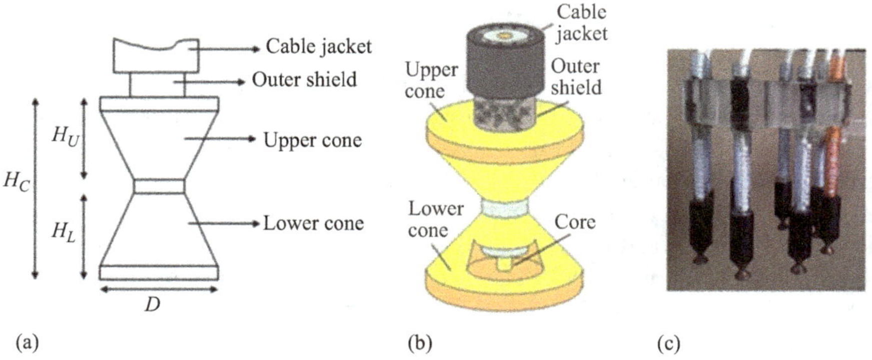

Figure 8.4 (a) Schematic of bi-conical antenna, (b) internal connection of coaxial cable with bi-conical antenna and (c) an array of eight equally distributed bi-conical antennas with coaxial cable connected with ferrite

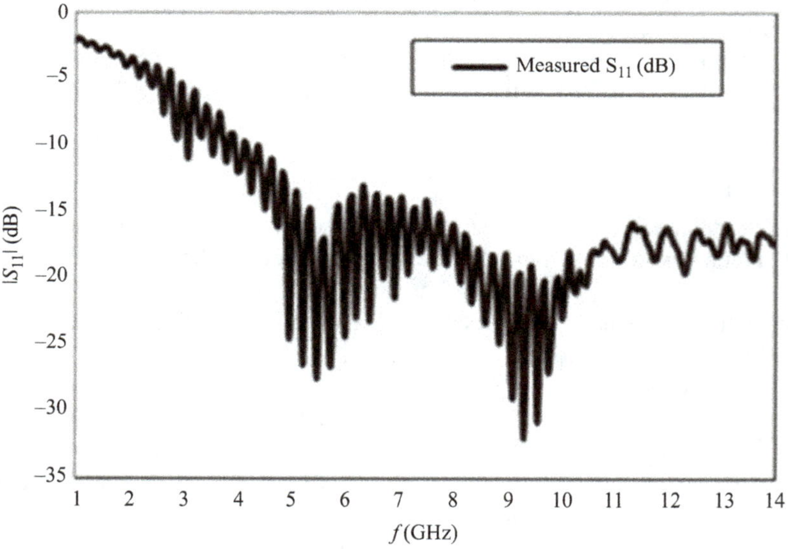

Figure 8.5 Measured reflection coefficient $|S_{11}|$ (dB) of bi-conical antenna

impedance matching of 50 Ω. To achieve high detection resolution and good clutter rejection, the antenna with an impedance bandwidth 4 to 10 GHz is typically employed [30]. Antenna measurements in free space are performed with an Agilent N5245A PNA-X Network Analyser, providing measurements up to 50 GHz. Figure 8.5 shows the measured reflection coefficient, S_{11} in free space. The bandwidth of the proposed antenna ranges from 3.73 to 14 GHz ($S_{11} < -10$ dB), with impedance matching greater than 115%. Hence, the frequency coverage of the proposed design is sufficiently wide and relatively stable for the desired bandwidth. During probing, two radar sensing elements are used. One sensor acts as a transmitter and the other as a receiver. For imaging purpose, the transmission coefficient (S_{21}), which represents the coupling between the two sensors are measured. As S_{21} increases, the coupling improves, yielding improved transmission.

Figure 8.6 shows the simulated and measured transmission coefficients of the bi-conical antenna. It can be seen that the magnitude of the simulated transmission coefficient is approximately –20 dB in the frequency range from 2 to 14 GHz, while the measured transmission coefficient is around –38 dB across the required frequency range. A maximum magnitude around 4.5 GHz is shown for both simulated and measured results. As suggested by earlier publications, oil-based immersion liquid provides the best performance in terms of matching media and coupling material [31]. Figure 8.7 shows the measured reflection coefficient of the antenna taken in oil over the frequency range from 0 to 5 GHz. Figure 8.8 shows the simulated radiation patterns of this sensor in oil.

Figure 8.7 shows the ripples appearing at high frequencies due to poor impedance matching, while the UWB characteristic of the antenna with a return loss can be observed to steadily decrease as the frequency increases. The ideal case for reflection coefficient is at –10 dB at which only 10% of the power is reflected back to the source. Hence, the antenna bandwidth defined at –10 dB cut-off point is approximately 3 GHz. In addition to S_{11} and S_{12}, antenna pattern is also important in the antenna design as it shows the ability of the bi-conical antenna to focus energy into the target rather than energy diverging out of the intended coverage

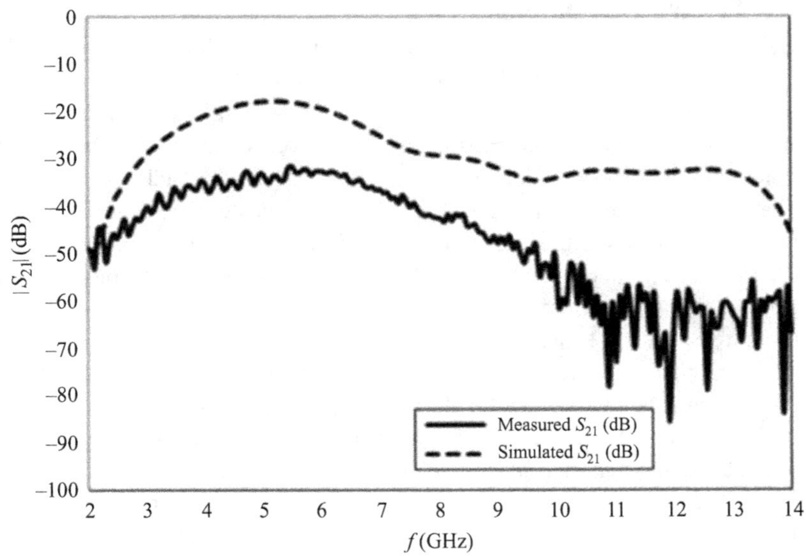

Figure 8.6 Simulated and measured transmission coefficient $|S_{21}|$ (dB) of bi-conical antenna

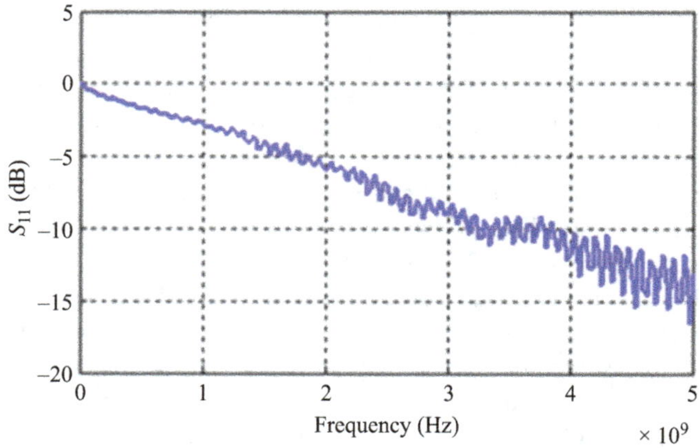

Figure 8.7 Measured reflection coefficient $|S_{11}|$ (dB) of bi-conical antenna in oil, for (0–5 GHz) frequency sweep

area. The results are shown in Figure 8.8. This figure clearly shows the formation of two resonant frequencies (3.62 GHz and 9.63 GHz) in *y–z* plane, indicating the directional properties of this type of antenna at low and high frequency regions. Also the resulting pattern appears like a rotated figure '8' at low frequency, while the pattern becomes butterfly-shaped with the direction of the main beam being shifted at high frequency in the *y–z* plane. This trend is in agreement with other omnidirectional antennas as the bi-conical antenna also belongs to this category.

8.3.1.2 Printed type antenna

Although the bi-conical antenna is a good design and exhibits important UWB characteristics, however, this type of sensor is relatively bulky and hence suitable for applications where the size of an object is significantly larger than the size of the antenna. For organ sensing a much more compact, planar-type, and smaller antenna is preferred. This has led to another design, which is referred to as the p-shaped slot radiator as shown in Figure 8.9. This design consists of a p-shaped slot radiator set in a ground plane on one side of a dielectric substrate Rogers RT6010 ($\varepsilon_r = 10.2$) and a simple 50 Ω micro-strip transmission line positioned along the *y*-axis on the other side.

In Figure 8.9, a few sensitive design parameters are investigated numerically to study antenna characteristics and behavioural performance across the entire band. Wide-slot antenna is well known for its wide impedance bandwidth. However, it is restricted by the degradation of the radiation patterns at higher frequencies. In this particular design the bandwidth of the antenna is determined by the shape and size of the slot. Figure 8.10 shows the reflection coefficient, S_{11} for different shapes of slot.

Clearly, the circular-shaped slot and the elliptical-shaped slot provide limited narrow bandwidth, averaging at 51% and 62% in the frequency range from 6.4–10.8 GHz, and 5.3–10.1 GHz, respectively. Meanwhile, the rectangular-shaped

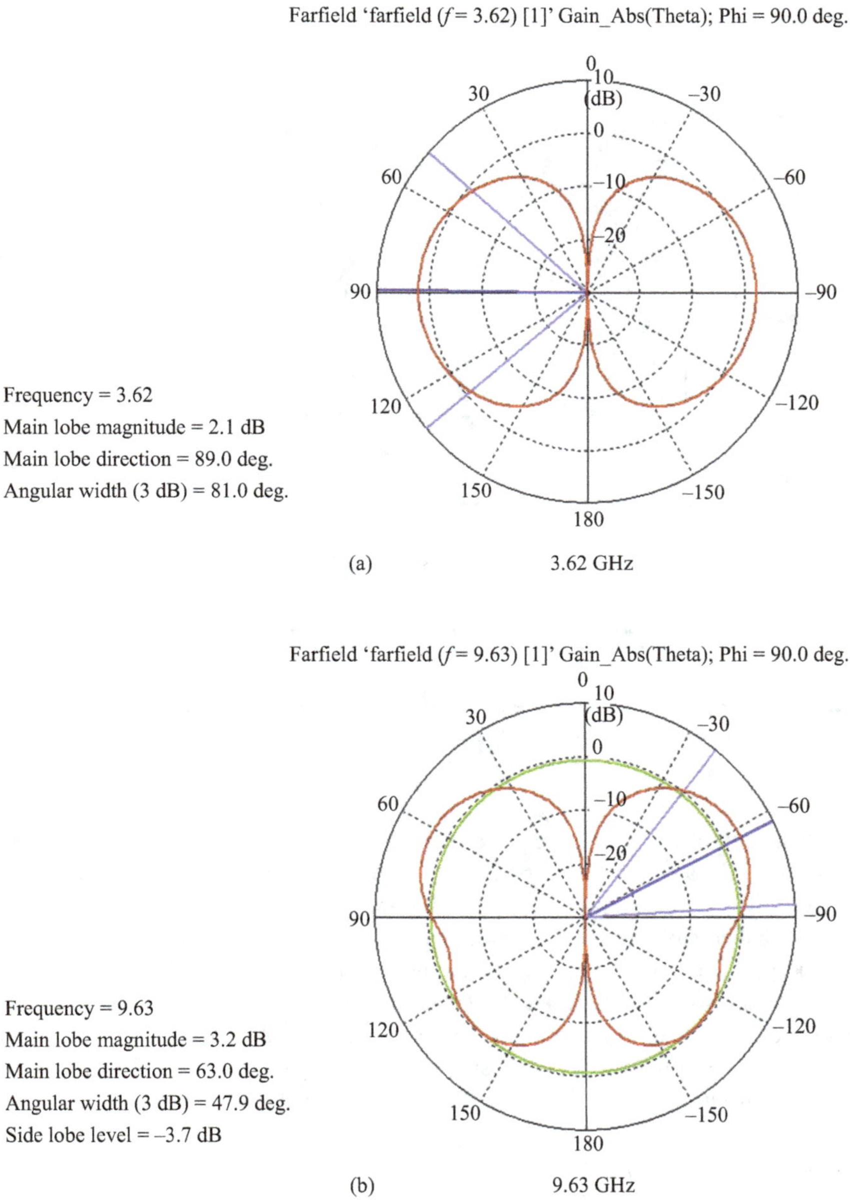

Figure 8.8 Simulated radiation patterns for bi-conical antenna at different frequencies in y–z plane: (a) 3.62 GHz and (b) 9.63 GHz

slot has an impedance bandwidth with large discrepancies where S_{11} rises above –10 dB from 6–8.5 GHz. Compared to these three designs, the p-shaped slot provides much wider bandwidth, averaging at 83% in the frequency range 4.5–10.9 GHz. The increase in the bandwidth is partly due to changes in the electric

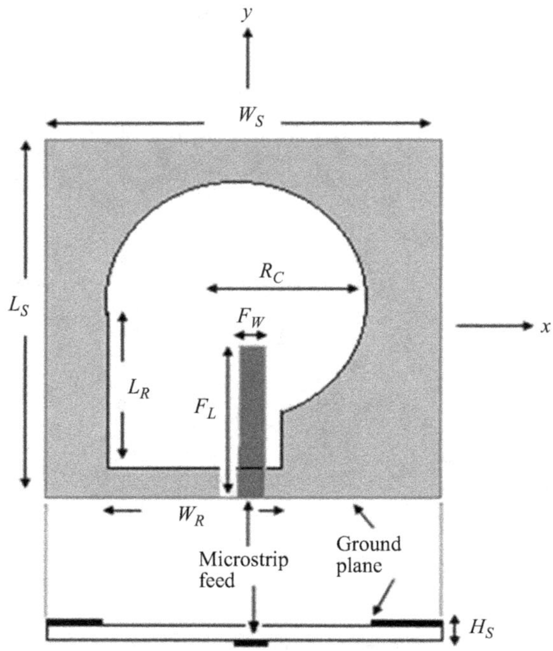

Figure 8.9 Geometry of proposed wide-slot antenna fed by micro-strip transmission line [32]

field distribution due to the presence of p-shaped slot, which reduces the longest current path. As a result, the impedance matching is significantly improved, especially at lower frequencies as evident in Figure 8.10. Hence, a micro-strip line fed with a p-shaped slot can provide wider bandwidth compared to other designs.

Another factor, which has increased the impedance bandwidth of the proposed design, is due to the simple and flat structure of the radiating micro-strip line. Compared to other designs the radiating element is formed by etching the micro-strip line directly onto the PCB substrate above the ground plane. In this case the size of the ground plane is determined by the frequency as follows [33]:

$$W_S = L_S = \frac{c}{2f_1}\sqrt{\frac{2}{\varepsilon_r + 1}} \tag{8.24}$$

where W_s and L_s are the width and length of the ground plane, respectively; f_1 is the lowest frequency of the operation, ε_r is the dielectric constant and c is the velocity of the light.

Given the dielectric of the substrate is 10.2, using (8.24), the width of the antenna is approximately 16 mm at 4 GHz. The length of the substrate is same as the width to achieve the symmetry of the structure. Figure 8.11 shows the simulated results of the proposed antenna with three different designs corresponding to three

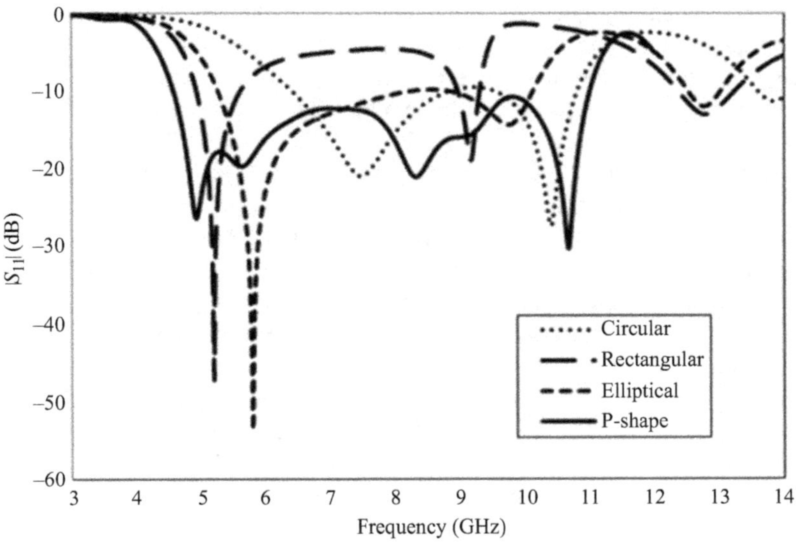

Figure 8.10 Simulated reflection coefficient for different slot shape

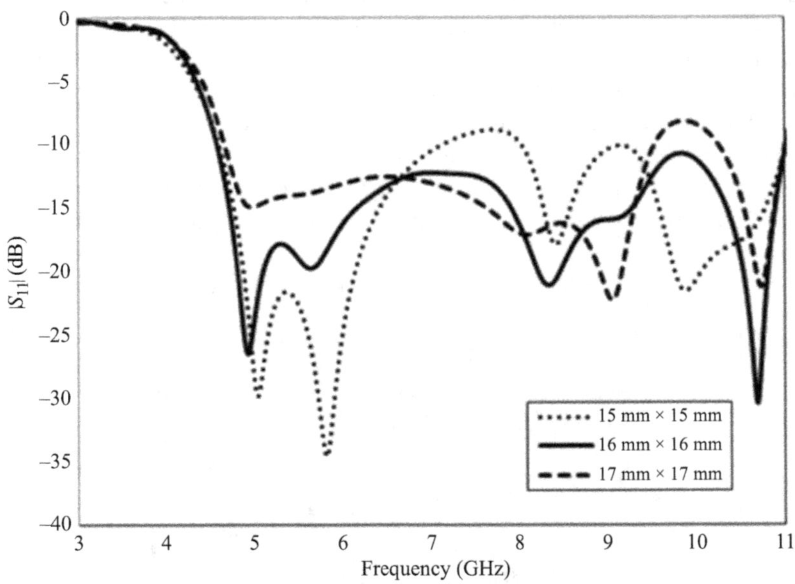

Figure 8.11 Simulated reflection coefficient for various substrate sizes

different sizes of ground plane: (1) 15 mm × 15 mm, (2) 16 mm × 16 mm and (3) 17 mm ×17 mm.

It can be seen that the lower resonant frequencies are significantly affected particularly in the first design, which is also the smallest antenna because the first and second resonant frequencies are associated with the size of the p-shaped slot. The surface current tends to concentrate strongly around the p-shaped slot, thus reducing the flow of currents to a ground plane, and hence lowering the antenna performance significantly. In general, the return loss for a second design at –10 dB ranges from 4.5 GHz to 10.9 GHz. In this frequency range the impedance bandwidth is approximately 83% compared to 80% and 78% for the first and third designs, respectively. Clearly, the second design provides a much better performance in terms of signal bandwidth while maintaining the antenna size to be as small and as compact as possible. Consequently, the second design is selected. The antenna is excited with micro-strip line having the length (F_L) of 7.5 mm and width (F_W) of 1.2 mm. The micro-strip feeder is designed on the substrate with size of 16 × 16 mm, and the thickness of the substrate (H_S) is 1.27 mm. The slot consists of a primary circular slot of radius 6 mm, and a secondary rectangular slot of dimensions $L_R \times W_R$. The antenna dimensions are as follows: $R_C = 6$ mm, $L_R = 6.5$ mm, $W_R = 7.5$ mm, $F_L = 7.5$ mm, $F_W = 1.2$ mm, $L_S = 16$ mm, $W_S = 16$ mm and $H_S = 1.27$ mm. Figure 8.12 shows an example of the fabricated antenna with an SMA connector.

Using the design as shown in Figure 8.12, experiments were carried out to study some important characteristics of the antenna. In the first experiment the antenna radiation pattern was measured in the anechoic chamber according to IEC 6100-4-3 standard. The anechoic facility consists of 19 GHz TDK HRN-0118 double-ridged horn transmitter, 40 GHz EMI Test Receiver, 40 GHz signal generator and antenna controller. Figure 8.13 shows the results comparing the measured and simulated radiation patterns, which are plotted in *y*-*z* plane.

The antenna exhibits a donut shape in *y*-*z* plane for all resonant frequencies. It can be seen from Figure 8.13 that there is a good agreement between measured and computed radiation patterns within some experimental errors. In theory, the slot

Figure 8.12 Actual p-shape wide-slot antenna

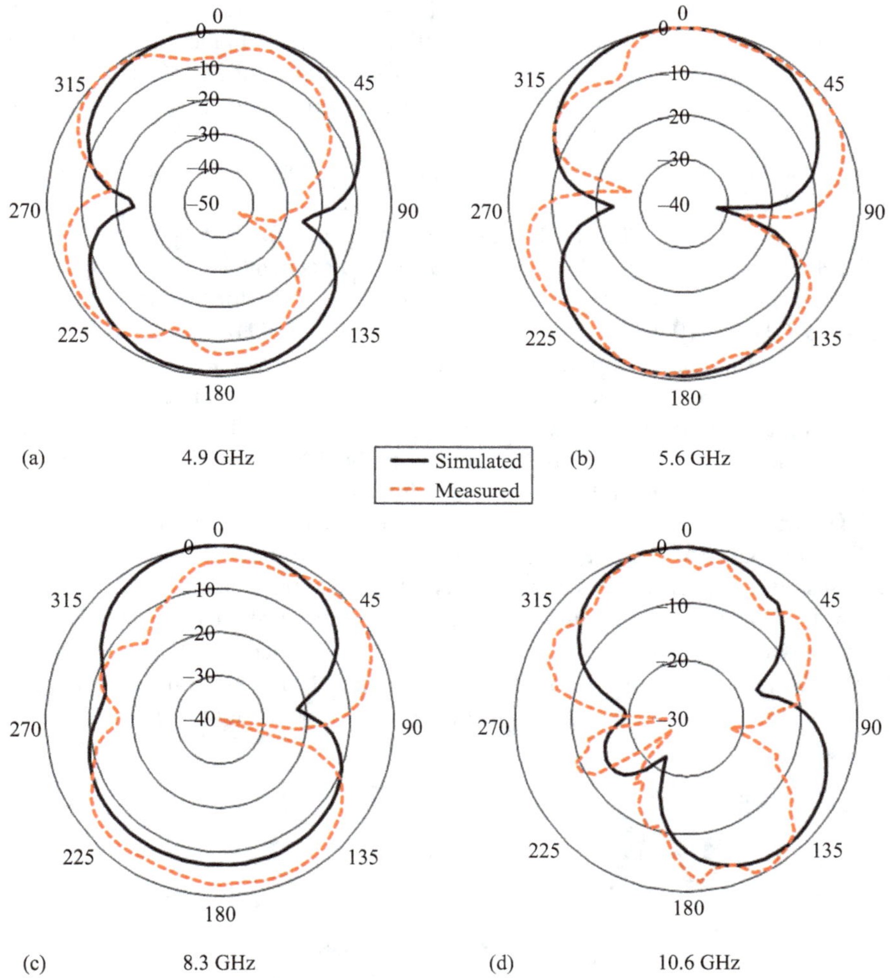

Figure 8.13 Simulated and measured radiation patterns at y-z plane at (a) 4.9 GHz, (b) 5.6 GHz, (c) 8.3 GHz and (d) 10.6 GHz

antenna is known as the complimentary dipole, in which the magnetic and electric fields are interchanged. Hence, it can be observed from Figure 8.13 that the antenna has the radiation patterns, which are typically the same as any typical dipole antenna at low frequencies. The pattern in the orthogonal plane (*x-y* plane) is omni-directional. Therefore, this antenna radiates its energy equally in all directions in the *x-y* plane. Figure 8.13 shows that the radiation patterns exhibit the typical nulls, particularly at lower frequencies and at angular positions of 90° and 270° from the *y*-axis. The nulls in the horizontal plane can be used to protect other transmitters from interference, as it is useful in suppressing interfering signals in a given direction. From the simulated *y-z* plane pattern, the antenna has a beam-width of

90° and 64° at low frequency of 4.9 GHz and high frequency of 10.6 GHz, respectively. The beam-width is significantly narrower for higher frequency than lower frequency, resulting in higher gain. Hence, the amount of power radiated at high frequency is more focused and directive compared to low frequency radiations.

The simulated reflection coefficient and S_{11} of the proposed wide-slot antenna are plotted in Figure 8.14, along with the equivalent measured results. It provides the evidence of wideband characteristic of the design, defined at threshold limit of –10 dB or less. It can be observed that a good agreement is obtained between simulated and measured data, averaging at 83% and 71% for frequency ranging from 4.5 GHz to 10.9 GHz and from 5.1 GHz to 10.7 GHz, respectively. Both simulated and measured results demonstrate that there are four major nulls in the response. Also, the difference between the two plots is likely due to a slight mismatch in the feeding cable, which is not properly modelled. There are some small discrepancies where S_{11} rises above –10 dB around 10 GHz in the measured data due to the fabrication tolerances of the feed line.

The transmission coefficient (S_{21}), which represents the coupling between the two sensors has also been investigated. It is desirable to have a flat magnitude of the transmission coefficient across the desired bandwidth, to achieve minimal distortion for UWB signals. Figure 8.15 compares the simulated and measured S_{21} of the proposed design. It can be seen that a good agreement is obtained between simulated and measured S_{21} over the frequency of interest. Furthermore, the simulated S_{21} is approximately –20 dB in the frequency range of 4–11 GHz, while

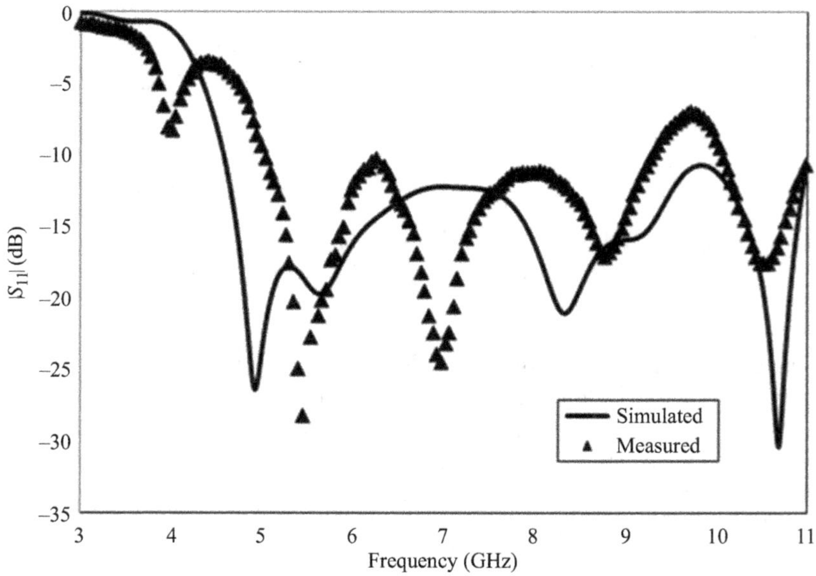

Figure 8.14 Reflection coefficients of the p-shaped wide-slot antenna, comparing simulated and measured values

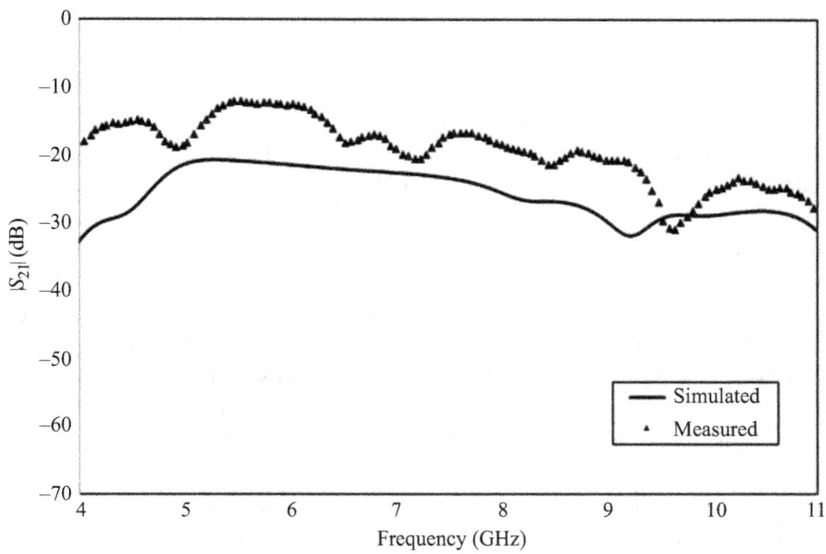

Figure 8.15 Transmission coefficients comparing simulated and measured values

the measured S_{21} is in the range of −10 dB to −20 dB, with variation less than 10 dB in this operating bandwidth. Although the antenna exhibits approximately flat magnitude, several slight fluctuations are observed at 4.9, 6.5, 7.3, 8.5 and 9.7 GHz in the measured data. These small dips are primarily due to the effect of the SMA connector in addition to errors in measurements. Nevertheless, these discrepancies do not distort the antenna's UWB performance significantly. In addition, the interference from multi-path signals increases in near-field zone, especially in conditions where the transmitting and receiving hardware are integrated. Figure 8.15 also shows that the measured and simulated S_{21} is optimum at frequency of around 5 GHz and 5.5 GHz, respectively. However, beyond 8 GHz, the measured S_{21} shows more variation, indicating the reflections of the signal at the terminating end of the antenna, thus causing more dispersion at higher frequencies.

Another important characteristic of the antenna is the shape and size of pulse because it tells general dielectric properties of the medium, particularly the scatterers. Theoretically, the pulse shape of the signals is affected as it travels from the transmitter to target and back to receiver. For this reason the shape of the incident pulses is geometrically different from the shape of the arriving pulses. Therefore, it is important for the antenna to transmit or receive signals with minimal distortion so that the image of the scatterer can be reconstructed as accurately as possible. In most UWB designs, the antennas emit high and low frequency components from different portions of the geometry, and so the radiated waveforms are highly dispersed. This undesirable characteristic is not suitable for impulse applications, such as for breast cancer detection. It is pertinently important for the design to fulfil the requirement of having phase centre that does not change with frequency so that the radiated emissions are minimally distorted.

To assess the performance of the antenna in transmission and reception of narrow pulses in distortionless condition, the time domain impulse response of the antenna was measured. Measurement was performed in a nearly lossless media using a vegetable oil as the homogeneous background. In this way the background interference especially the air-object interface is minimised. In the experiments, two antennas were located in the same height and separated by a fixed distance of 90 mm. The same experiment was repeated in a free space serving as reference for oil-based measurement. In both cases the pulse duration was set to 0.6 ns and the data were captured for 6 ns. The data were normalised and plotted in Figure 8.16.

It can be seen that the pulse distortion occurred at both the falling and rising edges. In the case of transmission, pulse distortions occur at 0.8 ns and 1.3 ns, with the distortion averaging at 4.1%. For the receiving case, the distortions occur at 2.1 ns and 3.3 ns, corresponding to an average distortion of 9.0%. Clearly the distortion is low in both cases and can be neglected. For the vegetable oil experiment, the pulse is delayed by 1.7 ns due to the reduced speed of the propagation. In this case the pulse distortion occurs at 4.1 ns and 5.4 ns. Overall, in these experiments the distortion level is consistently less than 18.7%. Although the distortion is high in oil than free space, the overall shape of size of pulse is still preserved. This result shows that the antenna retained the timing characteristics of the waveform before and after transmission. This is especially the case for sensing application with strong signals because the level of distortion is not sufficiently high to cause the pulse packets at the receiver be modified and transformed into a long envelope, thus altering the fall and rise times excessively. Hence, the design offers a significant advantage, in which the timing characteristic of a pulse is nearly preserved

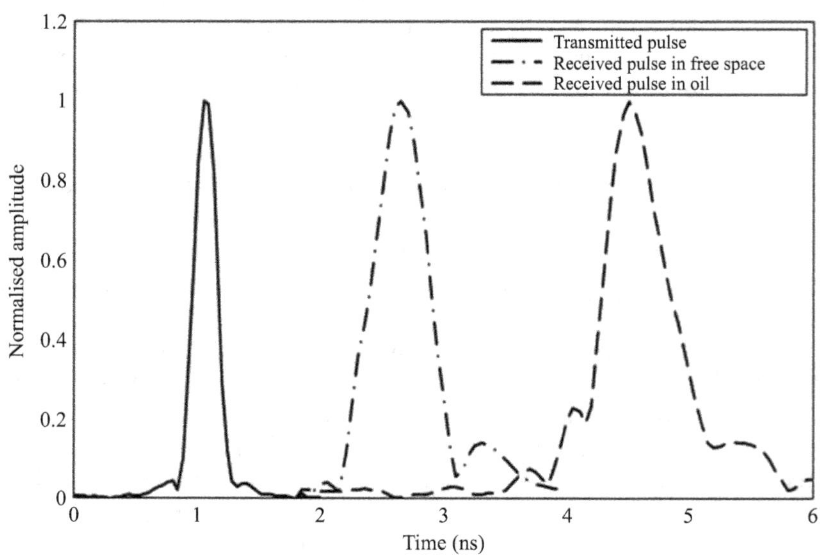

Figure 8.16 Measured response of the antenna

before and after transmission. Moreover the antenna performs much better in air compared to oil as evident from the distortion results in Figure 8.16. Therefore, such a design is practically feasible because the problems associated with the use of oil as coupling media can be eliminated.

Commonly in radar-based imaging, the ideal pulse produced by the antenna must be as short as possible with minimal late-time ringing, to ensure a sufficient amount of energy is transmitted, reflected and detected by a receiver. The pulse durations of the proposed antenna are measured for different transmitter-receiver positions in the circular array. It is discovered that the pulses produced by this antenna are relatively short, averaging at 0.56 ns when the transmitter and receiver are adjacently located. For other positions the pulse duration ranges from 0.6 ns to 0.7 ns. As expected the pulse duration increases as the distance between transmitter and receiver increases. On average the pulse duration recorded was 0.625 ns, which is relatively short compared to other designs [34].

Another factor used to evaluate the antenna is the fidelity factor [35], which is the maximum magnitude of the cross correlation between the received and incident pulses. Normalisation is generally needed in this case to compare the shape between received and incident pulses.

Assuming the time signal of receiver is much smaller than the time signal of transmitter, then the fidelity factor is defined as [36]

$$F = \max_{n} \int_{-\infty}^{\infty} \widehat{T}_S(t)\widehat{R}_S(t+\tau)dt \tag{8.25}$$

where $\widehat{R}_S(t)$ is the normalised E-field to unit energy and $\widehat{T}_S(t)$ is the input signal normalized to unit energy.

The result of the cross correlation ranges from 0 to 1. The value of 1 indicates the received pulse with no distortion and identical to the incident pulse, while the value of 0 indicates the received pulse is totally different than the transmitted pulse with severe distortion. Generally, a correlation with a value of less than 0.5 (50%) will deliver a high distortion [37], making the pulse almost unrecognisable. Figure 8.17 shows the computed fidelity factor as a function of the distance between transmitter and receiver.

It can be seen that the fidelity decreases as the distance increases. The fidelity factor is the highest for the adjacent antenna, averaging at 0.91 in most cases. It can be seen that the fidelity factor drops to 0.6 when the distance is increased to 90 mm. This distance corresponds to two opposing antennas in the circular array, which is also the longest sensing path. Although there is substantial reduction in the fidelity for two oppositely located antennas, the value is still reasonably good and acceptably high.

8.3.2 Data acquisition system

8.3.2.1 Limited view geometry

Several radar instrumentation systems were built to validate the methods and procedures discussed previously. The first system is based on limited view geometry,

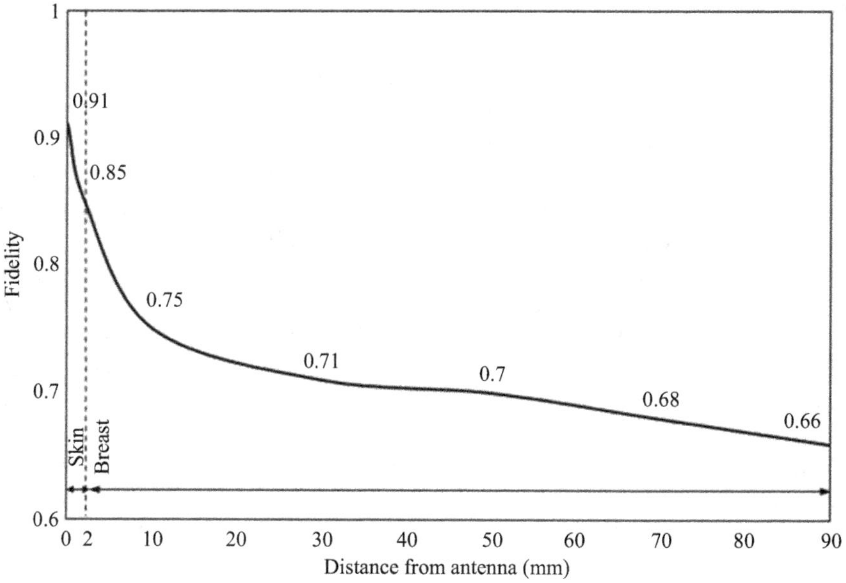

Figure 8.17 Fidelity factor as a function of distance between two antennas

simulating a typical cross-borehole reconstruction tomography. This data acquisition system was constructed in Manchester, UK and the details are provided elsewhere [38]. The instrumentation for automation comprises two stepper-motor driven antenna carriage assembly, each capable of moving in positive and negative directions with minimum step size of 1 mm. Figure 8.18(a) shows the recording system while the essential elements of the set-up are shown in Figure 8.18(b).

The antennas are positioned in a 62 × 47 × 30 cm size rectangular tank filled with tap water up to 22 cm in height. Both antennas are maintained at a constant height of approximately 18 cm below water surface. In this way, clutter caused by the reflections particularly at the water-tank and water-air interfaces can be minimised by appropriately time gating the arrival pulses. All antennas used in this set-up are bi-conical type as previously discussed. The region between the antennas of size 40 × 20 cm constitutes the space to be imaged. During operation, this region was remotely scanned by varying the transmitter and receiver locations at a predetermined step-size. For each transmitter location, signals were detected at different receiver locations. In this case the antenna was programmed to move at a discrete step-size of 2 cm, resulting in 400 unique measurements for 20 × 20 crosshole geometry.

8.3.2.2 Full view geometry

Figure 8.19(a) shows the schematic of data acquisition system for full-view measurements comprising of N antennas. It consists of a Cytec CXM 16 multiplexer that automatically switches the transmitting and receiving lines. All antennas are

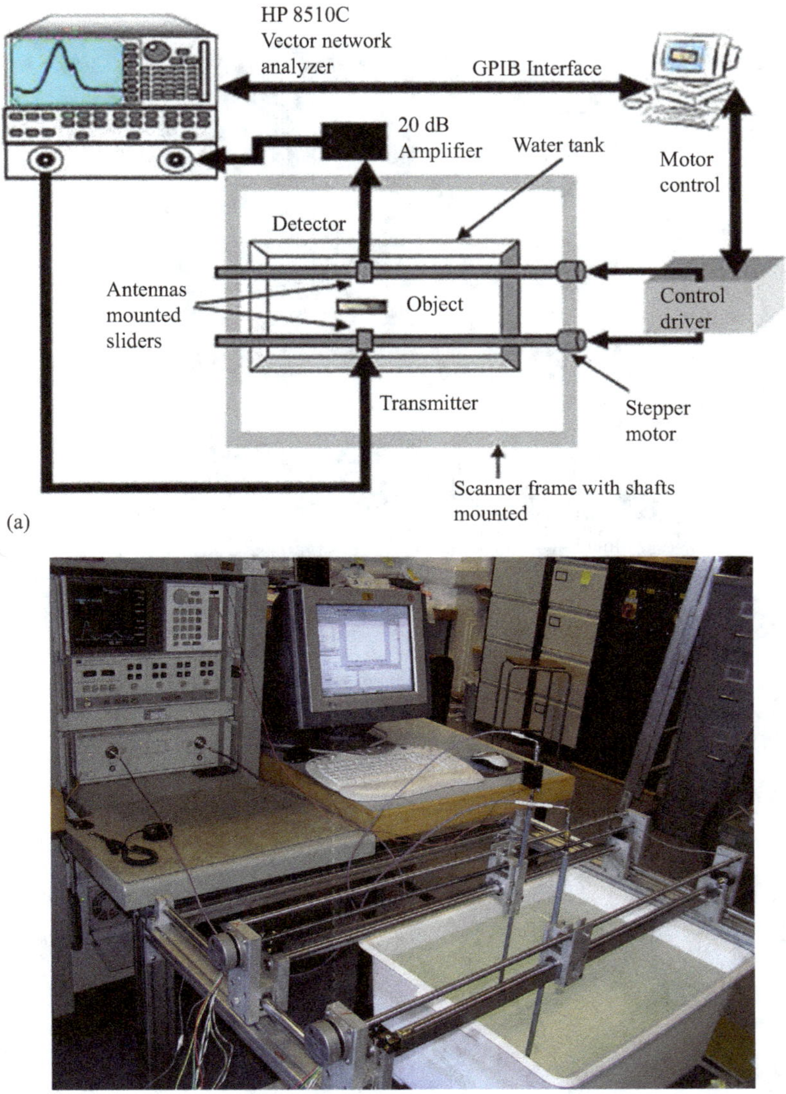

(a)

(b)

Figure 8.18 Experimental set-up for limited view geometry: (a) schematic and (b) actual system

transceivers and connected to an Agilent 91 E5071C 8.5 GHz VNA which provides time-domain measurements. A wide-band 20 dB amplifier is used to improve the received signal. Two-port calibration was performed to minimise insertion loss, impedance mismatch and electrical delay effects. The VNA and the multiplexer are controlled by a computer through a general purpose interface bus (GPIB) interface. Data are acquired by employing the multi-static mode, which allows a total of

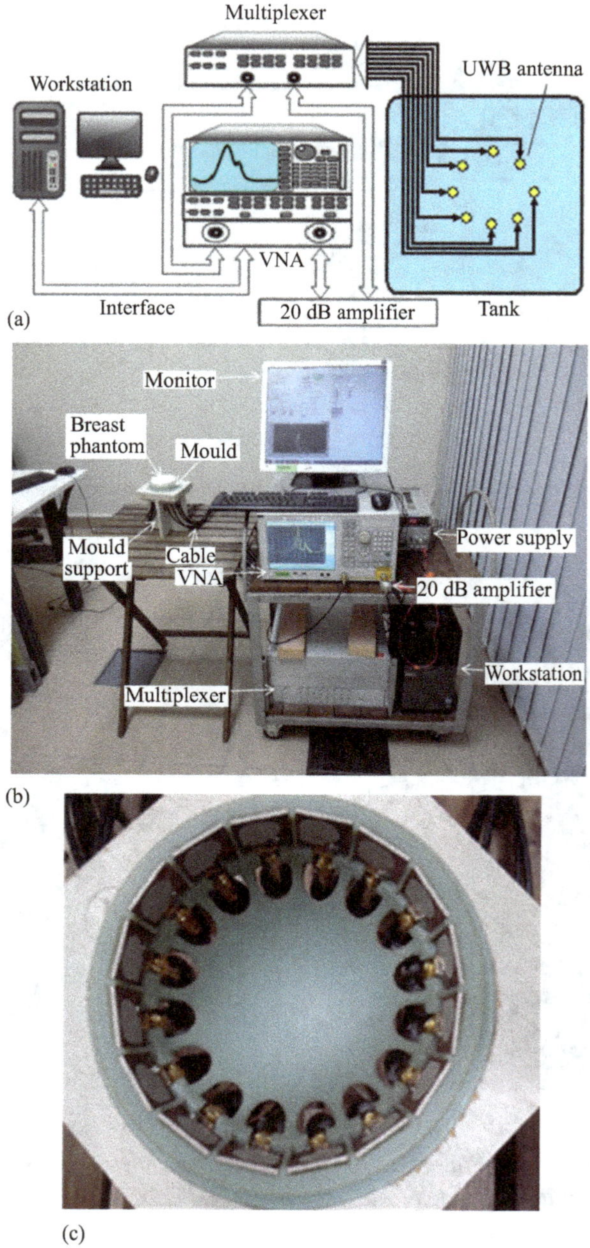

Figure 8.19 Experimental set-up for full view geometry: (a) schematic, (b) actual system and (c) 2D sensor array

$N(N-1)$ radar signals to be captured. Figure 8.19(b) shows the real set-up for $N = 16$. In this case the p-type design was used for sensing. As shown in Figure 8.19(c) the antennas are formed into a circular array comprising of 16 antennas, and placed equidistantly around the target, making an angle of 22.5° from the origin.

Each antenna is driven via a 50 Ω coaxial feeder, incorporating in-line ferrite torroid to ensure balanced antenna feeding. The sensor array is connected to a multiplexer by RG 54 cabling system. The switching of the multiplexer is controlled automatically by the PC workstation. This reduced the data acquisition time and prevents loss of information due to slow switching. LabView is used to configure the automated switching of the multiplexer and the data collection of VNA. It is a graphical control, test and measurement environment for data acquisition in academics and industries. The multiplexer pre-sets the switching time to control all of the antennas.

During measurement one antenna acts as transmitter whereas the other antennas act as receiver to collect the reflected signals. First, the multiplexer switches transmit (T_X) to 0. The multiplexer switches receive (R_X) from 1 to 15 to record the received signals to VNA. The process repeats until T_X reaches 15. Each transmitting antenna has 15 received signals, which results in 240 total number of measurements. All received signals are directed to the VNA to provide frequency and time domain measurements. The data are fed to an image reconstruction system, which produces approximate profile of the scatterers. A simple subtraction method is applied to remove the skin back-scattered signals. The received signals resulting from the homogeneous model (tumour free) are subtracted channel-by-channel from the received signals obtained from heterogeneous model (tumour), resulting in the calibrated signal.

As explained in Section 8.2.2.2, all calibrated signals are time-aligned, weighted and multiply in pair with two processing techniques: focusing quality and multiplication in pair. These products are summed and squared to create a synthetic focal point forming the image at each energy value of interest. Initially, vegetable oil was used as the background immersion matching medium to minimise interferences due to cluttering and multiple reflections. Later, it was discovered that the p-shaped antenna performs much better in free space compared to oil. Hence, in the later experiments, the target was placed directly onto the sensors instead of through the coupling medium. This set-up is more practical because it minimises the signal loss and risk of contamination.

8.4 Results and discussion

8.4.1 Limited view geometry

The validity of the inversion algorithm was first evaluated by simulation from limited view geometry as shown in Figure 8.1(a). In this case the modulated differentiated Gaussian pulse or popularly referred to as the Ricker wavelet is used to model the UWB wave. Figure 8.20 shows such a signal. During imaging such a

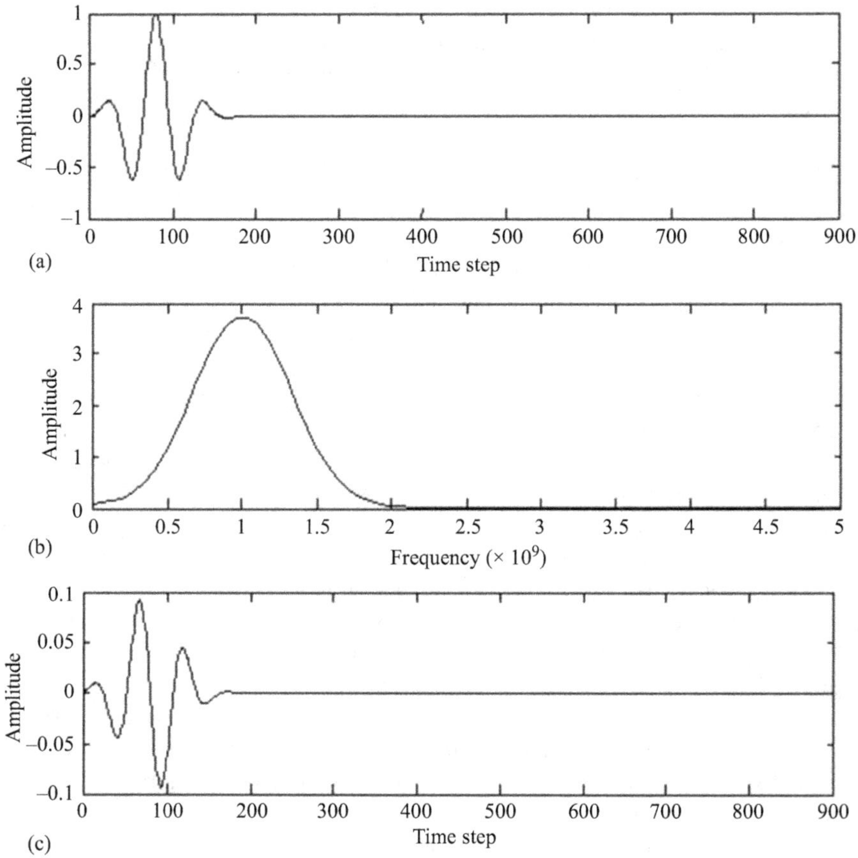

Figure 8.20 Example of cosine wave modulated Gaussian pulse with carrier frequency of 1 GHz: (a) original modulated signal, (b) spectrum of (a) and (c) differentiated signal of (a). The maximum time step in (a) and (c) is 801 points and time resolution of 16.23 ps

source is introduced into the grid by adding the signal to the *z*-component of the current density term in Maxwell's equations at the desired spatial location. When excited, the resulting E_z pulse travels through the grid and is recorded at the receiver locations. In this case both sources and receivers are actually line elements, extending from positive to negative infinity in the dimension perpendicular to the imaging plane. Using this signal as a source of excitation, a few numerical models are designed to investigate the validity of the inversion algorithm. Example of such a model is shown in Figure 8.21.

In Figure 8.21, the model is made-up from 8 cm diameter cylindrical target with $\varepsilon_r = 75$ and positioned at the center of the imaging space. The initial guess is that of the homogeneous medium. The reconstructed images are smoothed using threshold filters with cut-off points determined empirically. For comparison

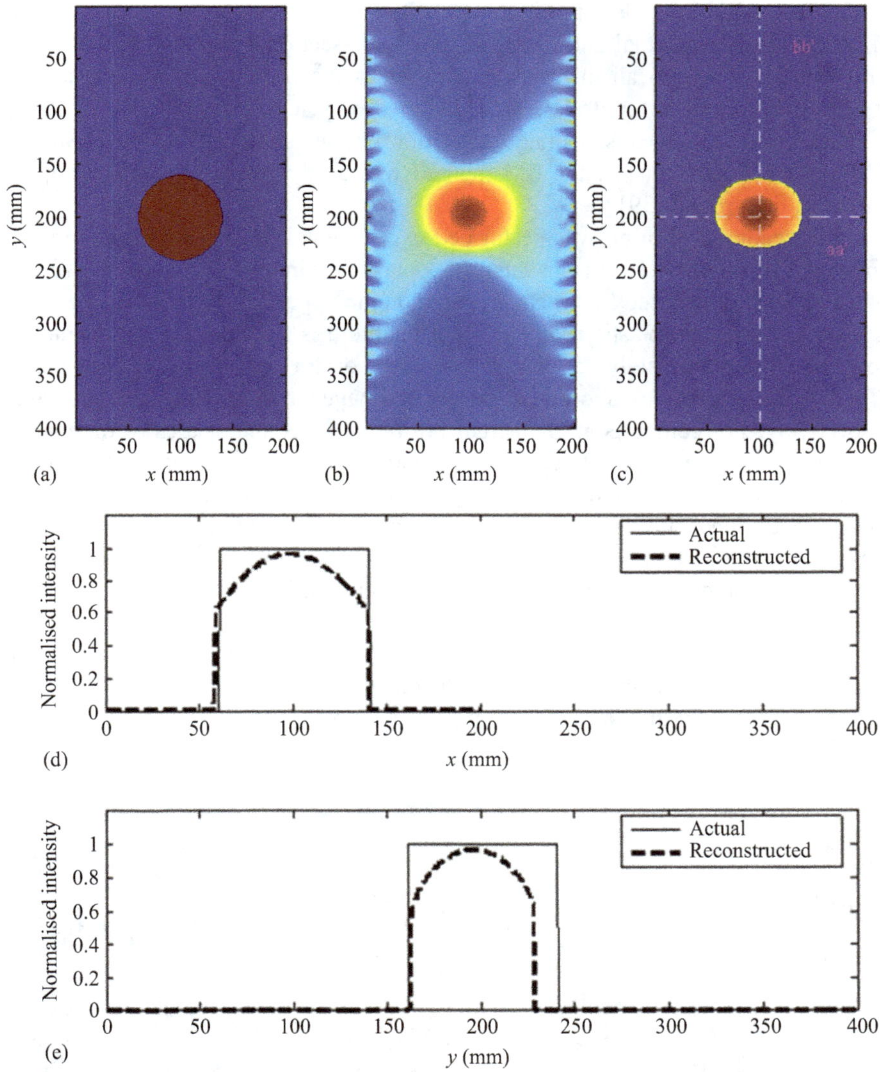

Figure 8.21 Limited view reconstruction of a single 8 cm diameter object located at centre: (a) actual, (b) original reconstructed, (c) filtered reconstructed and (d) and (e) are image profiles along the straight line aa′ and bb′, respectively

purpose both the original and filtered images are shown. Figure 8.21(a–c) shows the actual image, reconstructed profiles before and after filtering. Figure 8.21(d) shows the 1D profiles of Figure 8.21(c) drawn along straight lines *aa′* and *bb′*, respectively.

It can be seen that the size as well as the position of the target are correctly mapped. However, the shape of the object appeared slightly asymmetrical in the vertical direction as evident in Figure 8.21(d). The error calculated in terms of the difference between actual and reconstructed cross-sectional area is less than 15%. This is mainly due to an incomplete data-set as the present set-up assumes the region to be scanned for its lateral boundaries only. Consequently, the spatially truncated projections caused image distortion or the 'stretched' artefact. Similar finding was reported when the EM method was used to map tunnel-like objects buried in the ground [39].

The second validation was performed experimentally using the hardware set-up shown in Figure 8.18. In this case, two plastic phantoms ($\varepsilon_r \approx 6$) of different size and shape were used as targets. Located on the upper half of the imaging plane is a 2.5 cm diameter plastic rod. The other phantom is a rectangular rod, which is positioned approximately 5 cm below the origin of the imaging plane. Figure 8.22(b) and (c) show reconstructed images before and after filtering, respectively. For comparison, the actual positions, sizes and shapes of the targets are shown in Figure 8.22(a). Clearly, the important characteristics of the two targets are accurately reconstructed. Also, the general shape and size of the phantom are clearly distinguishable. However, there are some artefact in shape reconstruction particularly the rectangular rod, in which the four sharp corners appear smooth. This is consistent with the Born approximation used in the algorithm, which exhibits a low pass filtering effect.

Figure 8.21 and Figure 8.22 are some of the preliminary results from real experiments, which demonstrate the feasibility of imaging using UWB pulse. However, the reconstruction is far from perfect, partially because of the dispersion

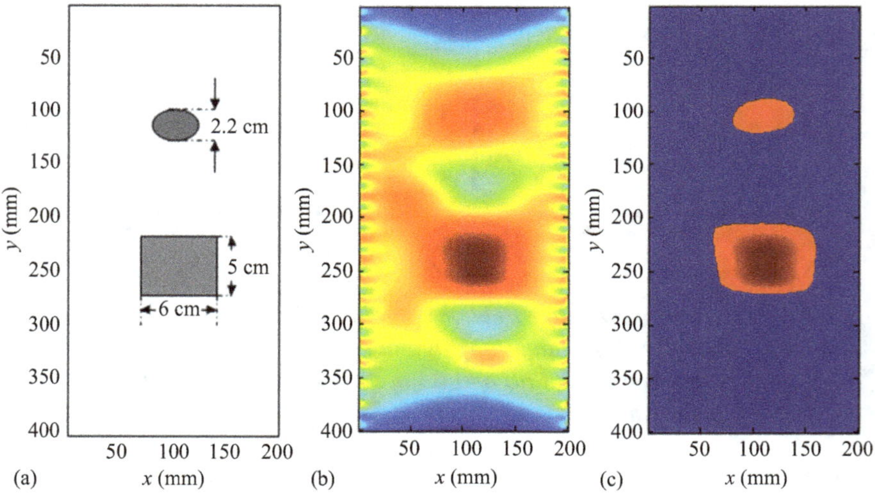

Figure 8.22 Reconstructed image of two plastic phantoms of different sizes and shapes immersed in tank filled with pipe water at 20 °C: (a) actual and (b) and (c) are the reconstructed profiles before and after filtering, respectively

effect, which is left unmodelled. In this case the reconstruction is terminated after a few iterations to avoid divergent. Overcoming these problems lead to conformal finite difference time-domain based image reconstruction algorithm using Debye dispersive model discussed previously. In so doing the domain is discretised into 54 × 54 cells of size 2.6 × 2.6 mm, terminated with perfectly matching layers (PML) as shown in Figure 8.1(b). The forward solver is then used to generate simulation data. Figure 8.23 shows the reconstructed results using the adjoint and Broyden reconstructions after 25 iterations each.

Using two quad-core Xeon processors, the adjoint gradient required less than 30 s compared to an hour for the CDF method, while the Broyden estimate is obtained in the range of 25 s. The reconstruction results show close resemblance where the 5 mm target has been successfully detected. Figure 8.24 shows the

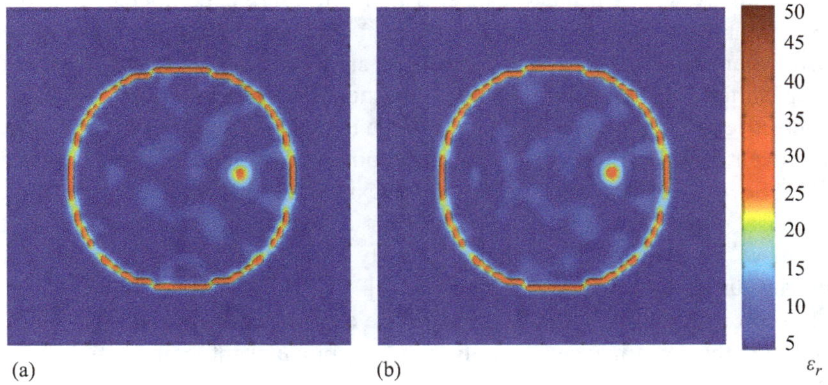

Figure 8.23 Full-view geometry numerical results obtained by (a) adjoint method and (b) central difference method

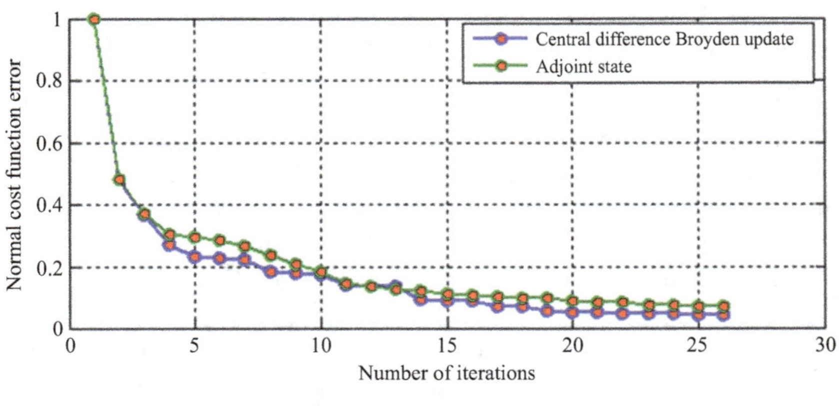

Figure 8.24 Iterative reduction in cost function error corresponding to image reconstruction in Figure 8.16

convergence trends of both approaches. Comparing both results, it is seen that both images indicate the presence of the target, but the geometry is distorted in the adjoint result. Examining the convergence trend, it can be seen that the adjoint solution cannot be enhanced with further iterations as the error has stabilised. Figure 8.24 also shows that more error reduction is obtained with the Broyden approach than the adjoint method.

More numerical results can be found in selected papers listed in the reference section. The remaining paragraphs in this section discuss important experiments using full-view set-up as shown in Figure 8.19, serving as references to results obtained by simulation.

The UWB system discussed in Section 8.3.2.2 specifically the full-view geometry was oriented for organ sensing, in particular the detection of cancerous tissues in human breast at the early stage. The system is built in such way that it can be used directly with real breast cancer patients in future experiment. Prior to this the performance of the system was investigated first using breast phantoms. Several realistic phantoms were designed, simulating an approximate anatomical structure of a typical female adult breast. In this connection previous experimental studies on the dielectric properties of female breast tissue in the microwave frequency suggest a significant contrast between breast fat, normal tissue and malignant tumours [40,41]. Several factors, such as the complexity of tissue structure and amount of adipose may contribute to the variance of reported properties. Table 8.1 summarises the permittivity and conductivity for breast fat, normal tissue and malignant tumours as reported elsewhere [41].

Several materials are used to represent the contrast of the breast tissue [42,43]. In assessing the performance of the algorithm, a simple phantom made from animal derived biological tissues and products is fabricated. The breast phantom is moulded into a hemisphere shape, with a diameter of 9 cm at the top. The skin is represented by using a 2–3 mm thick layer of chicken skin, while the interior tissue is represented by salted butter (0.60 g Na^+/100 g). The relative permittivity of salted butter ranges between 18 and 8.5 over the frequency range of 0 to 3 GHz, at 30 °C, with a conductivity of $\sigma = 0.45$ S/m [44]. Next, the high permittivity of cancerous tumours is modelled by using beef meat ($\varepsilon_r = 50$, $\sigma = 4$ S/m up to 5 GHz) covered by a thin plastic film layer and shaped into tubes of several diameters. The meat was compressed inside the tubes to remove any air pockets, before airtight sealing. Although the current breast phantom may not necessarily represent the actual human tissue, the structure offers good resemblance to the relative contrast in the human breast. Figure 8.25 shows the experimental breast

Table 8.1 Dielectric properties of female breast tissue [41]

Tissue	Permittivity	Conductivity (S/m)
Normal tissue	10–25	0.35–1.05
Fat	4–4.5	0.11–0.14
Malignant tumour	45–60	3.0–4.0

(a) (b) (c)

Figure 8.25 Experimental breast phantom made from chicken skin and butter: (a) top view, (b) side view and (c) 5 mm target placed to simulate early stage breast cancer [44]

phantom. In this case meat tubes with diameters of 5, 10 and 15 mm are used to represent different tumour sizes (see Figure 8.25(c)). Tubes were inserted vertically into the breast phantom, to obtain their cross-sectional image, and located 21 mm from the centre.

Prior to reconstruction the accuracy of Debye equation in modelling the dispersive model of UWB was first examined. Figures 8.26 shows the dispersive behaviour of butter and skin respectively, indicating a good agreement between calculated and measured data. It can be concluded that this dispersive properties of these materials have successfully been modelled using Debye's first order equation.

In addition to dispersion, the Jacobian or sensitivity was also examined prior to reconstruction. The Jacobians are mathematically independent, and therefore, may be computed in parallel provided the computational resources exist. The parallel calculations consume down to 1/8 of the total time in comparison with the time required to sequentially evaluate the Jacobians.

Figure 8.27 shows the isometric views of the normalised sensitivity distributions for $m = 1$ and $n = 4$ and 5 calculated for 1, 3 and 5 GHz excitation, respectively. Due to symmetrical properties of the model, other sensitivity distributions can be obtained by either coordinate transformations or calculations. In general UWB exhibits sensitivity patterns similar to sensitivity patterns of its electrical imaging counterparts, in that the sensitivity is relatively high near boundary compared with the central region. However, the sensitivity of the UWB tends to get narrower at high frequency, and hence increasing the capability of detecting smaller as well as bigger targets. This multi-resolution property of UWB means objects that may go undetected at one resolution may be detected at another. This is also one of the most attractive features of UWB compared to other monochromatic sensing techniques.

In the experiments, the breast phantom and sensors were first submerged in oil, and measurements were performed. For each antenna position, the S_{12} signals at receiver locations were sequentially recorded and transformed into the time domain via the internal chirp z transform implemented within the VNA. In this trial a total of 8 projections consisting of 56 unique measurements were obtained.

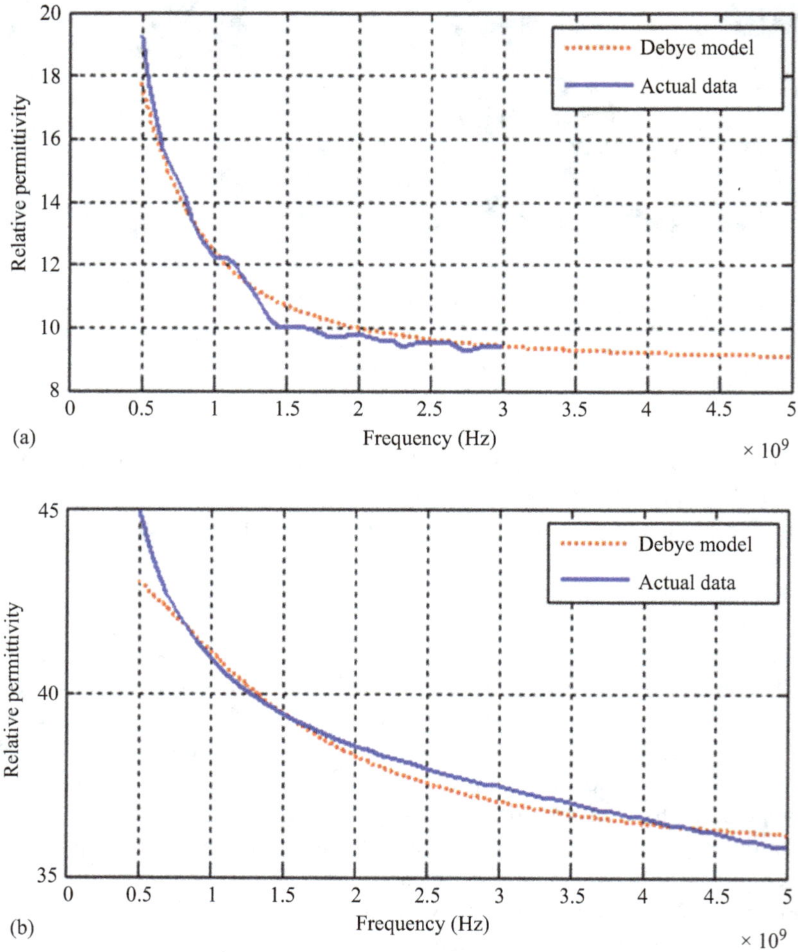

Figure 8.26 Variation of permittivity with frequency ranging from 0 GHz to 3 GHz using fitted Debye model for (a) salted butter and (b) dry skin

Image reconstruction using experimental data presents great difficulty due to the presence of noise. The ability of the computational model to accurately predict receiver response guarantees successful reconstruction. Several factors, such as antenna (i.e. sensor) gains, cable propagations delay, positioning errors and such like remain un-modelled. In addition, the model computes propagation in 2D only, and as a result, does not account the spherical wave propagation loss associated with the third dimension. Although the extension of 2D models to 3D is feasible to implement, the computational cost associated is considerably high. The magnitude of this cost accumulates when the 3D model is part of an extensive iterative process, such as the inverse solution.

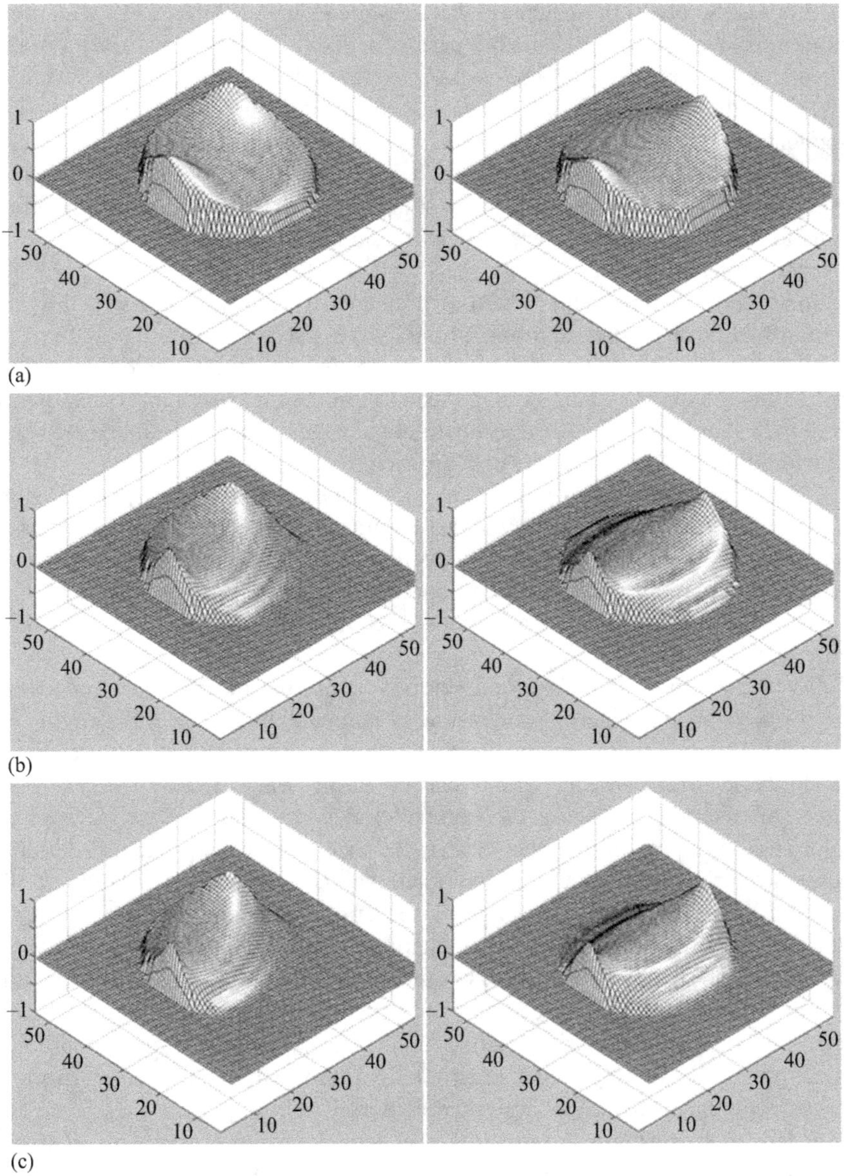

Figure 8.27 Sensitivity distributions corresponding to receivers 4 (left) and 5 (right) in response to exciting transmitter 1: (a) at 1 GHz, (b) at 3 GHz and (c) at 5 GHz

A correction (calibration) procedure is adopted to minimise systematic errors between the experimental system and the model, and ensure that the measured signals and calculated signals from the model are in close agreement for a defined set-up. This procedure also minimises errors inherited within the FDTD method.

The process involves in matching the homogeneous model results with those obtained from measurements for the case of a fixed homogeneous medium after normalising the amplitudes of both sets. Once the correction parameters have been established, they are applied to all data sets obtained using the defined setup. Two homogenous data sets corresponding to the homogeneous oil media and homogeneous breast phantom were obtained. It is observed that the best fit of data is achieved when $\varepsilon_\infty = 28$ for the chicken skin and $\varepsilon_\infty = 8$ for the butter. This could be due to the higher fatty content of chicken skin and lower room temperature for the salted butter at 22 °C.

For each experiment, the first four gradients were evaluated using the CDF equation, while Broyden approximations were used thereafter, resetting the Jacobian whenever the data residual error from two consecutive Broyden estimations is less than 5%. This strategy can improve the speed of convergence. Figure 8.28 shows the actual and reconstructed images obtained from experimental data while Figure 8.29 shows the convergence trend.

Clearly, the locations of targets are accurately determined. Despite the slight difference in size compared to the actual images, the difference in terms of reconstructed tumour size can be easily distinguished. Based on preliminary results, this algorithm is robust with respect to uncertainties in the measurement and noise artefact, where convergence is attained less than 15 iterations as evident in Figure 8.29.

Given the nature of the approximate inversion, the methods and procedures work well when the object's dimension is comparable to the wavelength. For 5 GHz UWB sensors, targets smaller than 9 mm in diameter are easily detected. The ability to detect smaller targets can be further enhanced by increasing the number of antennas, reducing the computational grid size cell to increase the resolution and improving the accuracy of the data acquisition system. Like other inverse scattering problems, the computational cost imposes limitation to the resolution and image quality. Decreasing the grid size would yield more accurate result but require more intense computation. Moreover, the reconstruction of inhomogeneity found within the breast tissue poses further challenges due to multi-scattering effects between the objects, and depending on the intensity of the cluttering, the local minima arise leading to divergence. Addressing this problem requires more robust image reconstruction algorithm, which has been met following the development of EDAS described earlier.

A few more phantoms simulating the heterogeneous breast with glandular properties are fabricated to evaluate the performance of EDAS, particularly when dealing with noisy signals. In the fabrication T1-weighted magnetic resonance images are considered corresponding to different types of human adult breasts. A typical human breast can be classified into one of the four possible types depending on the radiographic density: (1) fatty or Type-A, (2) mildly heterogeneous or Type-B, (3) densely heterogeneous breasts or Type-C and (4) very dense heterogeneous breasts or Type-D.

As shown in Figure 8.30, Type-A comprises of mostly fatty materials with glandular tissue contributing less than 25% of the total breast composition. The

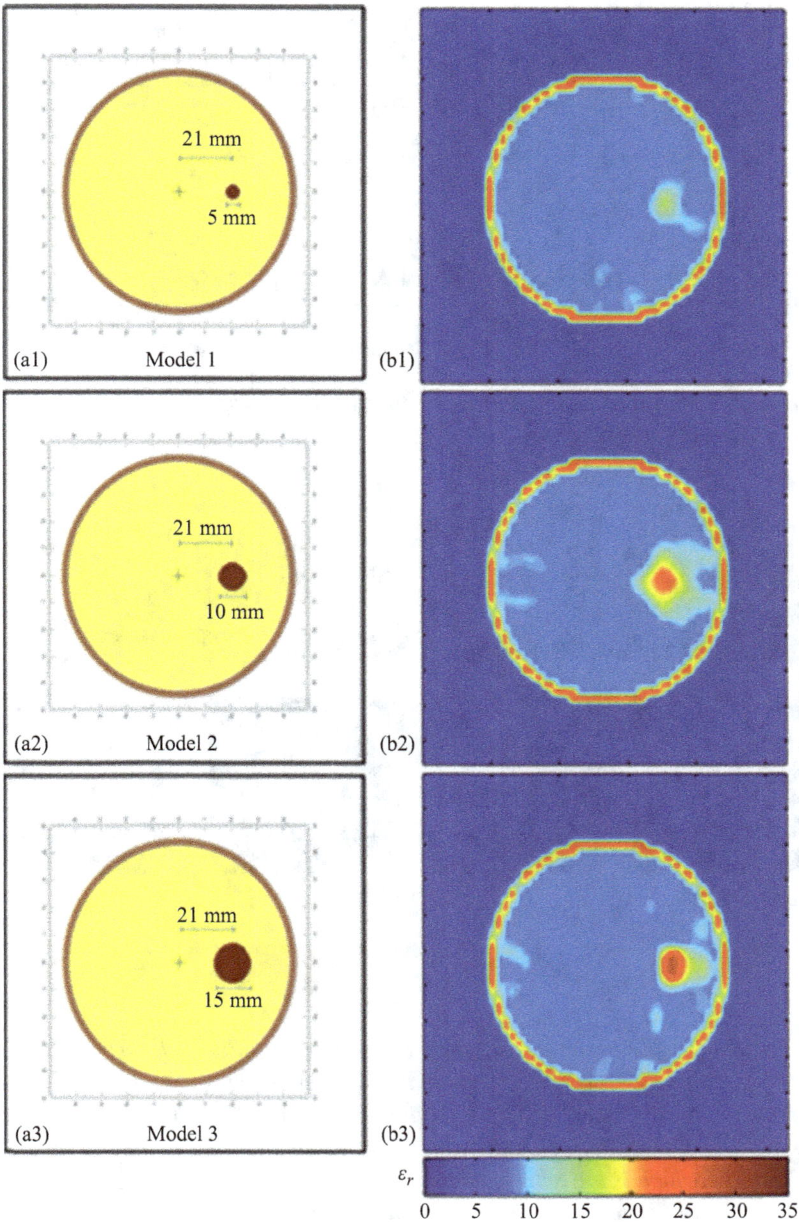

Figure 8.28 Experimental models simulating the growth of breast cancer at early stage: (a1), (a2) and (a3) actual images corresponding to tumour size of 5, 10 and 15 mm, respectively, and (b1), (b2) and (b3) show reconstructed images corresponding to each model

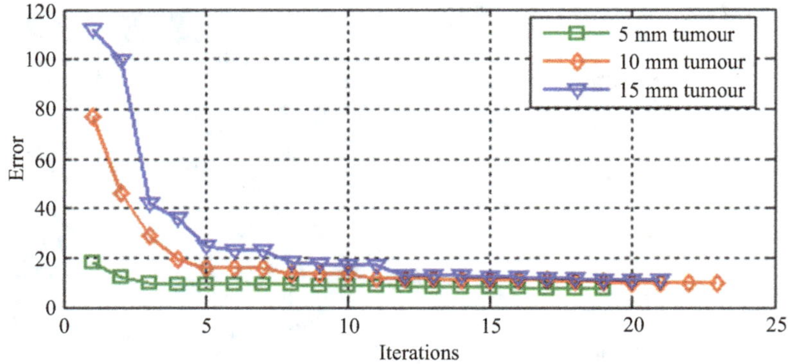

Figure 8.29 Iterative reduction in cost function error for experiments involving different tumour sizes

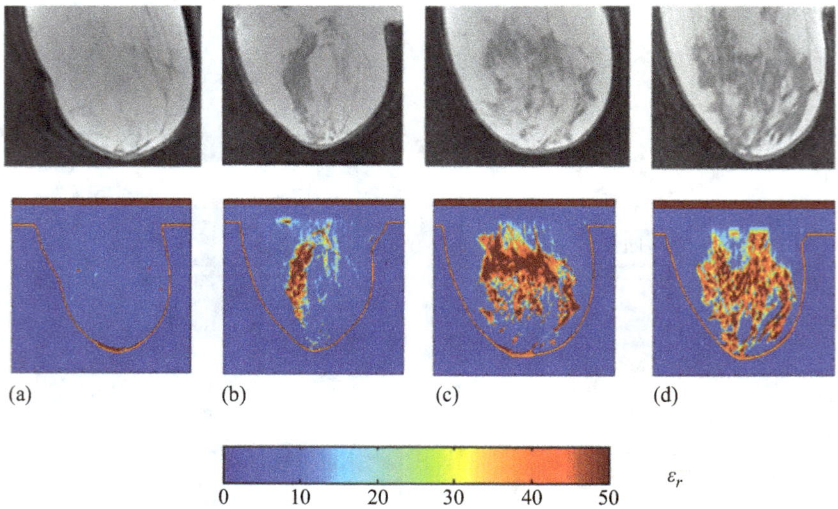

Figure 8.30 T1-MRI (upper) and their corresponding dielectric properties at 5GHz (lower) of breast classification showing: (a) Type-A which mostly fatty (<25% glandular tissue), (b) Type-B which is composed of scattered fibro-glandular tissue (25%–50% glandular), (c) Type-C which is the heterogeneously dense breast (51%–75% glandular) and (d) Type-D which is a very dense breast (>75% glandular) [45,46]

glandular composition lies in the range between 25% and 50% and 51% and 75% for Type-B and Type-C breasts, respectively. In contrast Type-D contains glandular tissues, which is more than 75% of the breast composition. Unlike the fabrication of Type-A or homogenous breast phantom, which requires materials of low water (that is fat tissue) and high water (i.e. tumour) only, the fabrication of the heterogeneously dense breast phantoms is slightly more complicated because of the presence of glandular structures in addition to tumour as well fatty tissues. In this work Type-A, Type-B and Type-C breasts are considered only whereas Type-D is reserved for future work. All the three types are fabricated using chemical derived materials with different level of heterogeneity. The chemicals are selected based on the electrical permittivity and conductivity covering the frequency of interest and dispersion characteristics as well as the electrical properties of fatty, glandular, skin and cancerous cells, whose values approximately match those of actual breast tissues [47]. The selection takes in consideration of the variations in dielectric properties of tissues among patients. In addition other factors, such as the low cost, the availability and the longer storage time, are also considered in selecting these materials.

In the fabrication the chemical compositions were varied so that the electrical characteristics (e.g. permittivity, conductivity and dispersion) of resulting mixtures closely matched with the breast tissues at the required signal bandwidth. During the experiment, an oil-gelatine mixture was used as a phantom material, based on heterogeneous dielectric permittivity of breast tissues available from the literature. Figure 8.31(a) shows three chemical-based breast models fabricated corresponding

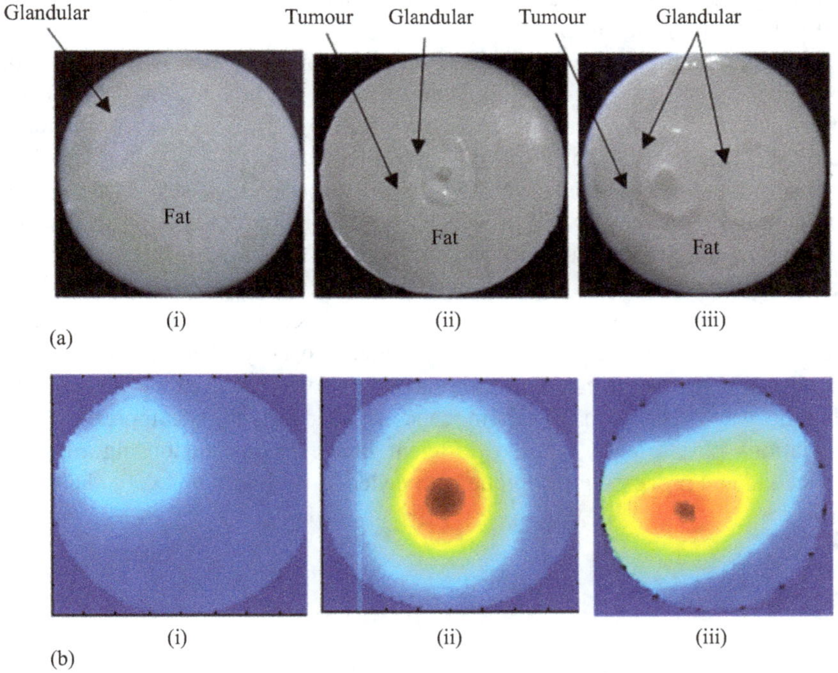

Figure 8.31 Realistic chemical-based heterogeneous-dense models corresponding to (i) Type-A, (ii) Type-B and (iii) Type-C breasts: (a) real and (b) reconstructed images, respectively [48]

to three different types of breasts. In this case Figure 8.31(a) (i), (ii) and (iii) represent Type-A, Type-B and Type-C breasts, respectively. All models are composed of 2 mm thick skin layer, 88 mm diameter fatty tissue and fibro-glandular with various sizes and structures. Their relative permittivities are fixed to 38, 15 and 32, respectively. Hence, these models capture the homogeneity and structural heterogeneity of normal breast tissue and integrate all realistic dispersive dielectric features from 0.5 to 20 GHz. A 5 mm target with relative permittivity of 54 is inserted in Type-B and Type-C models, simulating the cancerous tumour at the very early stage of development. The model in Figure 8.31(a) (i) contains both the fatty and fibro-glandular issues, thus simulating healthy breast but of heterogeneous type. This model serves as a reference to image reconstruction of cancerous models.

The acquired data are processed using EDAS, from which an image depicting the energy distribution inside the medium is reconstructed in 2D. The results are shown in Figure 8.31(b). It can be seen that important components like the fatty, the fibro-glandular and the cancerous tissues in the three models are detected and faithfully reconstructed. Visually, it is easy to distinguish between healthy and cancerous models because the response of UWB to the latter is more intense compared to the former. Hence, images reconstructed from Type-B and Type-C models contain areas with high intensity values as evident from Figure 8.31(b) (ii) and Figure 8.31(b) (iii), respectively.

In contrast, the image obtained from Type-A is relatively more uniform although the intensity of fibro-glandular region is slightly higher than the fatty background as clearly shown in Figure 8.31(b) (i). Close examination of Figures 8.31(b) (ii–iii) shows the presence of clutters, causing the difficulty in recovering the exact shape as well the size of the scatterers. This effect is more prominent in Figure 8.31(b) (iii), in which the glandular structures appear fused due to excessive image blurring and other artefact. This is partly attributed to the increasing difficulty in image reconstruction as the dielectric heterogeneity increases. Another problem is due to the low dielectric contrast of the heterogeneous tissues, decreasing from 3:6 for glandular-to-fatty tissues ratio to 1.7:1 for tumour-to-glandular tissues ratio. A similar finding was reported when radar-based technique was used to image inhomogeneous phantoms with complex dielectric distributions [26]. Clearly, a more efficient image reconstruction algorithm and advanced signal processing technique are needed to deal with this very challenging problem.

8.5 Summary

This chapter discussed methods for solving complicated inverse scattering problem arising from the propagation of UWB pulse in dielectric and dispersive media. Both qualitative and quantitative approaches are considered. Experimental results suggest that the Broyden method is robust with respect to uncertainties in the measurement and artefact. However, this technique is computationally intensive and requires longer reconstruction time. In contrast the single-step EDAS is faster and more resistant to noise.

The UWB antenna exhibits a wide impedance bandwidth averaging at 71% in the frequency range from 4.5 GHz to 10.9 GHz. Results from homogeneous breast phantoms show a 5 mm target corresponding to about 1/3 of the wavelength is easily detected both numerically and experimentally. The imaging system can also detect and resolve cancerous cells located inside the fibro glandular tissue of the heterogeneous breast phantom in spite of the presence of some artefact in the images. These anomalies are due to cluttering, which contaminated the measurement and introduced errors in the reconstruction. Further development on sensor and instrumentation is needed to improve the overall sensitivity of the system and SNR. The current work includes designing much more compact antenna based on dielectric resonant technology as a possible alternative to p-shaped radiator. Employing a much more advanced suppression technique like the phase coherence factor constitutes the future work in image reconstruction. Some preliminary works have already been performed and promising results have been published.

References

[1] Rekanos I T, Raisanen A, 'Microwave imaging in the time domain of buried multiple scatterers by using an FDTD-based optimization technique', *IEEE Trans. Magn.*, 2003, **39**, pp 1381–1384

[2] Abenius E, Strand B, 'Solving inverse electromagnetic problems using FDTD and gradient-based minimization', *Int. J. Numer. Methods Eng.*, 2006, **68**, pp 650–673

[3] Binajjaj S, Zanoon T, Abdullah M Z, 'Iterative and single-step solutions of multi-offset ultra wide band data in the time domain', *Procedings of IEEE Interntional Workshop on Imaging Systems and Techniques*, 11–12 May 2009, Shenzhen, China, pp 324–329

[4] Abdullah M Z, Binajjaj S, Zanoon T, Peyton A J, 'High-resolution imaging of dielectric profiles by using a time-domain ultra wideband radar sensors', *Measurement*, 2010, **44**, pp 859–870

[5] Xu L, Hagness S C, 'A confocal microwave imaging algorithm for breast cancer detection', *IEEE Microw. Wireless Comp. Lett.*, 2001, **11**, pp 130–132

[6] Nilavalan R, Gbedemah A, Craddock I J, Li X, Hagness S C, 'Numerical investigation of breast tumour detection using', *Electron. Lett.*, 2003, **39**, pp 1787–1789

[7] Kane Y, 'Numerical solution of initial boundary value problems involving Maxwell's equations in isotropic media', *IEEE Trans. Antennas Propag.*, 1966, **14**, pp 302–307

[8] Mur G, 'Absorbing boundary conditions for the finite-difference approximation of the time domain electromagnetic field equations', *IEEE Trans. Electromagn. Compat.*, 1981, **23**, pp 377–382

[9] Berenger J P, 'A perfectly matched layer for the absorption of electromagnetic waves', *J. Comput. Phys.*, 1994, **114**, pp 185–200

[10] Georgakopoulous S V, Birtcher C R, Balanis C A, Renaut R A, 'Higher order finite-difference schemes for electromagnetic radiation, scattering, and penetration, Part 1: Theory', *IEEE Antennas Propag. Mag.*, 2002, **44**, pp 134–142

[11] Supriyo D, Mittra R, 'A conformal finite-difference time-domain technique for modeling cylindrical dielectric resonators', *IEEE Trans. Microw. Theory Tech.*, 1999, **47**, pp 1737–1739

[12] Gabriel S, Lau R W, Gabriel C, 'The dielectric properties of biological tissues: III. Parametric models for the dielectric spectrum of tissues', *Phys. Med. Biol.*, 1996, **41**, pp 2271–2293

[13] Lazebnik M, Okoniewski M, Booske J H, Hagness S C, 'Highly accurate Debye models for normal and malignant breast tissue dielectric properties at microwave frequencies', *IEEE Microw. Wireless Comp. Lett.*, 2007, **17**, pp 822–824

[14] Sullivan D M, *Electromagnetic Simulation Using the FDTD Method*, IEEE Press, New York, 2000

[15] Debye P J W, *Polar Molecules*, The Chemical Catalog Company, Inc., New York, 1929

[16] Sullivan D M, 'Frequency-dependent FDTD methods using Z transforms', *IEEE Trans. Antennas Propag.*, 1992, **40**, pp 1223–1230

[17] Ashtari A, Noghanian S, Sabouni A, Aronsson J, Thomas G, Pistorius S, 'Using a priori information for regularization in breast microwave image reconstruction', *IEEE Trans. Biomed. Eng.*, 2010, **57**, pp 2197–2208

[18] Pastorino M, *Microwave Imaging*, Wiley, Hoboken, NJ, 2010

[19] Broyden C G, 'A class of methods for solving nonlinear simultaneous equations', *Math. Comput.*, 1965, **19**, pp 577–593

[20] Press W H, *Numerical Recipes: The Art of Scientific Computing*, Cambridge University Press, 2007

[21] Polak E, Ribierre G, 'Note sur la convergence de méthodes de directions conjuguées', *Revue Francaise Informat Reserche Opérationnelle*, 1969, **16**, pp 35–43

[22] Christiansen A V, Auken E, 'Optimizing a layered and laterally constrained 2D inversion of resistivity data using Broyden's update and 1D derivatives', *J. Appl. Geophys.*, 2004, **56**, pp 247–261

[23] Henriksson T, Joachimowicz N, Conessa C, Bolomey J, 'Quantitative Microwave imaging for breast cancer detection using a planar 2.45 GHz system', *IEEE Trans. Instrum. Meas.*, 2010, **59**, pp 2691–2699

[24] Hagness S C, Taflove A, Bridges J E, 'Two-dimensional FDTD analysis of a pulsed microwave confocal system for breast cancer detection: Fixed-focus and antenna-array sensors', *IEEE Trans. Biomed. Eng.*, 1998, **45**, pp 1470–1479

[25] Hollman K, Rigby K, O'Donnell M, 'Coherence factor of speckle from a multi-row probe', in *Proceedings of IEEE Ultrasonic Symposium*, 17–20 October 1999, CaesarsTahoe, NV, pp 1257–1260

[26] Klemm M, Leendertz J, Gibbins D, Craddock I, Preece A, Benjamin R, 'Microwave radar-based breast cancer detection: Imaging in inhomogeneous breast phantoms', *IEEE Antennas Wireless Propag. Lett.*, 2009, **8**, pp 1349–1352

[27] Armitage D W, Yin W, Abdullah M Z, Billal M, Peyton A J, 'Antenna design for ultra wide band electromagnetic tomography', in *Proceedings of 5th World Congress on Industrial Process Tomography*, 3–6 September 2007, Bergen, Norway

[28] Zanoon T F, Binajjaj S, Abdullah M, 'Electromagnetic tomography featuring ultra wide band sensor with conformal finite difference (CFDTD) modeling of dispersive media', in *IEEE Symposium on Industrial Electronics and Applications*, 4–6 October 2009, Kuala Lumpur, Malaysia, pp 377–382

[29] Samaddar S N, Mokole E L, 'Bi-conical antennas with unequal cone angles', *IEEE Trans. Antennas Propag.*, 1998, **46** (2), pp 181–193

[30] Valderas D, Sancho J I, Puente D, Ling C, Chen X, *Ultrawideband Antennas: Design and Applications*, Imperial College Press, 2011

[31] Sill J, Fear E, 'Tissue sensing adaptive radar for breast cancer detection: study of immersion liquids', *Electron. Lett.*, 2005, **41** (3), pp 113–115

[32] Tiang S S, Sadoon M S, Nik Anwar N S, Ain M, Abdullah M Z, 'Development of a compact wide-slot antenna for early stage breast cancer detection featuring circular array full-view geometry', *Int. J. Antennas Propag.*, 2014, 309321 (11 pages)

[33] Balanis C A, *Antenna Theory: Analysis and Design*, Wiley, New York, 1982

[34] Gibbins D, Klemm M, Craddock I J, Leendertz J A, Preece A, Benjamin R, 'A comparison of a wide-slot and a stacked patch antenna for the purpose of breast cancer detection', *IEEE Trans. Antennas Propag.*, 2010, **58**, pp 665–674

[35] Lamensdorf D, Susman L, 'Baseband-pulse-antenna techniques', *IEEE Antennas Propag. Mag.*, 1994, **36**, pp 20–30

[36] Allen O E, Hill D A, Ondrejka A R, 'Time-domain antenna characterizations', *IEEE Trans. Electromag. Compat.*, 1993, **35**, pp 339–346

[37] Chaudhary S S, Mishra R K, Swarup A, Thomas J M, 'Dielectric properties of normal and malignant human breast tissues at radiowave and microwave frequencies', *Indian J. Biochem. Biophys.*, 1984, **21**, pp 76–79

[38] Bilal M, Yin W, Abdullah M Z, Armitage D W, Peyton A J, 'Ultra wideband tomography experimental hardware and image reconstruction', in *Proceedings of of 5th World Congress Industrial on Industrial Process Tomography*, 3–6 September 2007, Bergen, Norway, pp 599–606

[39] Jia H, Takenaka T, Tanaka T, 'Time-domain inverse scattering method for cross-borehole radar imaging', *IEEE Trans. Geosci Remote Sens.*, 2002, **40** (7), pp 1640–1647

[40] Campbell A M, Land D V, 'Dielectric properties of female human breast tissue measured in vitro at 3.2 GHz', *Phys. Med. Biol.*, 1992, **37**, pp 193–210

[41] Sill J M, Fear E C, 'Tissue sensing adaptive radar for breast cancer detection: Preliminary experimental results', in *Proceedings of IEEE International Microwave Symposium Digest*, 2–17 June 2005, Long Beach, USA, pp 1789–1792

[42] Lai J C Y, Gunawan C B, Low K S, 'Homogeneous and heterogeneous breast phantoms for ultra-wideband microwave imaging applications', *Prog. Electromagn. Res.*, 2010, **100**, pp 397–415

[43] Ahmed J, Ramaswamy H S, Raghavan V G S, 'Dielectric properties of butter in the MW frequency range as affected by salt and temperature', *J. Food Eng.*, 2007, **82**, pp 351–358

[44] Zanoon T F, Abdullah M Z, 'Early stage breast cancer detection by means of time-domain ultra wide band sensing', *Meas. Sci. Technol.*, 2011, **22** (11), 114016

[45] Lazebnik M, Popovic D, McCartney L, *et al.*, 'A large-scale study of the ultrawideband microwave dielectric properties of normal, benign and malignant breast tissues obtained from cancer surgeries', *Phys. Med. Biol.*, 2007, **52**, 6093

[46] Zastrow E, Davis S K, Lazebnik M, Kelcz F, van Veen B D, Hagness S C, 'Development of anatomically realistic numerical breast phantoms with accurate dielectric properties for modeling microwave interactions with the human breast', *IEEE Trans. Biomed. Eng.*, 2008, **55**, pp 2792–2800

[47] Porter E, Fakhoury J, Oprisor R, Coates M, Popovic M, 'Improved tissue phantoms for experimental validation of microwave breast cancer detection', in *Proceedings of 4th European Conference on Antennas and Propagation*, 12–16 April 2010, Barcelona, Spain, pp 1–5

[48] Tiang S S, Sadoon M, Zanoon T F, Ain M F, Abdullah M Z, 'Radar sensing featuring bi-conical antenna for early stage breast cancer detection', *Prog. Electromagn. Res. B*, 2013, **46**, pp 299–316

Mohd Z Abdullah graduated from Universiti Sains Malaysia (USM) with a B. App. Sc. degree in Electronics in 1986 before joining Hitachi Semiconductor in Malaysia as a Test Engineer. In 1989 he commenced an MSc in Instrument Design and Application at University of Manchester Institute of Science and Technology (UMIST), UK. He remained in Manchester, carrying out research in Electrical Impedance Tomography at the same university, and received his PhD degree in 1993. His research interests include microwave tomography, digital image processing, computer vision and ultra-wide band sensing. He has published numerous research articles in international journals and conference proceedings. One of his papers was awarded The Senior Moulton medal for the best article published by the Institute of Chemical Engineering in 2002. At present, he is a Professor in the School of Electrical and Electronic Engineering at USM. He is a Fellow of IET and a Chartered Engineer.

Chapter 9

Synthetic aperture radar (SAR)

Guangcai Sun[1,2], Mengdao Xing[1,2] and Liang Guo[3]

List of abbreviations

BS	beam steering
CPI	coherent processing interval
CSA	chirp scaling algorithm
CZT	chirp Z transform
FFT	fast Fourier transform
FrFT	fractional Fourier transformation
ISAR	inverse synthetic aperture radar
LFM	linear frequency modulated
MTRC	migration through range cell
RCMC	range cells migration correction
RDA	range Doppler algorithm
SAR	synthetic aperture radar
SCFFT	scaled fast Fourier transform
SPM	stationary phase method
SRC	secondary range compression
TDJ-SPM	two-dimensional joint stationary phase method
TFD	time-frequency diagram
TOPS	terrain observation by progressive scans
UFA	unified focus algorithm

[1]National Laboratory of Radar Signal Processing, Xidian University, China
[2]Collaborative Innovation Center of information Sensing and Understanding, Xidian University, China
[3]School of Physics and Optoelectronic Engineering, Xidian University, China

9.1 Introduction

In 1951, Carl Wiley of Goodyear, an American company, discovered that side-looking radar can make use of the Doppler shift in echo to improve the azimuth resolution to the point that two-dimensional (2D) high-resolution imaging of observation objects can be realised by radar. This discovery greatly improves information perception and acquisition capability of radar and marks the birth of synthetic aperture radar (SAR) technology [1,2]. SAR is a system that processes transmitted and received signals to achieve high-resolution imaging of scene targets with small real antenna lengths and real apertures. SAR is an active system, which uses electromagnetic waves to observe surface characteristics. It is a technique of all-weather, long-distance, wide-amplitude and high-resolution imaging [2]. In military applications, SAR can be used for battlefield reconnaissance, military mapping, military target detection and other purposes to provide reliable information for strategic guidelines or tactical programs. SAR also has a wide range of civilian applications in agriculture, forestry, geology, oceanography, hydrology, flood detection, surveying and mapping, astronomy, disaster mitigation and disaster prevention, meteorology and so on. For over 60 years, the theory and system of SAR imaging have developed from coarse resolution imaging to high resolution imaging, single channel imaging to multi-channels and single sensor radar imaging to networked sensors processing [3–6]. The radar imaging processing technology has also been continuously extended to different application fields [7–10].

Due to the difference in imaging geometry, synthetic aperture imaging radar can be divided into SAR and inverse SAR (ISAR), but both make use of the principle of synthetic aperture. SAR commonly refers to a radar mounted on a motion platform (such as an aeroplane), which actively emits electromagnetic waves and receives echo signals to obtain geomorphological information [2]. ISAR commonly refers to a radar, which is stationary with the observed object in motion, and the radar obtains the electromagnetic scatter distribution of the target by processing the echo signal. Since both mechanisms make use of a virtual aperture formed by the relative motion between the radar and the observed target for synthetic aperture high-resolution imaging, the essential principles of SAR and ISAR are the same. This chapter introduces the basic concepts of SAR, the principle of 2D resolution, SAR imaging principle and imaging algorithms.

9.2 Principle of synthetic aperture

The basic principle of synthetic aperture technology is based on real aperture technology [2,4]. For real aperture antenna radar, the mechanism of 2D target resolution is to use a wide-band signal to distinguish the point target in the range direction, and use beam-forming to distinguish the azimuth direction, which is in parallel to the aperture direction of point target. According to the antenna theory, the antenna beam width is equal to the ratio of the radar wavelength λ to the antenna aperture D_a. Therefore, the azimuth resolution (cross-range resolution) of the real aperture radar can be approximated as

$$\rho_a = \frac{\lambda}{D_a} R_s \tag{9.1}$$

where R_s is the distance between the target and the antenna.

As can be seen from (9.1), the azimuth resolution is inversely proportional to the aperture length of the radar antenna. To obtain high resolution ρ_a, the antenna aperture must reach a longer length D_a, as shown in Figure 9.1. For example, if the antenna aperture is 300 wavelengths (about 10 m in the X-band) and the beam width is about 0.2°, the cross-range resolution at 30 km is about 100 m. Therefore, to improve the resolution to 1 m, the antenna aperture length needs to be increased 100 times, that is about 1,000 m, which is difficult to achieve. In particular, it is difficult to mount such a large radiation antenna on a flight platform. At the same time, the number of T/R components required for such a long antenna leads to high costs.

To overcome the limitation of the antenna aperture on the azimuth resolution, the concept of synthetic aperture is introduced into the imaging field. In principle, it is feasible to arrange a long linear array with small antennas (called array elements). To avoid blurring (i.e. no beam gratings lobe), the spacing distance of the array element should not exceed one-half wavelength [2]. If the target is fixed, a small radar can be set up to simplify the device, loading a single array element. All the array elements of the original real-aperture antenna send and receive signals at the same time. Now, a small radar is used to transmit and receive signals, and a straight track is laid, which is equivalent to placing the small radar on the trolley, pushing the trolley along the track step by step and recording the echoes of each step. These echoes contain the phase and amplitude information at the receiving end, and they are synthesised regarded as the array echoes. Obviously, the similar results to the actual array can be obtained, that is a high azimuth resolution can be obtained. In this way, although the actual antenna aperture is very short, the virtual antenna aperture is very long for each observed point. With the same principle, the radar can also be mounted on a plane or satellite to obtain a high-resolution images of a fixed ground scene. This is the basic principle of synthetic aperture.

The ability to achieve high azimuth resolution of ground scenes of a plane equipped with radar can also be explained by the Doppler effect [1,2]. As shown in Figure 9.2(a), radar moves along x-axis, and the angle between x-axis and the line from a radar to a target varies. Thus, the instantaneous Doppler also varies. To

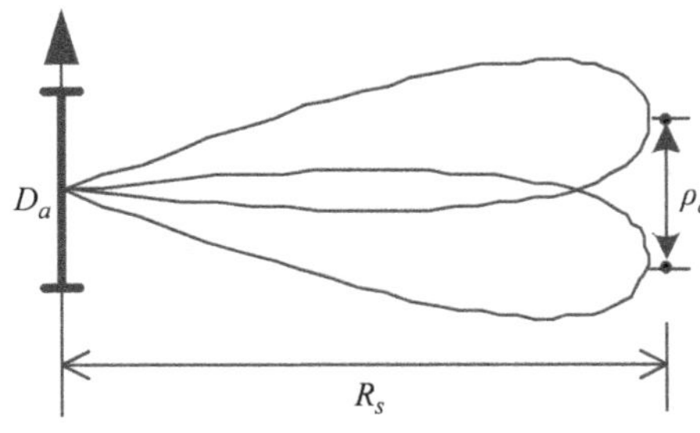

Figure 9.1 Real aperture antenna length and resolution

obtain high Doppler resolution, it is necessary to have a long coherent processing interval (CPI), that is the plane has to move a certain distance, during which the observation angle is constantly changing. The upper graph of Figure 9.2(b) shows the diagram of the radar echo phase of target O during the flight. The time, when the plane is closest to the target, is defined as the reference time (assumed to be 0). The corresponding distances before and after the reference time are longer than the closet distance, and also the phase of radar echo. From the distance change, one can calculate the expression of the phase, which is approximately a parabola. The time derivative of the above phase is Doppler, as shown in the lower part of Figure 9.2(b), showing the approximate line varying of Doppler with time (referred as time-frequency line). Figure 9.2(b) presents time-frequency lines of multiple point targets. They are all approximately linear frequency modulated (LFM) signals with the same frequency modulated rate and almost the same except for the shift in time. In other words, for different azimuth points with the same closest distance, their slant-range (or phase) expressions have the same form but with different reference times. Then, their signals in frequency domain can be written as the same expression multiplying with different linear-phase functions. Therefore, from the perspective of the linear system, their signals have the property of azimuth translation-invariance. This is a very important property and has important applications in later imaging algorithms.

It can also be seen from Figure 9.2(b) that the point targets on a parallel flight line have the same impulse response. Due to different frequency modulation rates, the impulse responses of targets on different parallel flight lines are different. The space-variant impulse response brings some problems to the calculation of image reconstruction.

In the above discussion, because the changing distance is approximately a wavelength and does not reach the range resolution, only the phase change which is caused by the distance from the target to the antenna phase centre is considered. If

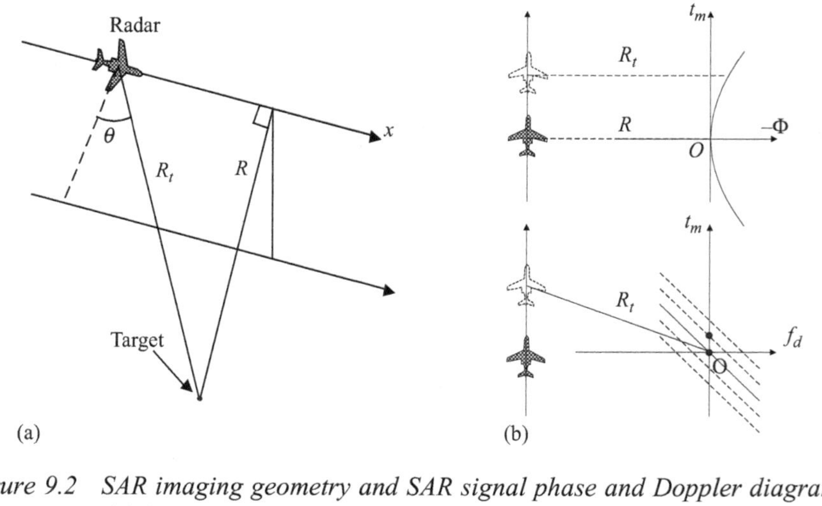

Figure 9.2 SAR imaging geometry and SAR signal phase and Doppler diagram: (a) SAR imaging geometry and (b) SAR phase history and time-frequency diagram

the distance variation can be compared with the range resolution, or even it is greater, then the migration through range cell (MTRC) should be considered, which will be discussed in detail later.

In Figure 9.3(a), the geometric relationship between the flight path *AD* and the point target *O* is drawn, where the synthetic apertures are arranged along the line *AD*. During the flight from *A* to *D*, signals are transmitted, received and recorded continuously. If the synthetic aperture length is not large, such as line *AB*, a small arc of AB' in Figure 9.3(a) can be used to approximate line *AB*, and then the received signal from *A* to *B* can be directly summed to reconstruct the target, which is called the non-focusing method.

As mentioned above, the synthetic aperture corresponding to the non-focusing method can only be a 'short straight line'. *A* question is how long the 'short straight line' should be. Suppose that the array elements are arranged symmetrically with *A* as the midpoint in Figure 9.3(a). If the signal phases on all array elements are approximately the same, the wave front can be regarded as a plane wave, but if the varying of signal phases is great enough, the wave front should be regarded as a spherical wave. If the point *A* is still used as the reference, the farther the plane is from the point *A*, the larger the phase difference is. For example, the phase of *C* and *D* is greater than *B*. When the phase difference is as large as $\pi/2$, such as *D* and *A*, the accumulated gain obtained by increasing the aperture and summing all the received signals is very small. In this case, the effective aperture length can be determined, that is line *AD*. At this time, the distance difference between *OA* and *OD* is just $\lambda/8$.

The above conclusion can be well explained by the superposition of the signal in the complex plane, as shown in Figure 9.3(b). The target is at point *A* as a reference. At this time, the phase of the received signal is 0. When the radar continues to move forward and reaches point *B*, due to the difference in wave path between the target and the radar position at point *A*, there is a phase difference between the received

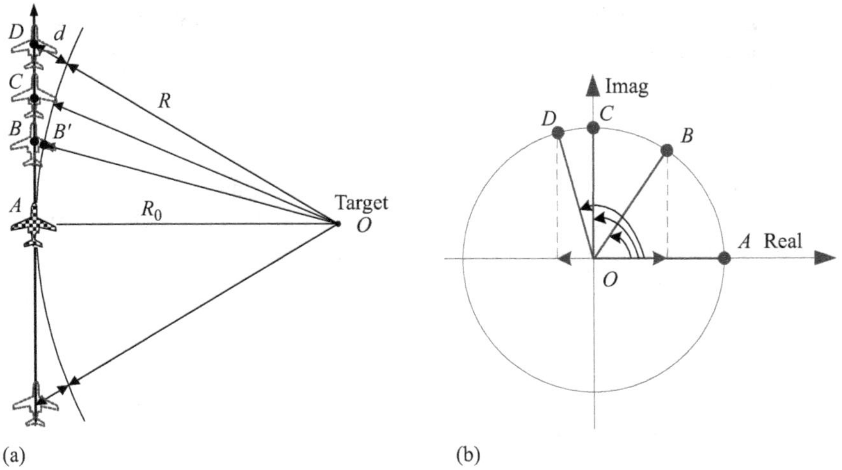

Figure 9.3 Complex interpretation of the effective length of a synthetic array: (a) effective length and (b) principle of adding complex plane of signal

signal at point B and the received signal at point A, and the position of signal at point B in the complex plane is shown in Figure 9.3(b). When the phase difference between point B and point A does not exceed $\pi/2$, the signal at point B still has a component on the real axis, that is it still contributes to the energy accumulation of the signal. When the radar moves to point C, the phase difference between point C and point A is $\pi/2$. In this case, the received signal at point C is orthogonal to the reference receiving signal, and thus on the real axis, it has no effect on the accumulation of the signal and exceeds the effective aperture. When the radar continues to move to point D, the phase difference between point D and point A exceeds $\pi/2$. At this time, the component of the receiving signal at point D on the real axis has side effects. The effective aperture length of non-focusing L_e equals to $\sqrt{Rl}$, where R is the target distance. By using (9.1) and considering signal two-way propagation, the cross-range (i.e. azimuth) resolution at this moment can be calculated as $\rho_a = \sqrt{R\lambda}/2$. For example, if wavelength l is 3 cm and the distance R is 30 km, then the effective aperture length of the non-focusing L_e is 30 m and azimuth resolution r_a is 15 m. If the distance is lengthened, the azimuth resolution will decrease.

If the synthetic aperture is longer, the signal between the equivalent elements has a larger phase difference. When the phase difference exceeds $\pm\pi/2$, the direct superposition of the signals does not produce a good resolution. In this case, only after phase compensation, the signals can be correctly reconstructed by summing up, which is equivalent to the focus in the optical system. Focus processing is the essence of the high-resolution synthetic aperture imaging technology. Although there are some differences between synthetic aperture imaging technology and real aperture technology, their resolution principles are the same. The principles of synthetic aperture imaging will be analysed in detail below.

9.2.1 Range resolution

Similar to the traditional radar, if SAR transmits and receives LFM signals, it can achieve high range resolution through pulse compression using matched filtering [5–10]. The slant-range resolution is determined by the bandwidth of the transmitted signal:

$$\rho_r = c/2B = c/2\gamma T_p \tag{9.2}$$

where c is the light speed, B is the bandwidth of the transmitted signal, γ is the frequency modulated rate of the transmitted signal and T_p is the pulse duration.

The resolution of ground distance ρ_g is obtained by

$$\rho_g = \rho_r/\sin\theta \tag{9.3}$$

where θ is the beam incident angle.

9.2.2 Azimuth resolution

Figure 9.4 shows the geometric relationship of side-looking SAR, where the radar beam centre line is perpendicular to the observing zone. It is assumed that the angular width of the radar beam in the azimuth dimension is β, the length of the synthetic

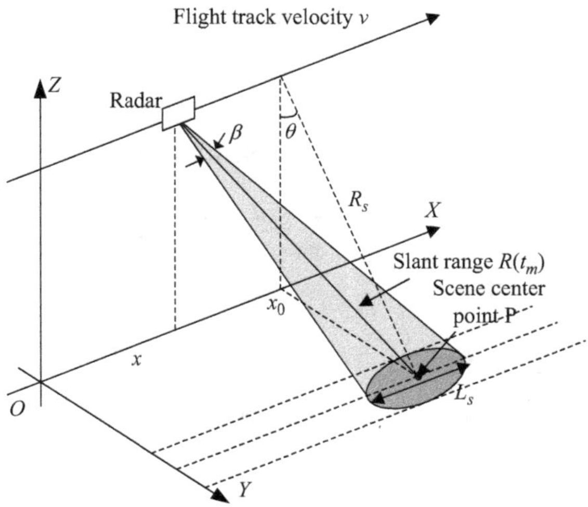

Figure 9.4 Geometric relationship diagram of side-looking SAR

aperture is L_s, the X coordinate of the radar platform is x, the forward movement velocity of the radar platform is v, the X coordinate of point target P is x_0, the slant-range distance from the radar to the target is $R(t_m)$ and the shortest distance is R_s.

From Figure 9.4, the instantaneous slant-range distance of the target point is

$$R(t_m) = \sqrt{{R_s}^2 + (x - x_0)^2} \tag{9.4}$$

Under ideal motion conditions, $x = vt_m$. When the target point and the carrier airplane satisfy the relationship $R_s \gg (x - x_0)$, (9.4) can be approximated as

$$R(t_m) \approx R_s + \frac{(vt_m - x_0)^2}{2R_s} \tag{9.5}$$

Since the azimuth resolution of SAR is mainly discussed here, the transmitted signal can be assumed as a single-frequency continuous wave $\exp(j2\pi f_c t)$, then the echo of the target at time t_m can be expressed as $A \cdot \exp(j2\pi f_c(t - 2R(t_m)/c))$, and the baseband echo is obtained by coherent detection:

$$s(t_m) = A \cdot \exp(j4\pi f_c R(t_m)/c) \tag{9.6}$$

where A is the amplitude of the echo.

When the radar platform moves relative to the target, the Doppler effect is generated and the Doppler frequency of the target echo in azimuth is

$$f_d = \frac{1}{2\pi}\frac{d}{dt_m}\varphi(t_m) = -\frac{2f_c}{c}\frac{d}{dt_m}R(t_m) \approx -\frac{2f_c v}{cR_s}(vt_m - x) \tag{9.7}$$

In (9.7), f_d is linear with t_m, which implies that under the condition of vertical side-looking view, the Doppler signal of the target in the azimuth time domain is a LFM signal, and when $t_m = x/v$ (i.e. when the radar is closest to the target), $f_d = 0$.

From (9.7), the Doppler frequency modulation rate γ_m of the echo can be obtained by

$$\gamma_m = -2v^2/\lambda R_s \tag{9.8}$$

The key to azimuth compression is to obtain Doppler centroid and frequency modulation rate correctly. Starting from Doppler bandwidth, the azimuth resolution is analysed below. Suppose that the time of the radar beam sweeping through the target (i.e. the synthetic aperture time) is T_s, and its relationship with platform speed and antenna length D is

$$T_s = \frac{L_s}{v} = \frac{\beta \cdot R_s}{v} = \frac{\lambda/D \cdot R_s}{v} \tag{9.9}$$

The Doppler bandwidth of the point target echo is

$$\Delta f_d = |\gamma_m| T_s = 2\beta v/\lambda \tag{9.10}$$

After the LFM signal passes through the matched filter, the main lobe width of the output signal envelope is

$$\tau_a = \frac{1}{\Delta f_d} = \frac{\lambda}{2v\beta} = \frac{D}{2v} \tag{9.11}$$

Then, the ultimate azimuth resolution of SAR is

$$\rho_a = \tau_a v = D/2 \tag{9.12}$$

As can be seen from the above derivation, the azimuth resolution is only related to the antenna size. From (9.12), the azimuth resolution can be improved by only reducing the real aperture length D, which is equivalent to expanding the antenna beam width and increasing CPI. However, this relationship must satisfy the approximation $R_s \gg L_s \geq (x - x_0)$. When D is infinitely reduced, the length of the synthetic aperture is $L_s \to \infty$ $(L_s = R_s \cdot \lambda/D)$. In the above approximation conditions, the condition that L_s is far less than the slant-range distance R_s cannot be satisfied, and then the Doppler frequency of the echo can no longer be approximated as an LFM signal but a signal containing many high-order phase terms, and the Doppler bandwidth cannot linearly increase with the increase of the beam width, and (9.12) is no longer valid. In the extreme case, the antenna beam width $\beta = \pi$, which is equivalent to that of the non-directional antenna, ρ_a in the extreme case approximately is

$$\rho_a = \frac{V}{\Delta f_d} = \frac{\lambda}{4} \tag{9.13}$$

9.3 Imaging principle and range Doppler algorithm

In Section 9.2, the basic principles of synthetic aperture technology are introduced. Obtaining high range resolution with wideband signal and high cross-range (azimuth) resolution with long synthetic array is the basis of the 2D high-resolution imaging [2]. The following will introduce the basic principles of synthetic aperture radar (SAR).

9.3.1 Basic principles of SAR imaging

SAR is an advanced active microwave equipment for the ground observation. It uses coherent radar system and a single moving antenna to simulate all antenna functions in a real linear antenna array [2–4]. The single antenna occupies the spatial position of the synthetic array in chronological order, stores the signals with the amplitude and phase information received at each position. Those signals are processed to present a reflection characteristic image of the ground object in the area.

SAR imaging has the ability to acquire high-resolution 2D landform images. Imaging SAR processing can be regarded as an input-output system, whose inputs are the scattering distribution of the target scene and outputs are echo data. According to the specific SAR signal model and the imaging algorithm, the response function of the SAR system can be obtained. Then, SAR imaging can be considered as an inverse problem of a given SAR system response function.

Let's take side-looking Stripmap SAR [3] as an example. To obtain a synthetic array, the radar moves along a straight line, transmitting and receiving broadband pulse signals periodically. Figure 9.5 shows radar receiving signals in the geometric plane, where t is the full time, $\hat{t}$ is the fast time or range time and t_m is the slow time or azimuth time. Then $\hat{t} = t - mT$, where m is an integer, T is the pulse repetition period and slow time is $t_m = mT$.

The radar moves along the x-axis with a speed of v, and the thick black line represents the trajectory of radar motion. The trajectory corresponds to the time interval $[-T_a/2, -T_a/2]$ and its centre A is also the centre of the full time and the slow time. Q is an arbitrary point in the scene. P_0 is the point on the centre line of the scene. B is the vertical projection of point P_0 on the flight path, called the close intersection point, and B is set as the origin of the coordinate $x(t)$ in the flight direction. In the $x(t)$

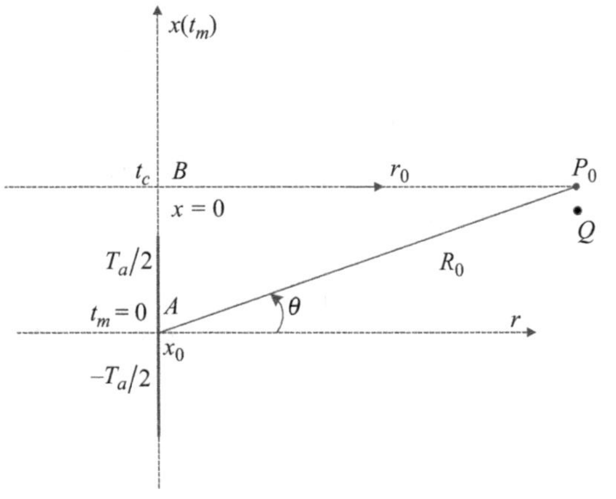

Figure 9.5 Geometry of radar collected signal

axis, the coordinate of point A is x_0, and the time of point B is t_c. The squint angle of the beam is θ, the slant range from the antenna phase centre to the strip centre line is R_0 and the slant range from the scene centre line to the flight path is r_0.

The radar requires the echo of each scattering point in the scene at fast time $\hat{t}$ and slow time t_m. Since the directionality of the radar antenna, a beam 'footprint' will be formed on the ground when the observation object is the ground scene. The illustrated targets are in the 'footprint', where the fast time $\hat{t}$ records the slant range of each target, the slow time records the position of the array elements. Through the position change of array elements, the change of echo phase with slow time can be known, that is the distance and Doppler that the beam 'footprint' sweeps over the target can be recorded. In practice, the Doppler of target echo is obtained by processing the phase history of slow time, and the high resolution requires a long coherent integration time.

From the above analysis, 2D data acquired by the synthetic aperture radar in the realistic three-dimensional (3D) space should be described by cylindrical coordinates. In other words, there is only range and no azimuth data in the normal plane of the flight path axis. However, the range of azimuth angle in the normal plane is limited, and only the area covered by the radar beam can be observed. The beam width of the real radar in elevation angle is not large, generally only a few degrees.

The recording of the 2D data can be vividly described through the data acquisition plane (see Figure 9.6). The data acquisition plane is constituted by the radial axis and the flight route axis. The radial axis is the line from the radar to the centre of the scene.

The objects outside the centre line of the scene are not in the data acquisition plane, while they are projected onto the data acquisition plane through a circular arc line centred on the radar, as shown in Figure 9.7. Essentially, what is acquired is the distance from these targets to the radar.

SAR imaging is essentially to extract the radar backscattering coefficients of the scattering unit on the surface of the Earth from the echo signals, and then

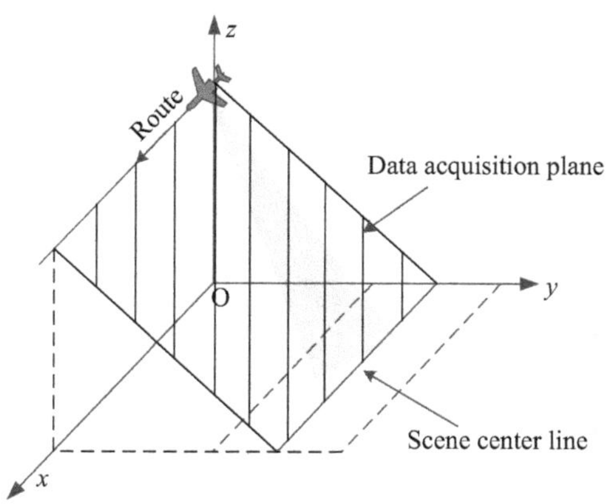

Figure 9.6 Data acquisition plane

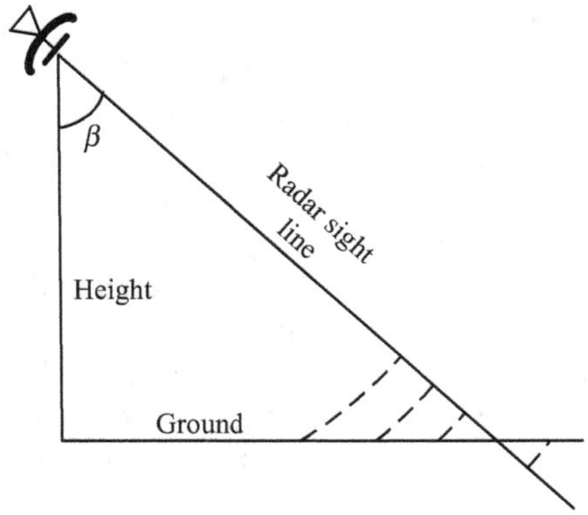

Figure 9.7 From data acquisition plane to imaging plane

displays the scattering points according to their range and azimuth positions. If x is used to denote the position of the azimuth and r is the position of the distance direction, the radar backscattering coefficient of each scattering unit can be written as $\sigma(x, r)$. Supposing the echo signal is represented by $s(t)$ and the impulse response function of the radar imaging system is $h(t)$, then the whole imaging process can be expressed as

$$\widehat{\sigma}(x, r) = s(t) \otimes h(t) \tag{9.14}$$

where $\otimes$ is the convolution operator.

The more accurate the imaging is, the closer $\widehat{\sigma}(x, r)$ will be to the true value $\sigma(x, r)$. Therefore, an SAR image is the real reflection of the scattering coefficient of the surface. SAR imaging usually involves the following steps:

1. Data acquisition. That is to store the baseband echo received by the radar during the flight of the plane.
2. Pulse compression. Since the matching function is spatial-variant in the range dimension, more attention should be paid on it.
3. 2D imaging for scene.

The above steps show the imaging process in the case of SAR working in the ideal condition. During the flight of the plane, if the unavoidable trembler and undulation are considered, the motion compensation must be implemented before the matched filtering. If the imaging result is geometrically distorted in some cases, the geometric distortion correction is required after imaging.

9.3.2 Range Doppler algorithm

Range Doppler algorithm (RDA) not only meets the requirements of efficient modular processing but also has the simplicity of one-dimensional (1D) operation [1,2]. The algorithm uses a range cells migration correction (RCMC) operation between two 1D operations to perform approximate separation of range and azimuth.

9.3.2.1 Property of azimuth-translation invariance

Azimuth-translation invariance means that the point targets with the same range all go through the same process, from entering the radar beam to leaving the radar beam. The only difference is that these points have a sequential order in time. After the signals are transformed into the azimuth-Doppler domain, their trajectories coincide. To correct a single target trajectory in the azimuth-frequency domain is equivalent to correct a group of target trajectories with the same nearest slant range. This is the key of the range-Doppler algorithm. As shown in Figure 9.8, the three curves correspond to the migration trajectories of points A, B and C, where A and B have the same nearest distance. When the signal is transformed into the azimuth-Doppler domain, the trajectories of points A and B coincide.

9.3.2.2 Expression of SAR signal

The geometry relationship diagram of side-looking SAR is shown in Figure 9.9. The vertical distance (or nearest distance) from the point target $P(X_0, R_B)$ to the flight route is R_B, and the slant range from the radar antenna phase centre to P at t_m is $R(t_m; R_B)$. It is supposed that the radar emits a linear FM (LFM) signal $s_t(\hat{t}) = a_r(\hat{t})\exp(j\pi\gamma\hat{t}^2)$, where γ is the frequency modulation rate. Then, the baseband echo of P in the range time and azimuth time domain ($\hat{t} - t_m$ domain) can be written as

$$s(\hat{t}, t_m; R_B) = a_r\left(\hat{t} - \frac{2R(t_m; R_B)}{c}\right) a_a(t_m) \exp\left[j\pi\gamma\left(\hat{t} - \frac{2R(t_m; R_B)}{c}\right)^2\right] \times \exp\left[-j\frac{4\pi}{\lambda}R(t_m; R_B)\right] \tag{9.15}$$

where a_r and a_a are the range and azimuth window functions, respectively; a_r is determined by the transmitted signal waveform and a_a is determined by the antenna pattern, $\lambda = c/f_c$ is the wavelength and $R(t_m; R_B) = \sqrt{R_B{}^2 + (vt_m - X_0)^2}$ is the instantaneous slant range between radar and P.

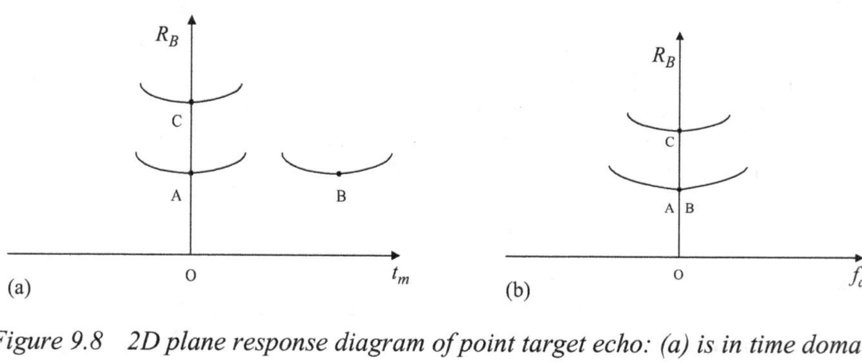

Figure 9.8 2D plane response diagram of point target echo: (a) is in time domain and (b) is in frequency domain

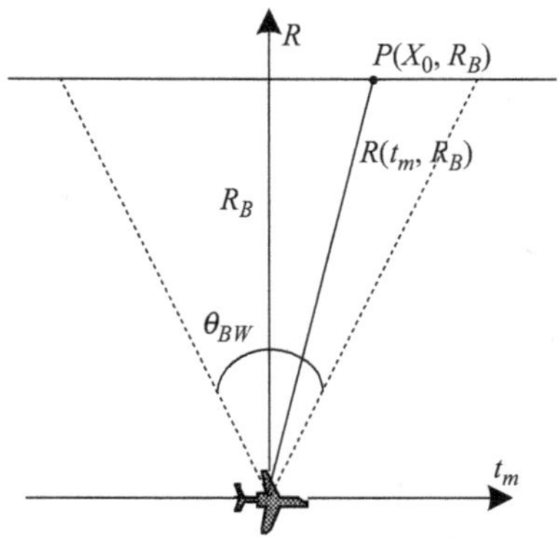

Figure 9.9 Geometry relationship diagram of side-looking SAR

9.3.2.3 2D spectrum

Since the signal has the property of azimuth-translation invariance, it is necessary to convert the azimuth signal into azimuth-Doppler domain for uniform matching processing. 2D spectrum is very important and should be derived accurately. The stationary phase method (SPM) is a common method for deriving 2D spectrum. In fact, it is a two-fold integrated form of the received signal. The 2D spectrum of the above equation can be calculated by referring to the two-dimensional joint stationary phase Method (TDJ-SPM) [2]:

$$\begin{aligned} S(f_r, f_a; R_B) &= \iint_{\hat{t}, t_m} s(\hat{t}, t_m; R_B) \exp(-j2\pi f_r \hat{t} - j2\pi f_a t_m) d\hat{t} dt_m \\ &= \iint_{\hat{t}, t_m} a_r(\hat{t} - \tau) a_a(t_m) \exp\left[j\pi\gamma(\hat{t} - \tau)^2\right] \\ &\quad \times \exp\left[-j\frac{4\pi}{\lambda} R(t_m; R_B)\right] \exp(-j2\pi f_r \hat{t} - j2\pi f_a t_m) d\hat{t} dt_m \end{aligned} \tag{9.16}$$

where f_r and f_a are the range frequency and azimuth frequency, respectively and $\tau = 2R(t_m; R_B)/c$ is the two-way delay.

The phase function in (9.16) is

$$\Theta(\hat{t}, t_m) = \pi\gamma(\hat{t} - \tau)^2 - \frac{4\pi}{\lambda} R(t_m; R_B) - 2\pi f_r \hat{t} - 2\pi f_a t_m \tag{9.17}$$

By generalising the principle of SPM for a 1D signal, 2D partial derivative of the phase function can be obtained and set to 0:

$$\frac{\partial\Theta}{\partial\hat{t}} = 2\pi\gamma(\hat{t} - \tau) - 2\pi f_r = 0 \tag{9.18}$$

$$\frac{\partial\Theta}{\partial t_m} = 2\pi\gamma(\hat{t} - \tau)\left(-\frac{2}{c}\right)\frac{\partial R(t_m; R_B)}{\partial t_m} - \frac{4\pi}{\lambda}\frac{\partial R(t_m; R_B)}{\partial t_m} - 2\pi f_a = 0 \tag{9.19}$$

Jointing (9.18) and (9.19), the following equations can be obtained:

$$\hat{t} = \frac{f_r}{\gamma} + \tau \tag{9.20}$$

$$\frac{\partial R(t_m; R_B)}{\partial t_m} = -\frac{c f_a}{2(f_c + f_r)} \tag{9.21}$$

On the other hand, the instantaneous slant range is a function of azimuth time, so its derivative can be written as

$$\frac{\partial R(t_m; R_B)}{\partial t_m} = \frac{v(vt_m - X_0)}{\sqrt{{R_B}^2 + (vt_m - X_0)^2}} \tag{9.22}$$

By combining (9.21) and (9.22), the following expression can be obtained:

$$\frac{vt_m - X_0}{\sqrt{{R_B}^2 + (vt_m - X_0)^2}} = -\frac{c f_a}{2v(f_c + f_r)} = \kappa \tag{9.23}$$

Solving (9.23), slow time can be obtained:

$$t_m = \frac{R_B}{v}\sqrt{\frac{\kappa^2}{1-\kappa^2}} + \frac{X_0}{v} \tag{9.24}$$

Then, the two-way delay and range time becomes

$$\tau = \frac{2R_B}{c}\sqrt{\frac{1}{1-\kappa^2}} \tag{9.25}$$

$$\hat{t} = \frac{f_r}{\gamma} + \frac{2R_B}{c}\sqrt{\frac{1}{1-\kappa^2}} \tag{9.26}$$

Substituting (9.24)–(9.26) into (9.16), the 2D spectrum can be obtained:

$$S(f_r, f_a; R_B) = \sigma' \mathrm{W}(f_r, f_a) \cdot \exp\begin{pmatrix} -j\pi\dfrac{f_r^2}{\gamma} - j2\pi f_a\left(\dfrac{\kappa}{v}R_B\sqrt{\dfrac{1}{1-\kappa^2}} + \dfrac{X_0}{v}\right) \\ -j\dfrac{4\pi R_B}{c}(f_c + f_r)\sqrt{\dfrac{1}{1-\kappa^2}} \end{pmatrix} \tag{9.27}$$

where σ is a constant derived from the TDJ-SPM derivation, and related to the signal intensity and the scattering characteristics of the target point and $\mathrm{W}(f_r, f_a)$ is a 2D frequency domain window function, which can be written as

$$\mathrm{W}(f_r, f_a) = \mathrm{rect}\left[\frac{f_r}{\gamma}\right]\mathrm{rect}\left[\frac{R_B}{v}\sqrt{\frac{\kappa^2}{1-\kappa^2}} + \frac{X_0}{v}\right] \tag{9.28}$$

Substituting (9.23) into (9.27), the 2D spectrum can be finally obtained as

$$S(f_r,f_a;R_B) = \sigma' W(f_r,f_a) \cdot \exp\begin{pmatrix} -j\pi\dfrac{f_r^2}{\gamma} - j2\pi f_a\dfrac{X_0}{v} \\ -j4\pi R_B\sqrt{\left(\dfrac{f_c+f_r}{c}\right)^2 - \left(\dfrac{f_a}{2v}\right)^2} \end{pmatrix} \tag{9.29}$$

By defining $\Phi(f_r,f_a;R_B) = -4\pi R_B\sqrt{\left(\frac{f_c+f_r}{c}\right)^2 - \left(\frac{f_a}{2v}\right)^2}$, its Taylor series expansion on f_r is as follows.

$$\Phi(f_r,f_a;R_B) = \phi_0(f_a;R_B) + \phi_1(f_a;R_B)f_r + \phi_2(f_a;R_B)f_r^2 + \cdots \tag{9.30}$$

where

$$\begin{aligned} \phi_0(f_a;R_B) &= -\frac{4\pi R_B\cos\theta}{\lambda} \\ \phi_1(f_a;R_B) &= -\frac{4\pi}{c}\frac{R_B}{\cos\theta} \\ \phi_2(f_a;R_B) &= \pi R_B\frac{2\lambda\sin^2\theta}{c^2\cos^3\theta} \end{aligned} \tag{9.31}$$

where $\sin\theta = f_a/f_{aM}$ and $\cos\theta = \sqrt{1-(f_a/f_{aM})^2}$.

By ignoring the third and higher order terms and substituting (9.30) into (9.29), a new 2D spectral expression can be obtained:

$$S(f_r,f_a;R_B) = \sigma' W(f_r,f_a) \cdot \exp\begin{pmatrix} -j\pi\dfrac{f_r^2}{\gamma_e(f_a;R_B)} - j\dfrac{4\pi}{c}\dfrac{R_B}{\cos\theta}f_r \\ -j2\pi f_a\dfrac{X_0}{v} - j\dfrac{2\pi R_B}{v}\sqrt{f_{aM}^2 - f_a^2} \end{pmatrix} \tag{9.32}$$

where $\frac{1}{\gamma_e(f_a;\,R_B)} = \frac{1}{\gamma} - R_B\frac{2\lambda\,\sin^2\theta}{c^2\,\cos^3\theta}$.

9.3.2.4 Range migration correction and range compression

If the signal is 2D separable, it is feasible and easy to eliminate the nonlinear phase term of the signal by the matched filtering. However, the signal is not 2D separable. When the echo is processed in the 2D frequency domain, the independent variable of the slant range also changed with t_m converted to be f_a, and the slant range expression is

$$R(f_a;R_B) = R_B/\sqrt{1-(f_a/f_{aM})^2} \approx R_B + \frac{1}{2}R_B(f_a/f_{aM})^2 \tag{9.33}$$

where the range migration is $\frac{1}{2}R_B(f_a/f_{aM})^2$.

To straighten the response curve with range migration in the $\hat{t}-f_a$ domain, $f_a = 0$ is taken as the reference (at this time $R(f_a;R_B) = R_B$) and echo at other f_a is moved forward along the $\hat{t}$-axis. This is easy to implement in the f_r domain by multiplied with a term $\exp\left[j2\pi\frac{R_B}{c}(f_a/f_{aM})^2 f_r\right]$. Therefore, to perform range migration correction and range pulse compression, the signal shown in (9.32) should be multiplied by the range migration correction function:

$$H_1(f_r, f_a; R_B) = \exp\left[j2\pi\frac{R_B}{c}(f_a/f_{aM})^2 f_r\right] \tag{9.34}$$

and distance pulse pressure function

$$H_2(f_r, f_a; R_B) = \exp\left[j\pi\frac{f_r^2}{\gamma_e(f_a; R_B)}\right] \tag{9.35}$$

9.3.2.5 Azimuth compression

By multiplying (9.32) by (9.34) and (9.35), the range migration correction and the range compression are performed. After transforming the result into the $\hat{t} - f_a$ domain, the azimuth compression function can be obtained:

$$H_3(\hat{t}, f_a; R_B) = \exp\left[j\frac{2\pi R_B}{v}\sqrt{f_{aM}{}^2 - f_a{}^2}\right] \tag{9.36}$$

By multiplying the signal in the $\hat{t} - f_a$ domain by (9.36) and performing the azimuth inverse Fourier transform, the compressed scene image can be expressed as

$$S(\hat{t}, t_m; R_B) = \sigma' \text{sinc}\left[\Delta f_r\left(\hat{t} - \frac{2R_B}{c}\right)\right] \text{sinc}\left[\Delta f_a\left(t_m - \frac{X_0}{v}\right)\right] \tag{9.37}$$

9.4 Imaging modes

With the development of SAR imaging requirements, SAR has developed a variety of working modes [3,6,7,11–13]. According to the steering direction of the SAR beam, the SAR imaging modes are divided into Stripmap, Spotlight, Sliding Spotlight, TOPS and Scan. Their working geometry is shown in Figure 9.10.

The Stripmap mode is the simplest, and the beam direction does not change during the radar movement. The Spotlight mode is that the beam always points to a point in a fixed scene. Sliding Spotlight and TOPS are that the antenna beam direction rotates around a virtual centre in the course of radar movement. The rotation centre of the Sliding Spotlight is farther than the reference centre line, while that of TOPS is on the side off course.

With the Scan mode, the radar beam illustrates one mapping zone in a short time and illustrates another mapping zone in the next time. Considering the synthetic aperture length of a single point, the following conclusions can be drawn. Under the same conditions, the resolution of the Spotlight mode is the highest, followed by Sliding Spotlight, Stripmap, TOPS and Scan modes. However, the order of their ability to cover the scene is opposite to the order of resolution ranking, and so these modes are the result of a compromise between mapping and resolution.

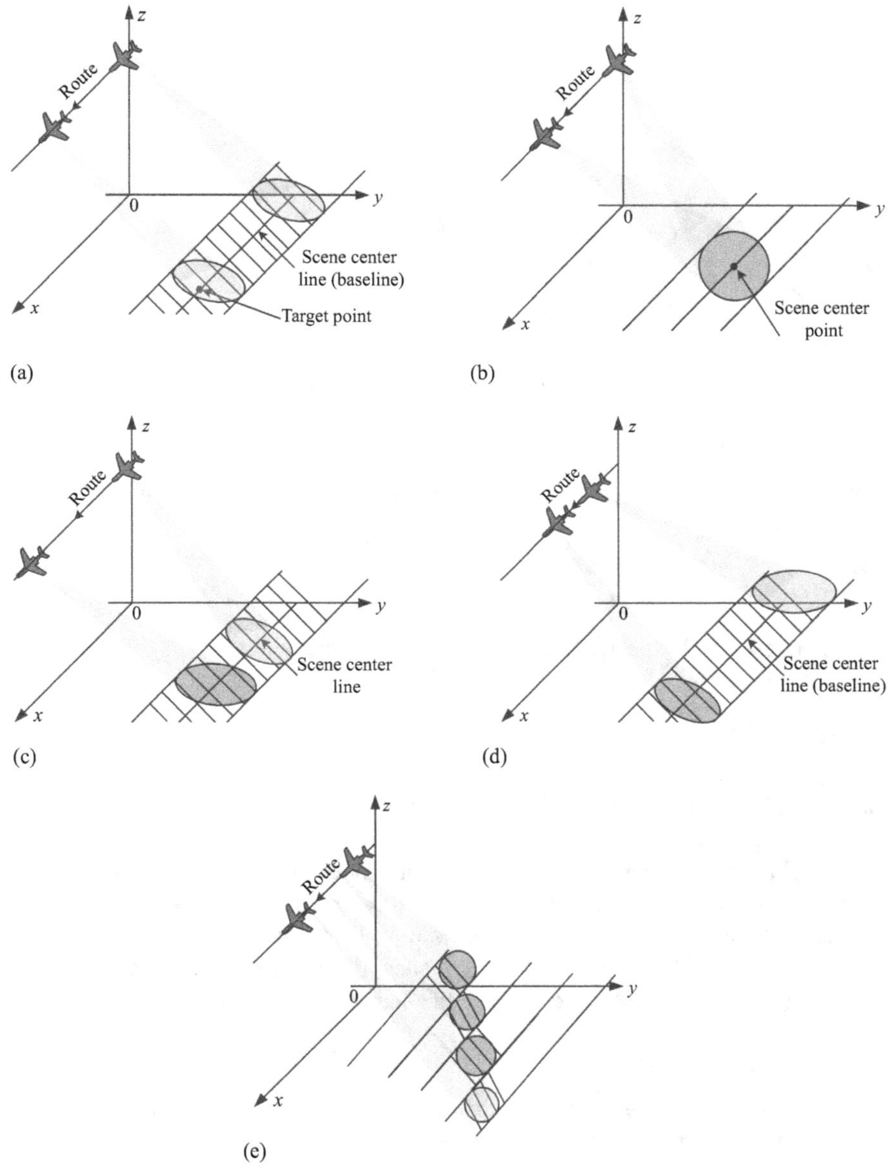

Figure 9.10 Geometry for multiple imaging modes: (a) Stripmap, (b) Spotlight, (c) Sliding Spotlight, (d) TOPS and (e) Scan

9.4.1 Multi-mode unified signal model

For a general analysis, some basic expressions are presented. Figure 9.11 shows the imaging geometry. R_{ref} is reference range, that is the nearest range from reference line (dashed line in Figure 9.11) to trajectory, v is the velocity of radar and t_a is the

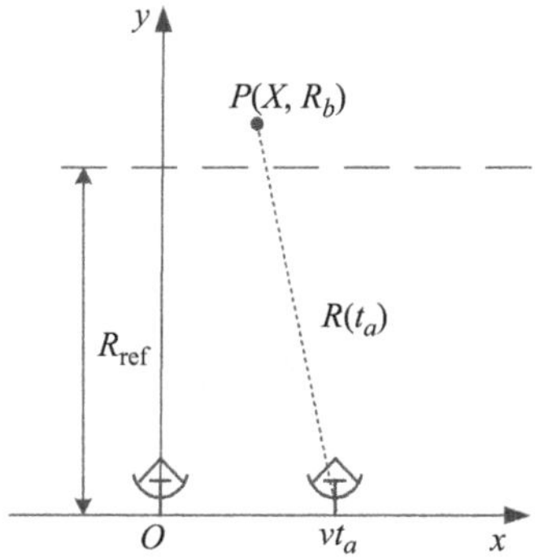

Figure 9.11 Imaging geometry

slow time. For an arbitrary point target $P(X, R_b)$, the instantaneous slant range can be expressed as

$$R(t_a) = \sqrt{{R_b}^2 + (vt_a - X)^2} \tag{9.38}$$

The phase of 2D spectrum can be directly written as

$$\Phi = -4\pi R_b \sqrt{\left(\frac{f_c + f_r}{c}\right)^2 - \left(\frac{f_a}{2v}\right)^2} - 2\pi f_a \frac{X}{v} \tag{9.39}$$

where f_c is the carried frequency, c is the light speed, f_r is the range frequency and f_a is the azimuth frequency.

The phases of 2D spectrums for different modes are the same. However, the varying of the azimuth bandwidth with slow time in each mode is different since the antenna beam is steered to different rotation centre. Due to the differences of the azimuth bandwidth of different SAR modes, if a unified focusing algorithm is to be obtained, it is necessary to analyse the signal properties for different modes.

9.4.2 Signal property

For Stripmap SAR, the antenna beam direction does not change and thus supporting areas of different point targets in Doppler domain are overlapped (see Figure 9.12(a)). However, the supporting areas in Doppler domain of Sliding Spotlight SAR, Spotlight SAR and TOPS SAR vary with slow times (see Figure 9.12(b–d)) since the antenna beam is steered to the rotation centre. In Figure 9.12, B_{inst} is the instantaneous bandwidth. It is determined by azimuth-aperture length and effective velocity. The dashed line in each time-frequency diagram (TFD) is the edge of time-frequency distribution of illustrated scene and the solid line is the time-frequency distribution of a certain point target.

Since the antenna pointing varies with slow time in beam steering (BS) SAR (generalising Spotlight SAR, Sliding Spotlight SAR and TOPS SAR), the

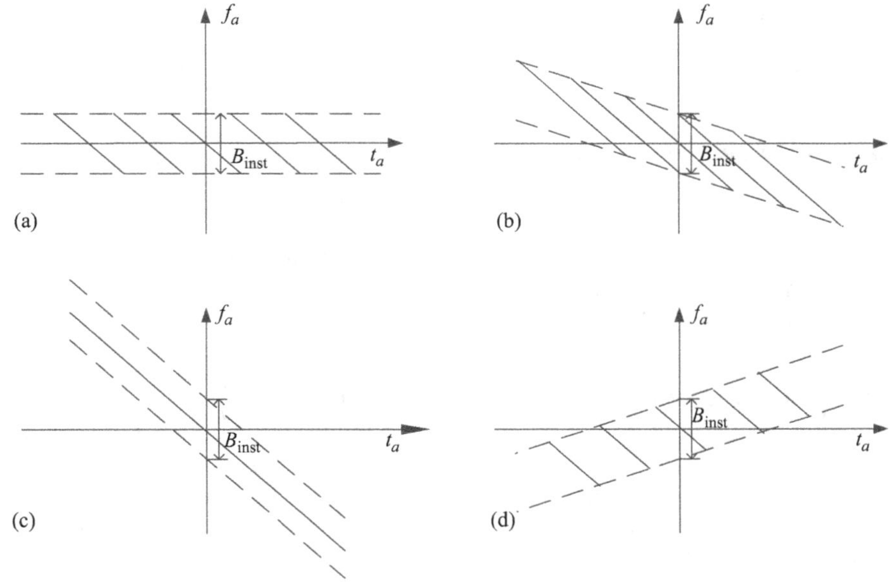

Figure 9.12 TFD of four modes: (a) Stripmap SAR, (b) Sliding Spotlight SAR, (c) Spotlight SAR and (d) TOPS SAR

bandwidth of azimuth signal may be much greater than B_{inst}. For a real application system, PRF is usually limited and slightly greater than B_{inst}. Therefore, the azimuth signal is usually under-sampled, which is one of the most important properties of BS-SAR. If a unified focusing algorithm is to be proposed, it should be able to avoid aliasing or to retrieve 2D spectrum.

The time-frequency distributions of BS-SAR are determined by the rotation-centre range, R_{rot}, which mainly determines the slope of the TFD. The rotation-centre range is the nearest range from the rotation centre to trajectory. Spotlight SAR can be directly regarded as a special case since the rotation centre is the same as the reference point. The concept of rotation-centre range can also be extended to Stripmap SAR by letting its rotation-centre range $+\infty$ or $-\infty$, that is $R_{\text{rot}} = +\infty$ or $R_{\text{rot}} = -\infty$. With such a rotation centre, the beam pointing is basically unchanged with slow time. Then, the slope of TFD for Stripmap SAR should be zero, which agrees with the actual case. With the above observations, these four modes can be regarded as special cases of BS-SAR and the sole representation is rotation centre or rotation-centre range, R_{rot}. Figure 9.13 shows the distribution of four SAR modes by using the concept of range-centre range. Stripmap SAR sits at the two ends of the coordinate. The rotation-centre range of TOPS SAR is negative while that of Sliding Spotlight SAR is positive. Spotlight SAR is a special case of Sliding Spotlight SAR. In Figure 9.13, four SAR modes can come down to one mode with a sole representation, that is R_{rot}.

Although four SAR modes can be unified conceptually, a unified focusing algorithm has not been studied so far, to the best of our knowledge. This chapter is

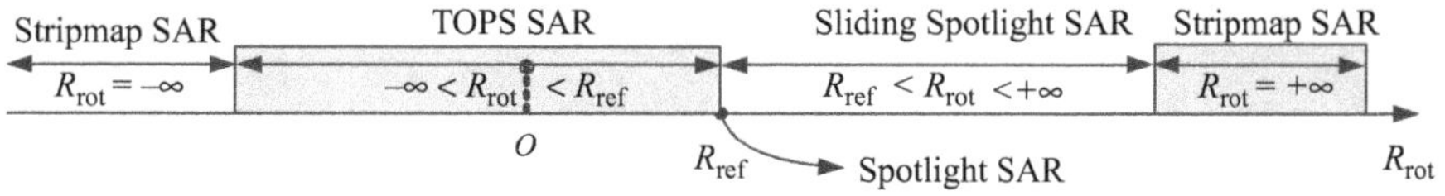

Figure 9.13 Distribution of four SAR modes on coordinate of rotation-centre range

to obtain a unified and parameterised focusing algorithm, in which the parameters are determined by rotation-centre range. The following is to introduce fractional Fourier transformation (FrFT) first.

9.5 Multi-mode unified imaging method

In this section, a unified focusing algorithm based on FrFT is shown, which can simultaneously process Stripmap, Spotlight, Sliding Spotlight and TOPS SAR mode data. In this algorithm, Stripmap mode and Spotlight mode can be considered as special forms of Sliding Spotlight SAR. The only characterisation of these modes is the distance between the rotation centres and radar. The focusing algorithm is parameterised into a rotation angle or a rotation factor, and the rotation angle is determined by the rotation centre distance, that is data of different modes can be focused by using different and appropriate rotation angles.

9.5.1 Fractional Fourier transformation (FrFT)

FrFT is a time-frequency analysis tool and is applied to SAR imaging recently [14–16]. FrFT can be defined as

$$X_p(u) = \mathrm{FrFT}_p(x(t)) = \int x(t)k_p(t,u)dt \tag{9.40}$$

where the subscript p is the order of FrFT and

$$k_p(t,u) = \begin{cases} A_0 \exp\left(j\left(\frac{1}{2}t^2 \cot\alpha - ut\csc\alpha + \frac{1}{2}u^2\cot\alpha\right)\right) & \alpha \neq n\pi \\ \delta(t-u) & \alpha = 2n\pi \\ \delta(t+u) & \alpha = (2n+1)\pi \end{cases} \tag{9.41}$$

where $A_0 = \sqrt{\frac{1-j\cot\alpha}{2\pi}}$, $\alpha = p\frac{\pi}{2}$ is the rotation angle and δ is the Dirac function.

When $\alpha \neq n\pi$, (9.41) can be simply expressed as

$$X_p(u) = A_0 \exp\left(j\frac{1}{2}u^2\cot\alpha\right)\int x(t)\exp\left(j\frac{1}{2}t^2\cot\alpha\right) \times \exp(-jut\csc\alpha)dt \tag{9.42}$$

From (9.42), one can see that the integration part is similar to the kernel of an FFT operation. When $p = 2n \pm 1$, (9.40) becomes Fourier transformation or

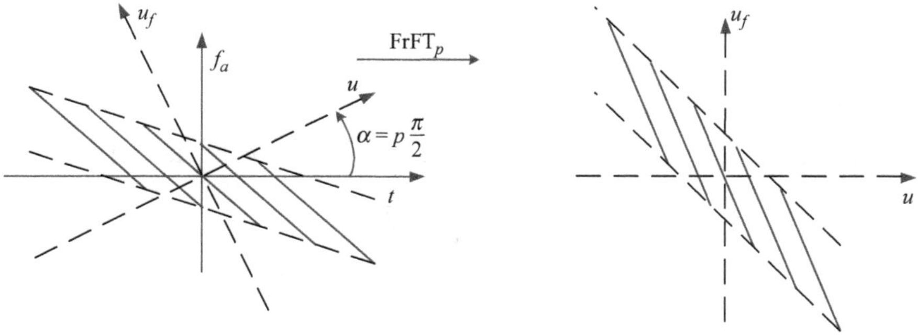

Figure 9.14 Rotation of time-frequency coordinate system in FrFT

inverse Fourier transformation of $x(t)$. FrFT can be regarded as a rotation of time-frequency coordinate system. Figure 9.14 shows a rotation of time-frequency coordinate. After FrFT, the slope of TFD is changed and it is the sampling interval in the new coordinate.

The kernel of an FFT operation in (9.42) can be used to determine the sampling intervals before and after FrFT. The sampling interval length Δ_u in u domain should satisfy

$$\Delta_u \csc \alpha = 2\pi \Delta_{f_a} \tag{9.43}$$

where Δ_{f_a} is the sampling interval length in f_a domain.

Then, (9.43) can be simplified as

$$\Delta_u = \frac{2\pi \sin \alpha}{M\Delta_t} \tag{9.44}$$

where Δ_t is the sampling interval in t and M is the number of sampling points.

When $p = 2n$, FrFT is simplified and sampling intervals satisfy

$$\Delta_u = (-1)^n \Delta_t \tag{9.45}$$

9.5.2 Algorithm description

For convenience, UFA is presented and then its applications analysed for four SAR modes [17]. UFA can be divided into three parts: azimuth scaling, RCMC and azimuth compression (see Figure 9.15). The azimuth scaling mainly consists of two FrFT elements, determined by two rotation factors, that is α and β, respectively. The RCMC is carried out by traditional Doppler domain algorithms, such as RDA, CSA, CZT, SCFFT, etc. The azimuth compression mainly consists of phase transformation from hyperbolic phase to quadratic phase, two FrFT elements and phase preserving compensation. Two FrFT elements in azimuth compression are determined by another two rotation factors, that is γ and χ, respectively. Four rotation factors are dependent on rotation-centre range. Therefore, data of four SAR modes can be focused by using UFA and selecting proper parameters.

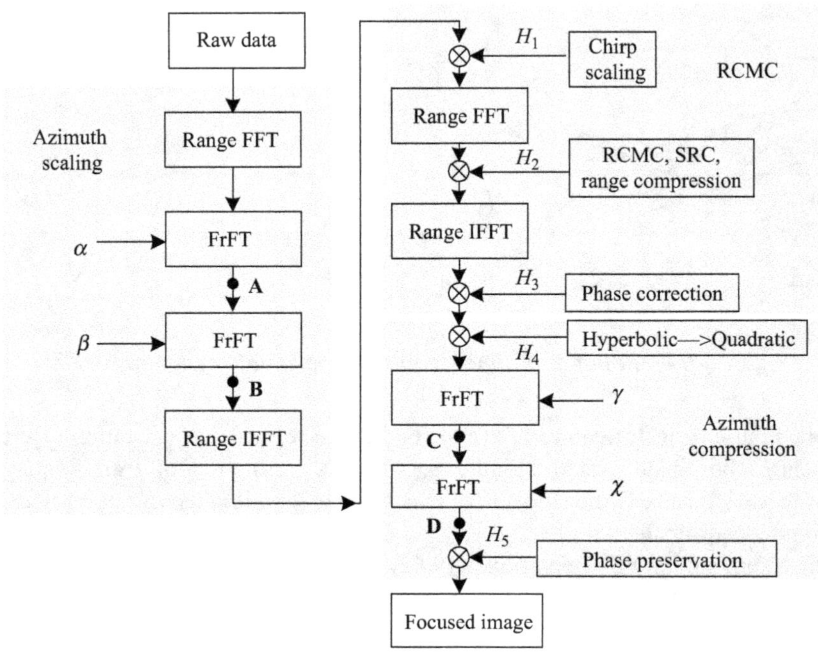

Figure 9.15 UFA for four SAR modes

To focus data, four rotation factors must satisfy some equations. The following is an important equation:

$$\alpha + \beta = \pi/2 \tag{9.46}$$

Equation (9.46) is denoted as the first equation. Also, γ and χ should satisfy another equation (9.64) below, which is analysed in Subsection 9.5.3.

Four FrFT elements are used in UFA. $x(u_{\mathbf{X}})$ is used to denote the signals at the stage $\mathbf{X} \in \{\mathbf{A}, \mathbf{B}, \mathbf{C}, \mathbf{D}\}$ in Figure 9.15 and $u_{\mathbf{X}}$ to denote its coordinate, respectively. Since H_1 to H_5 shown in Figure 9.15 are five functions in different FrFT domain, sampling interval length in $u_{\mathbf{X}}$ domain should be determined to construct these functions. Suppose the four rotation factors are fractional. According to (9.44), the sampling interval lengths in $u_{\mathbf{A}}$ and $u_{\mathbf{B}}$ domains can be expressed as

$$\Delta_{u_{\mathbf{A}}} = \frac{2\pi \sin\alpha}{M_t \Delta_{t_a}} \tag{9.47}$$

$$\Delta_{u_{\mathbf{B}}} = \frac{2\pi \sin\beta}{M_{\mathbf{A}} \Delta_{u_{\mathbf{A}}}} = \frac{M_t}{M_{\mathbf{A}}} \frac{\sin\beta}{\sin\alpha} \Delta_t \tag{9.48}$$

where M_t and $M_{\mathbf{A}}$ are the numbers of sampling points in t_a and $u_{\mathbf{A}}$ domains before each FrFT, respectively. M is the initial number of sampling points in t_a domain before the zero padding.

Note that zero padding is usually needed before each FrFT. The zero padding operation will be analysed later. Therefore, M_t and M are usually unequal. Then, in the azimuth compression, the sampling interval lengths in $u_{\mathbf{C}}$ and $u_{\mathbf{D}}$ domains can be expressed as

$$\Delta_{u_{\mathrm{C}}} = \frac{2\pi \sin\gamma}{M_{\mathbf{B}}\Delta_{u_{\mathrm{B}}}} = \frac{2\pi M_{\mathbf{A}} \sin\gamma \sin\alpha}{M_t M_{\mathbf{B}} \sin\beta\Delta_t} \tag{9.49}$$

$$\Delta_{u_{\mathrm{D}}} = \frac{M}{M_{\mathbf{A}}}\frac{M_{\mathbf{B}}}{M_{\mathbf{C}}}\frac{\sin\chi \sin\beta}{\sin\gamma \sin\alpha}\Delta_t \tag{9.50}$$

where $M_{\mathbf{B}}$ and $M_{\mathbf{C}}$ are the numbers of sampling points in $u_{\mathbf{B}}$ and $u_{\mathbf{C}}$ domains before FrFT, respectively.

Since TFD of the BS-SAR is much different with that of Stripmap SAR, Subsection 9.5.3 emphasizes on the application of UFA to four SAR modes, that is parameter set selection of $\{\alpha, \beta, \gamma, \chi\}$.

9.5.3 Applications for Sliding Spotlight and TOPS SAR

9.5.3.1 Expressions

Since the Spotlight SAR and Stripmap SAR can be regarded as special cases of Sliding Spotlight SAR, the signal expressions in different stages are first presented in Figure 9.15 for Sliding Spotlight SAR. The expressions for Stripmap SAR and Spotlight SAR can be obtained by similar derivations.

Two parallel red lines are used to denote boundaries of time-frequency distribution (see Figure 9.16). The green lines denote the TDF of point target. From

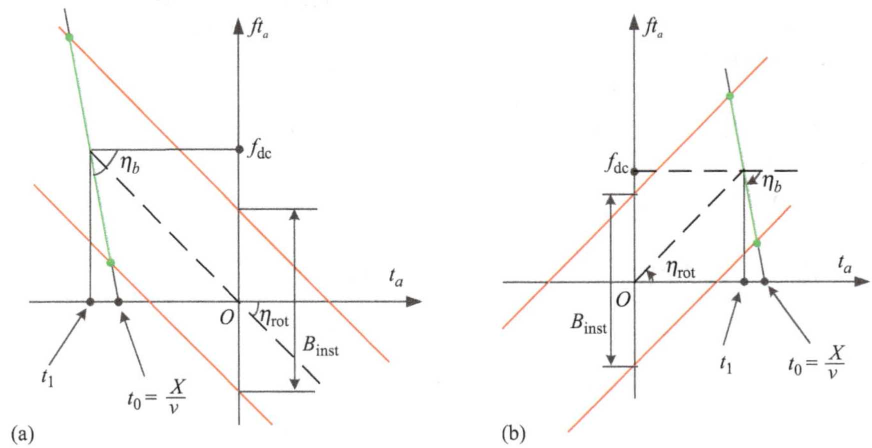

Figure 9.16 TFD of point targets in different modes: (a) Sliding Spotlight SAR and (b) TOPS SAR

Figure 9.16, one can express the echo signal after range compression in range-frequency and azimuth-time domain as

$$x(f_r, t_a) = \sigma W_r(f_r) w_{\text{azi}}\left(\frac{t_a - t_1}{T}\right) \exp\left(-j\frac{4\pi}{c}(f_c + f_r)R(t_a)\right) \tag{9.51}$$

where σ is the complex scattering coefficient, $W_r(f_r)$ is the range window function in range-frequency domain, w_{azi} is the azimuth window function in azimuth-time domain and T is the azimuth synthetic time.

$W_r(f_r)$ is determined by the transmitted waveform. T and w_{azi} depend on antenna beam pattern, antenna beam pointing, position of the point target, imaging geometry, etc. t_1 is the centre time of the signal projected to t_a axis (see Figure 9.16), t_0 is the azimuth time of the point target with nearest slant range and these two types of time satisfy

$$t_1 = t_0 \frac{\cos\eta_{\text{rot}} \sin\eta_b}{\sin(\eta_b + \eta_{\text{rot}})} \tag{9.52}$$

where η_b is the slope angle of time-frequency distribution of a point target and η_{rot} is the slope angle of time-frequency distribution of the illustrated scene.

According to Appendix A, η_{rot} and η_b can be expressed as

$$\tan\eta_b = 2\pi\frac{-2v^2}{\lambda R_b}; \quad \tan\eta_{\text{rot}} = 2\pi\frac{-2v^2}{\lambda R_{\text{rot}}} \tag{9.53}$$

Note that the positive value of slope angle indicates the counter-clockwise direction starting from t_a axis.

Four FrFT operations need to be performed in the focusing algorithm denoted in Figure 9.16. Four FrFT rotation operations can be represented in Figure 9.17. In Figure 9.17(a), two thick solid lines denoted by L_1 and L_2 can be used to compute the supporting areas of illustrated scene in $u_{\mathbf{X}}$ domain. The reason is that the supporting area of TFD projecting onto $u_{\mathbf{X}}$ axis can be expressed as the summation of the supporting areas of L_1 and L_2. These two thick solid lines will be used to compute the supporting areas and thus select rotational parameters in the following part. After the first FrFT in the azimuth scaling, the signal in range-frequency and azimuth-$u_{\mathbf{A}}$ domain can be expressed as

$$\begin{aligned} x(f_r, u_{\mathbf{A}}) = {} & \exp\left(j\frac{\cot\alpha}{2}u_{\mathbf{A}}^2\right)\int x(f_r, t_a)\exp\left(j\frac{\cot\alpha}{2}t_a^2\right) \\ & \times \exp(-ju_{\mathbf{A}}t_a\csc\alpha)dt_a \end{aligned} \tag{9.54}$$

Then, the signal at stage **B** can also be obtained

$$\begin{aligned} x(f_r, u_{\mathbf{B}}) = {} & \exp\left(j\frac{\cot\beta}{2}u_{\mathbf{B}}^2\right)\int x(f_r, u_{\mathbf{A}})\exp\left(j\frac{\cot\beta}{2}u_{\mathbf{A}}^2\right) \\ & \times \exp(-ju_{\mathbf{A}}u_{\mathbf{B}}\csc\beta)du_{\mathbf{A}} \end{aligned} \tag{9.55}$$

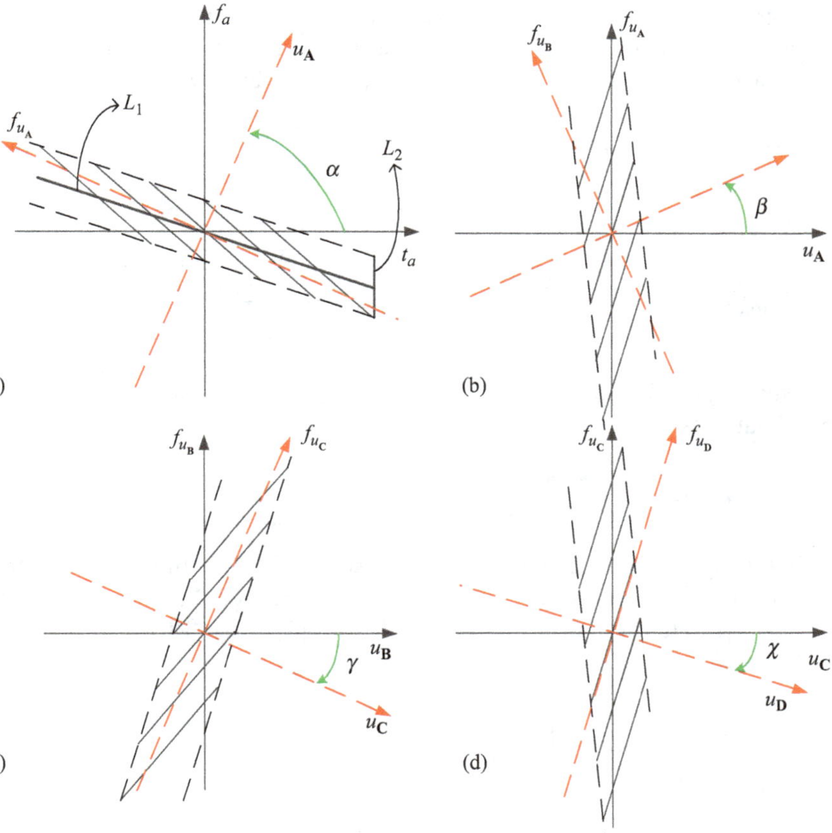

Figure 9.17 TFD of signal in four FrFT operations: (a) from t_a to $u_{\mathbf{A}}$, (b) from $u_{\mathbf{A}}$ to $u_{\mathbf{B}}$, (c) from $u_{\mathbf{B}}$ to $u_{\mathbf{C}}$ and (d) from $u_{\mathbf{C}}$ to $u_{\mathbf{D}}$

By substituting (9.54) and (9.46) into (9.55) and changing integral order, one can obtain

$$
\begin{aligned}
x(f_r, u_{\mathbf{B}}) &= \exp\left(j\frac{\cot\beta}{2}u_{\mathbf{B}}^2\right)\int x(f_r, t_a) \\
&\quad \times \exp\left(j\frac{\cot\alpha}{2}{t_a}^2 - j\frac{(t_a\csc\alpha + u_{\mathbf{B}}\csc\beta)^2}{2(\cot\alpha + \cot\beta)}\right)dt_a \\
&= \int x(f_r, t_a)\exp(-jt_a u_{\mathbf{B}})dt_a \\
&= X\left(f_r, f_a = \frac{u_{\mathbf{B}}}{2\pi}\right)
\end{aligned}
\tag{9.56}
$$

where $X\left(f_r, f_a = \frac{u_{\mathbf{B}}}{2\pi}\right)$ is the Fourier transform of $x(f_r, t_a)$ in azimuth-frequency domain.

From (9.56), one can see that signal in azimuth-frequency domain is obtained after two FrFT operations. $X(f_r,f_a)$ can be deduced from $x(f_r,t_a)$ by using stationary phase method [18]. For simplicity, $X(f_r,f_a)$ is directly expressed as

$$x(f_r, u_{\mathbf{B}}) = W_r(f_r) W_{\text{azi}}(u_{\mathbf{B}}) \exp\left(j\Phi\left(f_r, f_a = \frac{u_{\mathbf{B}}}{2\pi}\right)\right) \tag{9.57}$$

where

$$W_{\text{azi}}(u_{\mathbf{B}}) = w_{\text{azi}}\left(\frac{\frac{R_b}{v}\left(\frac{u_{\mathbf{B}}}{4\pi v}\right) - (t_0 - t_1)\sqrt{\left(\frac{f_c + f_r}{c}\right)^2 - \left(\frac{u_{\mathbf{B}}}{4\pi v}\right)^2}}{T\sqrt{\left(\frac{f_c + f_r}{c}\right)^2 - \left(\frac{u_{\mathbf{B}}}{4\pi v}\right)^2}}\right) \tag{9.57a}$$

Usually, f_a is far less than $2v/\lambda$, and thus the azimuth window function can be approximated as

$$W_{\text{azi}}(u_{\mathbf{B}}) = w_{\text{azi}}\left(\frac{\lambda R_b u_{\mathbf{B}} - 4\pi v^2 (t_0 - t_1)}{4\pi v^2 T}\right) \tag{9.57b}$$

Following azimuth scaling, the signal is sent to RCMC. In RCMC, the traditional imaging algorithms can be adopted to correct RCM. If CSA is used in RCMC, H_1 to H_3 can be directly obtained from [2]. After RCMC, the azimuth signal can be simply expressed as

$$x(u_{\mathbf{B}}) = W_{\text{azi}}(u_{\mathbf{B}}) \exp\left(-j2\pi\frac{R_b}{v}\sqrt{f_{\text{am}}{}^2 - \left(\frac{u_{\mathbf{B}}}{2\pi}\right)^2} - j2\pi\frac{u_{\mathbf{B}}}{2\pi}t_0\right) \tag{9.58}$$

where $f_{\text{am}} = 2v/\lambda$.

In (9.58), the phase function is a hyperbolic curve and it is not favourable to the azimuth scaling and compressing by using FrFT. A phase function can be used to transform the hyperbolic phase curve to a quadratic phase curve and the phase function:

$$H_4(u_{\mathbf{B}}) = \exp\left(j2\pi R_b\sqrt{f_{\text{am}}{}^2 - \left(\frac{u_{\mathbf{B}}}{2\pi}\right)^2}\Big/ v\right)\exp\left(j2\pi\frac{\lambda R_{\text{scl}}}{4v^2}\left(\frac{u_{\mathbf{B}}}{2\pi}\right)^2\right) \tag{9.59}$$

where R_{scl} is the scaling range.

By multiplying (9.59) to (9.58), the following equation can be obtained:

$$x(u_{\mathbf{B}}) = W_{\text{azi}}(u_{\mathbf{B}}) \exp\left(j\frac{\tan(\eta_{\text{scl}} + \pi/2)}{2}u_{\mathbf{B}}^2 - j2\pi\frac{u_{\mathbf{B}}}{2\pi}t_0\right) \tag{9.60}$$

where $(\eta_{\text{scl}} + \pi/2)$ is the angle of second term (LFM signal) of (9.59) in the FrFT domain.

η_{scl} should satisfy

$$\tan\eta_{\text{scl}} = 2\pi\frac{-2v^2}{\lambda R_{\text{scl}}}. \tag{9.60a}$$

Applying FrFT (the first FrFT in azimuth compression) to (9.59), one can obtain

$$x(u_{\mathbf{C}}) = \exp\left(j\frac{\cot\gamma}{2}u_{\mathbf{C}}^2\right)\int x(u_{\mathbf{B}})\exp\left(j\frac{\cot\gamma}{2}u_{\mathbf{B}}^2\right) \times \exp(-ju_{\mathbf{C}}u_{\mathbf{B}}\csc\gamma)du_{\mathbf{B}} \tag{9.61}$$

Then, the fourth FrFT is carried out and azimuth signal is compressed. The signal after the fourth FrFT can be written as

$$x(u_{\mathbf{D}}) = \exp\left(j\frac{\cot\chi}{2}u_{\mathbf{D}}^2\right)\int x(u_{\mathbf{C}})\exp\left(j\frac{\cot\chi}{2}u_{\mathbf{C}}^2\right) \times \exp(-ju_{\mathbf{C}}u_{\mathbf{D}}\csc\chi)du_{\mathbf{C}} \tag{9.62}$$

By substituting (9.61) into (9.62) and changing the integral order, we have

$$x(u_{\mathbf{D}}) = \exp\left(j\frac{1}{2}u_{\mathbf{D}}^2\cot\chi\right)\int x(u_{\mathbf{B}})\exp\left(j\frac{1}{2}u_{\mathbf{B}}^2\cot\gamma\right) \times \exp\left(-j\frac{1}{2}\frac{(u_{\mathbf{B}}\csc\gamma + u_{\mathbf{D}}\csc\chi)^2}{(\cot\gamma + \cot\chi)}\right)du_{\mathbf{B}} \tag{9.63}$$

Because it is expected to focus signal in $u_{\mathbf{D}}$ domain, the coefficient of the second order term of $u_{\mathbf{B}}$ should be zero. Then, we obtain

$$\gamma + \chi = \eta_{\mathrm{scl}} \tag{9.64}$$

Equation (9.64) is called as 'the second equation' in the focusing algorithm. By substituting (9.60) and (9.64) into (9.63), (9.63) can be rewritten as

$$\begin{aligned} x(u_{\mathbf{D}}) &= \exp\left(j\frac{\cot\eta_{\mathrm{scl}}}{2}u_{\mathbf{D}}^2\right)\int W_{\mathrm{azi}}(u_{\mathbf{B}}) \times \exp\left(-ju_{\mathbf{B}}\left(t_0 + u_{\mathbf{D}}\frac{\csc\gamma\csc\chi}{\cot\gamma + \cot\chi}\right)\right)du_{\mathbf{B}} \\ &= \exp\left(j\frac{\cot\eta_{\mathrm{scl}}}{2}u_{\mathbf{D}}^2\right)\mathrm{sinc}\left(\frac{R_{\mathrm{scl}}}{R_b}T\frac{\sec\eta_{\mathrm{scl}}}{2\pi}\left(u_{\mathbf{D}} + \frac{t_0}{\csc\eta_{\mathrm{scl}}}\right)\right) \end{aligned} \tag{9.65}$$

To have the algorithm for phase-tracking, the squinted time-frequency distribution needs to compensated as shown in Figure 9.17. For Sliding Spotlight SAR, TOPS SAR and Stripmap SAR, the slope of time-frequency distribution is denoted by L_1 and thus the compensation function can be expressed as

$$H_5(f_a) = \exp\left(-j\frac{\tan(\eta_1 - \eta_{\mathrm{scl}})}{2}u_{\mathbf{D}}^2\right) \tag{9.66a}$$

where η_1 is an angle and its definition is in Table 9.1. $\eta_1 - \eta_{\mathrm{scl}}$ is the angle of L_1 in $u_{\mathbf{D}}$ domain as given in Table 9.1.

Table 9.1 Lengths and angles of L_1 and L_2 in u_X domain

		Length	Angle
$u_{\mathbf{A}}$	L_1	$T\frac{\cos(\alpha-\eta_{rot})}{\cos\,\eta_{rot}}$	$\eta_{rot}-\alpha$
	L_2	$2\pi B_{inst}\sin\,\alpha$	$\frac{\pi}{2}-\alpha$
$u_{\mathbf{B}}$	L_1	$T\tan\,\eta_{rot}$	$\mathrm{atan}(-\cot\,\eta_{rot})$
	L_2	$2\pi B_{inst}$	0
$u_{\mathbf{B}}$	L_1	$T\tan\,\eta_{rot}$	$\mathrm{atan}(-\cot\,\eta_{rot}+\cot\,\eta_b-\cot\,\eta_{scl})\triangleq\eta_1$
	L_2	$2\pi B_{inst}$	$\mathrm{atan}(\cot\,\eta_b-\cot\,\eta_{scl})\triangleq\eta_2$
$u_{\mathbf{C}}$	L_1	$T\tan\,\eta_{rot}\frac{\cos(\eta_1-\gamma)}{\cos\,\eta_1}$	$\eta_1-\gamma$
	L_2	$2\pi B_{inst}\frac{\cos(\eta_2-\gamma)}{\cos\,\eta_2}$	$\eta_2-\gamma$
$u_{\mathbf{D}}$	L_1	$T\tan\,\eta_{rot}\frac{\cos(\eta_1-\gamma-\chi)}{\cos\,\eta_1}$	$\eta_1-\gamma-\chi$
	L_2	$2\pi B_{inst}\frac{\cos(\eta_2-\gamma-\chi)}{\cos\,\eta_2}$	$\eta_2-\gamma-\chi$

However, for Spotlight SAR, the slope should be denoted by L_2 and thus the compensation function should be expressed as

$$H_5(f_a)=\exp\left(-j\frac{\tan(\eta_2-\eta_{scl})}{2}u_{\mathbf{D}}^2\right) \tag{9.66b}$$

where η_2 is an angle defined in Table 9.1. $\eta_2-\eta_{scl}$ is the angle of L_2 in $u_{\mathbf{D}}$ domain as shown in Table 9.1.

9.5.4 Applied to Stripmap and Spotlight SAR

According to the above analysis, Stripmap and Spotlight SAR can be considered as a special form of Sliding Spotlight SAR. For the Stripmap, its centre of rotation range satisfies $R_{rot}=+\infty$ or $R_{rot}=-\infty$, therefore, $\eta_{rot}=0$. The centre of rotation range for the Spotlight satisfies $R_{rot}=R_{ref}$. According to their special types, the corresponding imaging parameters can be selected.

9.5.5 Real data results

9.5.5.1 Result of Sliding Spotlight SAR

Raw airborne data of an experimental Sliding Spotlight SAR at X-band were collected in 2008. Major system parameters of the SAR are given in Table 9.2. The data are focused by using our UFA with rotation factors as given in Table 9.2. R_{scl} is selected as the value of reference range. The focusing result is shown in Figure 9.18.

9.5.5.2 Result of TOPS SAR

Airborne data of an experimental Ku-band SAR with TOPS mode were collected on in 2011. Main system parameters of the SAR are given in Table 9.3. Using the studied focusing algorithm and selected rotation factors in Table 9.3, an image was obtained as shown in Figure 9.19.

Table 9.2 Parameters of an airborne x-band Sliding Spotlight SAR

Velocity	102.7 m/s
Bandwidth	600 MHz
PRF	312.5 Hz
IBW	258.9 Hz
Acquiring time	13.1 s
Rotation-centre range	79.87 km
Reference range	34.34 km
α	0.0188 rad
γ	–1.5520 rad

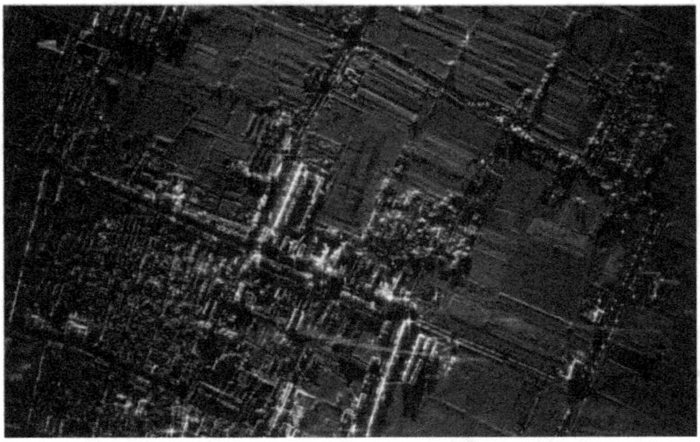

Figure 9.18 Truncated results from an airborne Sliding Spotlight SAR data (range $\times$ azimuth = 733.5 $\times$ 1,295.3)

Table 9.3 Parameters of an airborne ku-band tops SAR

Velocity	54 m/s
Bandwidth	40 MHz
Acquiring time	7 s
PRF	1,000 Hz
IBW	580 Hz
Rotation-centre range	–308 m
Reference range	12.12 km
α	3.141437 rad
γ	1.570641 rad

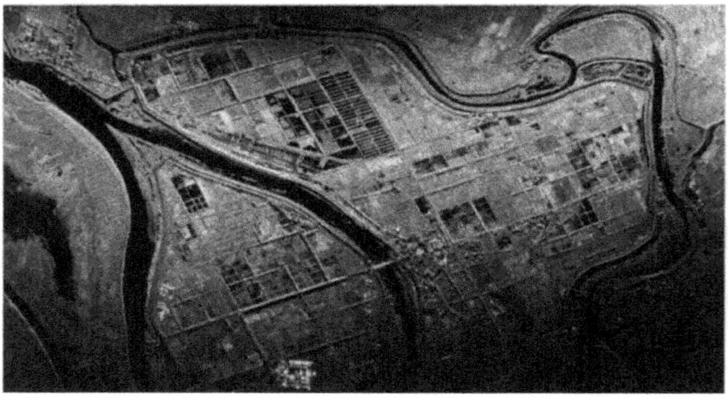

Figure 9.19 Truncated result from an airborne TOPS SAR data (range × azimuth = 5.66 km × 11.74 km)

Table 9.4 Parameters of an airborne x-band Spotlight SAR

Velocity	119.9 m/s
Bandwidth	600 MHz
PRF	416.66 Hz
IBW	258.9 Hz
Acquiring time	19.7 s
Rotation-centre range	40.84 km
Reference range	40.84 km
α	0.007064482 rad
γ	−1.563731845 rad

9.5.5.3 Result of Spotlight SAR

To assess the UFA for the Spotlight SAR, the raw data of an airborne X-band Spotlight SAR is processed. The raw data was collected in 2008, over a suburb of Yanliang City, China. The detailed parameters of this system are listed in Table 9.4. By using the proposed UFA, the raw data are focused, as shown in Figure 9.20. The horizontal direction is the azimuth, and the ordinate is the distance. The distance is getting further and further from top to bottom.

9.6 Summary

This chapter introduced the basic concepts of SAR and the principle of 2D target resolution. Obtaining high range resolution with wideband signal and high azimuth resolution with long synthetic array is the basis of the 2D high-resolution imaging. RDA is widely used in SAR imaging processing. Azimuth-translation invariance is the key of RDA. This chapter described the major processing step of RDA, including range compression, range migration correction and azimuth compression, and derives

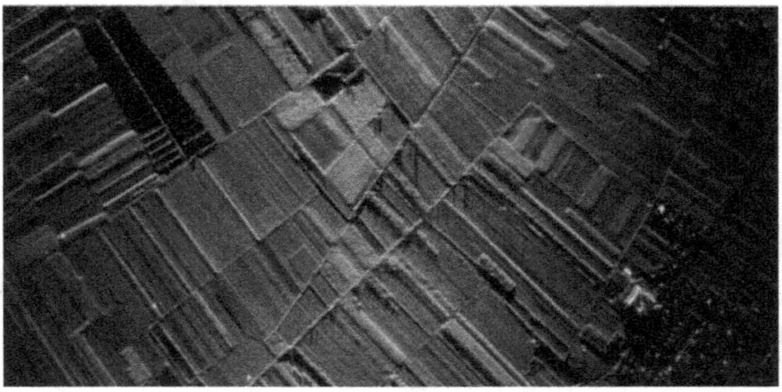

Figure 9.20 Truncated result from an airborne Spotlight SAR data (range × azimuth = 827 m × 1,909 m)

the signal expression of the key step. With the development of SAR imaging requirements, SAR has developed a variety of working modes. This chapter also introduced the working geometry of Stripmap mode, Spotlight mode, Sliding Spotlight mode, TOPS mode and Scan mode, established the unified signal model for multi-modes and analysed the signal property. In imaging processing for multi-modes, a unified focusing algorithm based on FrFT is shown, which can simultaneously process Stripmap, Spotlight, Sliding Spotlight and TOPS SAR mode data. The focusing algorithm is parameterised into a rotation angle or a rotation factor, which is determined by the rotation centre distance. Real data of different modes can be well focused by using different and appropriate rotation angles.

References

[1] Bamler R, 'A comparison of range-Doppler and wavenumber domain SAR focusing algorithms', *IEEE Trans. Geosci. Remote Sens.*, 1992, **30** (4), pp 706–713

[2] Bao Z, Xing M, Wang T, *Radar Imaging Technology*, Electronic Industry Press, 2005

[3] Lanari R, Zoffoli S, Sansosti E, Fornaro G, Serafino F, 'New approach for hybrid strip-map/spotlight SAR data focusing', *IEE Proc. - Radar, Sonar Nav.*, 2001, **148** (6), pp 363–372

[4] Cumming I G, Wong F H, *Digital Processing of Synthetic Aperture Radar Data Algorithms and Implementation*, Artech House, Norwood, MA, 2005

[5] Carrara W G, Majewski R M, Goodman R S, *Spotlight Synthetic Aperture Radar: Signal Processing Algorithms*, Artech House, Boston, MA, 1995

[6] Sun G, Xing M, Wang Y, Wu Y, Wu Y, Bao Z, 'Sliding spotlight and tops SAR data processing without sub-aperture', *IEEE Geosci. Remote Sens. Lett.*, 2011, **8** (6), pp 1036–1040

[7] Amein A S, Soraghan J J, 'Azimuth fractional transformation of the fractional chirp scaling algorithm (FrCSA)', *IEEE Trans. Geosci. Remote Sens.*, 2006, **44** (10), pp 2871–2879

[8] Raney R K, Runge H, Bamler R, Cumming I G, Wong F H, 'Precision SAR processing using chirp scaling', *IEEE Trans. Geosci. Remote Sens.*, 1994, **32** (4), pp 786–799

[9] Sun G, Jiang X, Xing M, Qiao Z, Wu Y, Bao Z, 'Focus improvement of highly squinted data based on azimuth nonlinear scaling', *IEEE Trans. Geosci. Remote Sens.*, 2011, **49** (6), pp 2308–2322

[10] Amein A S, Soraghan J J, 'A new chirp scaling algorithm based on the fractional Fourier transform', *IEEE Signal Process Lett.*, 2005, **12** (10), pp 705–708

[11] Lanari R, Franceschetti P, Tesauro M, Sansosti E (eds.), 'Spotlight SAR image generation based on strip mode focusing techniques', *1999 IEEE International Geoscience and Remote Sensing Symposium*, (Cat No. 99CH36293), 28 June–2 July 1999

[12] Lanari R, Tesauro M, Sansosti E, Fornaro G, 'Spotlight SAR data focusing based on a two-step processing approach', *IEEE Trans. Geosci. Remote Sens.*, 2001, **39** (9), pp 1993–2004

[13] Sun G, Xing M, Xia X, Wu Y, Bao Z, 'Beam steering SAR data processing by a generalized PFA', *IEEE Trans. Geosci. Remote Sens.*, 2013, **51** (8), pp 4366–4377

[14] Xiang-Gen X, 'On bandlimited signals with fractional Fourier transform', *IEEE Signal Process Lett.*, 1996, **3** (3), pp 72–74

[15] Sejdić E, Djurović I, Stanković L, 'Fractional Fourier transform as a signal processing tool: An overview of recent developments', *Signal Process.*, 2011, **91** (6), pp 1351–1369

[16] McBride A C, Kerr F H, 'On Namias' fractional Fourier transforms', *IMA J. Appl. Math.*, 1987, **39** (2), pp 159–175

[17] Sun G, Xing M, Xia X, Yang J, Wu Y, Bao Z, 'A unified focusing algorithm for several modes of SAR based on FrFT', *IEEE Trans. Geosci. Remote Sens.*, 2013, **51** (5), pp 3139–3155

[18] Born M, Wolf E, *Principles of Optics[M]*. Appendix III. Pergamon Press, New York, 1959

Guangcai Sun is an Associate Professor in the Department of Electronic Engineering and National Laboratory of Radar Signal Processing at Xidian University, China. Since 2007, his research has been focused on radar imaging technology, including imaging of several SAR modes and moving target detection and imaging. He is a co-author of a book and has more than 100 papers published and more than 20 Chinese patents granted. He won 2015 Xidian University Excellent Doctoral Dissertation Award and Shaanxi Province Outstanding Doctoral Dissertation Award. Dr Sun's research has been supported by various funding programs and research institutions, including National Science Foundation of China, Ministry of Education Doctoral Fund and Chinese Academy of Space Technology.

Mengdao Xing is a Professor and the Associate Dean for Research of the Department of Electronic Engineering and National Laboratory of Radar Signal Processing at Xidian University, China. His research interests include SAR, inversed SAR, sparse signal processing and microwave remote sensing. He has more than 200 journal papers (including 38 TGRS papers, 47 GRSL papers and 6 JSTAR papers). He also has authored and co-authored two books about SAR signal processing. His publications have been cited for more than 5,000 times (according to Google Scholar). He was rated as one of 'Most Cited Chinese Researchers' by Elsevier in 2014–2018. He has more than 40 granted Chinese patents. His PhD dissertation received honour nomination of National Excellent Doctoral Dissertation of China. Dr Xing's research has been supported for over $2 million grant by various funding programs, including the National Science Foundation of China, '973' Program, Program for New Century Excellent Talents in University and '863' Program. He is Associate Editor of *Journal of Radars*. He has served as a reviewer for many refereed journals.

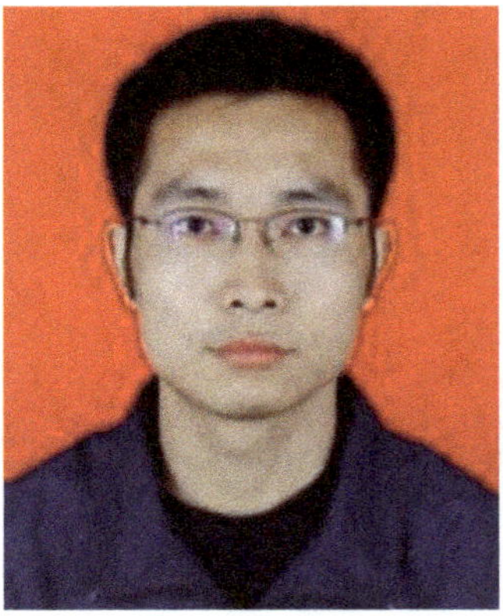

Liang Guo is an Associate Professor in the School of Physics and Optoelectronics Engineering at Xidian University, China. Since 2005, his research has been focused on Radar and Laser radar imaging technology, including imaging of synthetic aperture Lidar and moving target detection and imaging. He has published more than 30 papers, including journal papers on *IEEE Transactions on GRS, Optical Engineering* and *Optics and Laser Technology.* Dr Guo's research has been supported by various funding programs and research institutions. He received the First Prize of Awards of Science and Technology, Shaanxi Province in 2018.

Chapter 10

Electrical resistance tomography (ERT) sensors and applications

Huaxiang Wang[1]

List of abbreviations

ADC	analogue-to-digital converter
BEM	boundary element method
CCD	charge-coupled device
CMRR	common mode rejection ratio
DAC	digital-to-analogue converter
DAS	data acquisition system
DDS	direct digital synthesiser
EIT	electrical impedance tomography
ERT	electrical resistance tomography
FDM	finite difference method
FEM	finite element method
FPGA	field programmable gate array
ICA	independent component analysis
MAC	multiply accumulate
PDE	partial differential equation
PGA	programmable gain amplifier
SPD	symmetric positive definite
SVD	singular value decomposition
TSVD	truncated SVD
USG	superficial velocity of gas phase
USW	superficial velocity of water phase
VCCS	voltage-controlled current source

[1]Department of Automation, School of Automation and Information, Tianjin University, P.R. China

10.1 Introduction

Electrical resistance tomography (ERT) is a type of electrical impedance tomography (EIT). ERT is based on measuring resistive components of media with different electrical properties. Usually, an ERT sensor consists of many pin electrodes surrounding a domain of interest. A current is injected to a pair of electrodes, resulting in an electric field that depends on the conductivity distribution within the domain. The resulting potential changes are measured from the remaining electrodes. To take a set of measurements for one frame of image, all the electrodes are used in turn for current injection and potential measurements, which are affected by conductivity distribution. A particular difficulty with ERT is the so-called 'soft-field' effect, which results in low quality of reconstructed images and high measurement uncertainty. An ERT image can be reconstructed by solving the forward problem and inverse problem.

The advantages of ERT are no-radioactive, non-intrusive, on-line and fast imaging and of low-cost. ERT has been used in three areas: (1) medical imaging, (2) industrial process imaging and (3) geophysical surveying.

A typical ERT system with a 16-electrode sensor is shown in Figure 10.1. It consists of three units: a sensor, a measurement and data acquisition unit and an image reconstruction unit (i.e. computer).

10.1.1 Forward problem of ERT

Because ERT is a type of soft-field tomography, the inverse problem is ill-posed, which is extremely sensitive to the modelling error and measurement noise [1]. To reconstruct an image according to the injected current and potential measurements around the periphery, it is necessary to solve the forward and inverse problems. The forward problem of ERT is to estimate the electric field distribution to determine the geometric distribution and physical properties of the domain. In the forward step, the conductivity distribution is assumed to be known, and the boundary

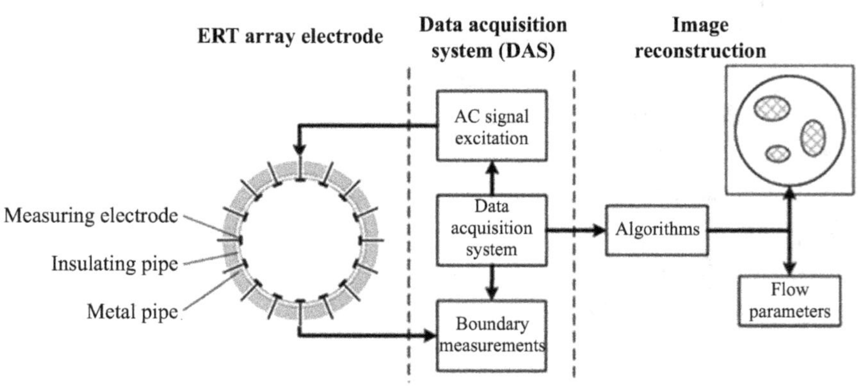

Figure 10.1 Typical ERT system with a 16-electrode sensor

voltage is calculated using finite element method (FEM). By using the FEM and the boundary conditions, the Poisson's equation over the horizontal slice of the vessel is solved and the voltages of all elements of the dissection domain are obtained. Finally, the sensitivity matrix is obtained from the measured voltages using FEM. Under the quasi-static assumption, the model consists of the following Laplace equation and the boundary conditions:

$$\nabla \cdot (\sigma \cdot \nabla u) = 0, (x, y) \in \Omega \tag{10.1}$$

$$\int_{e_l} \sigma \frac{\partial u}{\partial \overline{n}} dS = I_l, (x, y) \in e_l, l = 1, 2, \ldots, l \tag{10.2}$$

$$\sigma \frac{\partial u}{\partial n} = 0, (x, y) \in \partial\Omega \cup_{l=1}^{L} e_l \tag{10.3}$$

$$u + \xi_l \sigma \frac{\partial u}{\partial \overline{n}} = U_l, (x, y) \in e_l, l = 1, 2, \ldots, L \tag{10.4}$$

where σ is the conductivity distribution, u is the scalar potential distribution, $\overline{n}$ is the outward unit normal of the boundary $\partial\Omega$, ξ_l is the contact impedance, I_l is the injected currents, U_l is the corresponding potentials on the electrodes, L is the number of electrodes, e_l is the ith electrode and Ω is the computational domain.

To find a solution, the charge conservation law is used for the model and the potential reference point needs to be set as

$$\sum_{l=1}^{L} I_l = 0 \tag{10.5}$$

$$\sum_{l=1}^{L} U_l = 0 \tag{10.6}$$

Based on the abovementioned conditions, the forward problem can be solved by using the finite element method with the conductivity of the measured medium [2]. To solve (10.1)–(10.6), a numerical technique is preferable to analytical solution because the complexity of an ERT sensor makes it difficult to obtain an analytical solution. There are three commonly used methods for solving partial differential equations (PDE): finite difference method (FDM), FEM and boundary element method (BEM). FDM is often used in ERT for geophysical applications, because the simplicity of the domain in geophysical application makes the method suitable. However, in the case of dealing with complex geometries, which are common in biomedical models, a very fine level of discretisation is required to handle curved boundaries, making it difficult to use FDM in biomedical ERT.

BEM and FEM are two alternative approaches to FDM. BEM transforms any differential operator defined on a domain into an integral operator defined on a boundary. The object model is segmented in several regions, depending on its heterogeneity. The independent variable is assumed to be constant on each subdomain. Due to the attributes of BEM, this method is best suited for those applications, in which the domain is highly homogeneous. However, BEM involves in dense matrices. Consequently, it demands substantial computational load. FEM can be used to numerically obtain a solution to PDE by discretising the domain into a number of non-overlapping elements and approximating the piecewise solution within each element. It is widely used in ERT forward models because it is versatile with regards to the domain geometry, boundary conditions and the heterogeneity [3–5].

10.1.2 Inverse problem of ERT

With a solution to the forward problem, a physical model can be determined by describing the relationship between the potential of peripheral electrodes and the internal resistivity distribution and the injected currents on the boundaries. The conductivity distribution can be estimated by solving the inverse problem. To obtain the distribution of conductivity, the sensitivity theorem and the lead theorem are introduced for the impedance plethysmography, respectively [3]. According to the sensitivity theorem, in a domain V with conductivity $\sigma(x,y)$

$$\Delta U = J(\sigma)\Delta\sigma \tag{10.7}$$

where $\Delta\sigma$ is the variation of conductivity, $J(\sigma)$ is the sensitivity matrix, which changes with the variation of the conductivity, and ΔU is the change of the boundary voltage.

In general, the conductivity distribution is assumed to be uniform. The sensitivity is based on a homogeneous field background. The corresponding sensitivity matrix is J. Equation (10.7) can be written as

$$\Delta U = J\Delta\sigma \tag{10.8}$$

Because ERT is usually based on the adjacent current excitation, the sensitivity matrix can be defined as

$$J_{ij} = \int \frac{\nabla\phi_i}{J_i} \cdot \frac{\nabla\phi_j}{J_j} dxdy \tag{10.9}$$

where J_{ij} is the sensitivity coefficient of the jth electrode pairs to the ith electrode pairs, ϕ_i and ϕ_j are the potential distributions in the field of the ith electrode pairs and the jth electrode pairs when the exciting currents are I_i and I_j.

The nonlinear inverse problem is often simplified as a linear problem when the electrical property varies slightly. In this case, (10.8) can be written as

$$Sg = z \tag{10.10}$$

where S is the normalised sensitivity matrix, g is the vector of grey value and z is the normalised measurements.

Because of the ‘soft-field’ effect and the ill-posedness of the inverse problem, the solution is sensitive to measurement errors and noise. The existing image reconstruction algorithms can be divided into indirect methods, which are based on the sensitivity or Jacobian matrix and linearization, and direct methods, which are based on non-iterative and without linearization [6–10].

10.2 Optimised design of ERT sensor

FEM is widely used in ERT to solve the forward problem, because it is versatile with regards to the domain geometry, boundary conditions and the heterogeneity. COMSOL software can be used for both three-dimensional (3D) and two-dimensional (2D) numerical modelling. A 2D finite element mesh corresponding to cylindrical specimens is shown in Figure 10.2.

To improve the potential distribution and current density in the medium, and the sensitivity in the central area, it is necessary to find the optimum electrode size. A numerical model can be used to simulate the changes in conductivity of the material, with different size of electrodes, when a current is injected into a pair of adjacent electrodes. As an example, rectangular electrodes are selected initially for the FEM simulation using COMSOL. The influence of various width and length of the electrodes on the distribution of the sensing field is analysed. In ERT systems, the effective spatial resolution of the reconstructed images depends on the number of independent measurements. The more electrodes, the higher resolution is expected [2,11].

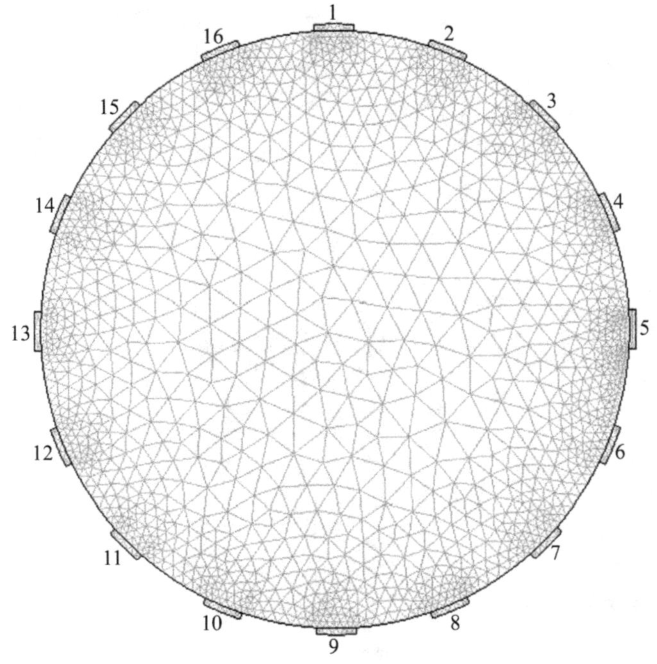

Figure 10.2 2D finite element mesh for cylindrical specimens

A 3D acrylic cylindrical object with conductive medium was meshed into many tetrahedral elements according to different requirements as shown in Figure 10.3, where the lower electrode plane is at $z = 8$ to $z = 9$ cm [12]. Yunus *et al.* analysed the potential distribution produced by a current source of 10 mA with 3D rectangular-shaped electrodes using the adjacent measurement strategy, including the spatial distribution and uniformity of the sensing field. Application of an AC current is preferable to prevent electrodes from polarisation. To minimise the capacitive-type electrode polarisation due to the ionic conduction in electrolytes, for example ERT is used to measure soil resistivity. At this time the measured medium is a water-soil mixture. The signal frequency for ERT should be in kHz range. Rasteiro *et al.* suggested a typical frequency range for ERT to be 20–150 kHz, at which the quasi-static conditions can be justified [13]. For the simulation in this part, 150 kHz is selected. In general, large electrodes would reduce the spatial resolution. In the ideal case, the excitation and sink electrodes should be as big as possible, while the measurement electrodes should be as small as possible [14]. However, it is difficult to implement different sizes of electrodes because the electrodes alternately function for excitation and measurement. Therefore, the choice of electrode size is a trade-off. As shown in Figure 10.4, larger electrodes

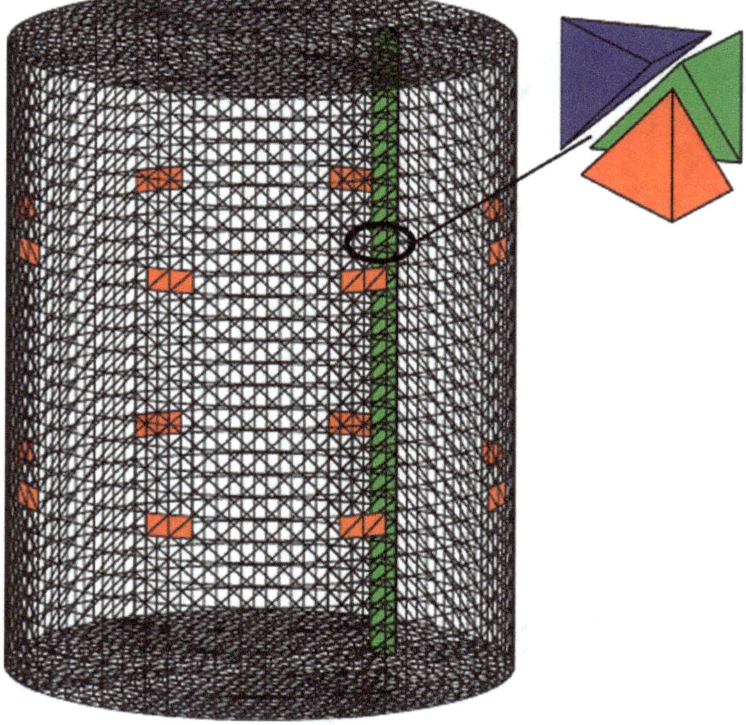

Figure 10.3 3D FEM mesh with three-tetrahedron wedge

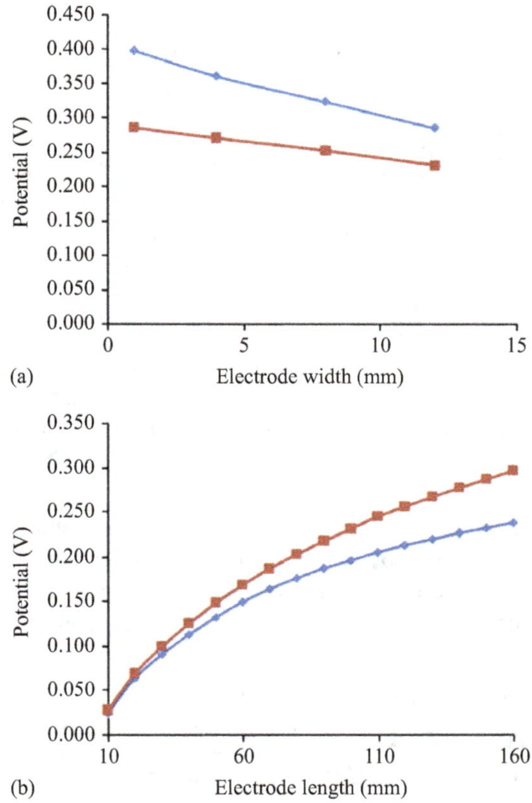

Figure 10.4 Potential produced by adjacent measurement strategy: (a) varying electrode width and (b) varying electrode length

can reduce the potential near the excitation and sink electrodes, because of less current density near the excitation and sink electrodes. The simulation results in Figure 10.4(b) show that longer electrodes can improve the evenness of the axial sensing field distribution [15,16].

With the adjacent excitation strategy, by comparing electrode heights of 10 mm and 90 mm, it can be seen from Figure 10.4(a) that long electrodes improve significantly the axial parallel distribution of the sensing field compared with the short electrodes in Figure 10.4(b). Wide electrodes improve the evenness of the field distribution, and hence improves signal strength. It also improves the ability of object detection, in particular in the central region. Simulation results are presented in Figure 10.5. However, with the adjacent measurement strategy proposed by Pinheiro *et al.*, it is suggested that to obtain maximum value, the optimal-sized electrodes should cover 80% and 60% of the surface for the opposite and adjacent protocol, respectively [17].

However, the electrode width must be selected carefully to avoid the signal reaching the asymptote level due to the number of electrodes and shunting effects

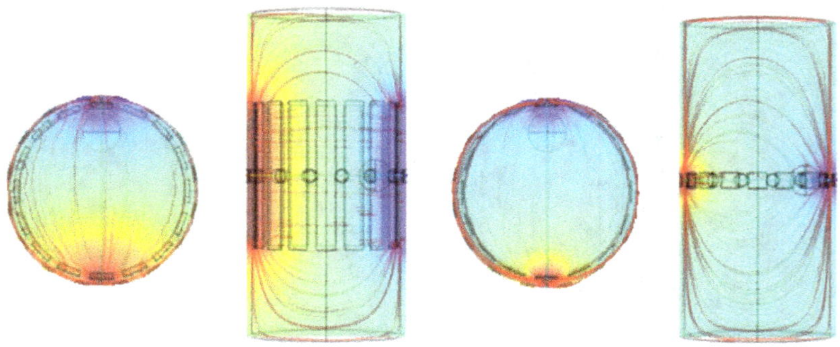

(a) (b)

Figure 10.5 Potential distribution at x–y plane and y–z plane with the adjacent strategy: (a) 12 mm (w) × 90 mm (h) and (b) 12 mm (w) × 10 mm (h)

in ERT. Numerical methods can prove that using the adjacent strategy, the optimum electrode width should cover 60% of the sensing surface and electrodes with greater height produce good axial field distribution. In general, the simulation results can be evaluated by the optimisation index of array electrodes.

10.2.1 Optimisation index of ERT sensor

10.2.1.1 Uniformity of field sensitivity

In ERT, the distribution of sensitive fields is usually non-uniform. Near the wall and the excitation electrodes, the sensitivity is high. In the central region, the sensitivity is low. The aim of optimising the electrodes is to make the sensing field be uniform. Therefore, the uniformity of the sensitive field is taken as an objective function for optimising the electrodes:

$$\boldsymbol{J}_{i,j}^{\text{avg}} = \frac{1}{n}\sum_{e=1}^{n} \boldsymbol{J}_{i,j}(e) \tag{10.11}$$

$$\boldsymbol{J}_{i,j}^{\text{dev}} = \left\{ \frac{1}{n-1}\sum_{e=1}^{n} \left[\boldsymbol{J}_{i,j}(e) - \boldsymbol{J}_{i,j}^{\text{avg}} \right]^2 \right\}^{1/2} \tag{10.12}$$

$$\boldsymbol{p}_{i,j} = \boldsymbol{J}_{i,j}^{\text{dev}} / \boldsymbol{J}_{i,j}^{\text{avg}} \tag{10.13}$$

$$P = \sum_{i,j} |\boldsymbol{p}_{i,j}| \tag{10.14}$$

where n is the number of the designed pixels, J is a sensitivity coefficient matrix, that is Jacobian matrix, $J_{i,j}^{\text{avg}}$ is the mean of sensitivity, $J_{i,j}^{\text{dev}}$ is the variance of sensitivity and P is a measure of field uniformity.

Obviously, the smaller the P value, the more uniform the sensitive filed distribution.

10.2.1.2 Correlation coefficient between real image and reconstructed image

Correlation coefficient between real image and reconstructed image is given by

$$a = \frac{\sum_{i=1}^{N} (\hat{g}_i - \bar{\hat{g}})(g_i - \hat{g})}{\sqrt{\sum_{i=1}^{N} (\hat{g}_i - \bar{\hat{g}})^2 \sum_{i=1}^{N} (g_i - \bar{g})^2}} \tag{10.15}$$

where g is the given field distribution, $\hat{g}$ is the reconstructed field distribution and $\bar{g}$ and $\bar{\hat{g}}$ are the average of g and $\hat{g}$, respectively.

The bigger a is, the stronger the correlation is, the higher the imaging accuracy is and the improved imaging quality can be achieved.

10.2.1.3 Resolution of data acquisition system

With different conductivity distributions in the sensitive field, the ratio of the norm of the resulting voltage difference and the excitation current norm is

$$d(\sigma_1, \sigma_2, I) = \frac{\|v_1(\sigma_1, I) - v_2(\sigma_2, I)\|}{\|I\|} \tag{10.16}$$

where I is the excitation current, v_1 and v_2 are the boundary voltage values obtained when the conductivity of the media in the internal field is σ_1 and σ_2, respectively.

10.2.2 Excitation and measurement strategies

After the sensor electrodes have been optimised, the next step is to choose excitation and measurement strategies, according to a specific application and the object to be measured. There are several strategies: adjacent, opposite and diagonal for a single-current source/sink-drive mode [18].

10.2.2.1 Adjacent strategy

Figure 10.6 shows that a current is applied to two adjacent electrodes and the voltages measured from other pairs of adjacent electrodes. The current is then applied to the next pair of electrodes and the voltages are measured. The procedure is repeated until all the independent measurements have been made. In Figure 10.6(a), the current is injected to electrode pair (1, 2) and the resulting boundary voltage differences are measured from electrode pairs (3, 4), (4, 5) and (15, 16). Voltages are not measured between pairs (16, 1), (1, 2) or (2, 3). In Figure 10.6(b), the current is injected to electrode pair (2, 3), and the voltage differences measured between pairs (4, 5), (5, 6) and (16, 1). Voltages are not measured between pairs (1, 2), (2, 3) or (3, 4).

The adjacent measurement strategy yields N^2 measurements, where N is the number of electrodes. However, among N^2 measurements, only $N(N-1)/2$ measurements are independent. To avoid electrode/electrolyte contact impedance

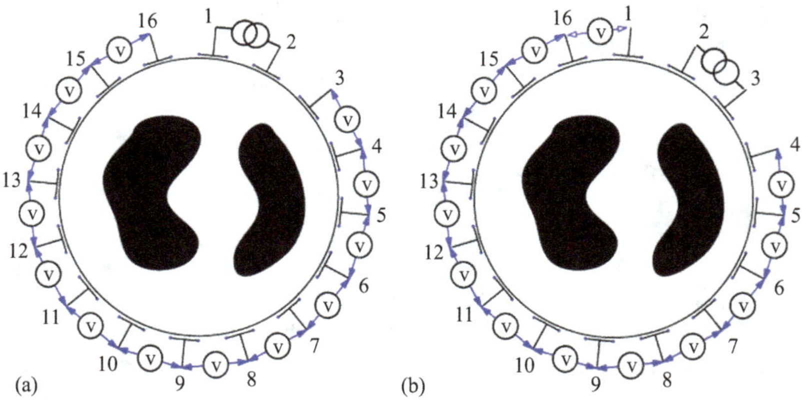

Figure 10.6 Adjacent current injection: (a) first of 16 drive pairs and (b) second of 16 drive pairs

problems, the voltage is not measured at a current-injecting electrode and therefore, the total number of independent measurements M is reduced to $N(N-3)/2$, that is the number of independent measurements is given by

$$L = \frac{N(N-3)}{2} \tag{10.17}$$

where N is the number of electrodes.

According to (10.17), for the 16-electrode ERT sensor and using the adjacent measurement strategy, 104 independent measurements can be obtained.

The main shortcoming of the adjacent strategy is that it has a non-uniform current distribution because most of the currents travel near the peripheral electrodes. Therefore, the current density in the central region is low, making this strategy sensitive to measurement error and noise. However, this strategy requires minimal hardware to implement and image reconstruction can be performed quickly with minimal computer resources required.

10.2.2.2 Opposite strategy

The opposite strategy is shown in Figure 10.7. In this strategy, a current is applied through diametrically opposed electrodes. The voltage reference is the electrode adjacent to the current-injecting electrode. Thus, for a particular pair of current-injecting electrodes, the voltages with respect to the reference are measured at all the electrodes except for the current-injecting ones. The next data set is obtained by switching the current to the next pair of opposite electrodes in the clockwise and the voltage reference electrode is changed accordingly. The voltages are again measured using the same procedure and the whole procedure is repeated until all independent measurements have been made.

Compared with the adjacent strategy, the opposite strategy is less sensitive to conductivity changes at the boundary because most of the currents flow through the

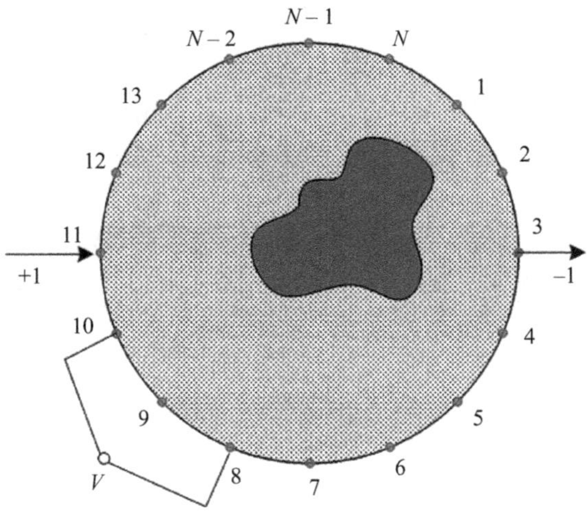

Figure 10.7 Opposite measurement strategy

central region. The opposite strategy yields a relatively good distinguishability due to the even distribution of currents. However, it suffers from a serious disadvantage. For the same number of electrodes N, the number of independent current projections is significantly less than that for the adjacent strategy. The number of independent measurements M is given by

$$M = \frac{N}{4}\left(\frac{3N}{2} - 1\right) \tag{10.18}$$

With a 16-electrode sensor, there are 92 independent measurements.

10.2.2.3 Diagonal strategy

This strategy is also known as the cross method. In this strategy, currents are injected between electrodes separated by large dimensions (see Figure 10.8). Compared with the adjacent strategy, this strategy can produce a more uniform current distribution in the region being imaged. For a 16-electrode sensor, the data collection procedure is as follows: electrode 1 is fixed as a current reference and electrode 2 as the voltage reference. A current is applied successively to electrodes 3, 5 and 15. For each of these current pairs, the voltages from all electrodes, except for the current electrodes, are measured with respect to electrode 2. The current reference is then changed to electrode 4 and the voltage reference to electrode 3 and current applied through electrodes 6, 8, 16 and 2. The voltages are measured on all other electrodes except for the current-injecting ones. Thus, for each pair of current-driven electrodes, 13 voltage measurements are obtained and seven different current electrode pairs are selected to give $7 \times 13 = 91$ data. A further 91 data can be obtained by changing the voltage and current reference. Therefore, a 16-electrode system yields a maximum of 182 data, out of which only 104 are independent.

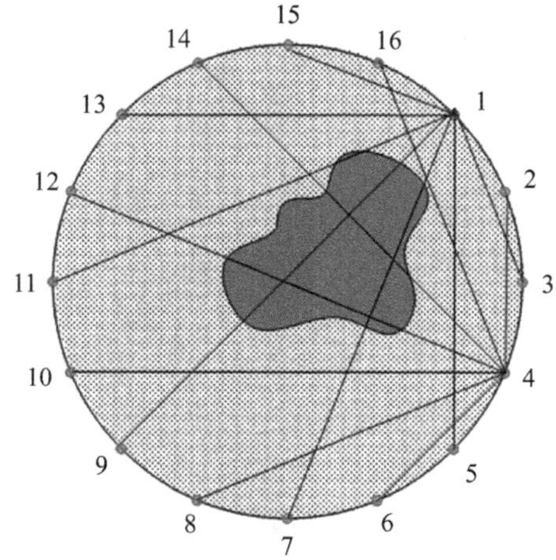

Figure 10.8 Diagonal measurement strategy

The diagonal strategy does not yield a high sensitivity in the periphery compared with the adjacent method. However, it has good matrix conditioning and sensitivity over the entire region and is not as sensitive to measurement error and thus can produce a high quality image.

10.3 Data acquisition system

Data collection system design is also a crucial part of an ERT system. A high-speed data acquisition system is desirable to meet the requirement of real-time imaging for industrial process monitoring and control. Moreover, the data acquisition system should have a high signal-to-noise ratio (SNR), because the measurement noise would be amplified during the process of solving the inverse problem and even dominate the reconstructed images.

Figure 10.9 shows an ERT system developed earlier [19]. It consists of (1) a current source and sink circuits, which must be accurately balanced, (2) current drive multiplexers (2 × 16 way), which select the current drive electrode pair, (3) a 16-channel amplifier with sample and holds, which simultaneously measures the differential voltages between adjacent measuring electrodes and (4) a processor, which controls the current pulse and measurement timing and the multiplexer.

This configuration can be used for current injection on any pair but restricts measurements to adjacent pairs as the electrodes are fixed to the amplifier inputs. The multiple amplifier configuration increases the attainable frame rate by a factor of 16 as all peripheral voltages are measured during each current injection cycle. The ERT system is controlled by a PC, which is used for data capture and real-time visualisation.

Depending on the spatial geometry and the disturbance, it may be beneficial to implement alternative current injection and measuring sequences, which are not

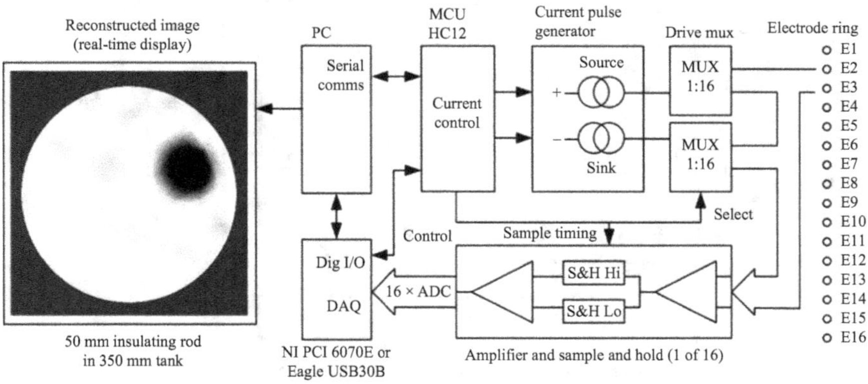

Figure 10.9 ERT system with a 16-electrode sensor

restricted to single plane measurements. To this end, a flexible and extendable multiplexing system was designed, which enables more elaborate sequences to be implemented by adding additional multiplexing modules connected in parallel to the electrode array, that is between the single plane instrument as shown in Figure 10.9 and the electrode rings. In its full implementation, the current may be applied between any of the 128 (8×16 rings) electrodes and voltage measurements made between either adjacent electrodes on the same layer or 'vertically adjacent' electrodes on different layers for 3D ERT.

ERT hardware can be designed with a field programmable gate array (FPGA). A typical architecture is shown in Figure 10.10. There are two basic strategies in the design. The first is to design the system as simple (for short development cycle and low cost) as possible. This can be achieved by implementing as many functions as possible in digital hardware rather than analogue hardware, mainly because of high noise level and limited dynamic range of analogue hardware and the availability of low-cost, high capacity FPGAs. The second strategy is to make the system as flexible as possible using FPGA-based modular design. Any module in the system can be updated or replaced without affecting other modules.

As shown in Figure 10.10, the developed platform consists of a low-cost, high-capacity FPGA, digital-to-analogue converter (DAC), analogue-to-digital converters (ADCs) and USB interface. Some data acquisition sub-systems (DAS) can be added to form a complete ERT system [20].

10.3.1 Sinewave generator

Single-frequency or sequentially applied multi-frequency ERT systems employ a pure sinewave. Alternative waveforms, in particular square waves, contain harmonics of the fundamental frequency, which affect the measurement accuracy. The AC signal source developed earlier is based on monolithic devices, which produce accurate waveforms via direct digital synthesisers (DDS), and has become

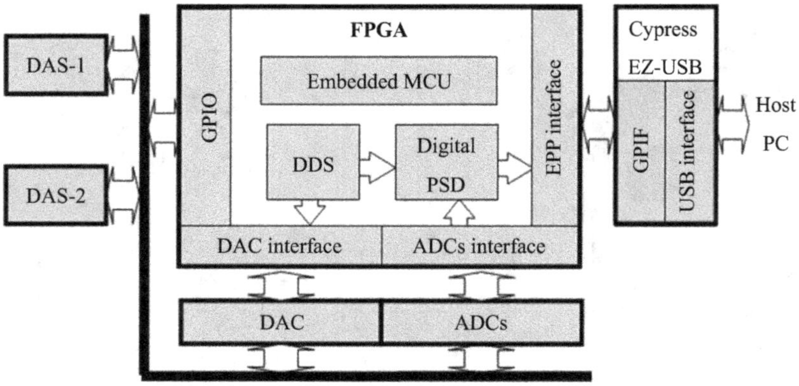

Figure 10.10 FPGA-based data acquisition platform

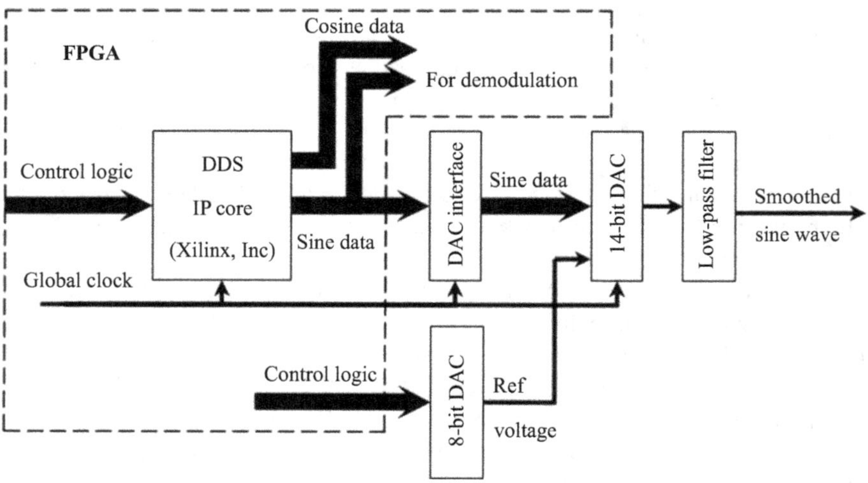

Figure 10.11 Structure of FPGA-based signal generator

commercially available, such as AD7008. DDS has two main advantages over their analogue predecessors: extremely fine resolution at low frequencies (up to 200 kHz) and the inherent stability, typically 100 ppm, of a reference crystal oscillator. An FPGA-based signal generator is shown in Figure 10.11. A DDS IP core (from Xilinx, Inc.) is implemented in FPGA. This IP core can produce sine and cosine signals synchronously. The sine data are converted to an analogue waveform via a DAC. A low-pass filter followed the DAC eliminates the staircase effect. The frequency and phase of the signal can be controlled via the DDS interface. The signal amplitude can be adjusted through a secondary DAC.

In addition to the excitation signal, sine and cosine signals are also used directly in the digital domain for demodulation. In this way, additional analogue circuits are not required to generate reference signals. Furthermore, over-zero signals and arbitrary waveform generation can also be realised by programming the FPGA without modifying external circuits. In this way, the system can have more functions than minimal hardware [20].

10.3.2 Signal conditioning circuits

A programmable gain amplifier (PGA) and an anti-aliasing filter are shown in Figure 10.12. The Nyquist criteria requires that the sampling frequency should be at least twice the signal bandwidth because otherwise information on the signal will be lost. If the sampling frequency is not satisfied, a phenomenon known as aliasing will occur. Though the excitation signal is a single frequency sinewave, noises of various frequencies are mixed in the signal. Therefore, an anti-aliasing filter ahead of the ADC is used to remove the unwanted frequencies. After being amplified and filtered, the signals are converted into the digital format via the ADC. As high-speed, high-resolution ADCs are expensive, sequential protocol is adopted in this design [20].

10.3.3 Voltage-controlled current source

Traditionally, a current is injected into a conductive medium on the premise, requiring a current source with high output impedance. As a result, the effects of electrode/electrolyte contact impedance can be negligible. The hardware required to construct a voltage-controlled current source (VCCS) is more complex than a voltage source especially at frequencies higher than 200 kHz. A simple current source is shown in Figure 10.13. In practice, the gain of an op-amp would not be infinite especially in the 10–100 kHz range used in many ERT systems. The open-loop gain should be sufficiently high. For example, the typical open-loop gain of μA741C and LM318 is 200 V/mV or 106 dB, to ensure that the output current error is well within the resolution limits of the ADC. The major drawback of this circuit

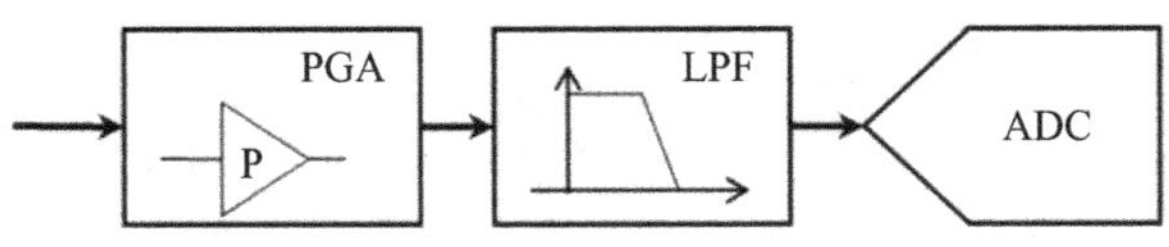

Figure 10.12 Signal conditioning circuits

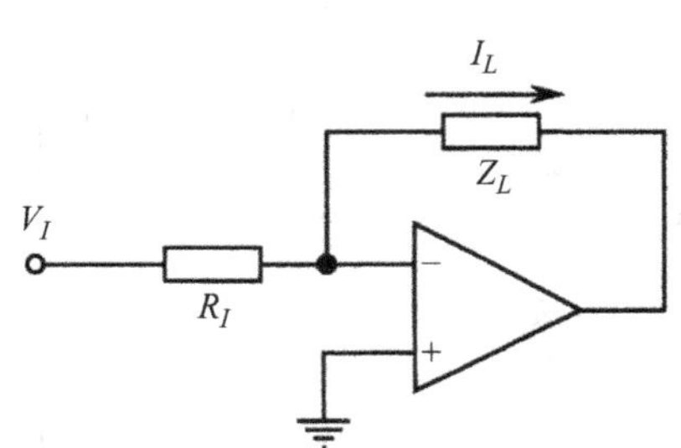

Figure 10.13 Simple VCCS with op-amp

is that it cannot be used in applications with an earthed load or for systems with more than one current source. In the majority of process applications. This is always the case because the pipelines and mixing vessels have some form of earth connection and paired source/sink current sources to reduce common-mode voltage.

Howland current source, modified Howland current source and a dual op-amp current source were developed. The Howland current source is shown in Figure 10.14.

By current analysis, the following equations can be obtained [21]:

$$\begin{cases} I_L \approx \dfrac{U_i - U_+}{R_1} + \dfrac{U_O - U_+}{R_2}, \\ U_+ \approx U_-, \\ \dfrac{U_-}{R_3} \approx \dfrac{U_o - U_-}{R_4} \end{cases} \tag{10.19}$$

When $R_1R_4 = R_2R_3$, the current flowing through the load resistor is

$$I_L \approx \frac{U_i}{R_1} \tag{10.20}$$

It can be seen from (10.20) that the load current I_L is independent on the load resistance R_L, indicating the output impedance of the current source is infinite if the op-amp is ideal. However, the op-amp limitations and resistor tolerances cause undesired behaviour, which exhibits instability at higher frequencies and in the presence of large load capacitance (>100 pF).

A modified Howland current source is shown in Figure 10.15.

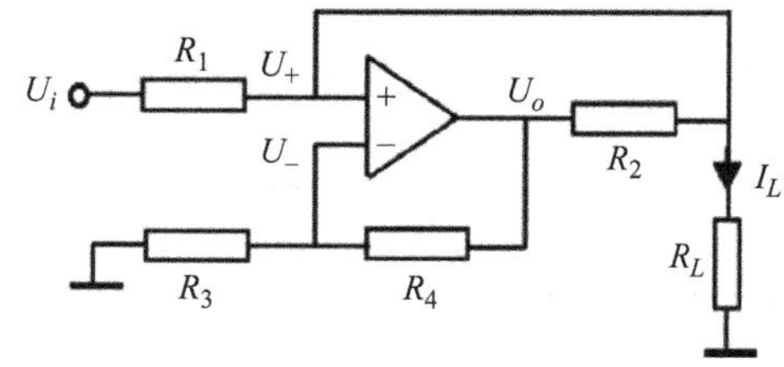

Figure 10.14 Howland current source

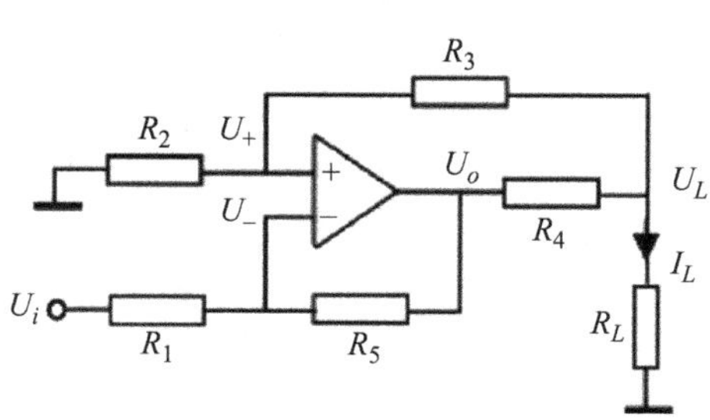

Figure 10.15 Modified Howland current source

By current analysis, the following equations can be written:

$$\begin{cases} I_L \approx \dfrac{U_o - U_L}{R_4} - \dfrac{U_L - U_+}{R_3}, \\ \dfrac{U_i - U_-}{R_1} \approx \dfrac{U_- - U_o}{R_5}, \\ \dfrac{U_L - U_+}{R_3} \approx \dfrac{U_+}{R_2}, \\ U_+ \approx U_- \end{cases} \tag{10.21}$$

When $R_1 = R_5$, $R_2 = R_3 + R_4$, the current flowing through the load resistor R_L is

$$I_L \approx -\frac{U_i}{R_4} \tag{10.22}$$

However, the modified Howland current source requires strict matching of resistors to achieve high output impedance and may suffer from instability over a wide frequency range. In the real implementation of this circuit, the presence of output capacitance acts in parallel with the finite output resistance, reducing the total output impedance. Although the output resistance is adjustable, the output and stray capacitance are still present, degrading the output impedance and hence the circuit performance.

The latest precision VCCS based on Wilson current mirrors has been developed. Its performance is superior to traditional current sources. AD844 (from Analog Devices, Inc.) is selected to construct the current source. AD844 has a wide bandwidth of 60 MHz and very high output slew rate of up to 2,000 V/μs. It differs from a conventional op-amp in that the two signal inputs have radically different impedance. The relationship between the output and inputs of AD844 is shown in Figure 10.16 [22].

The non-inverting input *Y* presents the usual high impedance, and input *X* is the common emitter node of a complementary pair of grounded base stages and presents an input resistance of about 50 Ω. The transfer function for AD844 is

$$\begin{pmatrix} I_y \\ V_x \\ I_z \end{pmatrix} = \begin{pmatrix} 0 & 0 & 0 \\ 1 & 0 & 0 \\ 0 & \pm 1 & 0 \end{pmatrix} \begin{pmatrix} V_y \\ I_x \\ V_z \end{pmatrix} \tag{10.23}$$

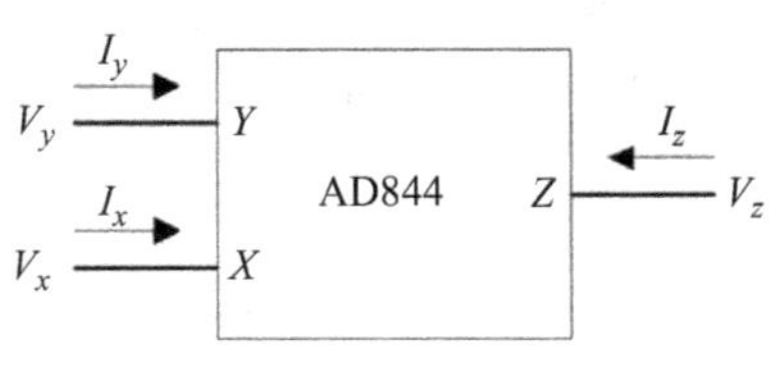

Figure 10.16 Relationship between inputs and outputs of AD844

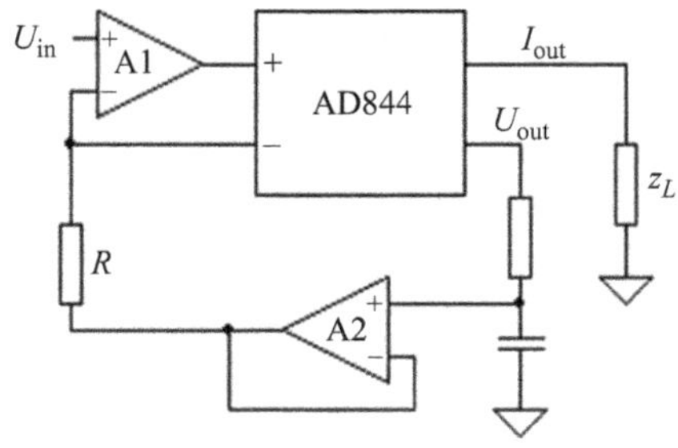

Figure 10.17 AD844 based VCCS

An AD844-based current source is shown in Figure 10.17. The op-amp A1 acts as input buffer and A2 acts as DC feedback to compensate the residual DC component. Test results show that this VCCS circuit can provide output impedance of 239 kΩ and phase shift of less than 10° at 1 MHz.

The output current can be calculated as

$$I_{out} = \frac{U_{in}}{R} \tag{10.24}$$

10.3.4 Differential amplifiers

In ERT, the resultant voltages due to current injection must be accurately measured. There is a wide dynamic range of differential voltages varying from millivolts to volts. Also, a high common mode rejection ratio (CMRR) is needed over the range of driving signal frequencies. Therefore, a differential amplifier should be able to accommodate such issues for accurate resistance measurement. AD8130 is a differential amplifier with CMRR of 70 dB at 10 MHz. AD8130 is stable at a gain of 1 and can be used for applications where lower gains are required. A high gain (10 or greater) version, AD8129, is also available for applications that requires high gain setting. A basic circuit for AD8130 is shown in Figure 10.18, in which $V_{out} = V_{in}$.

10.3.5 Signal conditioning circuit

PGA and anti-aliasing filter are essential parts in signal conditioning circuits. Figure 10.19 shows a design.

After being amplified and filtered, the signal is converted into the digital format via an ADC. The sinusoidal voltage sampled by the ADC is assumed to be of the form

$$V_m(i) = A \sin\left(\frac{2\pi i}{N} + \theta\right) \tag{10.25}$$

where N is the number of samples in one excitation period.

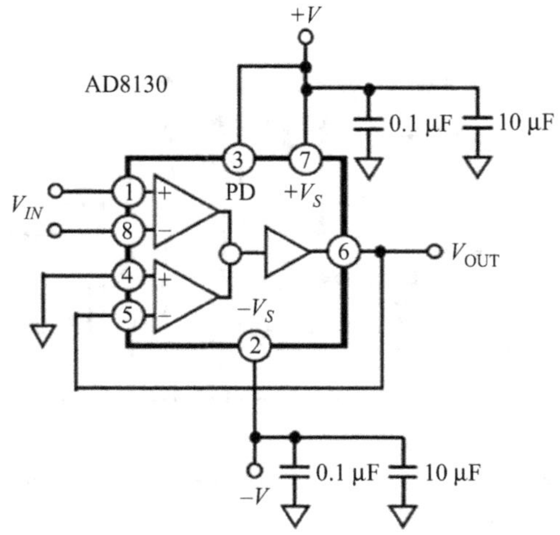

Figure 10.18 AD8130 based circuit with unity gain

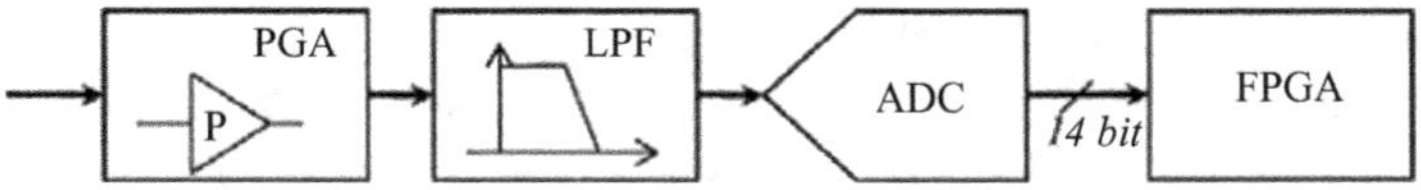

Figure 10.19 Signal conditioning circuit

Because high-speed, high-resolution ADCs are expensive, sequential protocol is adopted in this design.

10.3.6 Digital demodulation

Once the measured voltage $V_m(i)$ is attained, the in-phase (real) and quadrature (imaginary) components can be estimated as

$$V_r = \frac{2}{N}\sum_{i=0}^{N-1} \sin\frac{2\pi i}{N} = A\cos\theta \tag{10.26}$$

$$V_r = \frac{2}{N}\sum_{i=0}^{N-1} \cos\frac{2\pi i}{N} = A\sin\theta \tag{10.27}$$

As described in Section 10.3.1 about signal generator, excitation and reference signals are available in the digital domain, which is convenient for digital demodulation. The sampled data is forward to FPGA, in which digital phase-sensitive demodulations are performed for both in-phase and quadrature components, as illustrated in Figure 10.20.

Computationally intensive multiplication and accumulation operations are performed with dedicated multipliers, FPGA and other logics. No data buffers are

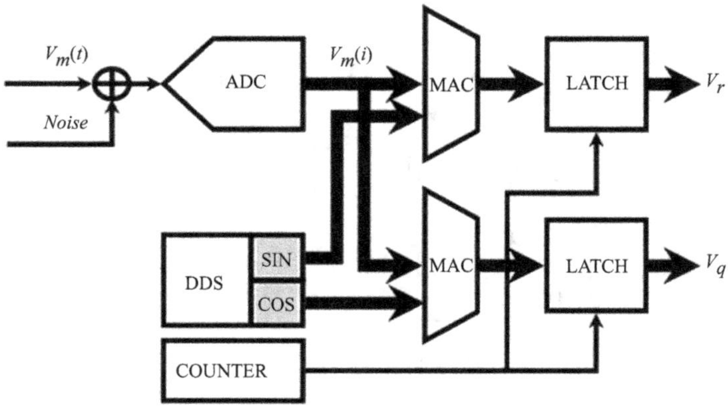

Figure 10.20 Discrete-time quadrature demodulation

required to regulate the data. Therefore, digital demodulation can be performed online to maximise the data acquisition rate. Once updated, the sample is forwarded to multiply-accumulates (MACs), where it can be multiplied by the reference signals and then moved to the accumulator. This procedure is of pipeline operation and would be carried out N times to produce a pair of results, V_r and V_q. Then the amplitude (A) and phase angle (θ) can be computed by

$$\begin{cases} A = \arg(V_r, V_q) = \sqrt{V_r^2 + V_q^2} \\ \theta = \arctan\left(\dfrac{V_q}{V_r}\right) \end{cases} \tag{10.28}$$

10.4 Image reconstruction

To determine the internal distribution from measurements acquired from an ERT sensor, different image reconstruction algorithms can be applied. In general, the algorithms can be used either on-line or off-line, depending on the time constraints and the type of image required. Image reconstructed algorithms have been published in many literatures [6,9,10,23–26]. The following is typical algorithms.

With the aid of physical modelling and FEM discretisation, a deterministic observation model of ERT can be written as

$$\mathbf{V} = \mathbf{U}(\boldsymbol{\sigma};\mathbf{I}) = \mathbf{R}(\boldsymbol{\sigma})\mathbf{I} \tag{10.29}$$

where $\mathbf{U}(\boldsymbol{\sigma};\mathbf{I})$is the forward model mapping $\boldsymbol{\sigma}$ and $\mathbf{I}$ to $\mathbf{V}$, and $\mathbf{R}(\boldsymbol{\sigma})$ is the model mapping $\boldsymbol{\sigma}$ to resistance. The model of $\mathbf{V} = \mathbf{R}(\boldsymbol{\sigma})\mathbf{I}$ depends nonlinearly on the conductivity $\boldsymbol{\sigma}$ and linearly on the current $\mathbf{I}$.

If changes in conductivity are small, the inverse problem can be solved with sufficient accuracy by considering the following linearized equation with the Jacobi matrix J, which relates the changes in conductivity ($\delta\boldsymbol{\sigma}$) to changes in the measured voltage ($\delta\mathbf{U}$):

$$\delta\mathbf{U} = \mathbf{U}'(\boldsymbol{\sigma}_0)\delta\boldsymbol{\sigma} = J\delta\boldsymbol{\sigma} \tag{10.30}$$

In ERT, the Jacobi matrix is made up of sensitivity maps. The forward problem can be simplified by linearization:

$$\boldsymbol{CX} = \boldsymbol{Y} \tag{10.31}$$

where $\boldsymbol{X} \in R^{n\times 1}$ is conductivity change, and $\boldsymbol{Y} \in R^{m\times 1}$ is the change between a measured value and the reference value and $\boldsymbol{C} \in R^{m\times n}$ is the sensitivity matrix.

10.4.1 Sensitivity/Jacobian matrix-based methods

10.4.1.1 Non-iterative methods

The non-iterative method is based on a one-step reconstruction. Examples of the non-iterative method include LBP method, sensitivity method and SVD method [8].

1. Linear back-projection. LBP is a simple image reconstruction algorithm and has commonly been used. For the l-electrode sensor, LBP can be represented as

$$x_k = \frac{\sum_{i=1}^{l-1}\sum_{j=i+1}^{l} y_{i\times(j-1)}c_{i\times(j-1),k}}{\sum_{i=1}^{l-1}\sum_{j=i+1}^{l} c_{i\times(j-1),k}} \tag{10.32}$$

   where x_k is the kth element of X.
2. Sensitivity method. If the inverse of C existed, (10.31) could be solved directly by

$$\boldsymbol{X} = \boldsymbol{C}^{-1}\boldsymbol{Y} \tag{10.33}$$

   However, the inverse of C does not exist and so other methods must be used. If C is considered to be a linear mapping from the conductivity vector space to the voltage vector space, C^T can be considered as a related mapping from the measured data vector space to distribution vector space, giving an approximated solution:

$$\boldsymbol{X} = \boldsymbol{C}^T\boldsymbol{Y} \tag{10.34}$$

   Although mathematically not accurate, the sensitivity algorithm is still widely used for online image reconstruction because of its simplicity. However, it produces poor images and can only provide qualitative information.
3. Singular value decomposition (SVD) method. Equation (10.30) can be simplified as

$$Ax = b \tag{10.35}$$

   where $A \in R^{m\times n}$, $b \in R^m$, $x \in R^n$ and $m < n$.

If a singular value system of A can be expressed as $\{\mathbf{u_k}, \mathbf{v_k}, \sigma_k\}$, where $\{\mathbf{u_k}\}_{k=1}^{m}$ is the left singular vector, $\{\mathbf{v_k}\}_{k=1}^{n}$ is the right singular vector and $\{\sigma_k\}_{k=1}^{n}$ is the singular value, then

$$\begin{aligned} \mathbf{A*Av_k} &= \sigma_k^2 \mathbf{v_k} \\ \mathbf{AA*u_k} &= \sigma_k^2 \mathbf{u_k} \\ \mathbf{A*u_k} &= \sigma_k \mathbf{v_k} \end{aligned} \tag{10.36}$$

The solution of (10.36) calculated by using SVD method is

$$\mathbf{x*} = \sum_{k=1}^{n} \frac{1}{\sigma_k} \langle \mathbf{b}, \mathbf{u_k} \rangle \mathbf{v_k} \tag{10.37}$$

If there exist some small singular values σ_k, this decision will produce a very large random component, which can completely submerge the components of x in that direction. Small singular values may be regarded as being zero, and to regard the components along such directions as free parameters, which are not determined by the data. The truncated SVD (TSVD) algorithm attempts to stabilise the solution by taking the small singular values:

$$\mathbf{x*} = \sum_{k=1}^{r} \frac{1}{\sigma_k} \langle \mathbf{b}, \mathbf{u_k} \rangle \mathbf{v_k} \tag{10.38}$$

where $r < n$.

The effectiveness of applying TSVD depends, to a great extent, on a proper choice of the truncated parameter r. The feasible approaches include L-curve method [24,25].

10.4.1.2 Iterative methods

1. Landweber method. The following objective function is formulated based on the residual error and resistivity distribution terms:

$$\phi_2(\rho_k) = \frac{1}{2} a \|R - \mathbf{W}\rho_k\|^2 + \frac{1}{2} \rho \|\Gamma \rho_k\|^2 \tag{10.39}$$

where $\|x\|^2 = x^T x$, α and β are constant step-lengths.

By differentiating (10.39) with respect to ρ_k, the following equation is obtained:

$$\nabla \phi_2(\rho_k) = a\mathbf{W}^T (R - \mathbf{W}\rho_k) + \beta \Gamma^T \Gamma \rho_k \tag{10.40}$$

To estimate the resistivity distribution, Landweber method can be obtained as follows:

$$\begin{aligned} \rho_{k+1} &= \rho_k + \nabla \phi_2(\rho_k) \\ &= \rho_k + a\mathbf{W}^T (R - \mathbf{W}\rho_k) + \beta \Gamma^T \Gamma \rho_k \end{aligned} \tag{10.41}$$

The reconstruction performance of Landweber method depends on appropriate step-lengths α and β. Typically, the step-length α is set to be $\alpha = 1$ or

$\alpha = 2$. Landweber method estimates unknown variables iteratively. In some cases, if optimal step-lengths are selected at every iteration, the reconstruction accuracy may be improved.

In addition, the sensitivity updating strategy can be used in Landweber method, and can be expressed by Figure 10.21, where inside of the dash line box is the direct Landweber method. For the modified Landweber method with updating the sensitivity matrix, the initial image is obtained by the direct Landweber method.

Landweber method with updated sensitivity matrix can be expressed as

$$\begin{cases} G_0 = G_m \text{ (direct Landweber iteration)} \\ G_{K+1} = G_K + \alpha S_{\text{up}}^T (V - S_{\text{up}} G_K) \end{cases} \tag{10.42}$$

where G_0 is the reconstructed image after m times iterations of direct Landweber method, which is used as the initial image for updating the sensitivity matrix, and S_{up} is the updated sensitivity matrix.

2. Conjugate gradient (CG) method. The CG method only works with a system with linear equations and a symmetric positive definite (SPD) coefficient matrix. Equation (10.30) can be transformed into

$$\mathbf{J}^{\mathrm{T}}\mathbf{J}\delta\sigma = \mathbf{J}^{\mathrm{T}}\delta\mathbf{U} \tag{10.43}$$

However, the coefficient matrix $\boldsymbol{J}^{\mathrm{T}}\boldsymbol{J} \in \boldsymbol{R}^{n\times n}$ or the number of pixels n in the reconstructed image dominates computation. To overcome these computational limitations, the computationally expensive system is transformed as

$$\mathbf{J}\mathbf{J}^{\mathrm{T}}\mathbf{y} = \delta\mathbf{U} \tag{10.44}$$

where $\delta\sigma = \mathbf{J}^{\mathrm{T}}\mathbf{y}$.

To improve the stability of the solution, the CG method is always applied to a Tikhonov system:

$$(\mathbf{J}\mathbf{J}^{\mathrm{T}} + \alpha\mathbf{L}\mathbf{L}^{\mathrm{T}})\mathbf{y} = \delta\mathbf{U} \tag{10.45}$$

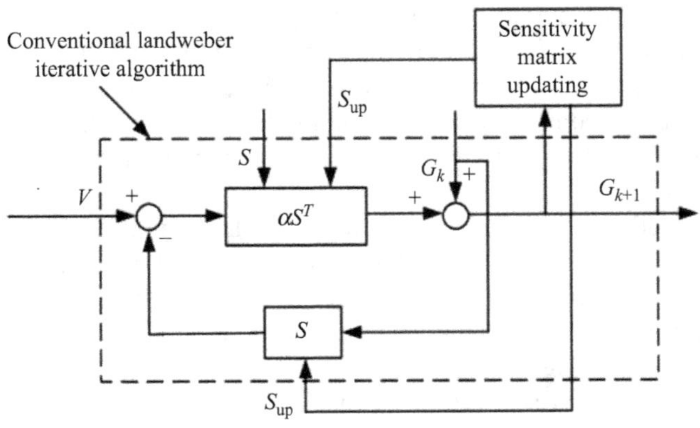

Figure 10.21 Landweber method with updated sensitivity matrix

If the following is assumed

$$\mathbf{A} = \mathbf{J}\mathbf{J}^{\mathrm{T}} + \alpha\mathbf{L}\mathbf{L}^{\mathrm{T}} \tag{10.46}$$

then a Tikhonov system can be transformed to

$$\mathbf{A}\mathbf{y} = \delta\mathbf{U} \tag{10.47}$$

Some intelligent algorithms are based on genetic algorithms, neural network and support vector machine algorithm.

10.5 Applications of ERT

Non-intrusive interior inspection through volume scanning first came to life in the medical arena for diagnosis of undesired tumours and lumps. This successful application later triggered applications of this technique in industry and stimulated a broad range of research into process imaging, such as multi-phase flow measurement in pipes, mixing vessels, reactors, digesters, hydro-cyclones and separators. Until now, the majority of ERT applications are limited to laboratory-scale and pilot-scale units. ERT measurements can assist in calculation of numerous critical process parameters, such as phase holdups, concentrations, flow regimes, flow rates, mixing time and detection of pathologies [27]. Some typical application examples are introduced as follows.

10.5.1 Multi-phase flow measurement

10.5.1.1 ERT sensor and system

Slug flow is one of the most common flow regimes, which occurs over a wide range of liquid and gas flows. It is characterised by the intermittent appearance of liquid slugs that may contain small bubbles, separated by a sequence of long bubbles. It is important to study the slug flow in a gas/liquid flow to achieve an in-depth understanding of the mixture flow and subsequent optimisation of industrial processes. The parameters of interest are (apart from hold-up and pressure drop) slug lengths and slug velocity. A dual-plane ERT system, as shown as Figure 10.22, is applied. Each plane consists of eight titanium electrodes, equi-spaced around the internal circumference of a Perspex pipe with the internal diameter of 85 mm, as shown in Figure 10.23.

Each electrode is 6 mm high and 4 mm wide. The axial distance L between up-stream and down-stream planes is 170 mm. The measured data are sent to a computer via an USB interface. Cross-sectional images of the phase distribution are reconstructed using an image reconstruction algorithm at 50 frames per second with 28 data in each frame [28].

A gas/water two-phase flow rig is shown in Figure 10.24. The flow rates of gas and water were controlled by adjusting the valves and measured by a Vortex flow meter and a Roots flow meter, respectively. The measured data were recorded and

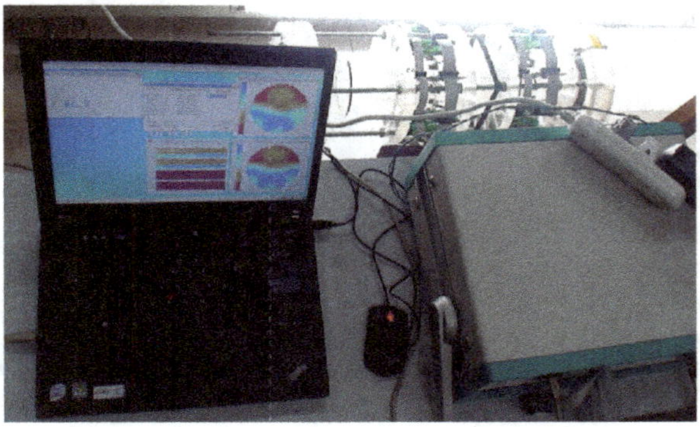

Figure 10.22 Dual-plane ERT system

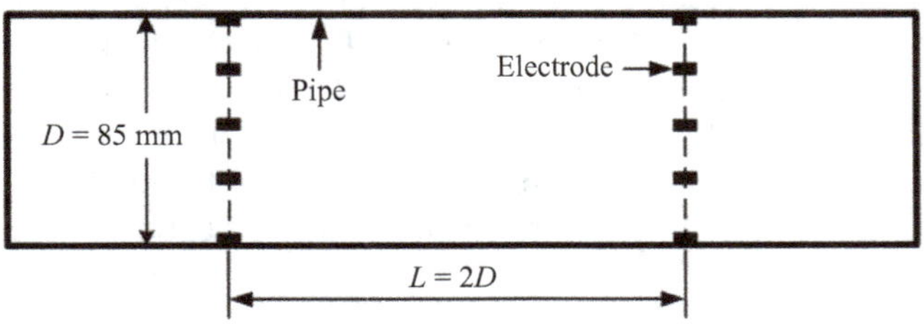

Figure 10.23 Mounting electrodes on pipe wall

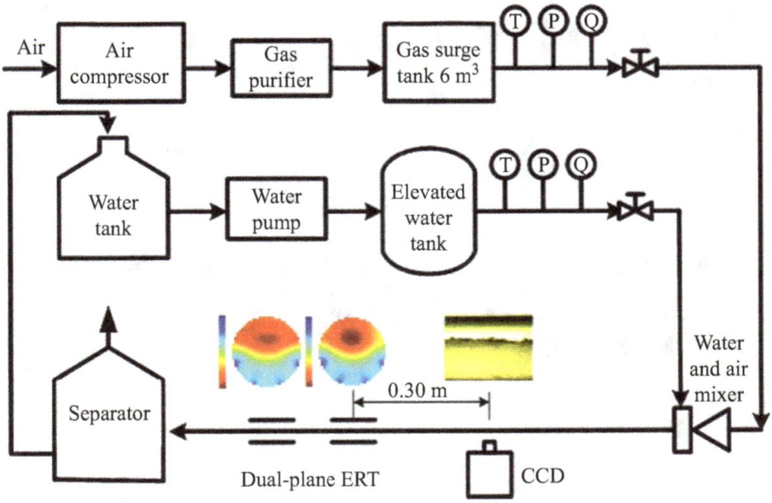

Figure 10.24 Experiment set-up

displayed on the computer. During the experiments, the following conditions are maintained:

- Gas temperature: 27 °C
- Gas pressure: 0.54–0.70 MPa
- Gas flow rate: 4.84–13.4 m^3/h
- Water temperature: 32 °C
- Water pressure: 0.19–0.22 MPa
- Water flow rate: 3.45–22.02 m^3/h

For the same water flow rate and the same gas flow rate, 2,000 frame ERT data were sampled for 40 s. The measured signals on 200 different flow conditions were acquired. For all slug flow, good agreement between the independent component with the maximal kurtosis and liquid slug fluctuation of slug flow was observed.

10.5.1.2 Stratified/slug flow

Figure 10.25 shows the reconstructed image in three representative locations for a stratified/slug flow (USW = 1.06 m/s, USG = 0.31 m/s). Figure 10.26(a) shows liquid flows along the bottom and gas on the top of a pipe. The liquid level ahead of a slug is relatively high. When liquid slug comes, liquid phase tends to move along the position closer to the top of pipe. In this case, the liquid hold-up is close to unity as shown in Figure 10.26(b). After the slug, the liquid level drops sharply behind its

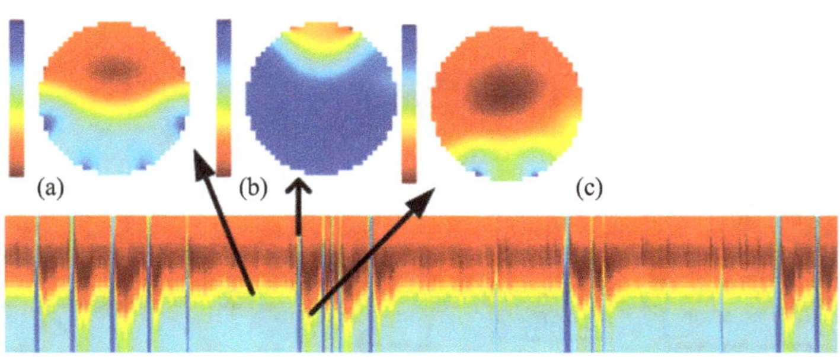

Figure 10.25 Reconstructed image using sensitivity algorithm in three representative positions

Figure 10.26 Liquid slug going through recorded by CCD: (a) frame 154, (b) frame 163 and (c) frame 175

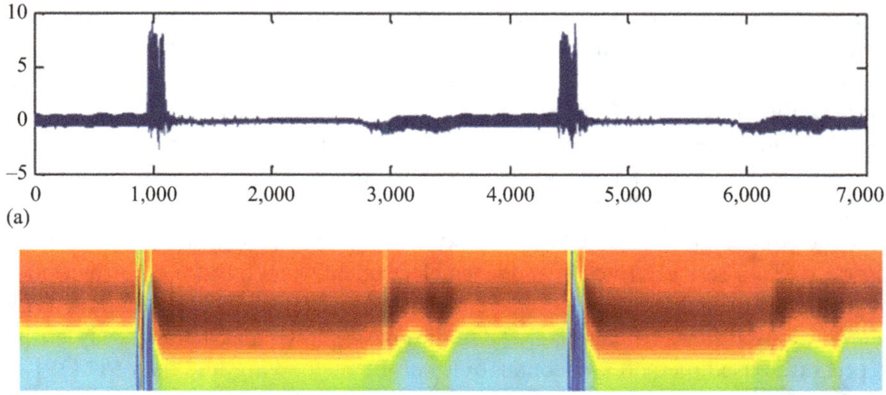

(a)

(b)

Figure 10.27 Comparison between estimated independent components and reconstructed ERT images for a wavy/slug flow: (a) independent component with the maximal kurtosis for 20 s and (b) reconstructed image series from side view for 20 s

tail as shown in Figure 10.26(c). Figure 10.26 shows a liquid slug going through, which is recorded by CCD. The frame frequency of CCD is 300 frames/s. The spent time for the liquid slug going through in this situation is about 70 ms. This corresponds to a very narrow pulse characteristics in reconstructed image series for 20 s. In Figure 10.26(c) the intermittent dark blue part represents the liquid slug going through, liquid filling almost the entire cross-section of pipe. Good agreement between the reconstructed images and CCD record for liquid slug information is observed. The visualisation capability of ERT, including the visualisation of slug size, shape, frequency and the height of the gas/liquid interface, shows the potential for enhancing understanding the flow dynamics of slug flow.

10.5.1.3 Wavy/slug flow

Figure 10.27(a) shows the independent component with the maximal kurtosis for 20 s for a wavy/slug flow (USW = 0.20 m/s, USG = 0.18 m/s). Figure 10.27(b) shows the reconstructed image series. The independent component with the maximal kurtosis corresponds to the intermittent blue part in Figure 10.27(b), which represents broader impulsive motion.

10.5.2 Volume fraction measurement of gas-liquid bubble column

Bubble columns are widely used in absorption, oxidation, fermentation, biochemical reactions and chemical wastewater treatment processes because of its advantages of simple structure, low cost and high efficiency. Recent studies on bubble column focus on local and global bubble patterns and distribution measurements. The quantitative measurement of radial and axial gas hold-up is particularly important. However, most industrial reactions in bubble columns are of high temperature and pressure, and usually non-transparent, all of which make conventional measuring method not applicable. Figure 10.28 shows the structure of gas/liquid bubble column.

The bubble column is filled with tap water and air is injected from the bottom. The air velocity can be controlled by a valve and measured by the flow meter under the bubble column. Big bubbles are smashed to smaller ones by the sieve plates. ERT electrodes are mounted on the two sides of sieve plate as shown in Figure 10.28. There are 16 equally spaced electrodes for each plane. The electrodes are made up of stainless steel. Differential pressure transducers and conductivity probes are used for verification. Early studies have shown that the relatively errors of pressure-based method and conductivity probe method are less than 5% [29].

Conductivity or gas hold-up distribution can be directly calculated and displayed on the host PC. The average value is calculated from every 200 images for different gas velocities. Figure 10.29 shows the experimental results: conductivity and gas hold-up images at different gas velocities. As the gas velocity rises, the conductivity at the centre region drops, and the gas hold-up increases.

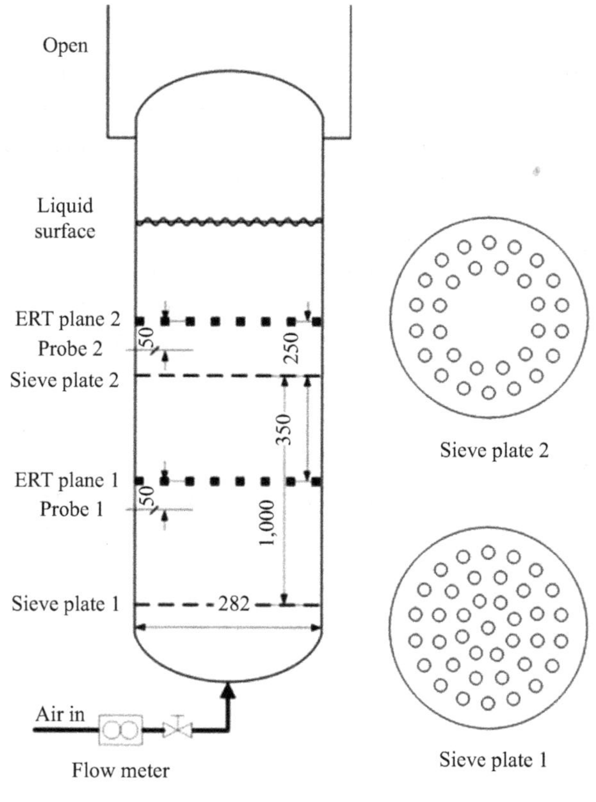

Figure 10.28 Bubble column and sieve plates (unit: mm)

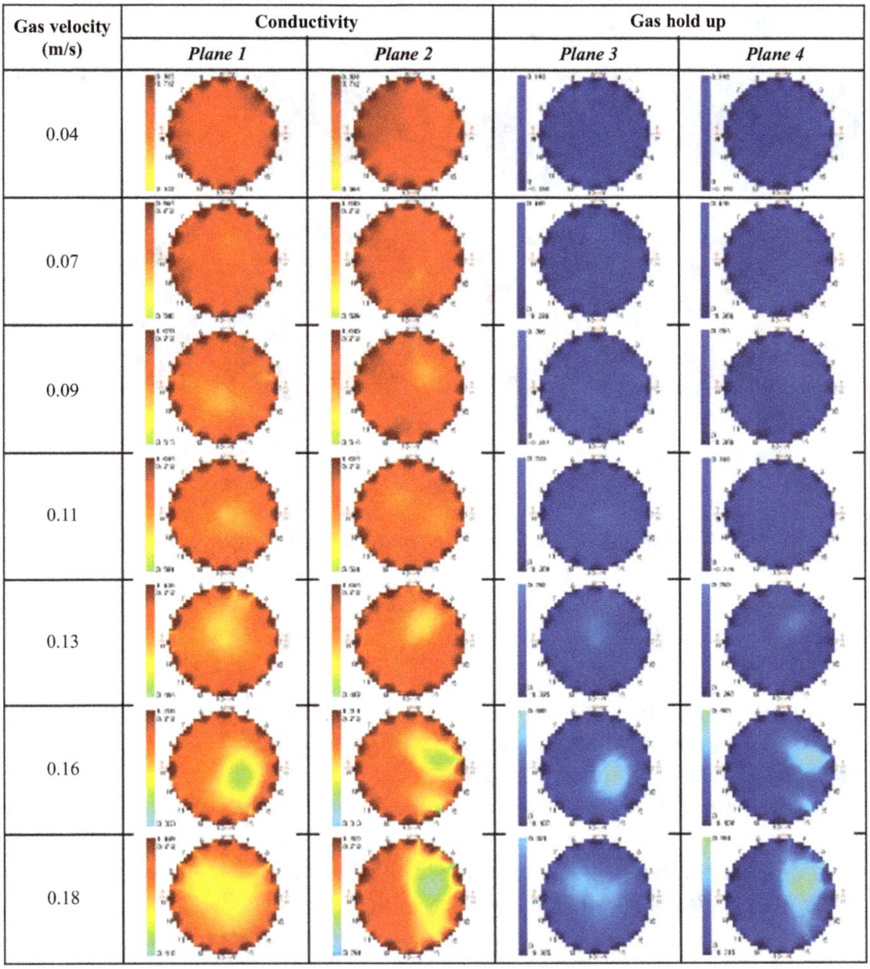

Figure 10.29 Conductivity and gas hold-up distributions

10.5.3 Fluidised bed reactor

ERT has been applied to visualise the flow pattern and distribution inside a radial flow packed bed for improving reactor performance with lower pressure drop. Sets of images representing slices through a packed bed can be obtained for an 8-plane × 16-electrode sensor, which produces the order 10^3 conductivity (mS/cm) measurements in 3D. Pulse injections of high conductivity tracer, both uniformly in the feed and localised, can be imaged as multiple images revealing the internal flow pattern. Differentiation of the motion of the tracer peak conductivity within pixels in the sensing planes and between the planes allows the local flow velocities and directions to be determined. This quantifies the flow pattern for uniformity and radial distributive properties.

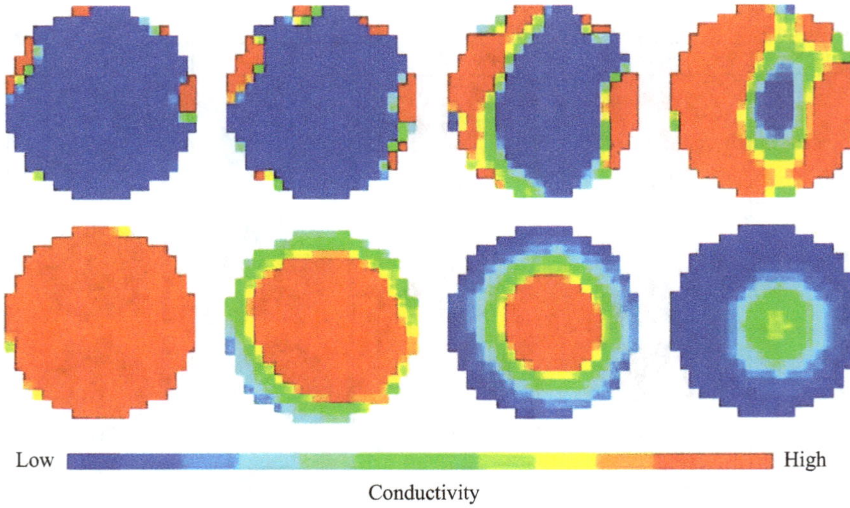

Figure 10.30 Time series of tomographic images from single, central measurement plane following direct injection of high conductivity tracer into top of the packed bed

A pulse of brine is injected into an aqueous flow at steady state with the feed, the flow and mixing of the tracer are captured in real time. The pulse progress is tracked as it enters the reactor and takes its tortuous but pre-determined path towards the exit. Figure 10.30 shows a series of reconstructed images from one measurement plane near the centre of the bed height. The successive images show the build-up of the pulse near the vessel wall, then the front moves towards the centre and the pulse is distributed right across the reactor section. The trailing edge of the pulse detaches from the wall and moves towards the axial exit zone. The desired flow pattern of axial flow near the wall, radial inward flow and exit via the central collector is well demonstrated. The results are similar to each other for each flow rate. The high conductivity tracer was detected initially around the periphery of plane 1, from where it moved down to planes 2, 3 and so on while simultaneously moving in towards the central collector.

Figure 10.31 shows 3D images for several frames of data from the measurement for a flow rate of 300 L/min at times, that is 3, 21, 39, 57, 75 and 93 s after the injection. The top view is an isometric view with a plan view underneath. These are processed by stacking the 2D images and performing a linear interpolation between planes to obtain the correct geometry. The simultaneous movement of the tracer down through the bed and towards the central collector can be seen [30].

10.5.4 Milk processing

ERT could be used in various stages of the milk processing for monitoring and control of crucial parameters, for example in 3D visualisation of milk holding tanks

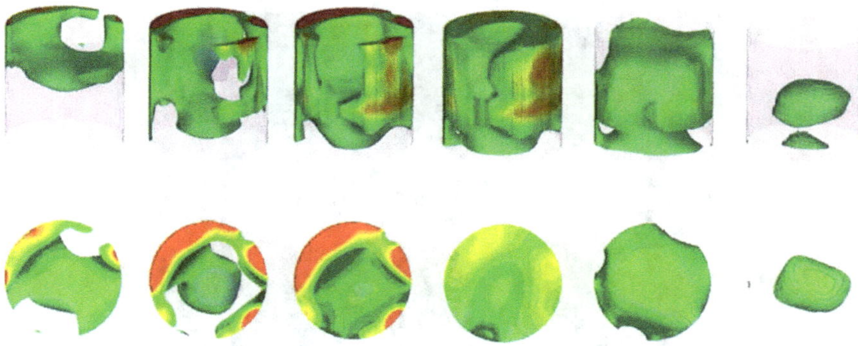

Figure 10.31 *Time series of '3D solids-body' images following direct injection of high conductivity tracer into top of packed bed*

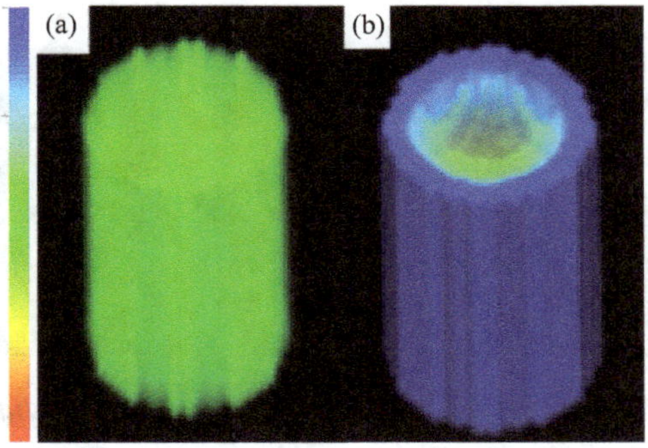

Figure 10.32 *Adulteration visualisation for whole milk solution: (a) desired state with no adulteration and (b) after adulteration*

for analysis of overall homogeneity or non-homogeneity, total solids content monitoring, adulteration, cream separation, aeration and object/powder lump detection, which are parameters requiring monitoring and control in the milk processing industry [31].

10.5.4.1 Adulteration detection

ERT can be used in detecting adulteration, desired solutions (12.7% total solids content skim and whole milk commensurate to raw milk) were set as the reference and measurements were taken after 1 L water was added and mixed into the milk solutions as shown in Figures 10.32 and 10.33.

10.5.4.2 Cream separation

ERT can be used in detecting cream separation. Whole milk solutions of 12.7%, 30% and 50% total solids content were prepared and left overnight without any agitation. Images were taken the next day using a reference image of the milk

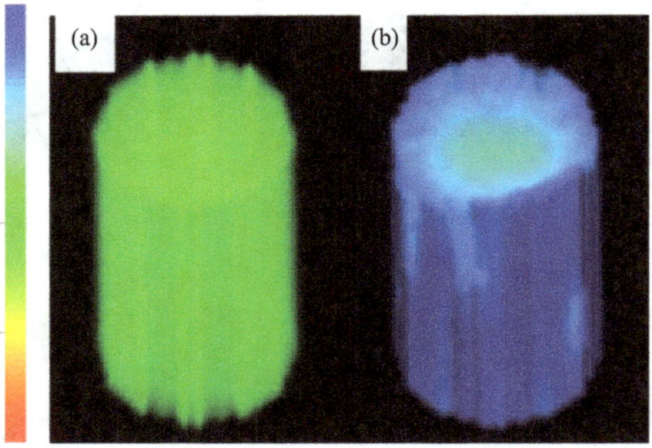

Figure 10.33 Adulteration visualisation for skim milk solution: (a) desired state with no adulteration and (b) after adulteration

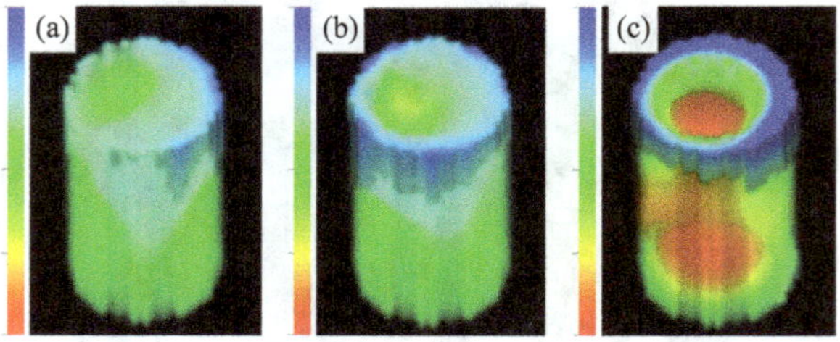

Figure 10.34 Overnight cream separation for whole milk solutions of different total solids content: (a) 12.7%, (b) 30% and (c) 50%

solution at a homogenous state. The resulting three-dimensional images are presented in Figure 10.34.

10.6 Summary

This chapter is about ERT sensors and systems as well as their applications, including optimal design of electrode arrays and sensor architecture based on an objective function and data acquisition system. DAS includes sinewave generators, signal conditioning circuits, VCCS, signal conditioning circuits and digital demodulation. Some typical applications are introduced, including mixers, pipes, reactors on process vessels and fluidised beds, with various aqueous conductive media, such as water, saline, pseudo plastic fluids and dairy products. Because ERT has low spatial resolution and high temporal resolution, ERT sensors should be optimised according to the characteristics of the measured object, and SNR and resolution of data acquisition system should be improved. To improve image reconstruction, prior information should be used, for example combining the characteristics of the tested object.

Research into ERT is currently aiming to improve the hardware and software and further enhance its abilities. To apply ERT in industry for process monitoring and control, higher accuracy, greater sampling rate and more robustness is required, which can enhance fundamental understanding of complicated processes and improve the design and operation of process equipment.

References

[1] Huang Z Y, Yan Y, Wang B L, Li H Q, 'Analysis of soft field in electrical resistance tomography sensors', *Chinese J. Sci. Instrum.*, 2001, **22** (6), pp 573–576

[2] Yunus F R M, Rahim R A, Aw S R, Ayob N M N, Goh C L, Pusppanathan M J, 'Simulation study of electrode size in air-bubble detection for dual-mode integrated electrical resistance and ultrasonic transmission tomography', *Powder Technol.*, 2014, **256**, pp 224–232

[3] Geselowitz D B, 'An application of electrocardiographic lead theory to impedance plethysmography', *IEEE Trans. Biomed. Eng.*, 1971, **18** (1), pp 38–41

[4] Mufti I R, 'A practical approach to finite-difference resistivity modeling', *Geophysics*, 1978, **43**, pp 930–942

[5] Polydorides N, 'Image reconstruction algorithms for soft-field tomography', *PhD Thesis*, University of Manchester Institute of Science and Technology (UMIST), UK, 2002

[6] Cui Z Q, Wang Q, Xue Q, *et al.*, 'A review on image reconstruction algorithms for electrical capacitance/resistance tomography', *Sensor Rev.*, 2016, **36** (4), pp 429–445

[7] Hu L, Wang H, Zhao B, Yang W, 'A hybrid reconstruction algorithm for electrical impedance tomography', *Meas. Sci. Technol.*, 2007, **18** (3), pp 813–818

[8] Isaksen Ø, 'A review of reconstruction techniques for capacitance tomography', *Meas. Sci. Technol.*, 1996, **7** (3), pp 325–337

[9] Lionheart W R, 'EIT reconstruction algorithms: Pitfalls, challenges and recent developments', *Physiol. Meas.*, 2004, **25** (1), pp 125–142

[10] Yang W, Peng L, 'Image reconstruction algorithms for electrical capacitance tomography', *Meas. Sci. Technol.*, 2003, **14** (1), pp R1–R13

[11] Seagar A D, Barber D C, Brown B H, 'Electrical impedance imaging', *IEE Proc. A: Sci. Meas. Technol.*, 1987, **134**, pp 201–210

[12] Graham B M, Adler A, 'Electrode placement configurations for 3D EIT', *Physiol. Meas.*, 2007, **28**, pp S29–S44

[13] Rasteiro M G, Silva R, Garcia F A P, Faia P, 'Electrical tomography: A review of configurations and applications to particulate processes', *KONA Powder Particle J.*, 2011, **29**, pp 67–80

[14] Wang H X, Wang C, Yin W L, 'Optimum design of the structure of the electrode for a medical EIT system', *Meas. Sci. Technol.*, 2001, **12**, pp 1020–1023

[15] Ma Y X, Wang H, Xu L A, Jiang C, 'Simulation study of the electrode array used in an ERT system', *Chem. Eng. Sci.*, 1997, **52**, pp 2197–2203

[16] Wang P, Guo B L, Nan L, 'Three-dimensional analysis on resistance tomography sensors and its application in inspection of milk flow', *8th IEEE International Conference on Control and Automation*, Xiamen, China, 9–11 June 2010

[17] Pinheiro P A T, Loh W W, Dickin F J, 'Optimal sized electrodes for electrical resistance tomography', *Electron. Lett.*, 1998, **34** (1), pp 69–70

[18] Dickin F, Wang M, 'Electrical resistance tomography for process applications', *Meas. Sci. Technol.*, 1996, **7**, pp 247–260

[19] Wilkinson A J, Randall E W, Long T M, Collins A, 'The design of an ERT system for 3D data acquisition and a quantitative evaluation of its performance', *Meas. Sci. Technol.*, 2006, **17**, pp 2088–2096

[20] Cui Z Q, Wang H X, Tang L, Zhang L F, Chen X Y, Yan Y, 'A specific data acquisition scheme for electrical tomography', *IEEE International Instrumentation and Measurement Technology Conference*, Vancouver Island, Canada, 12–15 May 2008

[21] Chen D X, Deng X, Yang W Q, 'Comparison of three current sources for single-electrode capacitance measurement', *Rev. Sci. Instrum.*, 2010, **81** (3), 034704

[22] Wang H X, Cui Z Q, 'Digital ECT/ERT dual modality system design', *MSc Thesis*, Tianjin University, China, 2007

[23] Isaksen O, Nordtvedt J E, 'A new reconstruction algorithm for process tomography', *Meas. Sci. Technol.*, 1998, **4** (12), pp 1464–1475

[24] Zhang L L, Xu G Q, Xue Q, Wang H X, Xu Y B, 'An iterative thresholding algorithm for the inverse problem of electrical resistance tomography', *Flow Meas. Instrum.*, 2013, **33**, pp 244–250

[25] Hansen P, 'Analysis of discrete ill-posed problems by means of the L-curve', *SIAM Rev.*, 1992, **34** (4), pp 561–580

[26] Wang Q, Wang H X, Cui Z Q, Xu Y B, Yang C Y, 'Fast reconstruction of electrical resistance tomography (ERT) images based on the projected CG method', *Flow Meas. Instrum.*, 2012, **27**, pp 37–46

[27] Sharifi M, Young B, 'Electrical resistance tomography (ERT) applications to chemical engineering', *Chem. Eng. Research and Design*, 2013, 9I, pp 1625–1645

[28] Xu Y B, Wang H X, Cui Z Q, Dong F, 'Application of electrical resistance tomography for slug flow measurement in gas/liquid flow of horizontal pipe', *IEEE International Workshop on Imaging Systems and Techniques*, Shenzhen, China, 11–12 May 2009

[29] Yang C Y, Wang H X, Cui Z Q, 'Application of electrical resistance tomography in bubble columns for volume fraction measurement', *IEEE International Instrumentation and Measurement Technology Conference*, Graz, Austria, 13–16 May 2012

[30] Bolton G T, Hooper C W, Mann R, Stitt E H, 'Flow distribution and velocity measurement in a radial flow fixed bed reactor using electrical resistance tomography', *Chem. Eng. Sci.*, 2004, **59**, pp 1989–1997

[31] Sharifi M, Young B, '3-Dimensional spatial monitoring of tanks for the milk processing industry using electrical resistance tomography', *J. Food Eng.*, 2011, **105**, pp 312–319

Huaxiang Wang is a Professor in the Department of Automation, School of Automation and Information, Tianjin University, Tianjin city, P.R. China. He has been working at Tianjin University since his graduation. He is a Senior Member of the IEEE, Member of International Advisory Committee of Industrial Process Tomography. He is in charge of a national key discipline 'Detection technique and automatic device'. His main research interests are in the area of sensing and information processing, process measurement and control system, electrical tomography and multi-phase flow measurement. He has published a monograph 'Electrical Tomography', many textbooks and nearly 300 papers. He has completed many research projects. He has many patents and awards for his technical achievements.

Chapter 11

Electrical tomography for medical applications

Zhen Ren[1] and Wuqiang Yang[1]

List of abbreviations

ARDS	acute respiratory distress syndrome
ART	algebraic reconstruction technique
CBCT	cone-bean computed tomography
CIN	cervical intraepithelial neoplasia
COPD	chronic obstructive pulmonary disease
FMMU	Fourth Military Medical University
FPCB	flexible printed circuit board
FTB	femoral test bed
ET	electrical tomography
ECT	electrical capacitance tomography
EEG	electroencephalogram
EIS	electrical impedance spectroscopy
EIT	electrical impedance tomography
EMG	electromyography
EMT	electro-magnetic tomography
ERT	electrical resistance tomography
fps	frames per second
IC	integrated circuit
MRI	magnetic resonant imaging
NOSER	Newton's one-step error reconstruction
PEEP	positive end-expiratory pressure
P-FCB	planar-fashionable circuit board
PMMA	polymethyl methacrylate
RCT	root canal treatment

[1]Department of Electrical and Electronic Engineering, The University of Manchester, UK

SNR	signal-to-noise ratio
TCI	Technology Commercialization International
THR	total hip replacement

11.1 Introduction

Electrical tomography (ET) is based on measuring electrical impedance, that is resistance, capacitance and inductance, and generating a cross-sectional image representing the distribution of electrical property, that is resistivity (or conductivity), permittivity and permeability. Accordingly, there are three ET modalities, that is electrical resistance tomography (ERT), electrical capacitance tomography (ECT) and electro-magnetic tomography (EMT). All three ET modalities may be called electrical impedance tomography (EIT). However, many people regard EIT being ERT only, which is a confusion.

ET is a non-intrusive imaging technique, that is it can generate an image of the inside of a subject being visualised/measured, without intrusion into the subject, like X-ray tomography. ECT and EMT are also non-invasive, that is without direct contact with a subject being visualised/measured. However, ERT is invasive because traditionally a current is injected into a subject being visualised/measured, which must be contacted by electrodes.

ET was first developed by a research group led by Professor Brown at University of Sheffield in the UK in 1980s for medical applications [1]. Almost at the same time, the Process Tomography Group led by late Professor Beck at University of Manchester Institute of Science and Technology (UMIST) in the UK started to develop ECT, followed by ERT and EMT, but mostly for industrial applications, such as to visualise and/or measure gas/oil/water flows [2], fluidised beds in the pharmaceutical industry [3] and for clean coal combustion [4], gas/solids pneumatic conveyors [5] and liquid/liquid mixing processes [6]. Since then, ET has attracted many researchers worldwide working in this research field.

While ET has found many industrial applications as mentioned above, ET does have potential to be used for medical imaging because various lesions are related to significant change in electrical impedance that can be detected and visualised by ET. In addition, ET has many advantages over other imaging modalities, such as X-ray tomography, including non-radiation, fast response and low cost, making ET an attractive technique for medical applications. So far, ET has found various medical applications, including lung imaging, mammography and brain function imaging. Some ET sensors and instruments have been commercialised and used in various clinical trials with encouraging results. Figure 11.1 shows the number of publications on ET for the three medical applications mentioned above over the past three decades. The increase in the number of publications year by year indicates that ET has attracted more and more attentions from research institutions and companies.

The chapter is organised into seven sections. Because many publications are available in ET principle, sensor and system design, and industrial applications, Section 11.1 of this chapter only introduces ET for medical applications, including

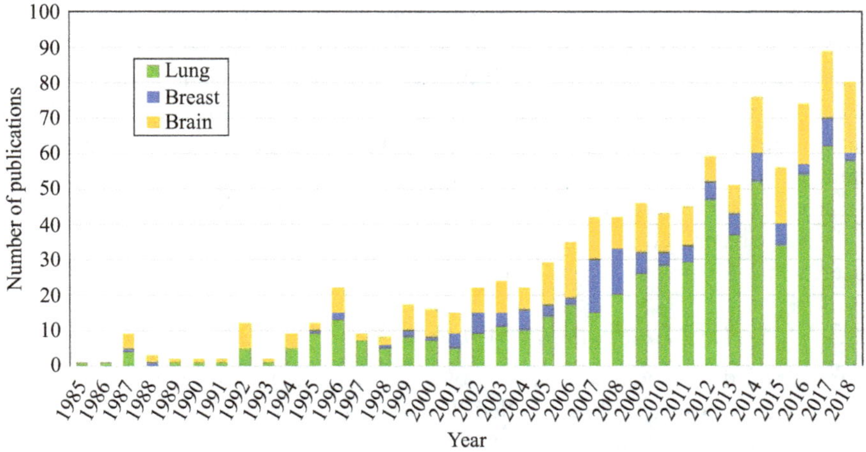

Figure 11.1 Number of publications on EIT for brain, breast and lung imaging in peer-reviewed journals from 1985 to 2018. Source: PubMed database of the US National Library of Medicine and National Institutes of Health

imaging of breast cancer, cervical, prostate and colorectal cancer, lung, brain and navigation surgery. Section 11.2 introduces ET mammography by different institutions. Section 11.3 summarises recent research into cervical, prostate and colorectal cancer detection by ET. Section 11.4 reviews ET systems for lung imaging and summarises the strength and limitation of current ET lung imaging systems. Section 11.5 is about ET for brain function imaging and neuro-science. Section 11.6 introduces two image-navigated-surgeries based on ET, which are root canal treatment and revision total hip replacement. Finally, Section 11.7 summarises the chapter.

11.2 Breast imaging

Among various types of cancer, breast cancer is the most dangerous for women with an incidence rate of 0.13%, which is much higher than lung and bronchus cancer with the second-highest incidence rate of 0.05%. Breast cancer is also the most cause of death in women. According to a recent cancer statistic report [7], over 268,670 new cases of breast cancer are estimated in the United States in 2018. It is not surprising that 99% cases of breast cancer are with women. The estimated deaths in the United States in 2018 are 41,400. According to this report, the number of breast cancer increases year by year. In contrast, the number of lung cancer and colon cancer show a decreasing tendency from 1995 due to the modern diagnosis techniques and therapy. In terms of cancer treatment, no matter what cancer, it follows one rule that the earlier the detection and treatment, the more lives can be saved. The purpose of breast imaging falls into detection of early stage of breast cancer.

ET has been used as an imaging technique for breast cancer screening. It is safe and easy to implement, whilst the commonly used imaging techniques, such as X-ray mammography has risk of radiation. Magnetic resonant imaging (MRI) mammography is barely used due to the economical reason. Research into breast cancer detection by electrical impedance measurement started as early as 1920s, when researchers launched various *in vitro* impedance measurement on normal and cancerous human breast tissue [8–10]. The early work established the fundamental for breast cancer detection by ET. Since then, a number of institutions started research into ET mammography systems.

11.2.1 TUGEM measuring system

Wtorek *et al.* discussed the possibility of using ET as a type of mammography and they built a measuring system called TUGEM (Technical University of Gdansk Electro-impedance Mammograph) for 3D phantom study [11,12]. The TUGEM system employs 64 electrodes, which are arranged into four layers and placed in a fixed position into a hemispherical measurement chamber. A DSP-based measuring unit provides two excitation strategies, either injecting currents for excitation and measuring voltages for detection, or applying voltage for excitation and measuring current for detection. Phantoms were set up with cucumber and potassium chloride (KCl) solution, Plexiglas and aluminium cylinders, and saline solution. All phantom measurements were performed at 10 kHz and images were reconstructed based on regulated correction frequency algebraic reconstruction technique (ART) [13]. Nowakowski *et al.* claimed that the TUGEM system has many advantages. For example, the use of fixed geometry can minimise problem caused by the uncertainty of sensor geometry. However, the fixed geometry doesn't fit all testing subject. Good contact between measurement electrodes and skin cannot be confirmed if an unfitted shape of testing subject is undertaken [14].

11.2.2 Duke University's system

Duke University designed a similar ET sensor but with more electrodes for breast cancer imaging [15]. The sensor consists of 128 electrodes, which are arranged in seven layers mounted on the surface of a cone shape chamber as shown in Figure 11.2.

A current-driven system was built with a function generator and a computer-controlled digital multi-meter. All measurements were made at 10 kHz. 3D images were generated using a regularised Gauss–Newton method. The system was evaluated by phantom studies for imaging a cylindrical target and a spherical target placed along the centreline of the applicator. Although good signal-to-noise ratio (SNR) of 90 dB and good special resolution were claimed, similar to the TUGEM system, no clinical trial can confirm its capability for breast cancer detection practically [11].

11.2.3 Dartmouth EIT system

Hartov *et al.* built a multi-channel multi-frequency electrical impedance spectroscopy (EIS) measurement system, which is Dartmouth's first-generation EIT

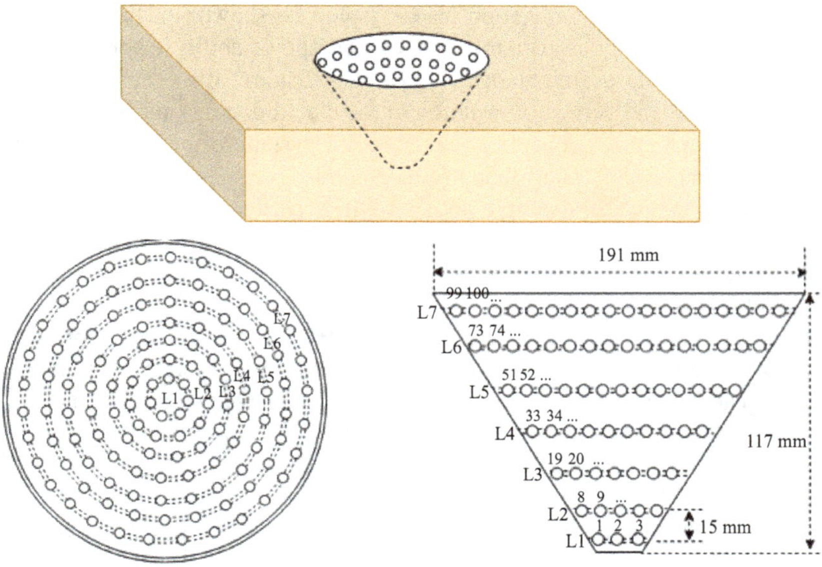

Figure 11.2 Duke University's system for breast cancer imaging. Graphic sketch (top) of ET sensor. Top-view (bottom-left) and side-view (bottom-right) of the graphic sketch of the sensor

system for breast imaging [16]. It was designed for diagnosis of tissue pathology since the most bio-tissues are frequency-dependent. The prototype has 32 channels, enabling the control of up to 32 electrodes, and provides a continuously selectable driving frequency up to 1 MHz. The system can be operated under either voltage source mode or current source mode and both modes have a good SNR. The performance of this EIT system was evaluated with phantoms, which were made of a plastic circular tank filled with 0.9% aqueous sodium chloride (NaCl) solution [17]. 2D absolute images were reconstructed using a Newton–Raphson method [18]. Initial clinical trial with this system was conducted with 25 breast examinations on 13 women participants [19]. In the *in vivo* experiment, examinations were performed under the voltage source mode with the maximum amplitude limited to 1 V, at ten different frequencies.

Before testing, a conductive gel was applied to the surface of the Ag-AgCl solid-disc electrode to improve coupling and to reduce the contact impedance. During testing, participants lied prone on an examination table with the breast to be pendant in an electrode array located below the table. An adjustable height of the array enables the ET sensor to provide information from different cross-sectional planes within the breast. After the experiment, 2D images were compared between mammographically normal cases, fluid-filled cyst and tumour cases. Although some encouraging results were obtained to identify some pathological conditions, the system has a number of practical limitations in the electrode array and exam reproducibility when it was used for breast imaging [20].

Dartmouth's second-generation EIT system is shown in Figure 11.3. It improves the design of the patient interface and shows ability to provide a consistent and repeatable examination [21]. The key feature of the new design is to enable both radial and vertical translation of the electrode array mechanically. In a later study, Kerner *et al.* deployed an updated EIT system for breast imaging of 26 female participants and found that useful diagnostic values can be probed at a high-frequency range, for example above 1 MHz, particularly in the permittivity image [21]. Although Dartmouth's first- and second-generation EIT systems for breast imaging showed some positive outcomes, both systems can produce 2D images only.

To obtain more useful information by EIT, Halter *et al.* further improved the system design and built Dartmouth's third-generation EIT system for breast imaging [22]. The system was developed based on DSP and evaluated by both phantom studies and clinical trials on 96 female volunteers [23]. The key improvements of this system are in two aspects: (1) it has more electrodes (four layers of 16 electrodes) enabling 3D imaging, and (2) broadband frequency spanning from 10 kHz to 10 MHz. The system showed a good capability of visualising different tissue types, particularly at a high-frequency range. A follow-up study was carried out to detect dynamic change within the breast that is associated with cyclic blood-volume changes. Nineteen women (ten with cancer and nine without) were imaged by EIT over the measurement of heartbeats in synchrony with pulse-oximetry acquisition [24]. Eight parameters related to EIT images and pulse-oximetry measurement were extracted and used to differentiate tumorous tissue from normal tissue. Seven out of eight parameters successfully traced the differences. The change in dynamic conductivity within a breast can provide additional physiological information, which is superior to those static and absolute breast imaging [24]. Most recent research at Dartmouth College focuses on mathematical model

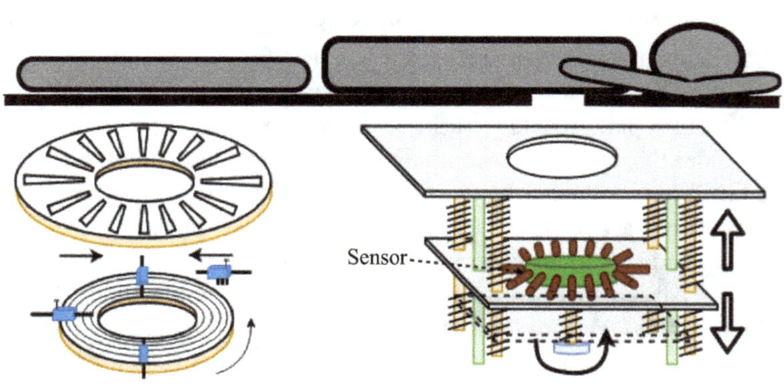

Figure 11.3 Dartmouth's second-generation EIT system for breast imaging. Graphic sketch (top) of the patient lying prone on the examination table. Graphic sketch of the radial (bottom-left) and vertical (bottom-right) translation stage located under the patient platform

and reconstruction algorithms to improve the spatial resolution and image accuracy [25,26].

11.2.4 Commercial EIT systems for breast cancer detection

A prototype 3D EIT mammography device was developed by Cherepenin *et al.* [14]. This device was patented by Technology Commercialization International (TCI), Inc., USA. The system consists of a compact array of 256 electrodes, 2 additional remoted driving electrodes, which are connected to a 10 kHz AC source and an output multiplexer for selection of measuring electrodes from the electrode array. The electrodes are arranged as a square matrix and supported on a rigid plane. Similar designs were adopted by Rensselaer Polytechnic Institute [27], Kyung Hee University [28] and University of Sussex [29,30] with some success in their phantom or clinical studies. During breast scanning, the electrode plane is pressed and flattened the breast towards the chest, to ensure the connection of electrodes with the breast and to reduce the thickness of the measured tissue layer. To compromise computational time, the system used a weighted back-projection method to reconstruct images, showing the region in parallel to the electrode array. Twenty-one female participants with breast cancer were examined by the TCI EIT mammography. Eighty-six percent of examinations were found to be fully or partially agreed with X-ray mammography [14].

The design of EIT mammography was updated for visualising mammary gland, which is associated with different hormonal status [31]. The key features of changes are (1) the shape of electrode array is changed to a round matrix, (2) a higher excitation frequency, from 10 kHz to 50 kHz, (3) transition to digital signal processing and (4) an updated weighted back-projection method. The second-generation system was evaluated by *in vivo* experiment with 57 female participants. The clinical results show that the updated system has good reproducibility and ability of visualising breast anatomic structure [31].

The proposed system shortly became a commercial EIT mammography product, as shown in Figure 11.4, for clinical use and is manufactured by Impedance Medical Technology (IMT) Company, LLC, Russia. The system has been used for

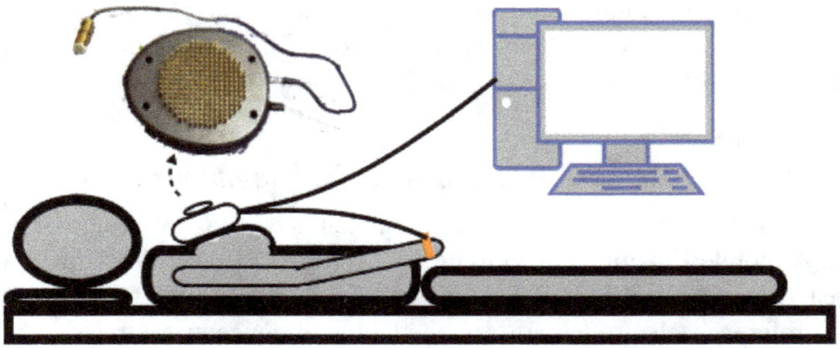

Figure 11.4 Commercial EIT mammography system MEIK device by IMT

various clinical activities worldwide [32,33]. According to a clinical survey in 2012 [34], diagnosis can achieve as accurate as 85.6%, based on 808 cases (i.e. examined patients). In terms of sensitivity, EIT mammography reaches 87%, slightly lower but close to X-ray mammography (89%) and ultrasonography (91%). Combining EIT with MMG or USG can increase the sensitivity to 96% and 98%, respectively. Although EIT shows encouraging results for early detection of breast cancer, it cannot replace MMG or USG, as it lacks of structural information, which is essential to a medical doctor [34].

Another commercial EIT mammography is TS2000 system from TransScan Medical Ltd, Israel. This system is based on T-Scan™ technology. It does not provide tomogram but an impedance map over the surface of a breast [35]. TS2000 consists of a hand-held reference electrode and a measuring probe pressed against a breast during the examination. The measuring probe contains electrodes arranged on a rectangular gird, which is similar to TCI first-generation system [14]. The system has three probes with different overall size and different number of electrodes: (1) small, 8 × 8 electrode array with total footprint of 32 × 32 mm, (2) large, 16 × 16 electrode array with footprint of 72 × 72 mm and (3) optimal, 16 × 16 electrode array with footprint of 32 × 32 mm. During examination, an AC voltage source with frequency spanning from 100 Hz to 100 kHz is applied to the hand-held electrode and the evoked current is measured from the electrodes on the measuring probe. Because cancerous tissue has a higher impedance than the normal tissue, enhanced current signals can be measured by the electrodes close to the lesion. In a grey-level image, the tumour can be observed as a focal white spot [35].

The TS2000 system has been used in clinical studies with 504 bioposied breast samples, which consists of 179 malignant and 325 benign findings [36]. The system was found to be a good adjunct to X-ray mammography with an improved specificity from 39% to 51% and enhanced sensitivity from 82% to 88%. Compared to ultrasonic and MRI, the overall sensitivity of EIT as an adjunct to mammography is higher than the sensitivity of ultrasonic but lower than MRI [37]. TS2000 was granted U.S. Food and Drug Administration (FDA)'s pre-market approval in 1999 for adjunctive use with X-ray mammography [36]. However, TS2000 has some limitations, such as (1) false positive by artefacts, for example air bubbles, bones, muscles and superficial skin lesions, which result in high conductivity measurement, (2) limited maximum depth of measurement (about 3–3.5 cm) and thus it cannot detect lesion in a deeper region or close to the chest wall and (3) no accurate localisation of positive lesion for biopsy [38].

11.2.5 Wearable smart device-based EIT mammography

The development of mobile technology stimulated new ideas for design of imaging systems. A mobile EIT integrated circuit (IC) for breast cancer detection was built by Korea Advanced Institute of Science and Technology [39]. Ninety electrodes plus two reference electrodes are arranged in five concentric circles, which are patterned on fabric using planar-fashionable circuit board (P-FCB), as shown in Figure 11.5.

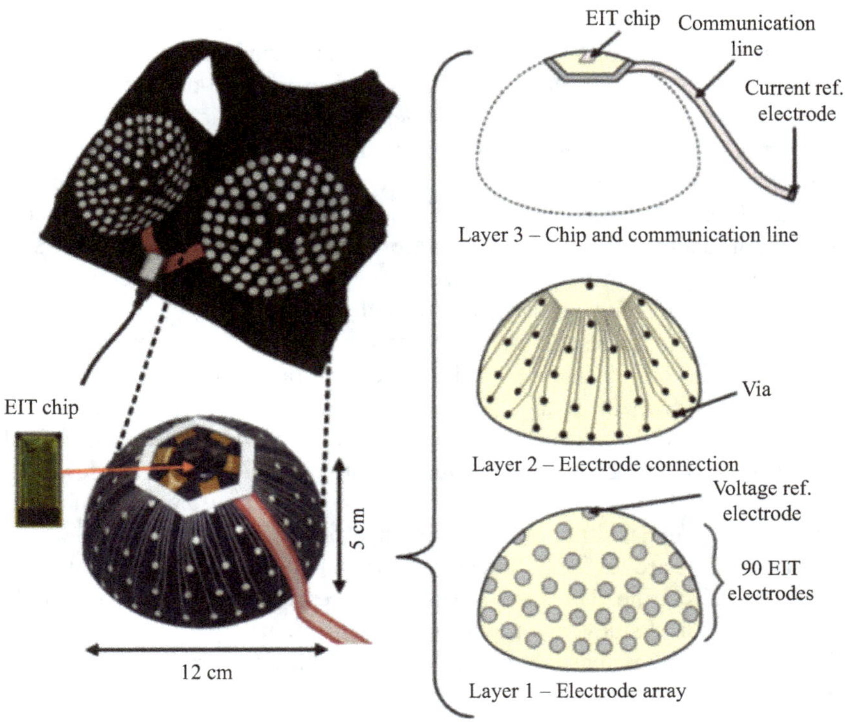

Figure 11.5 Wearable EIT mammography sensor by Korea Advanced Institute of Science and Technology [39]

The electrodes are driven by a current stimulator in a frequency range of 100 Hz to 100 kHz. 3D images are generated by a weighted back-projection algorithm and displayed on a portable smart device, for example a smartphone, which is connected to EIT IC via USB port. A good SNR of 90 dB is claimed. The fabric sensor design ensures the connection of electrodes to skin. The system was evaluated by a phantom model, using X-ray mammography as a reference. The result shows that the proposed system has good spatial resolution with a detectable tumour size of 5 mm, corresponding to stage 1 breast cancer [39]. Up to date, no *in vivo* experiment was carried out with this system.

11.3 Other cancer research: cervical, colorectal and prostates cancer

11.3.1 Cervical cancer

EIT mammography inspires research into various cancer detection. One of them is cervical cancer, which is the second most frequent cancer and the fourth common cause of death in women. Abdul *et al.* used a pencil probe to measure impedance of

8 points on the cervix of 176 female participants [40]. The probe is 5.5 mm in diameter and has four gold electrodes mounted on the surface at the top. Without generating an image, the probe measures cervical impedance-derived parameters, which are used to differentiate cervical intraepithelial neoplasia (CIN) from normal tissues and provides an immediate result. Sensitivity of 74% and specificity of 53% can be achieved from the clinical diagnosis. A commercial product called ZedScan from Zilico Ltd, UK is based on this study and is used as an adjunct to colposcopy. Clinical report claimed that 474 women were examined [41]. The result shows that as a real-time adjunct to colposcopy, ZedScan can improve colposcopic performance in terms of positive predictive value and specificity.

Korjenevsky led a research group in Moscow and launched a similar work and built a probe of 3 cm in diameter with 32-electrode array on a flat surface [42]. Measurements were made at an adjustable frequency from 1 to 50 kHz. Different from ZedScan, this system enables 3D images to be reconstructed with a back-projection algorithm. The prototype system can produce a 3D image in a few cross-sections with depth up to 8 mm, showing the cross-section anatomy of cervix. Cherepenin *et al.* updated the sensor design and built a new sensor array with 48 non-orthogonal electrodes [43]. This design can minimise noise and cross-talk and is easy in manufacturing. Three cross-sectional slices are available with depth of 2, 5 and 8 mm. Clinical study was conducted with 170 female participants [44]. The result shows that EIT is sensitive to high degree of squamous intraepithelial lesion and suspicion of invasion [44].

11.3.2 Colorectal cancer

Colorectal cancer is another high frequent type of cancer, which is the third-highest cancer death rate in both men and women, ranking after lung and bronchus cancer, breast cancer of females and prostate cancer of males. A recent study used ZedScan to investigate whether EIT can tell colorectal cancer from normal colonic mucosa [45]. An *ex vivo* experiment was carried out with 22 fresh colorectal specimens, of which 19 were histopathologically-proven colorectal cancer specimens. Promising results were obtained from the feasibility study, showing that electrical impedance of cancerous tissue is higher than the benign tissue under all scanning frequency from 76 Hz to 625 kHz. However, further investigation is needed to support such finding because *ex vivo* tissue cellular structure begins to change, resulting in the change in electrical properties of the tissues.

11.3.3 Prostate cancer

Approximately 164,690 new cases of prostate cancer are estimated in the United States in 2018, where 129,430 men are expected to die from it [7]. Traditional blood test for prostate cancer screening lacks sufficient specificity to diagnose the lesion. Ultrasound-guided biopsy may fail when the tumorous tissue is not sampled during the biopsy procedure.

Proven evidence shows that electrical impedance of malignant prostatic tissue is significantly different from that of benign tissues within the gland [46,47], implying the capability of EIT to serve as an imaging technique for biopsy guidance to help with the sampling problem. A research group at Dartmouth College designed a current-drive tetra polar EIT configuration coupled with a transracial ultrasound probe of 2.28 cm in diameter [48,49]. The EIT sensing array has 30 electrodes and is attached to the surface of the ultrasonic probe, which is arranged to open an acoustic window for ultrasonic energy transmission. Prior knowledge from ultrasound imaging is used for EIT reconstruction/parameter estimation for enhancing sensitivity along the sensing probe. The design of the probe was evaluated by numeral analysis with FEM modelling and reconstruction algorithms [50]. The proposed system can be used to image moderate contrast of 1 cm in diameter at a distance of 4.55 cm from the probe, which is sensitive enough for clinical use. So far, no clinical data were found with the proposed sensor and system.

11.4 Lung imaging

EIT has become a promising tool for monitoring lung function because the resistivity of lung tissue is more than five times larger than the other soft tissues in the thorax [51]. Because significant change in impedance occurs during inspiration and expiration, the pulmonary activity can be visualised and some ventilator-associated lung problem can be diagnosed by imaging the impedance variances. EIT is non-intrusive, fast response and does not require collaboration from patient. This is particularly useful when pulmonary activity in neonates is measured because there is no other imaging technique for pulmonary assessment [52]. In addition, the effect of aging can also be detected [53]. Generally, EIT is capable for monitoring several lung functions, such as

- *Pulmonary ventilation*: EIT can be used to measure spontaneous regional response to positive end-expiratory pressure (PEEP) [54–57] or a recruiting manoeuvre [58,59], providing dynamic information associated with distribution of tidal volume and intrathoracic gas volume [60–62].
- *Heart activity and lung perfusion*: EIT can be used to detect regional change in both air and blood content as well as the periodic changes in blood vessel dimension that is associated with heart–lung interactions [63–66].
- *Pulmonary function testing*: EIT can trace spatial and temporal heterogeneity in regional ventilation over time during spontaneous tidal breathing and forced ventilation manoeuvres. EIT can potentially be used as adjunct to conventional pulmonary function tests for diseases like acute respiratory distress syndrome (ARDS) [54,67–69], acute lung injury [70,71] and chronic obstructive pulmonary disease (COPD), including emphysema, chronic bronchitis and asthma [72–75], lung cancer [76] and cystic fibrosis [77].

Many review articles were published to introduce the principle of lung EIT system [78,79], various pulmonary applications [80–83] and the current status of clinical activity [52,84–87].

Since 2011, many commercial bedside EIT lung imaging system have been released to market, such as PulmoVista® 500 (Dräger, Germany), Swisstom BB^2 (Swisstom AG, Switzerland), Enlight 1800 (Timpel SA, Brazil) and Elisa 800^{VIT} (Heinen+Löwenstein GmbH & Co KG, Germany). Elisa 800^{VIT} is the first EIT system that integrates EIT function with a respirator [88]. All of those commercial bedside EIT systems employ similar configurations and functions. As an example, Figure 11.6 shows the Swisstom BB^2 system. For all those bedside EIT systems, a small current is injected to electrodes at a single frequency, range from 80 kHz to 150 kHz. Those systems all use a 32-electrode sensor on a belt, except for PulmoVista® 500, which uses a sensor of 16 electrodes plus a reference electrode.

In addition to those commercial bedside devices, many universities have developed commercial products or lab prototypes for pulmonary EIT research. Those systems provide flexibility of using different data acquisition strategies,

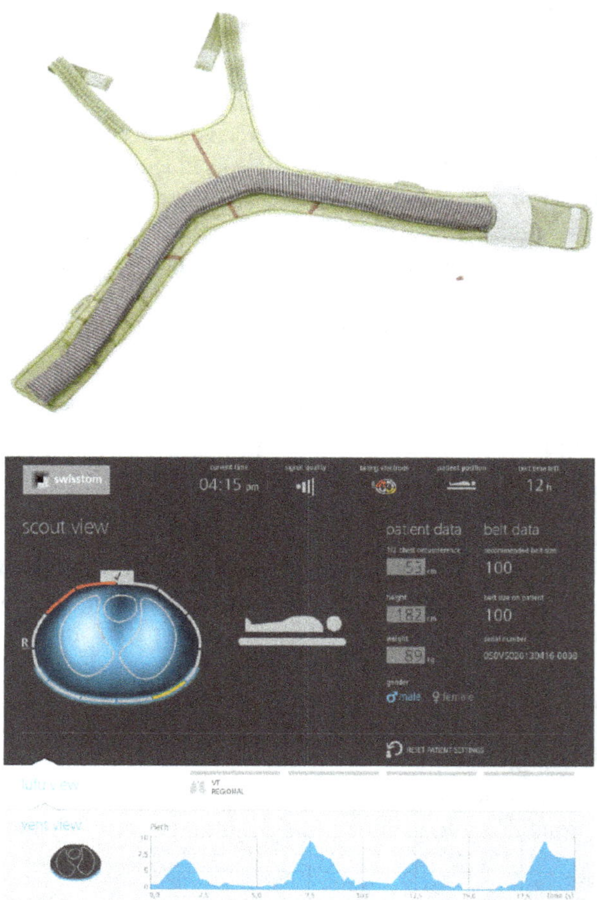

(a)

(b)

Figure 11.6 Swisstom BB^2 system: (a) sensor belt and (b) GUI [89] image reconstruction methods and quantitative analysis [90–97]. Most of those systems rely on single frequency excitation, such as Sheffield first and second-generation systems [90,91], Sheffield APT system DAS-01P [98], Rensselaer system ACT2 and ACT3 [92,93], Cardiff EIT system [99], Montreal EIT system [94], Keele tomograph system [95], Manchester fEITER system [100] and HUT EIT system [101]. The others use a multi-frequency strategy, for example Gottingen EITS system [96], which is the predecessor of Geo MF II (CareFusion, Germany) and PulmoVista® 500, Sheffield third-generation system [97] and its successive commercial system: Sheffield MK 3.5 [102], a wearable EIT system by a research group in South Korea [103] and Edinburgh multi-frequency EIT [104]. Table 11.1 summarises those commercial or research based lung EIT systems and lists their key features and the selected clinical activities or phantom studies.

An advantage of using multi-frequency excitation is that multi-frequency can offer a functional surveillance of tissues [52]. EIT systems provide either dynamic or statics images of lung. Dynamic lung images showing the relative impedance change with respect to a reference data set are generated because the algorithms used to reconstruct impedance change are less sensitive to noise compared with the algorithms for static images showing absolute impedance values [133]. In most cases, static images with a single frequency are difficult to obtain. On the other hand, using multi-frequency excitation, a quasi-static image can be generated because two data sets can be measured at different frequencies and an algorithm can be used for dynamic imaging. Additional physiological information can be obtained because the impedance of different tissues has different response to frequency [84]. The most recent developed systems, for example Wearable 3D EIT system [103] and Edinburgh multi-frequency EIT [104], offer two measurement modes: time-difference mode and frequency-difference mode. With frequency-difference mode, an optimal frequency range can be found for time-difference mode to achieve high image contrast for local lesions [103].

As the spontaneous response is one of the most significant features of a pulmonary EIT system, the most often used image reconstruction algorithms are back-projection [95,97,120] and Newton's one-step error reconstruction (NOSER) [126]. Although 2D 'slices' from a single layer of electrodes are commonly produced by most pulmonary EIT systems, 3D version of a volume provides more accurate information in the chest region. For many years, 3D EIT images are produced offline only, due to the limitation of the computational and analytical tools. In recent years, EIT systems with multiple layers of electrodes are reported to provide dynamic 3D information in the chest region with high SNR [94,101,103,104]. In the meantime, many 3D measurement strategies and image reconstruction methods are introduced with superior performance to the 2D slice [132,134].

For lung diagnosis, EIT has many advantages over classical pulmonary function tests, including

1. It is non-intrusive and does not require mouthpiece. Thus, it offers independency on conventional examination site and continuous assessment of

Table 11.1 List of pulmonary EIT systems

Devices	Organisation	Key features	Notes
PulmoVista®	Dräger, Germany	• 16 electrodes plus 1 reference electrode on silicone belt • 10–30 fps (4 options) • Single frequency selected in a range of 80–130 kHz • 2D image, Newton–Raphson method	• Commercial bedside device • Clinical activities [54,105–108] • Animal trails [109]
Swisstom BB^2	Swisstom AG, Switzerland	• 32 electrodes plus a reference electrode on textile belt • Up to 50 fps • Single frequency at 150 kHz • 2D image	• Commercial bedside device • Clinical activities [110,111]
Elisa 800^{VIT}	Heinen+Löwenstein GmbH & Co KG, Germany	• 32 electrodes plus a reference electrode on textile belt • Up to 50 fps • 2D image	• Commercial bedside device • The first fully integrated EIT system in an intensive care ventilator [84]
Enlight 1800	Timpel SA, Brazil	• 32 electrodes on silicon belt • Single frequency at 125 kHz • Up to 50 fps • 2D image	• Commercial bedside device • Clinical activities [112,113] • Animal trails [114]
Sheffield MK 3.5 [102]	Maltron International Ltd, UK	• Eight electrodes • Multi-frequency between 2 kHz and 1.6 MHz • 25 fps • SNR original specification at 60 dB and at 40 dB for small amplitude signals with saline solution • 2D image, back-projection	• Commercial device • University-based prototype (Sheffield University) • Clinical activities [115,116]

Sheffield APT system MK1 [90]	Sheffield University, UK	• 16 electrodes • Single frequency at 50 kHz • Up to 10 fps • 2D image, back-projection	• University-based prototype • Clinical activities [117–119] • Animal trails [120]
Sheffield APT system MK2 [121]		• 16 electrodes • Single frequency at 20 kHz • Up to 25 fps • 2D, back-projection	• University-based prototype • Clinical activities [91]
Sheffield APT system DAS-01P [98]		• 16 electrodes • Single frequency at 50 kHz • Up to 50 fps	• University-based prototype • Clinical activities [69,122–124]
Sheffield EITS system [97]		• 16 electrodes • Multi-frequency selected from 9.6 kHz to 614.4 kHz • 2D image, back-projection	• University-based prototype • Clinical activities [97,125]
Rensselaer system ACT2 [92]	Rensselaer Polytechnic Institute, USA	• 32 electrodes • Single frequency at 15 kHz • 2D image, NOSER	• University-based prototype • Clinical activities [126]
Rensselaer system ACT3 [93]		• 32 electrodes • Single frequency at 30 kHz • 2D image, NOSER	• University-based prototype • Animal trails [127]
Cardiff EIT system [99]	University Hospital of Wales, UK	• 16 electrodes • Single frequency at 40 kHz	• University-based prototype • Clinical activities [99]
Montreal EIT system [94]	Ecole Polytechnique et Universite de Montreal, Canada	• 64 electrodes (3 layers) • Single frequency selected from 800 Hz to 30 kHz • 2D image, regularised linear reconstruction	• University-based prototype • Clinical activity [128] • Animal trail [129,130]
	Keele University, UK	• 16 electrodes	• University-based prototype

(Continues)

Table 11.1 (Continued)

Devices	Organisation	Key features	Notes
Keele tomograph MK 1b [95]		• Single frequency at 20 kHz • 2D image, back-projection	• Clinical activity [95]
Goe MF II	CareFusion, Germany	• 16 electrodes and a reference electrode • Single frequency at 50 kHz • Up to 50 fps • 2D image, back projection	• Commercial device, University-based prototype (University of Göttingen) • Clinical activities [57,73,74,131]
Manchester fEITER system [100]	The University of Manchester, UK	• 32 electrodes • Single frequency at 10 kHz • Up to 100 fps • Low noise level	• University-based prototype, originally designed for brain function imaging • Clinical activities [132]
Wearable EIT system [103]	Korea Advanced Institute of Science and Technology, South Korea	• 48 electrodes (3 layers) on belt • Multi-frequency selected from 4 kHz to 128 kHz • SNR of 94 dB • Up to 10 fps • 3D image, Gauss–Newton iteration	• University-based prototype • Phantom study [103]
Edinburgh multi-frequency EIT [104]	The University of Edinburgh, UK	• 32 electrodes • Multi-frequency selected from 10 kHz to 1 MHz • Two modes: serial (up to 546 fps) and semi-parallel (up to 1014 fps) • SNR from 45.51 to 82.82 dB with 16-electrode sensor • 3D image at 120 fps, using Tikhonov regularisation	• University-based prototype • Phantom study [104]
HUT EIT system [101]	Hebei University of Technology, China	• 64 electrodes • Single frequency at 100 kHz • 3D image, Nodal TV regularisation	• University-based prototype • Clinical activities [101]

regional lung ventilation, reducing the risk of cross-contamination and makes patient more comfortable during examination [73,74,131].

2. Most of the parameters required in pulmonary function test can be measured with reasonable accuracy by indirect measurement with good reproductiveness [115,135].
3. EIT image is sensitive to small incident pneumothoraxes, but insensitive to other classical bedside parameters, such as tidal volume, respiratory system compliance or system arterial pressure [136].

However, EIT for lung imaging has some limitations, such as

1. Current 2D pulmonary EIT images can only present a slice of ventral-to-dorsal regional ventilation distribution and the sensitivity is limited to a single transverse thoracic plane of around 6 cm above and below the plane [54,105].
2. It can only provide low spatial resolution, compared to MRI and CT [73,74,78,82].
3. It is highly sensitive to the amount of air within the thorax and hence the change in air content can overlap with the change in body fluids [78].
4. It has limitation in studying regional lung function during large volume change that is related to the effects of chest expansion and organ/tissue movement [74].
5. Accuracy of EIT measurement is dependent on placement of the electrodes on the chest. Thus, the feasibility of long-term monitoring by EIT requires further investigation [80,82].

11.5 Brain function imaging and neuro-science

Brian function imaging or neuro-imaging is one of the most challenging, but attractive topics among researches into imaging techniques. Brain activities, such as evoking regional cerebral blood flow and volume changes, result in impedance change of the local active cortical area, which can be measured by EIT. While EIT can offer non-intrusive measurement and fast response for recording brain functions with a resolution of a few millisecond, the traditional image techniques, such as CT, MRI and PET, cannot permit long-time monitoring/imaging with fast response, and neither electroencephalogram (EEG) nor electromyography (EMG) has unique inverse solution to produce image [137].

Two types of electrodes are used for EIT brain function imaging. One is intracranial electrodes, which are placed on the exposed surface of cerebral cortex. The other is extracranial electrodes or scalp electrodes, which are placed on the subject's head. The intracranial electrodes use either ring [138], planar [137,139,140] or flexible cylindrical multi-electrode cuff structure [141], while scalp electrodes are normally used in EEG [142–146]. Practically, it is more challenging to obtain satisfactory brain function image by EEG electrodes, even with optimal head-net electrode array [147,148]. The main difficulty is the insufficient sensitivity to the small impedance change in the brain due to the diluted signal by diversion of current injected to the skull. The movement artefact in awake subjects and incorrect electrode position are also unavoidable factors that reduces overall SNR. In addition, the noise normally dominants the recorded impedance change and the expected impedance changes related to brain function would be hidden. By using intracranial electrodes, measurements can target a specific region. Thus, the sensitivity can be increased and movement artefact can be avoided. However, using intracranial electrodes, the EIT measurement is no longer non-intrusive and is limited to be used on human body [149].

Frequency plays a critical role in producing high quality EIT brain images. Different types of tissue have different response over frequency. Impedance change occurs due to a combination of cell swelling and blood flow/volume. The former occurs during intense brain activity, causing water to move from the extracellular to intracellular compartment and the extracellular space shrinks, and thus results in impedance increase. This can be detected in a low frequency range, for example a few hundred Hertz [139,150]. On the contrary, as resistivity of blood is much lower than that of brain tissue, the increase in blood volume lowers the overall impedance. Using a higher frequency can larger the contrast between blood and brain tissue [139].

Due to the technical limitation, single frequency EIT systems were adopted by research into EIT brain function imaging in the early years. A number of *in vivo* experiments were conducted with animal trails (e.g. rats, rabbits and piglets) and human subjects, most of which were operated in a frequency range between 10 kHz and 100 kHz [138,143,151,152], like many other EIT systems. Impedance changes related to brain activity or stimuli were obtained, implying the potential of EIT for monitoring brain function during stroke [138], cortical spreading depression [151], epilepsy [152] or auditory related activity [143]. However, a single frequency system can only produce time-difference image, which is difficult to use in clinical practice when there is no opportunity to take a previous reference image or to produce an absolute image [153,154].

Most recent studies employed multi-frequency EIT systems for brain function imaging [137,139,154–157]. Although multi-frequency EIT systems overcome some problems like imaging of an event without prior knowledge and insensitivity at a single frequency [153], few clinical activities were successfully undertaken using multi-frequency EIT systems, especially with scalp electrodes. A research group at University College London (UCL) used UCLH Mk2 multi-frequency EIT system, operating in a frequency range of 16 kHz –64 kHz, with scalp electrodes to distinguish between ischemic and haemorrhagic stroke [158]. *In vivo* experiments were conducted to image seven human subjects with brain tumours, arteriovenous malformations or chronic stroke. However, the impedance change of tissue within such a frequency range is not large enough to produce high quality EIT images for distinguishing brain tissues [158].

Oh *et al.* reported a low frequency EIT system [137], which allows to record neuronal depolarisation of a rat in a frequency range of 125 to 825 Hz. High quality data can be collected but with low-speed response [137]. Avery *et al.* introduced a highly versatile EIT system – ScouseTom – operating in a frequency range of

20 Hz to 20 kHz, for various medical purposes [142]. The system was evaluated with both scalp electrode for human study and epicortial electrode array in rats with high SNR over the frequency range. However, the system requires further investigation on clinical study with scalp electrodes.

A research group at Fourth Military Medical University (FMMU) in China developed a 32-channel multi-frequency EIT system for clinical study. The system has a broad frequency range from 1.6 to 380 kHz and presents high SNR over the frequency range [144]. Fu *et al.* employed the system to monitor regional cerebral edema of 23 patients during clinical dehydration treatment and successfully detected changes in brain fluid content non-intrusively and in real-time [156]. Li *et al.* used the same system for monitoring cerebral impedance changes of 42 patients during total aortic arch replacement [159]. The results indicated that EIT has the potential to provide quantitative information about regional cerebral perfusion. The FMMU group recently developed a new system using a high-precision digital synthesis method and a digital demodulation technique in the design of electrode interface and voltmeter, which presents high noise immunity to reduce shunt effect and stray capacitance [157]. Promising results were obtained from this system, but further investigation is needed, particularly for the clinical evaluation.

Low spatial resolution is the main limitation of EIT for brain function imaging. Although in principle, more electrodes can be used to produce images of higher spatial resolution to some extent, a large number of electrodes would result in a reduced measurement speed and reduced SNR. Thus, the number of electrodes is normally determined by a compromise of spatial resolution, imaging speed and SNR. Current EIT systems normally employ 16 or 32 electrodes [139,142,144,160]. Although some systems can support a greater number of electrodes, for example UCLH Mk2 and Mk 2.5 system with 64 channels [154,161], only half of those electrodes were used in practice [153,156]. To improve image quality, an accurate head model needs to be developed for calculating the sensitivity matrix. The use of mathematical model derived from an accurate MRI of the subject's head can improve the accuracy to present regional lesion in a specific subject [140,160,162,163]. In addition, a number of image reconstruction algorithms were introduced to enhance spatial resolution and to make a system robust to noise [140,160,164–166].

11.6 Image-based surgery

11.6.1 Endodontic therapy

Endodontic therapy, or root canal treatment (RCT), is common for elimination of infection and protection of a decontaminated tooth. In the traditional approach, a dentist makes treatment plan by the use of a reference image which shows the inner structure of teeth, to preliminarily access the tooth root. Most existing dental imaging technologies for this purpose are based on X-ray, either radiography or cone-bean computed tomography (CBCT), the latter is the most advanced technology in dental treatment. However, neither of them can provide real-time images, due to the radiation concerns and the limit in the imaging speed. Thus, during the treatment, there is no existing image guidance for the dentist to perform the RCT and hence it is difficult to ensure the quality of the treatment. While millions of RCT are carried out each year, the success rate is ranged between 68% and 85% [167]. It means that about 20% of treated teeth need re-treatment.

Ren and Yang reported the feasibility of using ECT to visualise tooth surface during RCT [168]. To confirm the feasibility, a tooth clamping tool was designed as shown in Figure 11.7. This tool can secure a multi-electrode capacitance sensor onto the gum [169,170]. The sensor employs a two-plate and open structure so that it can be easily placed onto each side of the gum in the mouth. Two types of sensor

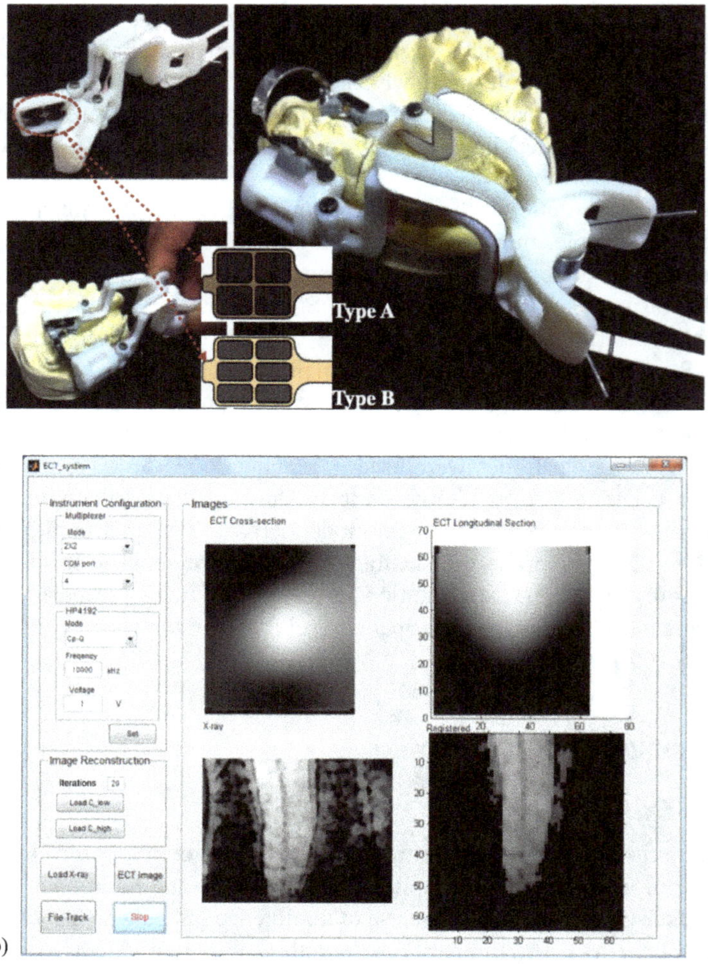

(a) (b)

Figure 11.7 (a) Tooth clamping tool and FPCB sensor and (b) GUI [169] heads on a flexible printed circuit board (FPCB) were designed. The performance of the designed sensor was fully discussed in [169]. The result shows that the two-plate sensor has similar performance to the traditional circular and closed-structure sensors. Although compared to the traditional sensor, this sensor has less sensitivity at the edge without covering electrodes, it is sensitive enough for imaging tooth surface because the tooth to be treated is normally placed in the central area [170]. An ECT system based on an impedance analyser (HP 4192A) was used. An *in vitro* experiment with a dried tooth was conducted [168]. The ECT image was fused with radiograph to acquire higher resolution of the inner structure as shown in Figure 11.7.

Then a grounded dental file was visualised on the fused tooth image by measuring capacitance between the file and electrodes [170]. A significant capacitance change occurs when the dental file approaching the apex of the tooth root, which is considered as the end-point. By the use of such point, the position of the file in the root can be estimated by mapping the relationship between the measured capacitance and the effective length of the file in the sensing area. Up to date, there is no clinical data of the use of ECT for RCT.

11.6.2 Revision total hip replacement

An aging population has led to a dramatic rise in the number of total hip replacement (THR) being carried out. According to the most recent report from National Joint Registry [171], over 90,000 THRs were recorded in the UK in 2016. Arthritis Care charity claimed that cases of osteoarthritis, which is one of the leading reasons for joint replacement, will double by 2030, leading to an even greater number of THR operations. During the THR procedure, there is always a potential of mechanical failure, which requires a revision [172]. In particular, 10 years after primary THR, many patients need revision surgery due to numbers of reasons such as deep infection, dislocation, fracture, wearing, loosening and hypersensitivity. For example, 5%–12% of the patients with cemented prostheses need to be replaced [172]. To replace a hip prosthesis, the bone cement must first be removed. This can be a difficult and intricate procedure. Conventionally, a hammer and a chisel are used to remove the bone cement manually. To do so, the lateral femur must be opened to access the operative field in the femur cavity. Currently, new surgery strategies with navigated drilling tools are being developed, attempting to achieve efficient removal of the bone cement. However, similar to RCT, a radiograph or X-ray CT image is taken before the operation starts as a reference. It is difficult to position the drilling tool very accurately because there is no real-time imaging and navigating tool available.

Brendle *et al.* developed an impedance controlled surgical instrumentation system to establish real-time control during the bone cement drilling process in revision THR [173]. The main purpose of this research is to find out the relationship between the thickness of the residual bone cement and the measured electrical impedance. By the use of the system, thickness of the residual bone cement can be estimated by measuring impedance, in that case, unintended cutting of bone can be avoided. A femoral test bed (FTB) was designed and used as a representation of the scenario for the adult human thigh. The FTB consists of 16 T-shaped electrodes placed inside of a cylindrical bowl, an off-centre anatomical femur and a carrier for polymethyl methacrylate (PMMA) dummy. In the experiment, NaCl solution was fully filled in the space between the dummy and the bowl wall as a substitute for human tissue. Electrical impedance was recorded by an Agilent E4890 LCR Meter operated in a frequency range between 200 kHz and 2 MHz. As a preliminary research, this report only focused on discussing the sensitivity of FTB to different thickness of bone cement.

In the later work by Brendle *et al.*, an updated bioimpedance-controlled surgical instrumentation was presented [174]. The key feature of this system is to use the drilling tool as a cutting tool and an electrode simultaneously. The change of impedance between the drilling tool and the electrodes on skin is dependent on the thickness of the residual bone cement. The system consists of an integrated surgical drilling tool and a rapid control system (MicroAutoBox, dSPACE GmbH, Germany). The control system records electrical impedance every 15 ms and the drilling tool will stop drilling when the measured impedance reaches a pre-set value. Experiment was carried out with agar-agar gel as a substitute for human tissue and PMMA as bone cement. The results show that the proposed system can provide a safe and accurate removal of bone cement during revision THR with residual thickness of 1 mm.

Inspired by Brendle's work, an ECT/ERT dual modality navigation system was developed at the University of Manchester, UK. The aim of this research is to obtain the 3D position of the drilling tool in femur. For this purpose, a single-plane ECT/ERT sensor with eight internal electrodes was used to generate 3D images by making use of the fringe effect and the prior knowledge of the drilling tool, like the shape and size [175]. The prior knowledge was used to adjust an original image generated by a linear back-projection algorithm.

A feasibility study was carried out by inserting a metallic rod into sensor filled with plastic beads. 3D imaging of the rod is implemented in three steps:

1. A 2D image is reconstructed and an elliptical fitting algorithm is applied to locate the cross-sectional centre of the rod on the 2D image.
2. Axial distance of the rod head from the origin is derived by calculating fringe effect in the related difference between measured normalised capacitance and normalised capacitance from the 2D simulation.
3. 3D image of the rod inside of the sensor is then reconstructed by fusing the results from (1) and (2) and prior knowledge of rod size and shape.

A follow-up study on ECT/ERT for revision THR was reported by Ren and Yang [176]. Similar to Brendle's work [174], the metallic rod is considered as a drilling tool and an electrode at the same time. The work aims to measure electrical impedance between the rod and electrodes to locate the cross-sectional and axial (depth) position of the rod in the femur. In the experiment, the ECT/ERT sensor was filled with saline solution. A cemented bone was placed in the sensor, either near the centre or at edge (i.e. close to electrodes), as shown in Figure 11.8. The

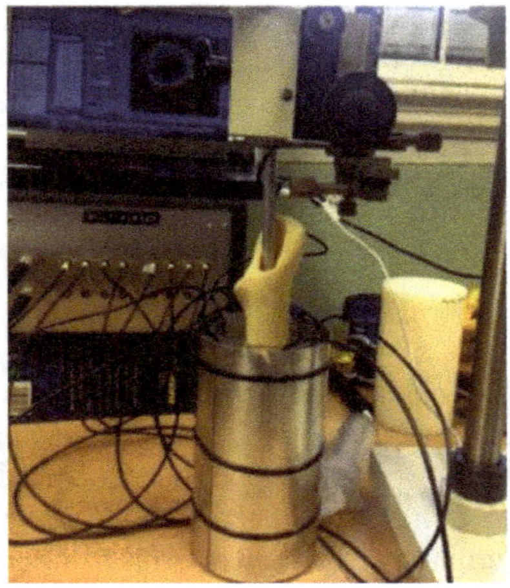

(a)

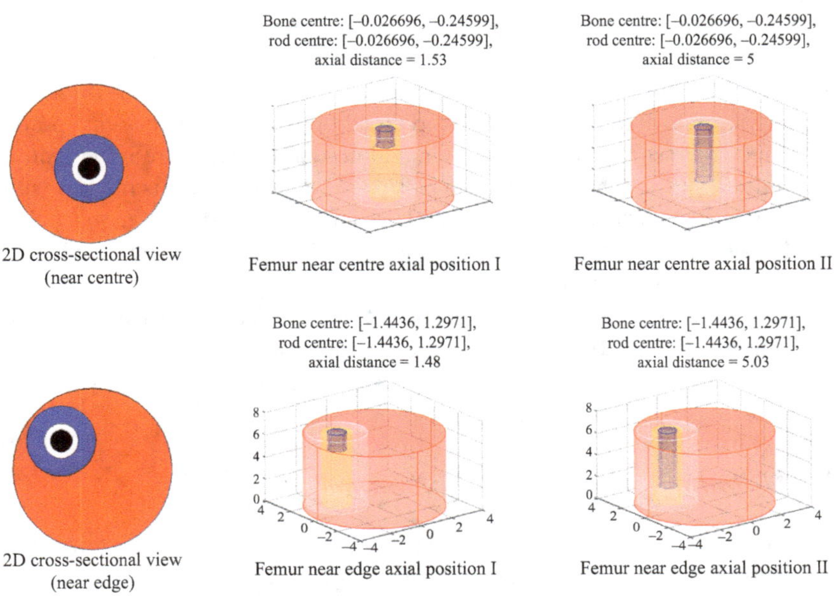

(b)

Figure 11.8 ECT/ERT system for revision THR: (a) experiment set-up and (b) tracing metal rod in cemented bone by ET [176]

bone has central hollow, which enables the rod to go into the shaft. The metallic rod, as a substitute for the drilling tool, was controlled by an XYZ stand. The cross-sectional position of the femur was calculated by a weighted mean method based on 2D cross-sectional images. The depth of the rod was estimated by mapping the relationship of axial position of the rod and the measured impedance between the rod and each electrode. The 3D image of revision THR procedure can be then reconstructed by making use of 3D position of the rod and the prior knowledge of the femur [176]. The ECT/ERT system by Ren and Yang presents high accuracy in estimation of the axial position of the drilling tool with the average absolute error less than 1 mm [176]. The feasibility of using EIT for revision THR surgery requires further investigation as there is no clinical data currently available.

11.7 Summary

EIT has good capability of obtaining images with advantages of non-intrusive, fast-response and low cost. It is expected to be an ideal imaging technique for medical applications, particularly for long-time healthcare monitoring. However, as a new imaging technique, it encounters a number of challenges when EIT is applied for clinical applications. Current medical EIT systems show common problems of low spatial resolution, insufficient sensitivity and inaccurate localisation, which may be partially improved by using novel image reconstruction algorithms and mathematical forward models from high resolution image, such as MRI and CT.

Developing multi-frequency systems is a trend for further research into medical EIT as they can provide time-difference and frequency-difference information. Operating at multi-frequencies can find optimal operation frequency that produces high image contrast to distinguish lesion and normal tissue. More importantly, by using multi-frequency the problem can be solved when an object is imaged without prior knowledge or when it is impossible to obtain baseline dataset in most clinical cases. Further investigation on multi-frequency EIT is needed to improve hardware design and frequency selection.

Most medical EIT systems rely on current-driven for safety reason. However, a voltage-driven EIT system may be more desirable as it is technically easier to implement with high precision and a broader frequency range.

References

[1] Brown B H, Barber D C, Seagar A D, 'Applied potential tomography: Possible clinical applications', *Clin. Phys. Physiol. Meas.*, 1985, **6** (2), pp 109–121

[2] Li Y, Yang W Q, Xie C G, Huang S M, Wu Z P, Tsamakis D, Lenn C, 'Gas/oil/water flow measurement by electrical capacitance tomography', *Meas. Sci. Technol.*, 2013, **24** (7), 074001 (12pp)

[3] Yang W Q, Wang H G, 'Application of electrical capacitance tomography in pharmaceutical manufacturing processes', *IEEE International Instrumentation and Measurement Technology Conference*, Auckland, New Zealand, 20–23 May 2019 (accepted)

[4] Wang H G, Li Y, Qiu G Z, Song G L, Yang W Q, 'Measurement of gas-solids flow in loop seal and external heat exchanger in a circulating fluidized bed', *Powder Technol.*, 2014, **266**, pp 249–261

[5] Yang W Q, Liu S, 'Role of tomography in solid flow measurement', *Flow Meas. Instrum.*, 2000, **11** (3), pp 237–244
[6] Mann R, Dickin F J, Wang M, *et al.* 'Application of electrical resistance tomography to interrogate mixing processes at plant scale', *Chem. Eng. Sci.*, 1997, **52** (13), pp 2087–2097
[7] Siegel R L, Miller K D, Jemal A, 'Cancer statistics, 2018', *CA: Cancer J. Clin.*, 2018, **68** (1), pp 7–30
[8] Fricke H, Morse S, 'The electric capacity of tumors of the breast', *J. Cancer Res.*, 1926, **10** (3), pp 340–376
[9] Surowiec A, Stuchly S, Barr J, Swarup A, 'Dielectric properties of breast carcinoma and the surrounding tissues', *IEEE Trans. Biomed. Eng.*, 1988, **35** (4), pp 257–263
[10] Jossinet J, 'Variability of impedivity in normal and pathological breast tissue', *Med. Biol. Eng. Comput.*, 1996, **34** (5), pp 346–350
[11] Wtorek J, Stelter J, Nowakowskim A, 'Impedance mammograph 3D phantom studies', in *Electrical Bioimpedance Methods: Applications to Medicine and Biotechnology*, *Ann. New York Acad. Sci.*, 1999, **873** (1), pp 520–533
[12] Nowakowski J W A, Stelter J, 'A Technical University of Gdansk Electroimpedance Mammorgraph', in *Proceedings of the IX Int. Conference on Electrical Bioimpedance*, Heidelberg, Germany, 26–30 September 1995, pp 434–437
[13] Kocikowski M, Nowakowski A, 'Rcfart-3d reconstruction algorithm for EIT', in *Proceedings of the X Int. Conference on Electrical Bioimpedance*, Barcelona, Spain, 5–9 April 1998, pp 425–428
[14] Cherepenin V, Karpov A, Korjenevsky A, *et al.*, 'A 3D electrical impedance tomography (EIT) system for breast cancer detection', *Physiol. Meas.*, 2001, **22** (1), pp 9–18
[15] Ye G, Lim K H, George R T, Ybarra G A, Joines W T, Liu Q H, '3D EIT for breast cancer imaging: System, measurements, and reconstruction', *Microwave Opt. Technol. Lett.*, 2008, **50** (12), pp 3261–3271
[16] Hartov A, Mazzarese R, Reiss F, *et al.*, 'A multi-channel continuously selectable multi-frequency electrical impedance spectroscopy measurement system', *IEEE Trans. Biomed. Eng.*, 2000, **47** (1), pp 49–58
[17] Kerner T E, Williams D B, Osterman K S, Reiss F R, Hartov A, Paulsen K D, 'Electrical impedance imaging at multiple frequencies in phantoms', *Physiol. Meas.*, 2000, **21** (1), pp 67–77
[18] Paulsen K D, Jiang H, 'An enhanced electrical impedance imaging algorithm for hyperthermia applications', *Int. J. Hyperthermia*, 1997, **13** (5), pp 459–480
[19] Osterman K S, Kerner T E, Williams D B, Hartov A, Poplack S P, Paulsen K D, 'Multi-frequency electrical impedance imaging: Preliminary in-vivo experience in breast', *Physiol. Meas.*, 2000, **21** (1), pp 99–109
[20] Kerner T E, Hartov A, Soho S K, Poplack S P, Paulsen K D, 'Imaging the breast with EIS: An initial study of exam consistency', *Physiol. Meas.*, 2002, **23** (1), pp 221–236

[21] Kerner T, Paulsen K, Hartov A, Soho S, Poplack S, 'Electrical impedance spectroscopy of the breast: Clinical imaging results in 26 subjects', *IEEE Trans. Med. Imaging*, 2002, **21** (6), pp 638–645

[22] Halter R, Hartov A, Paulsen K D, 'Design and implementation of a high frequency electrical impedance tomography system', *Physiol. Meas.*, 2004, **25** (1), pp 379–390

[23] Halter R, Hartov A, Paulsen K, 'A broadband high-frequency electrical impedance tomography system for breast imaging', *IEEE Trans. Biomed. Eng.*, 2008, **55** (2), pp 650–659

[24] Halter R J, Hartov A, Poplack S P, *et al.*, 'Real-time electrical impedance variations in women with and without breast cancer', *IEEE Trans. Med. Imaging*, 2015, **34** (1), pp 38–48

[25] Murphy E K, Mahara A, Halter R J, 'Absolute reconstructions using rotational electrical impedance tomography for breast cancer imaging', *IEEE Trans. Med. Imaging*, 2017, **36** (4), pp 892–903

[26] Murphy E K, Mahara A, Wu X, Halter R J, 'Phantom experiments using soft-prior regularization EIT for breast cancer imaging', *Physiol. Meas.*, 2017, **38** (6), pp 1262–1277

[27] Mueller J L, Isaacson D, Newell J C, 'A reconstruction algorithm for electrical impedance tomography data collected on rectangular electrode arrays', *IEEE Trans. Biomed Eng.*, 1999, **46** (11), pp 1379–1386

[28] Oh T I, Lee J, Seo J K, Kim S W, Woo E J, 'Feasibility of breast cancer lesion detection using a multi-frequency trans-admittance scanner (TAS) with 10 Hz to 500 kHz bandwidth', *Physiol. Meas.*, 2007, **28** (7), pp S71–S84

[29] Zhang X, Wang W, Sze G, Barber D, Chatwin C, 'An image reconstruction algorithm for 3D electrical impedance mammography', *IEEE Trans. Med. Imaging*, 2014, **33** (12), pp 2223–2241

[30] Zhang X, Chatwin C, Barber D C, 'A feasibility study of a rotary planar electrode array for electrical impedance mammography using a digital breast phantom', *Physiol. Meas.*, 2015, **36** (6), pp 1311–1335

[31] Cherepenin V, Karpov A, Korjenevsky A, *et al.*, 'Three-dimensional EIT imaging of breast tissues: System design and clinical testing', *IEEE Trans. Med. Imaging*, 2002, **21** (6), pp 662–667

[32] Zain N M, Chelliah K K, 'Breast imaging using electrical impedance tomography: Correlation of quantitative assessment with visual interpretation', *Asian Pacific J. Cancer Prevent.*, 2014, **15** (3), pp 1327–1331

[33] Chakraborti D K L, Selvamurthy D W, 'Clinical application of electrical impedance tomography in the present health scenario of India', *J. Phys.: Conf. Ser.*, 2010, **224** (1), 012069 (4pp)

[34] Raneta O, Ondru D, Bella V, 'Utilisation of electrical impedance tomography in breast cancer diagnosis', *Klin Onkol*, 2012, **25** (1), pp 36–41

[35] Assenheimer M, Laver-Moskovitz O, Malonek D, *et al.*, 'The t-SCAN™ technology: Electrical impedance as a diagnostic tool for breast cancer detection', *Physiol. Meas.*, 2001, **22** (1), pp 1–8

[36] Zou Y, Guo Z, 'A review of electrical impedance techniques for breast cancer detection', *Med. Eng. Phys.*, 2003, **25** (2), pp 79–90

[37] Malich A, Boehm T, Facius M, *et al.*, 'Differentiation of mammographically suspicious lesions: Evaluation of breast ultrasound, MRI mammography and electrical impedance scanning as adjunctive technologies in breast cancer detection', *Clin. Radiol.*, 2001, **56** (4), pp 278–283

[38] Malich A, Fritsch T, Anderson R, *et al.*, 'Electrical impedance scanning for classifying suspicious breast lesions: First results', *Eur. Radiol.*, 2000, **10** (10), pp 1555–1561

[39] Hong S, Lee K, Ha U, Kim H, Lee Y, Kim Y, Yoo H J, 'A 4.9 mΩ sensitivity mobile electrical impedance tomography IC for early breast-cancer detection system', *IEEE J. Solid-State Circuits*, 2015, **50** (1), pp 245–257

[40] Abdul S, Brown B, Milnes P, Tidy J, 'The use of electrical impedance spectroscopy in the detection of cervical intraepithelial neoplasia', *Int. J. Gynecol. Cancer*, 2006, **16** (5), pp 1823–1832

[41] Tidy J A, Brown B H, Healey T J, *et al.*, 'Accuracy of detection of high-grade cervical intraepithelial neoplasia using electrical impedance spectroscopy with colposcopy', *Int. J. Obstet. Gynaecol.*, 2013, **120** (4), pp 400–411

[42] Korjenevsky A, Cherepenin V, Trokhanova O, Tuykin T, 'Gynecologic electrical impedance tomograph', *J. Phys.: Conf. Ser.*, 2010, **224** (1), 012070 (4pp)

[43] Cherepenin V, Gulyaev Y, Korjenevsky A, Sapetsky S, Tuykin T S, 'An electrical impedance tomography system for gynecological application GIT with a tiny electrode array', *Physiol. Meas.*, 2012, **33** (5), pp 849–862

[44] Trokhanova O V, Chijova Y A, Okhapkin M B, Korjenevsky A V, Tuykin T S, 'Possibilities of electrical impedance tomography in gynecology', *J. Phys.: Conf. Ser.*, 2013, **434** (1), 012038 (4pp)

[45] Pathiraja A, Ziprin P, Shiraz A, *et al.*, 'Detecting colorectal cancer using electrical impedance spectroscopy: Anex vivo feasibility study', *Physiol. Meas.*, 2017, **38** (6), pp 1278–1288

[46] Halter R, Hartov A, Heaney J, Paulsen K, Schned A, 'Electrical impedance spectroscopy of the human prostate', *IEEE Trans. Biomed. Eng.*, 2007, **54** (7), pp 1321–1327

[47] Halter R J, Schned A, Heaney J, Hartov A, Schutz S, Paulsen K D, 'Electrical impedance spectroscopy of benign and malignant prostatic tissues', *J. Urol.*, 2008, **179** (4), pp 1580–1586

[48] Borsic A, Halter R, Wan Y, Hartov A, Paulsen K D, 'Sensitivity study and optimization of a 3d electric impedance tomography prostate probe', *Physiol. Meas.*, 2009, **30** (6), pp S1–S18

[49] Wan Y, Halter R, Borsic A, Manwaring P, Hartov A, Paulsen K, 'Sensitivity study of an ultrasound coupled transrectal electrical impedance tomography system for prostate imaging', *Physiol. Meas.*, 2010, **31** (8), pp S17–S29

[50] Borsic A, Halter R, Wan Y, Hartov A, Paulsen K D, 'Electrical impedance tomography reconstruction for three-dimensional imaging of the prostate', *Physiol. Meas.*, 2010, **31** (8), pp S1–S16

[51] Brown B, 'Electrical impedance tomography (EIT): A review', *J. Med. Eng. Technol.*, 2003, **27** (3), pp 97–108

[52] Chatziioannidis I, Samaras T, Nikolaidis N, 'Electrical impedance tomography: A new study method for neonatal respiratory distress syndrome?' *Hippokratia*, 2011, **15** (3), pp 211–215

[53] Frerichs I, Braun P, Dudykevych T, Hahn G, Genée D, Hellige G, 'Distribution of ventilation in young and elderly adults determined by electrical impedance tomography', *Respir. Physiol. Neurobiol.*, 2004, **143** (1), pp 63–75

[54] Franchineau G, Bréchot N, Lebreton G, *et al.*, 'Bedside contribution of electrical impedance tomography to setting positive end-expiratory pressure for extracorporeal membrane oxygenation–treated patients with severe acute respiratory distress syndrome', *Am. J. Respir. Crit. Care Med.*, 2017, **196** (4), pp 447–457

[55] Meier T, Luepschen H, Karsten J, *et al.*, 'Assessment of regional lung recruitment and derecruitment during a PEEP trial based on electrical impedance tomography', *Intensive Care Med.*, 2007, **34** (3), pp 543–550

[56] Dargaville P A, Rimensberger P C, Frerichs I, 'Regional tidal ventilation and compliance during a stepwise vital capacity manoeuvre', *Intensive Care Med.*, 2010, **36** (11), pp 1953–1961

[57] Becher T H, Bui S, Zick G, *et al.*, 'Assessment of respiratory system compliance with electrical impedance tomography using a positive end-expiratory pressure wave manoeuvre during pressure support ventilation: A pilot clinical study', *Critical Care*, 2014, **18** (6), p 679

[58] Frerichs I, Dargaville P A, van Genderingen H, Morel D R, Rimensberger P C, 'Lung volume recruitment after surfactant administration modifies spatial distribution of ventilation', *Am. J. Respir. Crit. Care Med.*, 2006, **174** (7), pp 772–779

[59] Odenstedt H, Lindgren S, Olegard C, *et al.*, 'Slow moderate pressure recruitment manoeuvre minimizes negative circulatory and lung mechanic side effects: Evaluation of recruitment manoeuvres using electric impedance tomography', *Intensive Care Med.*, 2005, **31** (12), pp 1706–1714

[60] Grychtol B, Elke G, Meybohm P, Weiler N, Frerichs I, Adler A, 'Functional validation and comparison framework for EIT lung imaging', *PLoS ONE*, 2014, **9**(8), 103045, [Online], 2018, Available from: https://journals.plos.org/plosone/article?id=10.1371/journal.pone.0103045

[61] Mauri T, Bellani G, Confalonieri A, *et al.*, 'Topographic distribution of tidal ventilation in acute respiratory distress syndrome', *Critical Care Med.*, 2013, **41** (7), pp 1664–1673

[62] Becher T, Kott M, Schdler D, *et al.*, 'Influence of tidal volume on ventilation inhomogeneity assessed by electrical impedance tomography during controlled mechanical ventilation', *Physiol. Meas.*, 2015, **36** (6), pp 1137–1146

[63] Frerichs I, Pulletz S, Elke G, *et al.*, 'Assessment of changes in distribution of lung perfusion by electrical impedance tomography', *Respiration*, 2009, **77** (3), pp 282–291

[64] Smit H J, Noordegraaf A V, Roeleveld R J, *et al.*, 'Epoprostenol-induced pulmonary vasodilatation in patients with pulmonary hyper-tension measured by electrical impedance tomography', *Physiol. Meas.*, 2002, **23** (1), pp 237–243

[65] Li Y, Tesselaar E, Borges J B, Bhm S H, Sjberg F, Janerot-Sjberg B, 'Hyperoxia affects the regional pulmonary ventilation/perfusion ratio: An electrical impedance tomography study', *Acta Anaesth. Scand.*, 2014, **58** (6), pp 716–725

[66] Schibler A, Pham T M T, Moray A A, Stocker C, 'Ventilation and cardiac related impedance changes in children undergoing corrective open heart surgery', *Physiol. Meas.*, 2013, **34** (10), pp 1319–1327

[67] Pulletz S, Kott M, Elke G, Schdler D, Vogt B, Weiler N, Frerichs I, 'Dynamics of regional lung aeration determined by electrical impedance tomography in patients with acute respiratory distress syndrome', *Multidiscipl. Respir. Med.*, 2012, **7** (1), p 44

[68] Victorino J A, Borges J B, Okamoto V N, *et al.*, 'Imbalances in regional lung ventilation', *Am. J. Respir. Critical Care Med.*, 2004, **169** (7), pp 791–800

[69] Kunst P W, Noordegraaf A V, Raaijmakers E, *et al.*, 'Electrical impedance tomography in the assessment of extravascular lung water in noncardiogenic acute respiratory failure', *Chest*, 1999, **116** (6), pp 1695–1702

[70] Hinz J, Gehoff A, Moerer O, *et al.*, 'Regional filling characteristics of the lungs in mechanically ventilated patients with acute lung injury', *Eur. J. Anaesth.*, 2007, **24** (5), pp 414–424

[71] Wolf G K, Grychtol B, Frerichs I, *et al.*, 'Regional lung volume changes in children with acute respiratory distress syndrome during a derecruitment maneuver', *Critical Care Med.*, 2007, **35** (8), pp 1972–1978

[72] Vogt B, Pulletz S, Elke G, *et al.*, 'Spatial and temporal heterogeneity of regional lung ventilation determined by electrical impedance tomography during pulmonary function testing', *J. Appl. Physiol.*, 2012, **113** (7), pp 1154–1161

[73] Vogt B, Zhao Z, Zabel P, Weiler N, Frerichs I, 'Regional lung response to bronchodilator reversibility testing determined by electrical impedance tomography in chronic obstructive pulmonary disease', *Am. J. Physiol.-Lung Cell. Mol. Physiol.*, 2016, **311** (1), pp L8–L19

[74] Frerichs I, Zhao Z, Becher T, Zabel P, Weiler N, Vogt B, 'Regional lung function determined by electrical impedance tomography during bronchodilator reversibility testing in patients with asthma', *Physiol. Meas.*, 2016, **37** (6), pp 698–712

[75] Smit H J, Vonk-Noordegraaf A, Marcus J T, *et al.*, 'Pulmonary vascular responses to hypoxia and hyperoxiain healthy volunteers and COPD patients measured by electrical impedance tomography', *Chest*, 2003, **123** (6), pp 1803–1809

[76] Serrano R E, de Lema B, Casas O, *et al.*, 'Use of electrical impedance tomography (EIT) for the assessment of unilateral pulmonary function', *Physiol. Meas.*, 2002, **23** (1), pp 211–220

[77] Zhao Z, Mller-Lisse U, Frerichs I, Fischer R, Mller K, 'Regional airway obstruction in cystic fibrosis determined by electrical impedance tomography in comparison with high resolution CT', *Physiol. Meas.*, 2013, **34** (11), pp N107–N114

[78] Bodenstein M, David M, Markstaller K, 'Principles of electrical impedance tomography and its clinical application', *Critical Care Med.*, 2009, **37** (2), pp 713–724

[79] Lundin S, Stenqvist O, 'Electrical impedance tomography', *Curr. Opin. Crit. Care*, 2012, **18** (1), pp 35–41

[80] Frerichs I, Becher T, Weiler N, 'Electrical impedance tomography imaging of the cardiopulmonary system', *Curr. Opin. Crit. Care*, 2014, **20** (3), pp 323–332

[81] Nguyen D T, Jin C, Thiagalingam A, McEwan A L, 'A review on electrical impedance tomography for pulmonary perfusion imaging', *Physiol. Meas.*, 2012, **33** (5), pp 695–706

[82] Walsh B K, Smallwood C D, 'Electrical impedance tomography during mechanical ventilation', *Respir. Care*, 2016, **61** (10), pp 1417–1424

[83] Muders T, Luepschen H, Putensen C, 'Impedance tomography as a new monitoring technique', *Curr. Opin. Crit. Care*, 2010, **16** (3), pp 269–275

[84] Frerichs I, 'Electrical impedance tomography (EIT) in applications related to lung and ventilation: A review of experimental and clinical activities', *Physiol. Meas.*, 2000, **21** (2), pp R1–R21

[85] Adler A, Amato M B, Arnold J H, *et al.*, 'Whither lung EIT: Where are we, where do we want to go and what do we need to get there?', *Physiol. Meas.*, 2012, **33** (5), pp 679–694

[86] Frerichs I, Amato M B P, van Kaam A H, *et al.*, 'Chest electrical impedance tomography examination, data analysis, terminology, clinical use and recommendations: Consensus statement of the Translational EIT Development Study Group', *Thorax*, 2016, **72** (1), pp 83–93

[87] Leonhardt S, Lachmann B, 'Electrical impedance tomography: The holy grail of ventilation and perfusion monitoring?', *Intensive Care Med.*, 2012, **38** (12), pp 1917–1929

[88] Gibas C, Grnewald A, Bchner S, Brck R, 'An EIT system for mobile medical diagnostics', in Gimi B, Krol A (eds.), *Medical Imaging 2018: Biomedical Applications in Molecular, Structural, and Functional Imaging*, SPIE, 2018

[89] Swisstom AG, 'Swisstom BB^2 product information', [Online], Available from: http://www.swisstom.com/wp-content/uploads/GB_2ST100-112_002_ Product_Brochure.pdf, 2015

[90] Brown B H, Seagar A D, 'The Sheffield data collection system', *Clin. Phys. Physiol. Meas.*, 1987, **8** (4A), pp 91–97

[91] Leathard A D, Brown B H, Campbell J, Zhang F, Morice A H, Tayler D, 'A comparison of ventilatory and cardiac related changes in EIT images of normal human lungs and of lungs with pulmonary emboli', *Physiol. Meas.*, 1994, **15** (2A), pp A137–A146

[92] Newell J, Gisser D, Isaacson D, 'An electric current tomograph', *IEEE Trans. Biomed. Eng.*, 1988, **35** (10), pp 828–833

[93] Cook R, Saulnier G, Gisser D, Goble J, Newell J, Isaacson D, 'ACT3: A high-speed, high-precision electrical impedance tomograph', *IEEE Trans. Biomed. Eng.*, 1994, **41** (8), pp 713–722

[94] Guardo R, Boulay C, Murray B, Bertrand M, 'An experimental study in electrical impedance tomography using back-projection reconstruction', *IEEE Trans. Biomed. Eng.*, 1991, **38** (7), pp 617–627

[95] Taktak A, Spencer A, Record P, Gadd R, Rolfe P, 'Feasibility of neonatal lung imaging using electrical impedance tomography', *Early Human Devel.*, 1996, **44** (2), pp 131–138

[96] Osypka M, Gersing E, Meyer-Waarden K, 'Komplexe elektrische impedanztomografie im frequenzbereich von 10 Hz bis 50 kHz', *Z. Med. Phys.*, 1993, **3** (3), pp 124–132

[97] Brown B H, Barber D C, Wang W, *et al.*, 'Multi-frequency imaging and modelling of respiratory related electrical impedance changes', *Physiol. Meas.*, 1994, **15** (2A), pp A1–A12

[98] Smith R, Freeston I, Brown B, 'A real-time electrical impedance tomography system for clinical use-design and preliminary results', *IEEE Trans. Biomed. Eng.*, 1995, **42** (2), pp 133–140

[99] Griffiths H, Leung H T L, Williams R J, 'Imaging the complex impedance of the thorax', *Clin. Phys. Physiol. Meas.*, 1992, **13** (A), pp 77–81

[100] McCann H, Ahsan S T, Davidson J L, Robinson R L, Wright P, Pomfrett C J D, 'A portable instrument for high-speed brain function imaging: FEITER', in *2011 Annual International Conference of the IEEE Engineering in Medicine and Biology Society*, Boston, MA, USA, 30 August–3 September 2011

[101] Zhang S, Xu G, Zhang X, *et al.*, 'Computation of a 3-d model for lung imaging with electrical impedance tomography', *IEEE Trans. Magn.*, 2012, **48** (2), pp 651–654

[102] Wilson A J, Milnes P, Waterworth A R, Smallwood R H, Brown B H, 'Mk3.5: A modular, multi-frequency successor to the mk3a EIS/EIT system', *Physiol. Meas.*, 2001, **22** (1), pp 49–54

[103] Kim M, Jang J, Kim H, *et al.*, 'A 1.4-m Ω-sensitivity 94-dB dynamic-range electrical impedance tomography SoC and 48-channel hub-SoC for 3-D lung ventilation monitoring system', *IEEE J. Solid-State Circuits*, 2017, **52** (11), pp 2829–2842

[104] Yang Y, Jia J, 'A multi-frequency electrical impedance tomography system for real-time 2D and 3D imaging', *Rev. Sci. Instrum.*, 2017, **88** (8), 085110

[105] Lehmann S, Leonhardt S, Ngo C, *et al.*, 'Electrical impedance tomography as possible guidance for individual positioning of patients with multiple lung injury', *Clin. Respir. J.*, 2016, **12** (1), pp 68–75

[106] Hsu Y L, Tien A J, Chang M Y, *et al.*, 'Regional ventilation redistribution measured by electrical impedance tomography during spontaneous

breathing trial with automatic tube compensation', *Physiol. Meas.*, 2017, **38** (6), pp 1193–1203

[107] Bikker I G, Preis C, Egal M, Bakker J, Gommers D, 'Electrical impedance tomography measured at two thoracic levels can visualize the ventilation distribution changes at the bedside during a decremental positive end-expiratory lung pressure trial', *Crit. Care*, 2011, **15** (4), p R193

[108] Zhao Z, Peng S Y, Chang M Y, *et al.*, 'Spontaneous breathing trial after prolonged mechanical ventilation monitored by electrical impedance tomography: An observational study', *Acta Anaesth. Scand.*, 2017, **61** (9), pp 1166–1175

[109] Wolf G K, Gómez-Laberge C G, Rettig J S, *et al.*, 'Mechanical ventilation guided by electrical impedance tomography in experimental acute lung injury', *Crit. Care Med.*, 2013, **41** (5), pp 1296–1304

[110] Khodadad D, Nordebo S, Seifnaraghi N, Waldmann A D, Muller B, Bayford R, 'Breath detection using short-time Fourier transform analysis in electrical impedance tomography', in *2017 32nd General Assembly and Scientific Symposium of the International Union of Radio Science*, 19–26 August 2017, Montreal, Canada

[111] Zhou L, Harrach B, Seo J K, 'Monotonicity-based electrical impedance tomography for lung imaging', *Inverse Problems*, 2018, **34** (4), pp 1–25

[112] Rosa R G, Rutzen W, Madeira L, *et al.*, 'Use of thoracic electrical impedance tomography as an auxiliary tool for alveolar recruitment maneuvers in acute respiratory distress syndrome: Case report and brief literature review', *Rev. Bras. Ter. Intensiva*, 2015, **27** (4), pp 406–411

[113] Lopes F A, Souza L A M, Bernardi J T N, *et al.*, 'Pendelluft diagnosed from ventilator weaning indexes obtained through bioelectrical impedance tomography: A case report', *Sao Paulo Medical J.*, 2017, **135** (3), pp 302–308

[114] Borges J B, Suarez-Sipmann F, Bohm S H, *et al.*, 'Reginal lung perfusion estimated by electrical impedance tomography in a piglet model of lung collapse', *J. Appl. Physiol.*, 2012, **112** (1), pp 225–236

[115] Nebuya S, Kitamura K, Kobayashi H, Noshiro M, Brown B, 'Measurement accuracy in pulmonary function test using electrical impedance tomography', *13th International Conference on Electrical Bioimpedance and the 8th Conference on Electrical Impedance Tomography*, 29 August–2 September 2007, Graz, Austria, in *IFMBE Proceedings*, Springer, 2007, pp 539–542

[116] Nebuya S, Noshiro M, Yonemoto A, *et al.*, 'Study of the optimum level of electrode placement for the evaluation of absolute lung resistivity with the Mk3.5 EIT system', *Physiol. Meas.*, 2006, **27** (5), pp S129–S137

[117] Harris N D, Suggett A J, Barber D C, Brown B H, 'Applications of applied potential tomography (APT) in respiratory medicine', *Clin. Phys. Physiol. Meas.*, 1987, **8** (4A), pp 155–165

[118] Harris N D, Suggett A J, Barber D C, Brown B H, 'Applied potential tomography: A new technique for monitoring pulmonary function', *Clin. Phys. Physiol. Meas.*, 1988, **9** (4A), pp 79–85

[119] Campbell J H, Harris N D, Zhang F, Brown B H, Morice A H, 'Clinical applications of electrical impedance tomography in the monitoring of change in intrathoracic fluid volumes', *Physiol. Meas.*, 1994, **15** (2A), pp A217–A222

[120] Hahn G, Sipinkova I, Baisch F, Hellige G, 'Changes in the thoracic impedance distribution under different ventilatory conditions', *Physiol. Meas.*, 1995, **16** (3A), pp A161–A173

[121] Brown B H, Sinton A M, Barber D C, Leathard A D, McArdle F J, 'Simultaneous display of lung ventilation and perfusion on a real-time EIT system', in *Proceeding of 14th Annual International Conference of the IEEE Engineering in Medicine and Biology Society*, 29 October–1 November 1992, Paris, France

[122] Kunst P W A, Noordegraaf A V, Hoekstra O S, Postmus P E, de Vries P M J M, 'Ventilation and perfusion imaging electrical impedance tomography: A comparison with radionuclide scanning', *Physiol. Meas.*, 1998, **19** (4), pp 481–490

[123] Kunst P W A, Noordegraaf A V, Straver B, *et al.*, 'Influences of lung parenchyma density and thoracic fluid on ventilator EIT measurements', *Physiol. Meas.*, 1998, **19** (1), pp 27–34

[124] Zhang J, Patterson R, 'Non-invasive determination of absolute lung resistivity in adults suing electrical impedance tomography', *Physiol. Meas.*, 2010, **31** (8), pp S45–S56

[125] Hampshire A R, Smallwood R H, Brown B H, Primhak R A, 'Multifrequency and parametric EIT images of neonatal lungs', *Physiol. Meas.*, 1995, **16** (3A), pp A175–A189

[126] Cheney M, Isaacson D, Newell J C, Simske S, Goble J, 'NOSER: An algorithm for solving the inverse conductivity problem', *Int. J. Imag. Syst. Technol.*, 1990, **2** (2), pp 66–75

[127] Newell J C, Edic P M, Ren X, Larson-Wiseman J L, Danyleiko M D, 'Assessment of acute pulmonary edema in dogs by electrical impedance imaging', *IEEE Trans. Biomed. Eng.*, 1996, **43** (2), pp 133–138

[128] Adler A, Guardo R, Berthiaume Y, 'Impedance imaging of lung ventilation:Do we need to account for chest expansion?', *IEEE Trans. Biomed. Eng.*, 1996, **43** (4), pp 414–420

[129] Adler A, Amyot R, Guardo R, Bates J H T, Berthiaume Y, 'Monitoring changes in lung air and liquid volumes with electrical impedance tomography', *J. Appl. Physiol.*, 1997, **83** (5), pp 1762–1767

[130] Adler A, Shinozuka N, Berthiaume Y, Guardo R, Bates J H T, 'Electrical impedance tomography can monitor dynamic hyperinflation in dogs', *J. Appl. Physiol.*, 1998, **84** (2), pp 726–732

[131] Luepschen H, Meier T, Grossherr M, Leibecke T, Karsten J, Leonhardt S, 'Protective ventilation using electrical impedance tomography', *Physiol. Meas.*, 2007, **28** (7), pp S247–S260

[132] Crabb M G, Davidson J L, Little R, *et al.*, 'Mutual information as a measure of image quality for 3D dynamic lung imaging with EIT', *Physiol. Meas.*, 2014, **35** (5), pp 863–879

[133] Barber D C, 'Quantification in impedance imaging', *Clin. Phys. Physiol. Meas.*, 1990, **11** (4A), pp 45–56

[134] Wagenaar J, Adler A, 'Electrical impedance tomography in 3d using two electrode planes: Characterization and evaluation', *Physiol. Meas.*, 2016, **37** (6), pp 922–937

[135] Reifferscheid F, Elke G, Pulletz S, *et al.*, 'Regional ventilation distribution determined by electrical impedance tomography: Reproducibility and effects of posture and chest plane', *Respirology*, 2011, **16** (3), pp 523–531

[136] Morais C C A, Santiago R R D S, de Oliveira Filho J R B, *et al.*, 'Monitoring of pneumothorax appearance with electrical impedance tomography during recruitment maneuvers', *Am. J. Respir. Crit. Care Med.*, 2017, **195** (8), pp 1070–1073

[137] Oh T, Gilad O, Ghosh A, Schuettler M, Holder D S, 'A novel method for recording neuronal depolarization with recording at 125–825 Hz: Implications for imaging fast neural activity in the brain with electrical impedance tomography', *Med. Biol. Eng. Comput.*, 2011, **49** (5), pp 593–604

[138] Holder D S, 'Electrical impedance tomography with cortical or scalp electrodes during global cerebral ischaemia in the anaesthetised rat', *Clin. Phys. Physiol. Meas.*, 1992, **13** (1), pp 87–98

[139] Dowrick T, Blochet C, Holder D, 'In vivo bio-impedance measurement of healthy and ischaemic rat brain: Implications for stroke imaging using electrical impedance tomography', *Physiol. Meas.*, 2015, **36** (6), pp 1273–1282

[140] Aristovich K Y, dos Santos G S, Packham B C, Holder D S, 'A method for reconstructing tomographic images of evoked neural activity with electrical impedance tomography using intracranial planar arrays', *Physiol. Meas.*, 2014, **35** (6), pp 1095–1109

[141] Aristovich K, Donegá M, Blochet C, *et al.*, 'Imaging fast neural traffic at fascicular level with electrical impedance tomography: Proof of principle in rat sciatic nerve', *J. Neural Eng.*, 2018, **15** (5), 056025

[142] Avery J, Dowrick T, Faulkner M, Goren N, Holder D, 'A versatile and reproducible multi-frequency electrical impedance tomography system', *Sensors*, 2017, **17** (2), p 280

[143] Davidson J L, Wright P, Ahsan S T, Robinson R L, Pomfrett C J D, McCann H, 'fEITER – a new EIT instrument for functional brain imaging', *J. Phys.: Conf. Ser.*, 2010, **224** (1), 012025 (4pp)

[144] Xuetao S, Fusheng Y, Feng F, Ruigang L, Xiuzhen D, 'High precision multi-frequency electrical impedance tomography system and preliminary imaging results on saline tank', in *IEEE 27th Annual Conference in Engineering in Medicine and Biology*, Shanghai, China, 1–4 September 2005, pp 1492–1495

[145] Manwaring P K, Moodie K L, Hartov A, Manwaring K H, Halter R J, 'Intracranial electrical impedance tomography', *Anesth. Analgesia*, 2013, **117** (4), pp 866–875

[146] Goren N, Avery J, Dowrick T, *et al.*, 'Multi-frequency electrical impedance tomography and neuroimaging data in stroke patients', *Sci. Data*, 2018, **5**,

180112, [Online], Available from: https://www.nature.com/articles/sdata2018112

[147] Tidswell A T, Bagshaw A P, Holder D S, *et al.*, 'A comparison of headnet electrode arrays for electrical impedance tomography of the human head', *Physiol. Meas.*, 2003, **24**, pp 527–544

[148] Fabrizi L, Sparkes M, Horesh L, *et al.*, 'Factors limiting the application of electrical impedance tomography for identification of regional conductivity changes using scalp electrodes during epileptic seizures in humans', *Physiol. Meas.*, 2006, **27** (5), pp S163–S174

[149] Holder D S, 'Electrical impedance tomography of brain function', in *2008 World Automation Congress*, 28 September–2 October 2008, Hawaii, USA, 10411809 (6pp)

[150] Holder D S, Gonzalez-Correa C A, Tidswell T, Gibson A, Cusick G, Bayford R H, 'Assessment and calibration of a low-frequency system for electrical impedance tomography (EIT), optimized for use in imaging brain function in ambulant human subjects', *Ann. New York Acad. Scie.*, 1999, **873** (1), pp 512–519

[151] Boone K, Lewis A M, Holder D S, 'Imaging of cortical spreading depression by EIT: Implications for localization of epileptic foci', *Physiol. Meas.*, 1994, **15** (2A), pp A189–A198

[152] Holder D S, Rao A, Hanquan Y, 'Imaging of physiologically evoked responses by electrical impedance tomography with cortical electrodes in the anaesthetized rabbit', *Physiol. Meas.*, 1996, **17** (4A), pp A179–A186

[153] Malone E, dos Santos G S, Holder D, Arridge S, 'Multi-frequency electrical impedance tomography using spectral constraints', *IEEE Trans. Med. Imaging*, 2014, **33** (2), pp 340–350

[154] McEwan A, Romsauerova A, Yerworth R, Horesh L, Bayford R, Holder D, 'Design and calibration of a compact multi-frequency EIT system for acute stroke imaging', *Physiol. Meas.*, 2006, **27** (5), pp S199–S210

[155] Oh T I, Koo H, Lee K H, *et al.*, 'Validation of a multi-frequency electrical impedance tomography (mfEIT) system KHU mark1: Impedance spectroscopy and time-difference imaging', *Physiol. Meas.*, 2008, **29** (3), pp 295–307

[156] Fu F, Li B, Dai M, *et al.*, 'Use of electrical impedance tomography to monitor regional cerebral edema during clinical dehydration treatment', *PLoS ONE*, 2014, **9** (12), 0113202, [Online], 2018, Available from: https://journals.plos.org/plosone/article?id=10.1371/journal.pone.0113202

[157] Shi X, Li W, You F, *et al.*, 'High-precision electrical impedance tomography data acquisition system for brain imaging', *IEEE Sensors J.*, 2018, **18** (14), pp 5974–5984

[158] Romsauerova A, McEwan A, Horesh L, Yerworth R, Bayford R H, Holder D S, 'Multi-frequency electrical impedance tomography (EIT) of the adult human head: Initial findings in brain tumours, arteriovenous malformations and chronic stroke, development of an analysis method and calibration', *Physiol. Meas.*, 2006, **27** (5), pp S147–S161

[159] Li Y, Zhang D, Liu B, *et al.*, 'Noninvasive cerebral imaging and monitoring using electrical impedance tomography during total aortic arch replacement', *J. Cardiothor. Vascul. Anesth.*, 2018, **32** (6), pp 2469–2476

[160] Aristovich K Y, Packham B C, Koo H, dos Santos G S, McEvoy A, Holder D S, 'Imaging fast electrical activity in the brain with electrical impedance tomography', *NeuroImage*, 2016, **124**, pp 204–213

[161] Yerworth R J, Bayford R H, Cusick G, Conway M, Holder D S, 'Design and performance of the UCLH mark 1b 64 channel electrical impedance tomography (EIT) system, optimized for imaging brain function', *Physiol. Meas.*, 2002, **23** (1), pp 149–158

[162] Tidswell T, Gibson A, Bayford R H, Holder D S, 'Three-dimensional electrical impedance tomography of human brain activity', *NeuroImage*, 2001, **13** (2), pp 283–294

[163] Tizzard A, Horesh L, Yerworth R J, Holder D S, Bayford R H, 'Generating accurate finite element meshes for the forward model of the human head in EIT', *Physiol. Meas.*, 2005, **26** (2), pp S251–S261

[164] Li Y, Rao L, He R, *et al.*', A novel combination method of electrical impedance tomography inverse problem for brain imaging', *IEEE Trans. Magn.*, 2005, **41** (5), pp 1848–1851

[165] Clay M, Ferree T, 'Weighted regularization in electrical impedance tomography with applications to acute cerebral stroke', *IEEE Trans. Med. Imaging*, 2002, **21** (6), pp 629–637

[166] Kumar S P, Sriraam N, Benakop P, Jinaga B C, 'Reconstruction of brain electrical impedance tomography images using particle swarm optimization', in *Proceedings of 2010 5th International Conference on Industrial and Information Systems*, 29 July–1 August 2010, Mangalore, India, pp 339–342

[167] Elemam R F, Pretty I, 'Comparison of the success rate of endodontic treatment and implant treatment', *ISRN Dentistry*, 2011, **2011**, pp 1–8

[168] Ren Z, Yang W Q, 'Visualisation of tooth surface by electrical capacitance tomography', *Biomed. Phys. Eng. Exp.*, 2017, **3** (1), 015021

[169] Ren Z, 'Exploration of medical applications of electrical capacitance tomography', *PhD thesis*, The University of Manchester, 2015

[170] Ren Z, Yang W, 'A miniature two-plate electrical capacitance tomography sensor', *IEEE Sens. J.*, 2015, **15** (5), pp 3037–3049

[171] National Joint Registry, '14th Annual Report 2017 National Joint Registry for England, Wales, Northern Ireland and the Isle of Man', [Online], 2017, Available from: http://www.njrreports.org.uk/Portals/0/PDFdownloads/NJR%2014th%20Annual%20Report%202017.pdf

[172] Springer B D, Fehring T K, Griffin W L, Odum S M, Masonis J L, 'Why revision total hip arthroplasty fails', *Clin. Orthop. Relat. Res.*, 2008, **467** (1), pp 166–173

[173] Brendle C, Niesche A, Korff A, Radermacher K, Leonhardt S, 'Femoral test bed for impedance controlled surgical instrumentation', *Acta Polytech.*, 2012, **52** (5), pp 17–21

[174] Brendle C, Rein B, Niesche A, *et al.*, 'Electrical bioimpedance-controlled surgical instrumentation', *IEEE Trans. Biomed. Circuits Syst.*, 2015, **9** (5), pp 743–750

[175] Sun J, Ren Z, Yang W, '3D imaging with single-plane electrical capacitance tomography sensor', *Electron. Lett.*, 2015, **51** (3), pp 222–224

[176] Ren Z., Yang W., 'Development of a navigation tool for revision total hip surgery based on electrical impedance tomography', *IEEE Trans. Instrum. Meas.*, 2016, **65**, pp 2748–2757

Zhen Ren received her PhD degree from The University of Manchester, UK in 2015. She is currently a postdoctoral scholar in the Department of Radiology in the University of Chicago. She was a postdoctoral research associate in the School of Chemical Engineering and Analytical Science in The University of Manchester, from 2015 to 2017. Her research interests include dynamic contrast-enhanced magnetic resonance imaging, breast cancer research, electrical impedance tomography, image guided surgery/therapy, industrial process control and image analysis. She is a member of IEEE and a referee for 12 journals. She was recipient of the Outstanding Organization Award from IEEE International Conference on Imaging Systems and Techniques, Manchester, UK in 2012, and the Outstanding Service Award from the IEEE International Conference on Systems, Man and Cybernetics, Manchester, UK in 2013.

Wuqiang Yang is a Fellow of the IEEE, Fellow of the IET and Fellow of the Institute of Measurement and Control. He spent 13 years at Tsinghua University in Beijing, as an undergraduate, MSc and PhD student and then as a lecturer. Since 1991, he has been working with UMIST and University of Manchester in the UK. He became a Professor in the Department of Electrical and Electronic Engineering in 2005. His main research interests include electrical capacitance tomography (ECT), instrumentation and multiphase flow measurement. He has published 400 papers and edited a book 'Sensor array', which is published by Intech. He is an Associate Editor of *IEEE Trans. on Instrum. and Meas. and IET Sci., Meas. and Technol.*, editorial board member of several other journals (including *Meas. Sci. and Technol.*

and Sensor Review), guest editor of many journal special issues and visiting professor at several other universities. He is one of the key organisers of the annual *IEEE International Conference on Imaging Systems and Techniques*. He was an IEEE IMS Distinguished Lecturer (2010–2016) and a JSPS Invitation Fellow (2016) and is a JSPS Bridge Fellow (2020). His biography has been included in Who's Who in the World since 2002.

Chapter 12

Electro-magnetic tomography (EMT) sensors

Wuliang Yin[1]

List of abbreviations

AGC	auto gain control
BEM	boundary element method
CGS	conjugate gradients stabilised
DDS	direct digital synthesiser
DSP	digital signal processor
ECT	electrical capacitance tomography
EMT	electro-magnetic tomography
ERT	electrical resistance tomography
FEM	finite element method
FPGA	field-programmable gate array
MIT	magnetic inductance tomography
PEC	perfect electric conductor
SNR	signal-to-noise ratio

12.1 Introduction

Magnetic inductance tomography (MIT) is also called electro-magnetic tomography (EMT) or electro-magnetic induction tomography. It has been developed for nearly three decades for industrial process and biomedical imaging applications. The basic principle is to place a set of coils around an object being imaged, measure the mutual inductive coupling between the coils, and reconstruct cross sectional images using the measured data. The mutual inductive coupling changes with the electromagnetic, geometrical and positional properties of the object(s). Because MIT is sensitive to both permeability and conductivity, an image reconstruction algorithm can be used to obtain the permeability/conductivity distribution in the imaging space according to the mutual induction measurements.

[1]The Department of Electrical & Electronic Engineering, The University of Manchester, UK

The operating principle of an EMT system is as follows. An excitation current is multiplexed to a transmitter coil (one out of n) and the remaining $n - 1$ coils act as receivers. The induced voltage signals from the receiving coils are connected to a measurement channel, which converts each signal to a complex number, that is with in-phase (real) and quadrature (imaginary) components relative to a synchronous reference waveform. Then the next coil is selected as the transmitter with the remaining coils as receivers. This process continues until all the coils have been excited. Therefore, one frame of data consists of $n \times (n - 1)$ measurements, covering all the transmitter/receiver coil combinations. However, there are only $n \times (n - 1)/2$ independent measurements as redundant measurements occur due to reciprocity. During the measurement process, the excitation currents are fed into an independent measurement channel, which provides the in-phase and quadrature components of the injected currents relative to the same synchronous reference waveform. Mutual impedance/inductance between the coils can then be calculated.

The governing equation for MIT is a diffusion equation, which can be derived from Maxwell equations:

$$\nabla \times (v\nabla \times \mathbf{A}) + j\omega\xi(\mathbf{A} + \nabla V) = J_S \tag{12.1}$$

$$j\omega\nabla \cdot (\xi(\mathbf{A} + \nabla V)) = 0 \tag{12.2}$$

where $\xi = \sigma + j\omega\varepsilon$, σ is the conductivity, $v = 1/\mu$ is the inverse of the permeability, J_s is the current density and $\mathbf{A}$ is the magnetic vector potential and V is the electrical potential.

In contrast with electrical capacitance tomography (ECT) and electrical resistance tomography (ERT), where the governing equation is the Laplace equation, the governing equation for EMT is a diffusion equation, where the frequency term f is present.

In an MIT sensor, coils are normally arranged with their axes towards the object space as shown in Figure 12.1. A typical set of parameters are the diameter of the imaging space 150 mm, the diameter of the coil 50 mm, the number of the turns of the coil 50 and the sensor former made of non-conducting plastics.

Other alternative sensor designs have been reported. A planar sensor for the detection of conductivity inhomogeneity on a metallic plate was reported in [1], where the coils are arranged with their axes perpendicular to the object space. A coil array with an angle to the object space was investigated for object recognition in [2] (Figure 12.2).

In addition to coil angles, MIT sensors with different geometrical scales are needed to address particular problems. A relationship has been developed between sensors of different geometrical scales to provide sensor design guidance in process scaling-up and scaling-down [3].

The dimensional scaling transformation is defined by assuming

$$x = kx', y = ky', z = kz' \tag{12.3}$$

where x, y and z are coordinates in the original geometry while x', y' and z' are coordinates in the new scaled geometry.

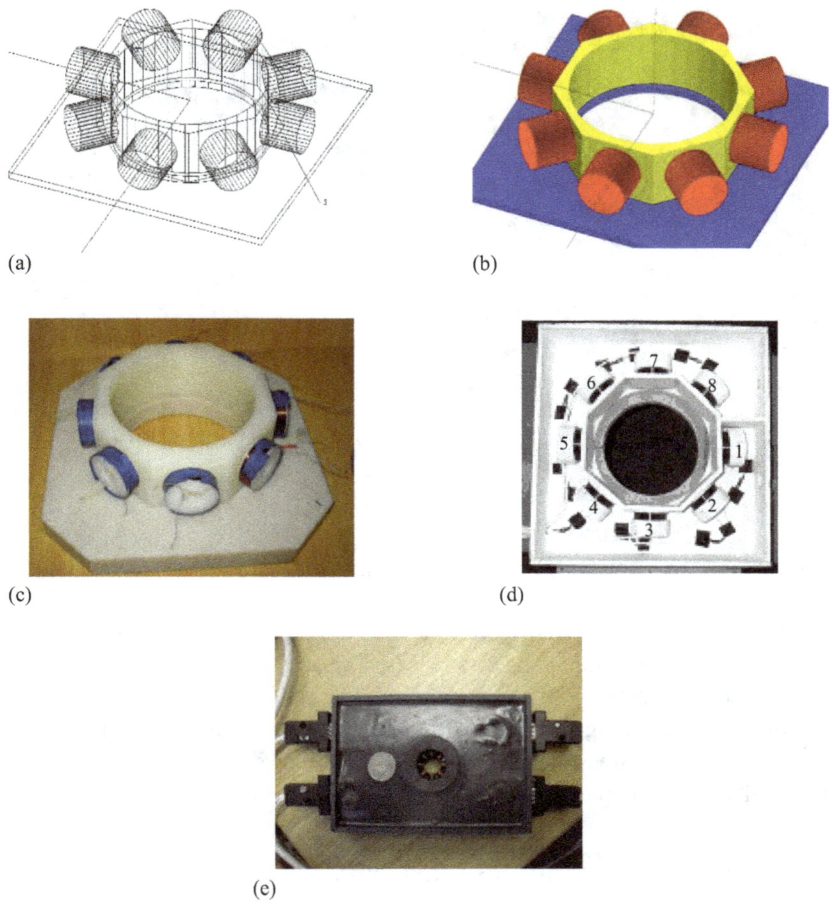

Figure 12.1 (a) Simulation model of an eight coil MIT sensor, (b) 3D illustration of an eight coil MIT sensor, (c) photo of a lab version MIT sensor, (d) photo of an industrial version MIT sensor and (e) photo of a miniature MIT sensor

If $k > 1$, the new system is a deflated (by $1/k$ times) version of the original system. If $k < 1$, the new system is an inflated version (by $1/k$ times).

Let magnetic vector potential $\mathbf{A}'(x', y', z')$ be a solution to (12.1) and defining $\mathbf{A}'(x', y', z') = \mathbf{A}(kx', ky', kz')$. Because $\frac{1}{\mu}\nabla^2\mathbf{A} = \frac{1}{\mu k^2}\nabla^2\mathbf{A}'$, (12.1) becomes

$$\frac{1}{\mu}\nabla^2\mathbf{A}' - j\omega\sigma k^2\mathbf{A}' = -k^2\mathbf{J} \tag{12.4}$$

It indicates that if the current density $\mathbf{J}$ and the frequency increases by k^2 time, then the magnetic potential in the new system is a just a geometrically scaled version of the original system. For an inflated system ($k < 1$), if the current density

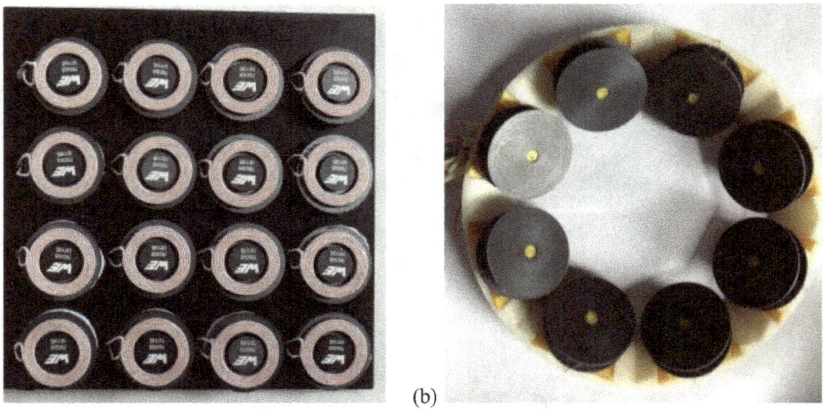

(a) (b)

Figure 12.2 Two alternative sensor designs: (a) planar sensor and (b) angled sensor

reduces by $\frac{1}{k^2}$ times, then the current remains the same considering the area increases by $\frac{1}{k^2}$ times. From $\mathbf{B}' = \nabla \times \mathbf{A}' = k\nabla \times \mathbf{A} = k\mathbf{B}$, the magnetic flux density **B** reduces by $\frac{1}{k}$ times. However, the area the flux flows through increases by $\frac{1}{k^2}$ times, and hence the inductance increases. The induced voltage and the inductance can also be expressed in terms of a line integration of **A**, where an inflated system increases the integral line length and results in an increase in inductance. The properties of a range of sensors with scaling relations can be predicted based on the results from one particular sensor.

12.2 Forward problem and sensitivity maps

In some special cases, such as axis-symmetry or simple plate geometry, (12.1) can be solved using analytical methods, but in most cases, numerical methods such as finite element methods (FEM) and boundary element methods (BEM) are suitable. Examples are given below for these methods.

12.2.1 Analytical solution for calculating sensitivity maps

The analytical solution of an air-cored coil placed above a uniform conducting plate was given by Cheng [4,5]. The magnetic vector potential in the space in this case can be obtained from the analytical solution. The induced voltage in a receiving coil can be calculated by line-integrating the magnetic vector along the coil. Figure 12.3 shows the schematic diagram and geometry of the model for calculating the mutual inductance of a double air coil next to a uniform conducting plate. The base of the coil is at a height of l_1 above the surface and the top of the coil is at l_2. The coil parameters of importance are number of turns N, coil radius r_0 and height $L = l_2 - l_1$. Note that the space below the sample conductor is free space. In the forward model, it is treated as a layer with infinite thickness.

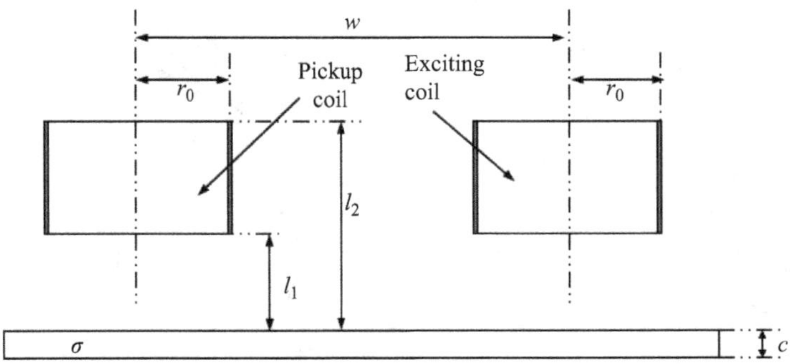

Figure 12.3 Schematic and geometry of double air-coil sensor

In the region between l_1 and l_2, the vector potential can be expressed as

$$A(r,z) = \frac{1}{2}\mu\frac{I\cdot n}{l_2 - l_1}\int_0^\infty \frac{1}{\alpha}J_1(\alpha r_0)J_1(\alpha r)\{2 - e^{\alpha(z-l_2)} - e^{-\alpha(z-l_1)}$$
$$+ e^{-\alpha z}(e^{-\alpha l_1} - e^{-\alpha l_2})\frac{(\alpha_1+\alpha)(\alpha_1-\alpha) - (\alpha_1+\alpha)(\alpha_1-\alpha)e^{2\alpha_1 c}}{-(\alpha_1-\alpha)(\alpha_1-\alpha) + (\alpha_1+\alpha)(\alpha_1+\alpha)e^{2\alpha_1 c}}\Bigg\}\, d\alpha \tag{12.5}$$

where $\alpha_1 = \sqrt{\alpha^2 + j\omega\sigma\mu\mu_0}$, μ and σ are the relative permeability and conductivity of the conducting plate, respectively, I is the current flowing in the coil, n is the number of turns in the coil, r_0 is the radius of the coil, l_1 and l_2 are the height of the bottom and top of the coil and c is the thickness of the plate.

The above formula is obtained by solving differential equations describing a coil above a stratified conductor, where α is a spatial frequency variable and $J_1(x)$ is a first-order Bessel function of the first kind.

The voltage induced in a receiving coil is obtained by integrating the vector potential produced by the exciting coil along the receiving coil (see Figure 12.4). The mutual impedance between the double air coils can then be obtained by dividing the voltage induced in the pickup coil by the current flowing through the exciting coil.

The mutual inductance between an exciting and a receiving coil as shown in Figure 12.4 is

$$M = \frac{1}{2}\frac{\mu n^2 r_0}{(l_2 - l_1)^2}\int_0^{2\pi}\int_0^\infty \cos\left[\theta + \tan^{-1}\left(\frac{r_0\sin\theta}{w - r_0\cos\theta}\right)\right]$$
$$\times \frac{1}{\alpha}J_1(\alpha r_0)J_1\left(\alpha\sqrt{r_0^2\sin^2\theta + (w - r_0\cos\theta)^2}\right)$$
$$\times\left\{2(l_2 - l_1) + \frac{1}{\alpha}\left[2e^{-\alpha(l_2-l_1)} - 2 + (e^{-\alpha l_1} - e^{-\alpha l_2})^2\right.\right.$$
$$\left.\left.\times\frac{(\alpha_1+\alpha)(\alpha_1-\alpha) - (\alpha_1+\alpha)(\alpha_1-\alpha)e^{2\alpha_1 c}}{-(\alpha_1-\alpha)(\alpha_1-\alpha) + (\alpha_1+\alpha)(\alpha_1+\alpha)e^{2\alpha_1 c}}\right]\right\} d\alpha\, d\theta \tag{12.6}$$

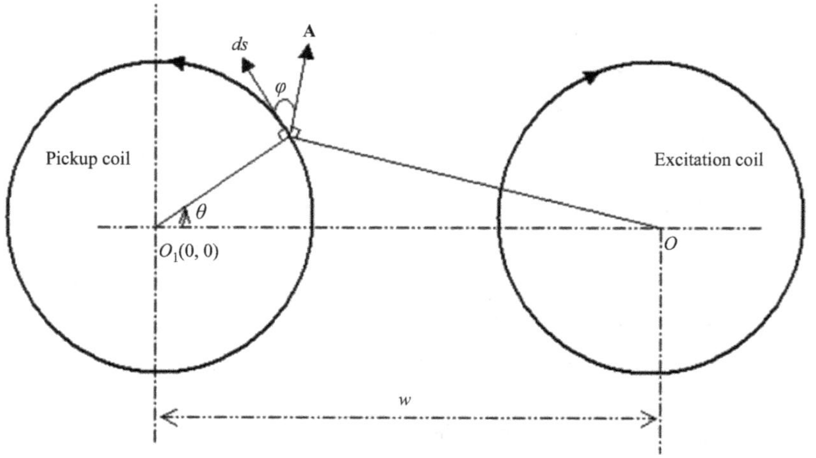

Figure 12.4 Calculation of mutual inductance between two air coils (line-integration of magnetic potential along the receiving coil)

The inductance difference for the cases: a uniform conducting plate and free space is

$$\Delta M = M - M_{\text{air}} = \frac{1}{2}\frac{\mu n^2 r_0}{(l_2 - l_1)^2}\int_0^{2\pi}\int_0^{\infty}\cos\left[\theta + \tan^{-1}\left(\frac{r_0\sin\theta}{w - r_0\cos\theta}\right)\right]\frac{1}{\alpha^2}J_1(\alpha r_0)$$
$$\times J_1\left[\alpha\sqrt{r_0^2\sin^2\theta + (w - r_0\cos\theta)^2}\right](e^{-\alpha l_1} - e^{-\alpha l_2})^2$$
$$\times \frac{(\alpha_1+\alpha)(\alpha_1-\alpha) - (\alpha_1+\alpha)(\alpha_1-\alpha)e^{2\alpha_1 c}}{-(\alpha_1-\alpha)(\alpha_1-\alpha) + (\alpha_1+\alpha)(\alpha_1+\alpha)e^{2\alpha_1 c}} d\alpha d\theta. \tag{12.7}$$

To calculate the sensitivity map for the sensor, the imaging space was divided into a number of rectangular elements as shown in Figure 12.5.

The *k*th sensitivity map (*k*th column of the Jacobian matrix) is expressed as the dot product of the electrical fields produced by the excitation coil and that by the receiving coil as if it is excited by a unit current:

$$\mathbf{S}_k = -\omega^2 \mathbf{A}_i \cdot \mathbf{A}_j \tag{12.8}$$

where $\mathbf{A}_i$ and $\mathbf{A}_j$ are the vector potentials when coils *i* and *j* are excited with unit current, respectively. They were obtained using (12.5).

Figure 12.6 shows three typical sensitivity maps for the planar EMT at frequency 10 kHz and 100 kHz. Other sensitivity maps for coil pairs 1-5, 1-6, 2-3, etc. can be obtained either by rotating or mirroring.

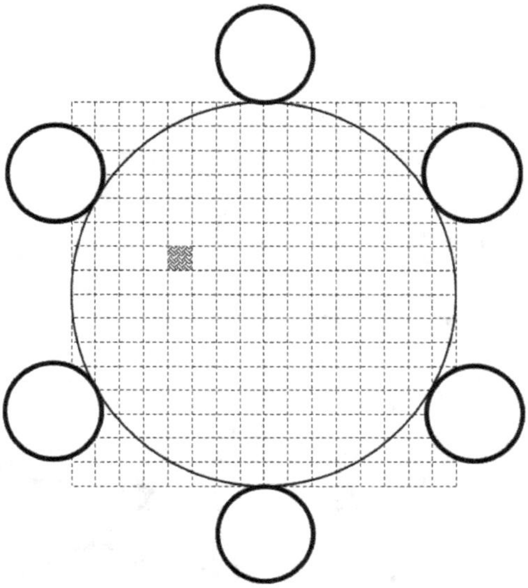

Figure 12.5 Illustration of coil arrangement and pixels (elements) arrangement for calculating sensitivity matrix

12.2.2 Simulation with a custom edge-element FEM solver

In this section, the canonical edge-element FEM solver – Galerkin formulations-based FEM solver is presented. Galerkin's equations were formulated over 10 decades ago and are still the most common solver for calculating the global stiffness matrix in FEM [6]. First, the original Galerkin's equations applied in electromagnetics (EM) are transformed into matrix form, as shown in the following:

$$\int_{\Omega_c} \nabla \times \mathbf{N}_i \cdot \nu \nabla \times \mathbf{A}^{(n)} d\Omega + \int_{\Omega_c} j\omega\sigma \mathbf{N}_i \cdot \mathbf{A}^{(n)} d\Omega + \int_{\Omega_c} j\omega\sigma \mathbf{N}_i \cdot \nabla V^{(n)} d\Omega$$
$$= \int_{\Omega_c} \nabla \times \mathbf{N}_i \cdot \nu_0 \nabla \times \mathbf{A}_s d\Omega \qquad (12.9)$$
$$i = 1, 2, ..., 6$$

$$\int_{\Omega_c} j\omega\sigma \nabla L_i \cdot \mathbf{A}^{(n)} d\Omega + \int_{\Omega_c} j\omega\sigma \nabla L_i \cdot \nabla V^{(n)} d\Omega = 0$$
$$i = 1, 2, ..., 4 \qquad (12.10)$$

where L_i is the elemental interpolation of ith node corresponding to its nth element, $\mathbf{N}_i$ is a vector interpolation of the ith edge corresponding to its nth edge element, $\mathbf{A}^{(n)}$ is the edge vector potential of the nth element, $V(n)$ is the electrical potential on the four vertices for the nth element in the meshed domain (including the tested plate subdomains and the surrounding air subdomains), $\mathbf{A}_s$ is the applied edge vector potential of the nth element, ν is the reluctivity (the reciprocal of the permeability) of the target, σ is the conductivity of the target, ν_0 is the reluctivity (the

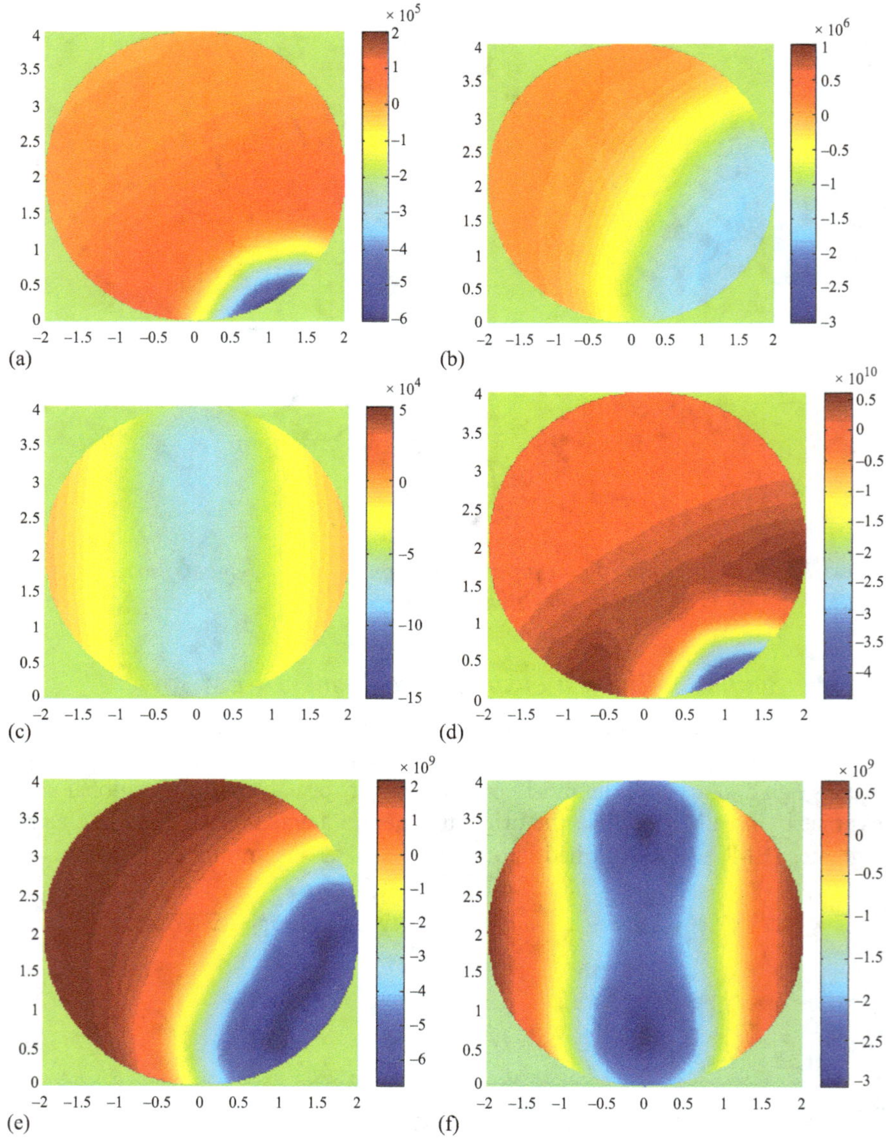

Figure 12.6 Sensitivity maps for coil pairs 1-2, 1-3 and 1-4 at 10 kHz (a)–(c) and 100 kHz (d)–(f)

reciprocal of the permeability) of the air, ∇ is the gradient operation and $\nabla\times$ is the curl operation.

For element n, the stiffness matrix Q is:

$$Q^{(n)} = \begin{pmatrix} K^{(n)}(6 \times 6) & L^{(n)}(6 \times 4) \\ M^{(n)}(4 \times 6) & N^{(n)}(4 \times 4) \end{pmatrix} \tag{12.11}$$

Combining (12.11) with (12.9) and (12.10), the Galerkin form of the whole system becomes

$$\sum_{n=1}^{N0} Q^{(n)} \cdot \begin{pmatrix} \mathbf{A}^{(n)} \\ V^{(n)} \end{pmatrix} = \sum_{n=1}^{N0} \begin{pmatrix} K^{(n)} & L^{(n)} \\ M^{(n)} & N^{(n)} \end{pmatrix} \cdot \begin{pmatrix} \mathbf{A}^{(n)} \\ V^{(n)} \end{pmatrix} = B \tag{12.12}$$

where matrix K is composed of matrices K_1 and K_2, L is the matrix form of the second term in (12.9), which is related to the electric potential and the Maxwell–Wagner effect, M and N are the matrices form of the first and second term in (12.10), B is the matrix form of the right side term of (12.9) and (12.10), which represents the Dirichlet boundary condition and $N0$ is the total number of elements.

The electric field of an arbitrary element can then be derived from the vector potential and electric potential in (12.12) by using V-A formulation incorporating Coulomb gauge [7]:

$$\mathbf{E}^{(n)} = -j\omega\mathbf{A}^{(n)} - j\omega\nabla V^{(n)} \tag{12.13}$$

where $\mathbf{A}^{(n)}$ is the vector sum of the vector potential on all the edges of each tetrahedral element and $V^{(n)}$ denotes the electric potential on all the vertices of each tetrahedral element.

The change in mutual inductance between the transmitter and the receiver can be obtained by applying the equation presented by Mortarelli [8] or Auld and Moulder [9]. In both analyses, they started from the Lorentz reciprocity relation and arrived at the same generalized equation that can be applied to any pair of coils [8,9]:

$$\Delta L = \frac{1}{j\omega I^2}\int_c \mathbf{E}_a \cdot \mathbf{J}_b dv = \frac{1}{j\omega I^2}\int_c \mathbf{E}_a \cdot \mathbf{E}_b \cdot (\sigma_a - \sigma_b) dv \tag{12.14}$$

where ΔL is the change in inductance caused by the difference between medium a and b.

To develop an efficient FEM solver that can simulate the eddy currents induced by an excitation coil in biological tissues in any sectional views for normal and pathological conditions (such as internal and peripheral strokes), a custom FEM formulation based on the Rayleigh–Ritz–Galerkin method under low frequency is presented in (12.16).

Because the angular frequency ω is small under the low-frequency condition, the term $\int_{\Omega_c} j\omega\sigma\mathbf{N}_i \cdot \mathbf{A}^{(n)} d\Omega$ is much smaller than that of $\int_{\Omega_c} \nabla \times \mathbf{N}_i \cdot \nu\nabla \times \mathbf{A}^{(n)} d\Omega$, which suggests the secondary field produced by the eddy currents are very small compared to the primary field. This approximation is called the weakly coupled approximation. Then, ignoring the ω terms, (12.9) becomes

$$\int_{\Omega_c} \nabla \times \mathbf{N}_i \cdot \nu\nabla \times \mathbf{A}^{(n)} d\Omega \approx \int_{\Omega_c} \nabla \times \mathbf{N}_i \cdot \nu_0 \nabla \times \mathbf{A}_s d\Omega \qquad i = 1, 2, ..., 6 \tag{12.15}$$

Assuming the reluctivity of the target ν is approximately equal to that in the air ν_0, the estimated edge vector potential $\mathbf{A}^{(n)}$ is approximately the same as the primary edge vector potential. Therefore, $\mathbf{A}^{(n)} \approx \mathbf{A}_s$. Consequently, the electric potential on all the vertices of each tetrahedral element can be calculated by the following:

$$\int_{\Omega_c} j\omega\sigma\nabla L_i \cdot \mathbf{A}_s d\Omega + \int_{\Omega_c} j\omega\sigma\nabla L_i \cdot \nabla V^{(n)} d\Omega \approx 0 \quad i = 1, 2, \ldots, 4 \tag{12.16}$$

The primary field $\mathbf{A}_s$ due to the current flowing in transmitter can be obtained based on the Biot–Savart law [10]. The electric field can be obtained once V is known:

$$\mathbf{E}^{(n)} = -j\omega\mathbf{A}_s{}^{(n)} - j\omega\nabla V^{(n)} \tag{12.17}$$

Because only scalar electric field $\nabla V^{(n)}$ remains in (12.16) under low frequency, the eddy current computation time will be much reduced when compared to vector electric field in (12.13).

Compared with the typical FEM solver in (12.9) and (12.10), the proposed simplified FEM solver in (12.16) can significantly reduce the computation burden as only one polynomial needs to be solved. In addition, the proposed simplified FEM solver has eliminated the frequency terms in typical FEM solver in (12.9), which would require a larger region to be defined to encompass the original model to precisely compute the eddy currents affected by the skin effect. Therefore, the overall mesh size is much larger than 2 million when calculated by the typical FEM solver.

Based on the simplified FEM algorithm as given by (12.16), the Sparse Reverse Cuthill–McKee orderings technique and the incomplete LU (ILU) decomposition method have been implemented in the proposed FEM solver to reorder and decompose the sparse polynomial matrix ($j\omega\sigma\nabla \mathrm{d}L_i$ term in (12.16)) so that most of the elements are distributed close to the diagonal of the stiffness matrix, which is a sparse polynomial matrix. This method has been shown to be very efficient in accelerating the convergence of the solving process [11]. A 2M realistic human brain model's stiffness sparse matrix elements distributions before and after the reordering and decomposing are presented in Figure 12.7(a) and (b). After reordering and decomposing the sparse stiffness matrix, Bi-conjugate Gradients Stabilised (CGS) iterative method was utilized to find the final numerical solution of the system.

As an example of the FEM solver, a 2 million-element realistic human brain with a dimension of 172 × 211 × 192 mm is simulated (see Figure 12.8). The mesh statistics of the realistic human brain mesh model (University College London, UK) is presented in Table 12.1. The conductivity of brain tissues is listed in Table 12.2.

The brain model is excited by a coil of 62.5 mm in radius positioned on its top with a distance of 147 mm to the brain centre. A current of 1 A at 10 Hz is applied to the coil. A receiver coil is also located on its top with a distance of 145 mm to the brain centre. The sensor setup and parameters (both transmitter and receiver) are shown in Figure 12.9 and Table 12.3. The simulation results are shown in Figure 12.10.

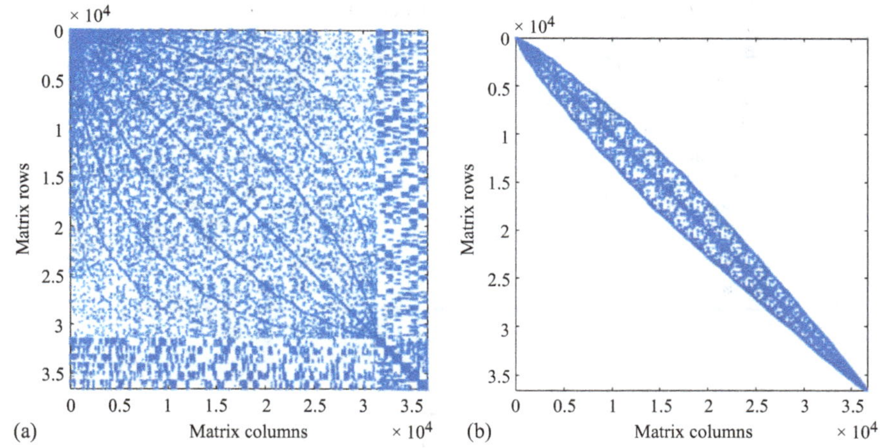

Figure 12.7 2M realistic human brain model's stiffness sparse matrix elements distributions: (a) before re-orderings and decomposition and (b) after re-orderings and decomposition

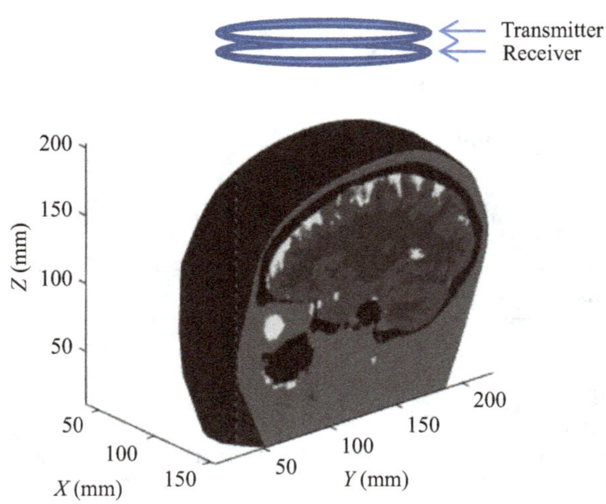

Figure 12.8 2M-elements realistic human brain model

Table 12.1 Mesh statistics for 2M elements realistic human brain model

Parameters	Values
Number of elements	2,047,408
Number of nodes	349,305
Minimum intermodal spacing (mm)	0.9801E–6
Average intermodal spacing (mm)	0.5094

Table 12.2 Electrical properties of brain tissues

Tissues	Conductivity (S/m)
White matter	0.15
Grey matter	0.3
CSF	1.79
Dura mater	0.44
Skull	0.018
Air cavities	0.001
Scalp	0.44

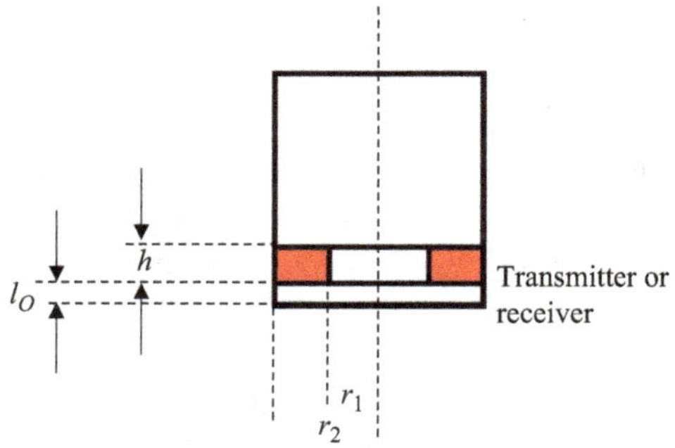

Figure 12.9 Sensor configuration

Table 12.3 Coil parameters

Sensor parameters	Values
r_1 (mm)	62.5
r_2 (mm)	65
lo (lift-off) (mm)	2
h (height) (mm)	2.2
Number of turns N_1 (transmitter) = N_2 (receiver)	80

12.2.3 Simulation with a custom BEM solver

For a typical MIT system for industrial applications, the driving frequency is normally in the range of kHz to several hundred kHz. For metallic object, when the driving frequency is high, the conductivity of the metal can be treated as infinite, the skin depth approaches to zero, the metal behaves essentially like a perfect electric conductor (PEC) and at this thin skin depth limiting case, the incident magnetic field is almost totally reflected, namely the magnetic field only exists on the surface of the metal. Hence, it is not necessary to discretise the volume inside the object and it is sufficient to discretise only on the surface. In this case, the traditional versions of FEM are not efficient for the calculation of the sensitivity

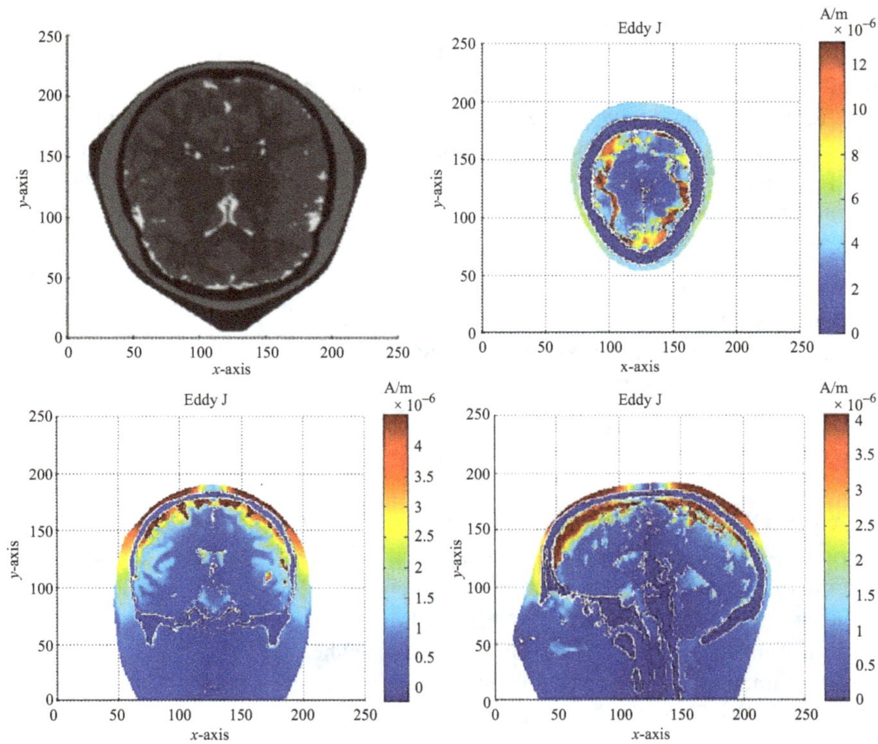

Figure 12.10 Eddy current simulations for a 2M-elements brain model

and the forward problem due to the requirement for large number of elements to reach an acceptable computational precision. In contrast, BEM based on integral formulations becomes effective because meshes are only required on the surface of the object.

Investigation on how to deal with the computation of MIT with integral equations has been carried out by several workers [12,13]. The magnetic scalar potential formulations are used to calculate the sensitivity maps of a MIT due to a small PEC perturbation. Beginning with Ampere's law, applied to the external region of the target, and Gauss's magnetic divergence law gives [14]:

$$\nabla \times \mathbf{H}^e = \sigma^e \mathbf{E}^e + i\omega\varepsilon \mathbf{E}^e \tag{12.18}$$

$$\nabla \cdot \mu \mathbf{H} = \nabla \cdot \mu^e \mathbf{H}^e = 0 \tag{12.19}$$

The exterior (air) environment is assumed to have the magnetic permeability of free space. The superscript *e* indicates the quantities in the external environment. The total magnetic field consists of two parts: the primary magnetic field generated by the excitation coil and the scattered magnetic field radiated by the eddy currents, that is $\mathbf{H}^e = \mathbf{H}^{pr} + \mathbf{H}^{sc}$, where the superscripts *pr* and *sc* indicate the primary field and the scattered field, respectively.

Assuming that the magnetic field is quasi-static in the external region, and hence the displacement current can be negligible. Also, the space exterior to the target is non-conductive so that the electric field is almost zero and the region investigated does not include the source current. So the exterior total magnetic field is irrotational and thus can be represented efficiently using a simple scalar potential. Together with the divergence theorem, the magnetic scalar potential φ^e satisfies the Laplace equation:

$$\nabla^2 \varphi^e = 0 \tag{12.20}$$

The Laplace equation can be converted to a boundary integral equation, with the jump condition on the normal magnetic field across the surface in the case of PEC, it produces the final expression:

$$c\varphi^e(\mathbf{r}) + \int_\Gamma \left[\varphi^e(\mathbf{r}') \frac{\partial g(\mathbf{r}, \mathbf{r}')}{\partial \mathbf{n}} \right] d\Gamma = \varphi^{pr}(\mathbf{r}) \tag{12.21}$$

where $g(\mathbf{r}, \mathbf{r}') = \frac{1}{4\pi} \frac{1}{|\mathbf{r} - \mathbf{r}'|}$ is the Green's function for the Laplace equation, φ^{pr} is the magnetic scalar potential of the primary magnetic field, vectors $\boldsymbol{r}$ and $\boldsymbol{r}'$ are the observation point and source point, respectively, Γ is the surface of the object with the unit normal vector $\boldsymbol{n}$ and c depends on the location of observation point and is 1/2 on a smooth boundary of Γ, 1 in Γ and 0 elsewhere [13].

For numerical purposes, the object surface is modelled using planar triangular patches [15–18]. Then the integration can be transformed into equations in discrete collocation points. A local coordinate system [16,19] is introduced to facilitate the calculation of 3D numerical integrations based on Green's function or its gradient on a plane triangle. The new coordinate system parallels with triangular element and has origin on the first vertex of the triangle. One edge of the triangle is set as the abscissa, as shown in Figure 12.11.

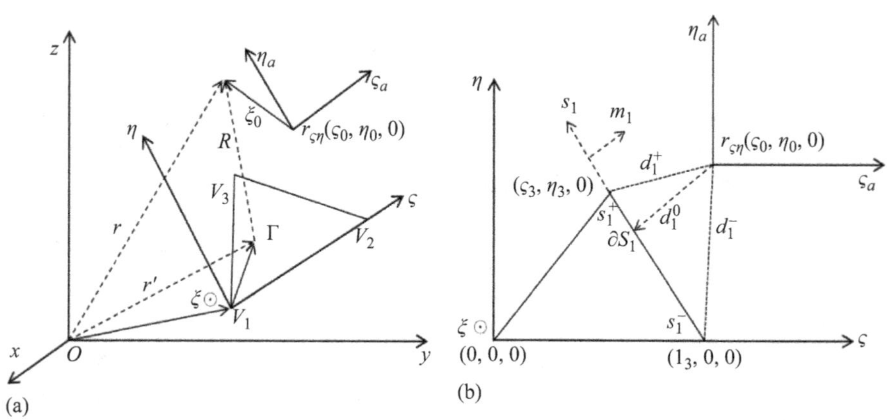

Figure 12.11 (a) Relationship between the local coordinate system and the global coordinate system and (b) parameters of the local coordinate system

Over the surface, $\varphi^e(\mathbf{r}')$ is interpolated with simple linear function using the linear nodal function and the magnetic scalar potential on the vertex of triangular patches. A transformation matrix is introduced to transform the local number into the global number. Therefore, the final integral equation is given by the coefficient matrix and the integral equation could be simplified to the matrix form. By calculating the linear simultaneous equations, the magnetic potentials and magnetic field can be obtained.

The magnetic field generated by the coil can be easily derived from the Biot–Savart law. Note that $\mathbf{H}^{pr} = -\nabla\varphi^{pr}$, the potential of the primary magnetic field φ^e can be derived, as well as ultimate result for the exterior magnetic field $\mathbf{H}^e$.

The voltage on the receiving coil could be calculated according to Maxwell's equations. Sensitivity distributions of the measured space S_{en} are calculated using the formula:

$$S_{en} = -\frac{U^e - U^{pr}}{U^{pr}} = -\frac{U^{sc}}{U^{pr}} \tag{12.22}$$

where U^e, U^{sc} and U^{pr} are the total voltage, voltage produced by the scattering field and voltage produced by the primary field, respectively.

The induced voltage of the receiving coil can be obtained by moving the target object in the space between the coils with small steps each time and the sensitivities can be deduced. Assume a sinusoidal current is applied to the excitation coil and the frequency is $f = 10$ MHz. Simulations were carried out with one excitation coil and four receiving coils. The five coils are of the same size with radius $ra = 1$ m, as shown in Figure 12.12. Four typical sensitivity maps are shown in Figure 12.13.

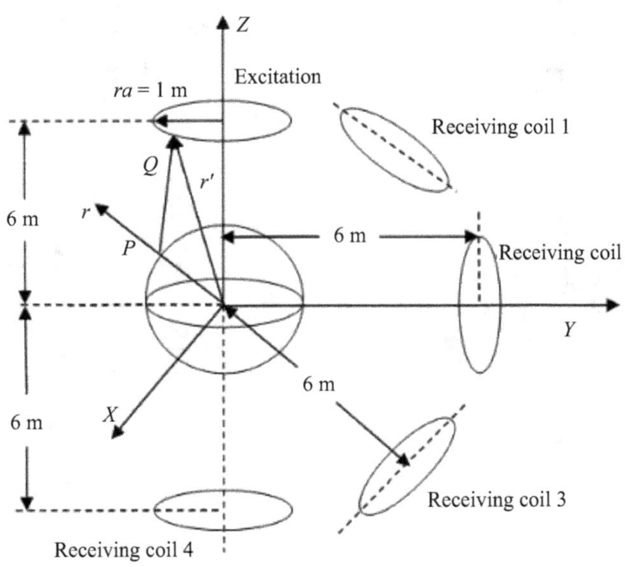

Figure 12.12 Positions of coils and target

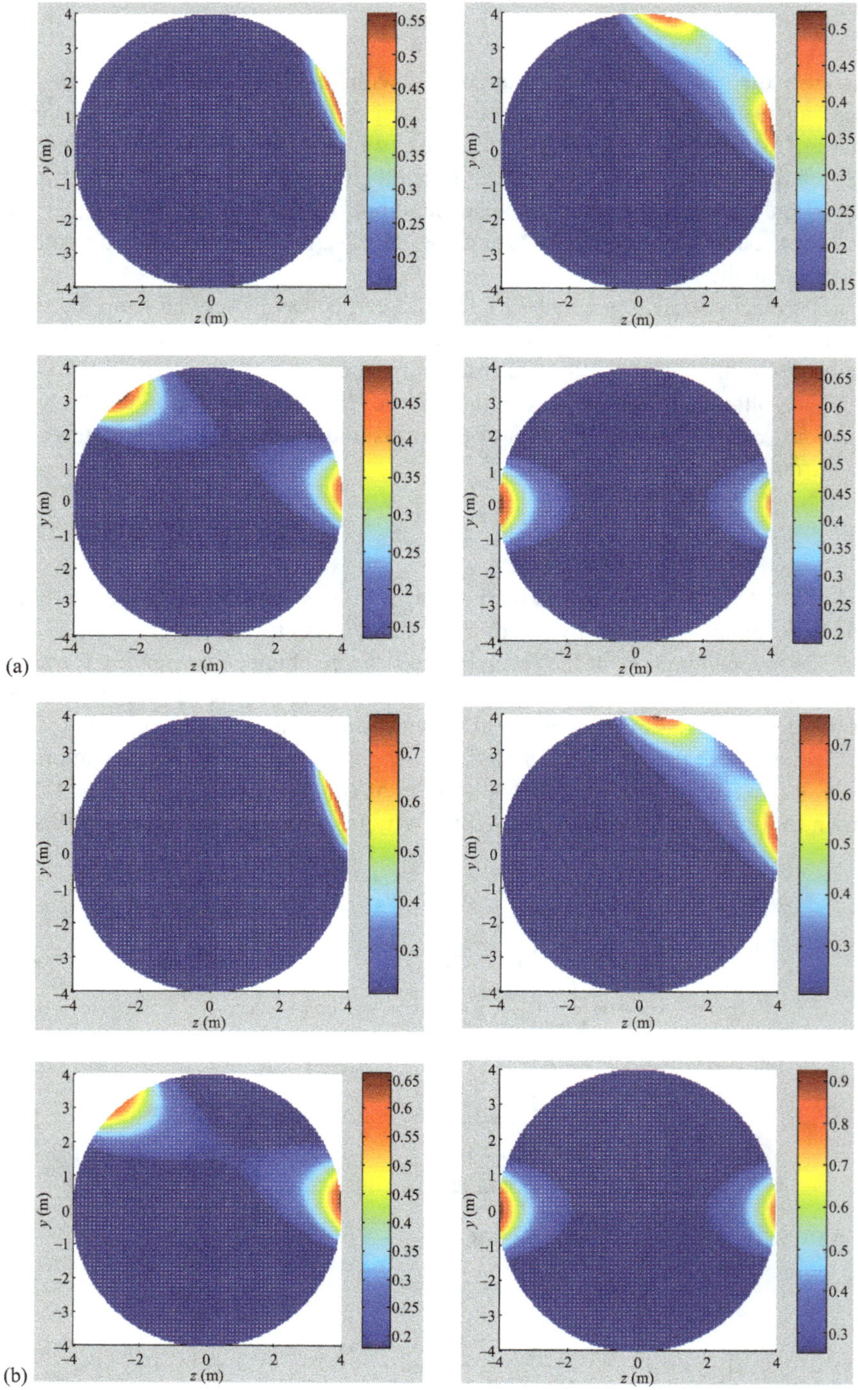

Figure 12.13 *Four typical sensitivity maps obtained by BEM: (a) for a sphere perturbation, (b) for a cube perturbation, (c) for a rectangle perturbation, (d) for an ellipsoid perturbation and (e) for a cylinder perturbation*

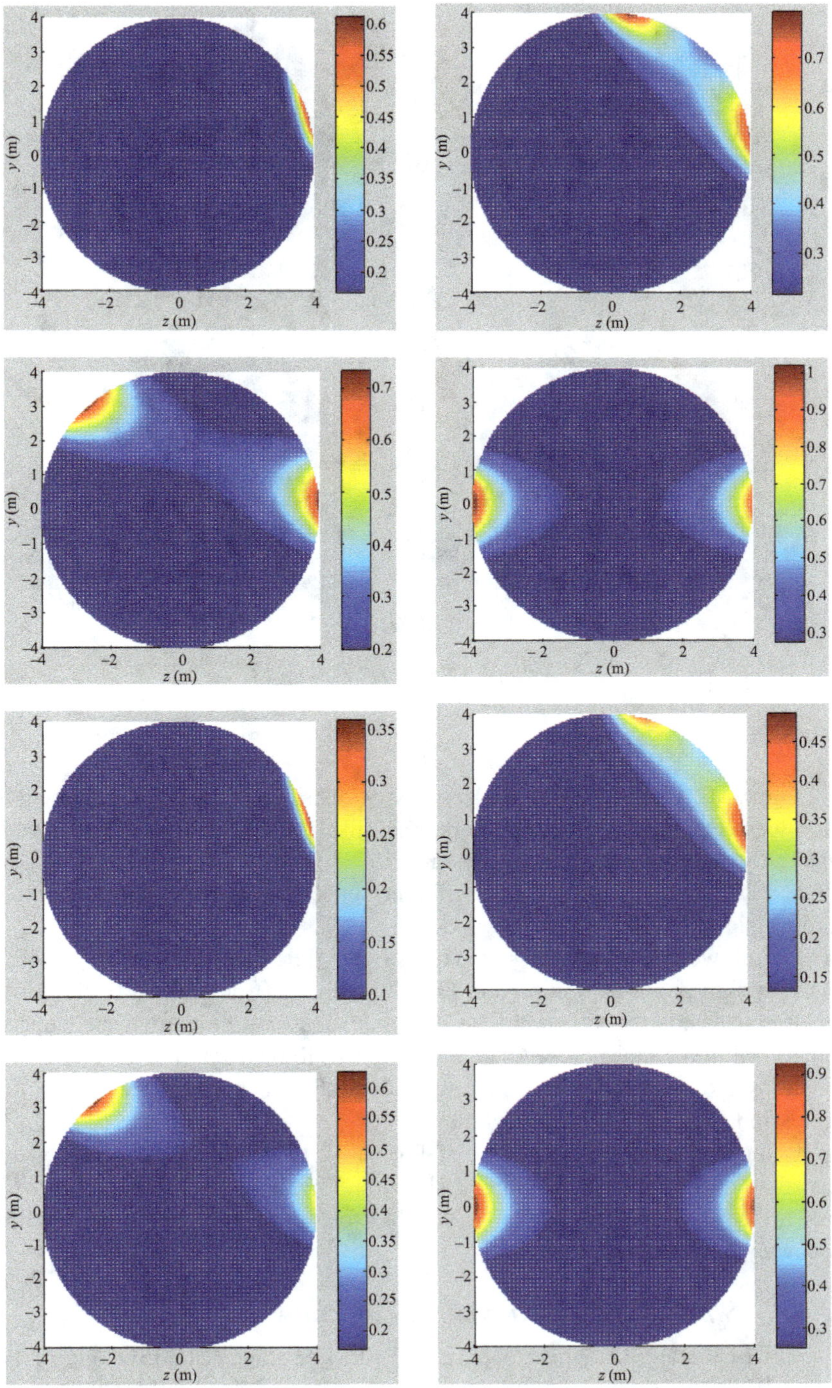

(c)

(d)

Figure 12.13 (*Continued*)

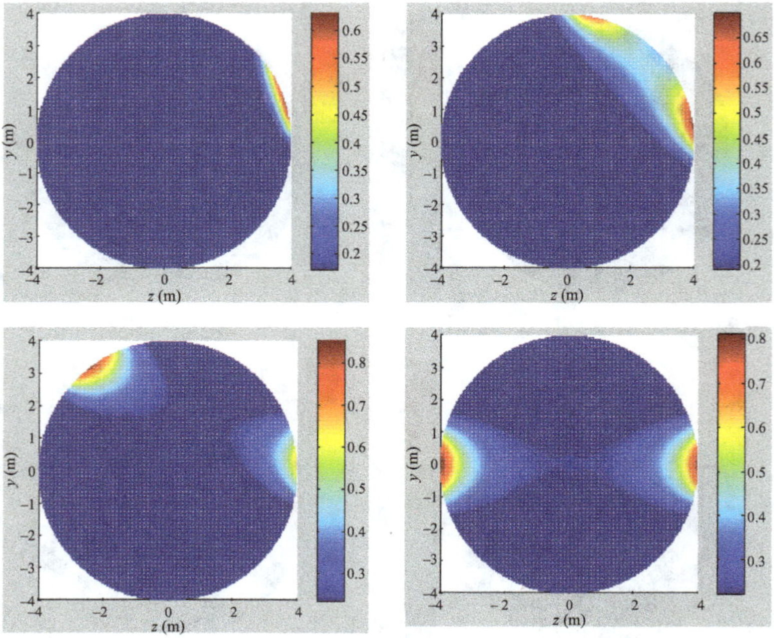

(e)

Figure 12.13 (Continued)

Figure 12.13 shows that the sensitivity distributions are in agreement with what were reported in previous literatures, displaying typical saddle shapes. The sensitivity maps are generally the same for five different perturbation shapes, which indicate that objects with different shapes but same volume is likely to produce similar sensitivities.

For a lower frequency range, the transmitted field penetrates the target slightly. The thin skin depth approximation (TSA) [6,12,14,20] is applied to describe the penetration inside the target surface. For this situation, the normal magnetic field component in the global coordinate system needs to be taken into account and new equations are needed by choosing several meshes to apply a weighted integral expression of the divergence relation.

Using the same configuration of Figure 12.12, simulations were conducted according to the TSA assumption, as shown in Figure 12.14. From the results, it is easy to see that the TSA model is suitable for the calculation of the broadband frequency MIT system.

12.2.4 Simulation with a commercial software

A simulation example using a commercial FEM package (COMSOL) is shown in Figure 12.15. An eight coil sensor model is shown in Figure 12.15(a) and (b), while

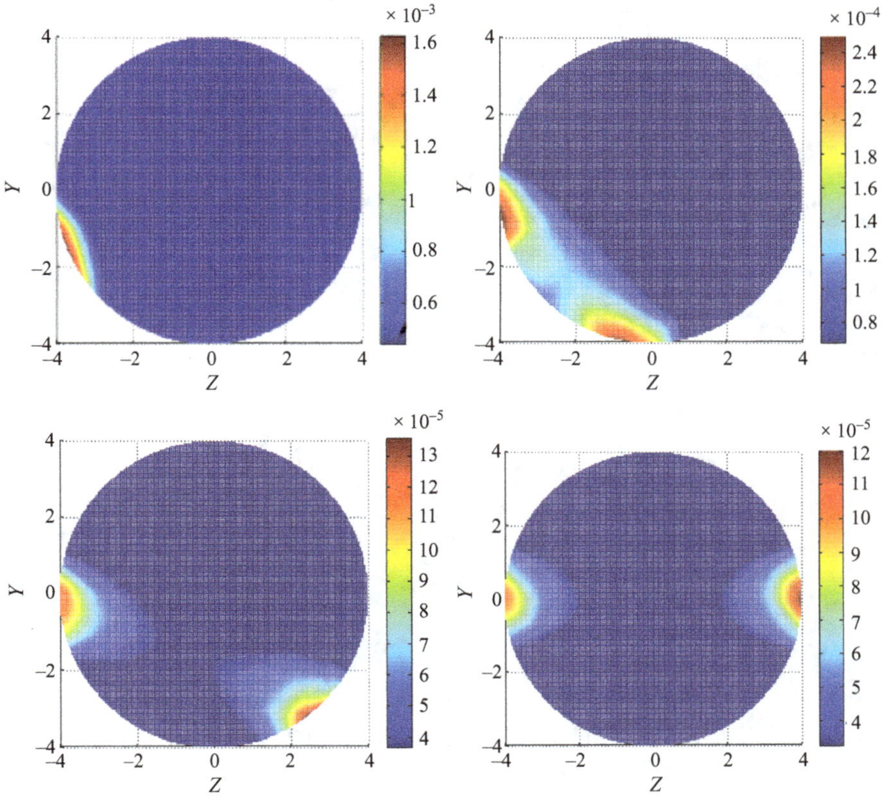

Figure 12.14 Four typical sensitivity maps obtained by TSA for a sphere perturbation

the eddy currents induced in the cylindrical object located in the centre of the imaging space is shown in Figure 12.15(c) and (d).

12.3 MIT hardware system design

The measurement of mutual inductances between the coils is normally accomplished by applying a sinusoidal signal to the excitation coil and measuring the response voltage at the receiving coil. The amplitude and phase of the excitation current signal and those of the receiving voltage are obtained by the demodulation process.

The hardware comprises mainly two parts, for example the front end circuits and the FPGA module as shown in Figure 12.16. Due to the fact that signal generation and demodulation function are implemented digitally in the FPGA, front

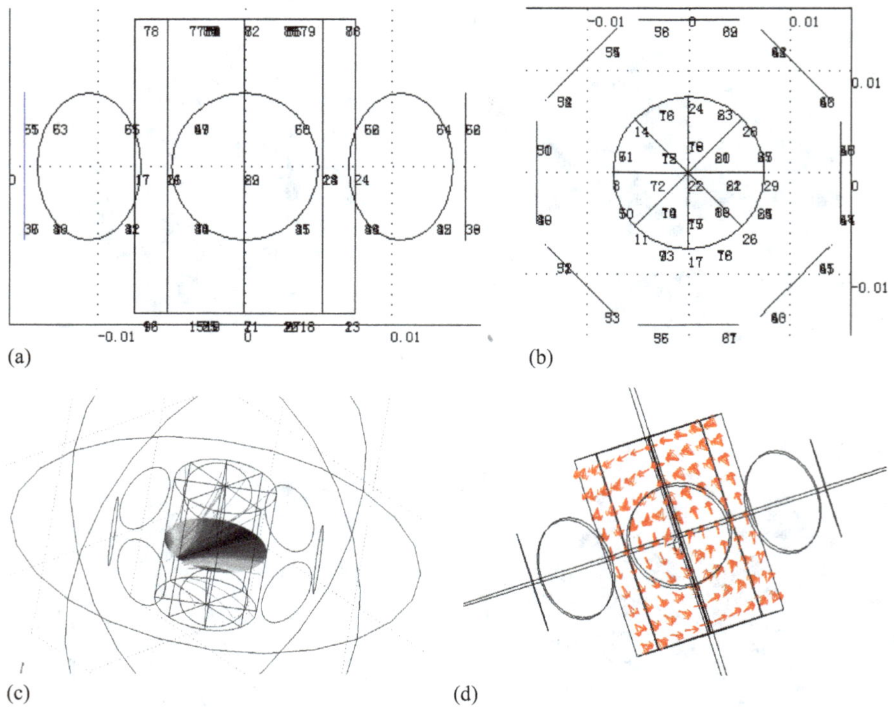

Figure 12.15 (a–b) A FEM model for an eight-coil MIT sensor and (c–d) eddy current distributions obtained when one coil is excited

end circuits are relatively simple, that is only amplification and switching. The MIT instrument operates in the spectrum from 10 kHz to 500 kHz and is equipped with the latest FPGA Zynq module, capable of digital demodulation at a rate of 100 k/s and has an Ethernet link to a PC. The parameters such as amplification gain, frequency and sampling rate can be set from the PC and provide good compatibility with varieties of sensor designs and applications. The signal-to-noise ratio (SNR) of the output signal is on average above 70 dB.

12.4 Typical applications

MIT has been applied in the imaging of molten steel flow through the submerged entry nozzle (SEN) during continuous casting. In a lab, a scaled (approx. 10:1) experimental rig consisting of a tundish, stopper rod, nozzle and mould was used. Argon gas was injected through the centre of the stopper rod and the behaviour of the two phase GaInSn/Argon flow was studied. The experiments were performed with GaInSn as an analogue for liquid steel, which has similar conductive properties as molten steel and allows the measurements at room temperature. In a pilot

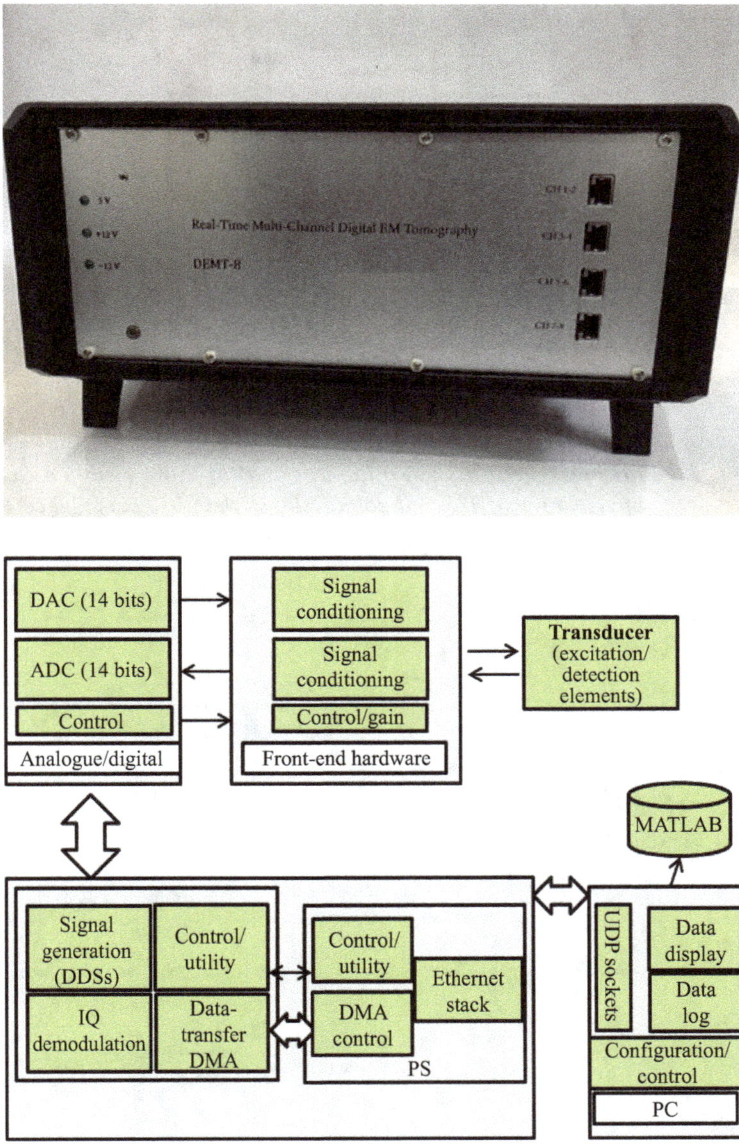

Figure 12.16 (a) System photo and (b) system block diagram

plant, imaging of molten steel was carried out (Figure 12.17). The system operated with a sinusoidal excitation waveform with a frequency of 20 kHz and a capture rate of 40 frames per second.

Further results show that the ability of the EMT system to distinguish different flow scenarios (see Figure 12.18).

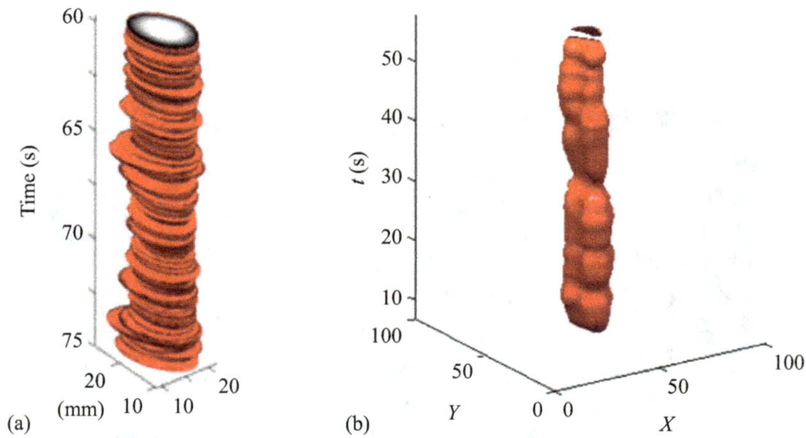

Figure 12.17 *(a) Flow of GaInSn in a lab test and (b) flow of molten steel in an industrial test. The red colour shows the 20% iso-surface of the maximum conductivity*

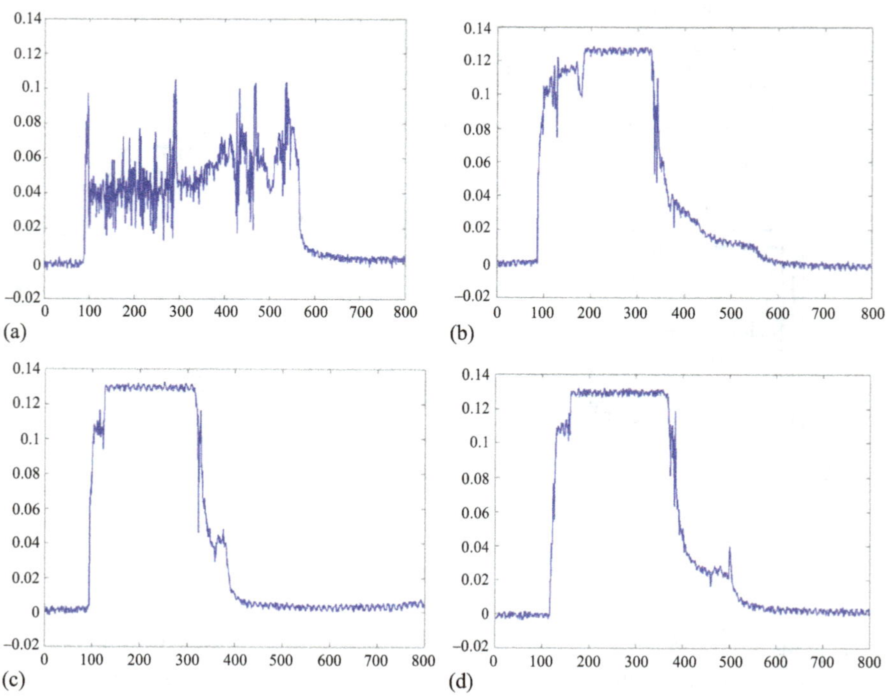

Figure 12.18 *Different flow regime of GaInSn due to the stopper position: (a) 12.2 mm, (b) 15.3 mm, (c) 18 mm and (d) 21.4 mm stopper lift*

12.5 Summary

EMT has found increasing applications in industrial processes and biomedical applications, such as imaging molten steel flow and detecting brain oedema. Further developments of the technique would benefit from advances in fast forward and inverse solvers, low cost yet high performance systems and new emerging application areas such as in landmine detection, food inspection and security applications.

References

[1] Yin W, Peyton A J, 'A planar EMT system for the detection of faults on thin metallic plates', *Meas. Sci. Technol.*, 2006, **17** (8), pp 2130–2135

[2] Tao Y, Yin W, Zhang W, Zhao Y, Ktistis C, Peyton A J, 'A very-low-frequency electromagnetic inductive sensor system for workpiece recognition using the magnetic polarizability tensor', *IEEE Sensors J.*, 2017, **17** (9), pp 2703–2712

[3] Yin W, Chen G, Chen L, Wang B, 'The design of a digital magnetic induction tomography (MIT) system for metallic object imaging based on half cycle demodulation', *IEEE Sensors J.*, 2011, **11** (10), pp 2233–2240

[4] Cheng C C, Dodd C V, Deeds W E, 'General analysis of probe coils near stratified conductors', *Int. J. Nondestruct. Testing*, 1971, **3**, p 109

[5] Terzija N, Yin W, Gerbeth G, *et al.*, 'Electromagnetic inspection of a two-phase flow of GaInSn and argon', *Flow Meas. Instrum.*, 2011, **22** (1), pp 10–16

[6] Bíró O, 'Edge element formulations of eddy current problems', *Comput. Methods Appl. Mech. Eng.*, 1999, **169** (3–4), pp 391–405

[7] Zeng Z, Udpa L, Udpa S S, Chan M S C, 'Reduced magnetic vector potential formulation in the finite element analysis of eddy current nondestructive testing', *IEEE Trans. Magn.*, 2009, **45** (3), pp 964–967

[8] Mortarelli J R, 'A generalization of the Geselowitz relationship useful in impedance plethysmographic field calculations', *IEEE Trans. Biomed. Eng.*, 1980, **11**, pp 665–667

[9] Auld B A, Moulder J C, 'Review of advances in quantitative eddy current nondestructive evaluation', *J. Nondestruct. Eval.*, 1999, **18**, pp 3–6

[10] Hayashi N, Isaka K, Yokoi Y, 'ELF electromagnetic environment in power substations', *Bioelectromagnetics*, 1989, **10**, pp 51–64

[11] Bridson R, Tang W P, 'A structural diagnosis of some IC orderings', *SIAM J. Sci. Comput.*, 2001, **22**, pp 1527–1532

[12] Pham M H, Peyton A J, 'A model for the forward problem in magnetic induction tomography using boundary integral equations', *IEEE Trans. Magn.*, 2008, **44** (10), pp 2262–2267

[13] Engleder S, Steinbach O, 'Boundary integral formulations for the forward problem in magnetic induction tomography', *Math. Methods Appl. Sci.*, 2011, **34** (9), pp 1144–1156

[14] Sun K, O'Neill K, Shubitidze F, Haider S A, Paulsen K D, 'Simulation of electromagnetic induction scattering from targets with negligible to moderate penetration by primary fields', *IEEE Trans. Geosci. Remote Sens.*, 2002, **40** (4), pp 910–927

[15] Morrison J A, 'Integral equations for electromagnetic scattering by perfect conductors with two-dimensional geometry', *Bell Syst. Tech. J.*, 1979, **58** (2), pp 409–425

[16] Graglia R D, 'On the numerical integration of the linear shape functions times the 3-D Green's function or its gradient on a plane triangle', *IEEE Trans. Antennas Propag.*, 1993, **41** (10), pp 1448–1455

[17] Graglia R, 'Static and dynamic potential integrals for linearly varying source distributions in two- and three-dimensional problems', *IEEE Trans. Antennas Propag.*, 1987, **35** (6), pp 662–669

[18] Rao S, Wilton D, Glisson A, 'Electromagnetic scattering by surfaces of arbitrary shape', *IEEE Trans. Antennas Propag.*, 1982, **30** (3), pp 409–418

[19] Zhao Q, Chen G, Hao J, Xu K, Yin W, 'Numerical approach for the sensitivity of a high-frequency magnetic induction tomography system based on boundary elements and perturbation method', *Meas. Sci. Technol.*, 2013, **24** (7), 074004

[20] Zhao Q, Yin W, 'The multi-frequency responses and sensitivity calculation of broadband magnetic induction tomography system using the boundary element method', in *Proceedings of 2014 IEEE International Conference on Imaging Systems and Techniques*, Santorini, Greece, 14–17 October, 2014, pp 48–52

Dr Wuliang Yin received his BSc and MSc degrees in Electronic Measurement and Instrumentation from Tianjin University, China, in 1992 and 1995, and PhD degree in Automotive Electronics from Tsinghua University, China, in 1999. Since 2005, he has been with the School of Electrical and Electronic Engineering, The University of Manchester, UK, and currently is a Senior Lecturer in Electromagnetic Sensors and Instrumentation. He has published more than 200 refereed journal and conference papers and has more than 10 patents granted. His research interests include advanced sensors and instrumentation, automotive electronics and electromagnetics. Dr Yin was the recipient of 2000 Science and Technology Award from the Chinese Ministry of Education, 2014 and 2015 Williams Award from The Institute of Materials, Minerals and Mining, 2016 Best Graduate Student Paper from the IEEE I2MTC, Excellent Paper Award – Agilent Measurement Forum in 2000 and National Scientific and Technical Progress Award by the Chinese Government in 1999. He is an IEEE Senior Member since 2007 and an EPSRC college member since 2017. He is leading and involved in a number of EPSRC and EU funded projects.

Chapter 13
Micro-sensors for cell and blood imaging

Jiafeng Yao[1] *and Masahiro Takei*[2]

List of abbreviations

ECT	electrical capacitance tomography
EDL	electrical double layer
EEC	electrical equivalent circuits
EIS	electrical impedance spectroscopy
EIT	electrical impedance tomography
FEM	finite-element method
GVSPM	generalised vector sampled pattern matching
IC	image correlation
IE	image error
Iter	iteration times
LBP	linear back-projection
MEMS	micro-electro-mechanical systems
PC	personal computer
PCB	printed circuit board
PLI	projected Landweber iteration
RBC	red blood cell
TK	Tikhonov regularisation

13.1 Introduction

Cell and blood imaging is a key process for developing biomedical measurement devices. Visualisation for measuring cell concentration distribution can potentially improve the control of cells in micro-fluidic flow [1], and also it is useful for

[1]College of Mechanical & Electrical Engineering, Nanjing University of Aeronautics & Astronautics, Nanjing, China
[2]Department of Mechanical Engineering, Chiba University, Chiba, Japan

validating computational models of cells distribution in micro-fluidics [2]. Electrical impedance tomography (EIT) is one of the visualisation technique to measure the electrical conductivity or permittivity distribution within a medium using electrical measurements from a series of electrodes on the surface of the object [3]. EIT uses low frequency (0–10 MHz), low voltage (0–5 V) and low current (0–4 mA) [4]. It has advantages, such as no radioactive source, non-invasive, fast and low cost [5]. EIT has been widely used for biological applications [5,6].

The diameter of an EIT sensor in biological applications is usually between 2.5 cm and 10 cm [7]. EIT has an advantage to be scaled down to suit a specific application [8], for example micro EIT (μEIT) to visualise cell distribution in micro-fluidics. A numerical EIT simulation conducted by Meir and Rubinsky preliminarily proved that EIT could be a viable method to monitor the spatial and temporal dynamics of a single cell electroporation process [9]. However, few studies have been applied a miniaturised micro-fluidic device to visualise cell distribution in a microchannel flow, due to the distinct properties and phenomena with greater influence at the micro-scale than at the macro-scale by miniaturisation [10].

Biophysical study with the use of a micro-scale device reveals that impedance measurements are frequency dependent [11]. At relatively low-frequency, the cell/fluid solution is primarily affected by the electrical currents flowing around and between the cells (as para-cellular current), while in relatively high frequency, the solution is primarily affected by the electrical currents flowing across the cell membranes (as transcellular current) [12]. The phenomena present new challenges during a measurement process in selection of optimal frequency. Furthermore, much more noise is introduced when a device is in micro scale, including capacitance between cables, electrical double layer around cells and contact impedance [13].

The quality of the reconstructed images from the μEIT could be improved from two aspects: hardware and software. In hardware, the electrodes used in μEIT must be sufficiently small and uniform with a fixed shape and size, which are advantageous in terms of sensitivity with respect to a local change in conductivity.

In software, the image reconstruction algorithm used and the excitation frequency should be optimised to improve the image quality. Various image reconstruction algorithms have been developed to find the distribution of materials from boundary measurements [14]. An algorithm for large-scale EIT can be adopted for micro-scale for observing cellular behaviour in vitro [15]. Linear back-projection (LBP) algorithm is a popular method with advantages of low computation, simple and fast response, but it causes smearing effect when the reconstructed images overlapped due to the summation of the back projected signals for each pixel [16]. Tikhonov regularisation (TK) is another popular method for solving linear equations or linear least-squares problems with a severely ill-conditioned matrix [17]. This method solves the given problem by a penalised least squares problem. Landweber iteration algorithm is widely used with the advantage of high image quality. In most cases, it produces the best images [18]. However, the Landweber iteration algorithm cannot satisfy the real-time measurement because of the time consumption of the iteration process. To improve the convergence time, the projected Landweber iteration (PLI) algorithm was proposed [19]. An optimal image reconstruction algorithm should be selected based on a μEIT device since each algorithm has advantages and disadvantages. Regarding the optimal frequency, at low frequencies, the change in impedance is very insensitive to cell spreading, as the impedance is mostly dominated by the electrode-electrolyte interface (i.e. Warburg impedance). At mid frequencies, the impedance is biphasic and increases slowly to a point and then starts decreasing. At high frequencies, the capacitance becomes significant [20]. Thus, to measure cell impedance on a planar electrode, it is important to record the impedance signal at the optimal frequency [21].

13.2 Micro-sensors with different structures

In 2006, York *et al.* developed a micro electrical capacitance tomography (ECT) sensor with 8 to 16 electrodes and a diameter of 750 μm [22]. The sensor is made with a CMOS integrated circuit, which contains a signal source and measures voltage by switching electronic devices through eight measurement channels. The sensor consists of a hole and an array of electrodes around it. The average diameter of the round hole is 0.75 mm. The sensor is made on a ceramic substrate by ultrasonic drilling. The sensor chip is connected to a host computer through a customised data acquisition system based on microcontroller, field programmable logic device and wide shift register. This provides fast capture of up to 750 frames of data prior to uploading to the host computer. Data capture rates of about 6,000 frames per second have been achieved for the eight-electrode sensor. Initial tests are reported for the static case involving 200 μm diameter rods that are placed in the sensor and for the dynamic case using the dose from an inhaler.

Sun *et al.* developed a miniaturised electrical impedance imaging sensor in 2010. The sensor is a circular array with 16 electrodes of 6 mm in diameter [23]. It can be used to visualise a production process of cells. The electrochemical impedance spectroscopy (EIS) characteristics of single cell were analysed using high-speed micro-fabrication technology, and the electrical characteristics of different cells were compared and analysed. A single cell equivalent circuit analysis method was established for particle models of different cell size and wall thickness. The equivalent circuit includes cell membrane capacitance and cytoplasmic resistance. The dielectric constant and conductivity of a single cell are measured. Based on the established equivalent circuit model, PSPICE circuit simulation was carried out to analyse the experimental data from the mechanism point of view.

Yang *et al.* developed a miniature EIT sensor suitable for cell culture, and established a 3D imaging model of the sensor [24]. The miniature EIT sensor can realise 3D visualisation of cell culture process. At the same time, to improve the image resolution, a new joint regularisation algorithm of 3D Laplacian sparseness was proposed. The 3D imaging performance evaluation of the sphere in vertical position and horizontal position is realised. In addition, the imaging performance of human breast cancer cell spheres and triangular breast cancer cell granules was experimentally verified. The results show that this method can improve the imaging quality of 3D cell culture process. This work accelerated real-time quantitative imaging of 3D cell culture to assess cell dynamics.

In 2007, Chai *et al.* developed a planar electrode array microchip using CMOS and used it to observe the cell activity in vitro [15]. They proposed a linear theory for EIT imaging, and tested the proposed EIT method using the sensor chip and electrolyte solution. First, the detection model is established, and the target is placed in the EIT chip to study whether the EIT sensor can recognise the location of the target. Subsequently, the EIT method is developed to detect the survival status of biological cells.

13.3 Noise reduction methods for micro-sensors

In micro scale, due to the enormous effect of contact impedance on the interface between solution and electrodes, the image reconstruction algorithms in conventional macro EIT sensors is not suitable for micro-sensors. Many researchers are interested in the effect of contact impedance on EIT sensing.

13.3.1 Noise from macro- to micro-sensors

Boyle and Adler's research in 2011 proved that the contact impedance between the electrode and the solution can produce 53% voltage drop, which greatly reduces the EIT image quality [25]. In 2007, Kim *et al.* proposed a contact impedance compensation algorithm to improve EIT image quality [26]. Previous studies shows that the impact of contact impedance can be reduced by improving the image reconstruction algorithm, to improve the imaging quality of EIT. However, little research has been done on the effect of contact impedance on μEIT from the perspective of electrochemical principle. In addition, the quantitative impact of algorithm on contact impedance becomes a key problem.

Image reconstruction for EIT is key to solve the inverse problem. Yang gave a detailed overview of non-iterative and iterative algorithms in 2002 [18]. Traditional EIT algorithms can be used for μEIT. Tikhonov regularisation (TK) is a commonly used algorithm for solving linear equations or linear least square problems with severely ill-conditioned matrices. It replaces a given problem with a penalty least squares problem [27]. In addition, Projection Landweber iteration (PLI) improves the problem of time consumption in the iteration process. PLI has been widely used because of its high image quality. In addition, generalised vector sampling pattern matching (GVSPM) is applied to solve the ill-posed inverse problem without empirical relaxation to achieve stable convergence, and GVSPM can obtain high quality images in fewer iterations [28].

In addition to the impact of contact impedance, the principle of μEIT is similar to that of traditional macro-EIT [29]. At relatively low frequencies, the cell/liquid solution is mainly affected by the peri-cellular current, while at relatively high frequencies, the solution is mainly affected by the extracellular current flowing through the cell membrane [12]. Because the contact impedance also depends on the current frequency, it is necessary to measure the contact impedance and the battery solution impedance at different applied current frequencies. In μEIT, the contact impedance is much larger than the traditional macro-sensor, which is the key point affecting the imaging accuracy of μEIT. However, little research has been done to analyse the effect of contact impedance on EIT.

13.3.2 Contact impedance

Measurements results are affected by contact impedance in the interface of electrode and solution. To quantitatively evaluate the influence caused by contact impedance, a complex coefficient ξ, which is called contact impedance effect ratio, is defined by (13.1), where ξ_{Re} is the real part of ξ, ξ_{Im} is the imaginary part of ξ, $|\xi|$ is the magnitude value of ξ, R_{C} [Ω] is the real part of contact impedance Z_{C} [Ω] and X_{C} [Ω] is the imaginary part of Z_{C}.

Figure 13.1 shows that when a current is injected to an electrode, the contact impedance Z_{C} occurs on the interface between electrode and solution.

$$\begin{aligned}
\xi &= \xi_{\text{Re}} + j\xi_{\text{Im}} \\
\xi_{\text{Re}} &= \frac{dR_{\text{C}}}{\kappa^{-1}R_{\Omega}}, \\
\xi_{\text{Im}} &= \frac{dX_{\text{C}}}{\kappa^{-1}X_{\Omega}}, \\
|\xi| &= \sqrt{\xi_{\text{Re}}^{2} + \xi_{\text{Im}}^{2}}
\end{aligned} \tag{13.1}$$

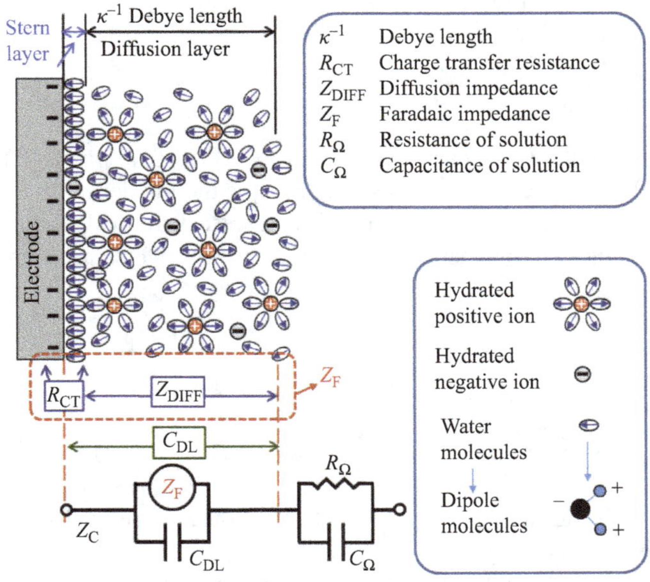

Figure 13.1 Electrical properties on electrode surface and electrical equivalent circuit

Under an assumption of the equivalent electrical circuit, the contact impedance Z_C [Ω] is expressed by

$$Z_C = 1/(1/Z_F + j\omega C_{DL}) \tag{13.2}$$

where Z_F [Ω] is the Faraday impedance and C_{DL} [F/m^2] is the capacitance of the electrical double layer (EDL) [30].

The Faraday impedance is represented by a series combination of charge transfer resistance R_{CT} [Ω] and diffusion impedance Z_{DIFF} [Ω]. Here, Z_{DIFF} is treated as semi-infinite diffusion at the frequency range from f = 4 Hz to f = 5 MHz. Z_F is expressed by

$$Z_F = R_{CT} + Z_{DIFF} = \frac{R_G T}{n_z F i_0} + \frac{R_G T}{n_z{}^2 F^2 c_0 A (2\pi f D j)^{1/2}} \tag{13.3}$$

where R_G is the gas constant R_G = 8.31 J/(K mol), T [K] is the absolute temperature, n_z is the integer charge number, F is the Faraday constant F = 96,500 C/mol, i_0 [A] is the electrode current, c_0 [mol/m^3] is the charge concentration, A [m^2] is the area surface of the electrode, c_0 [mol/m^3] is the charge concentration, f [Hz] is frequency of exciting alternating current (AC) and D [m^2/s] is the ions diffusion coefficient.

The EDL capacitance C_{DL} in (1) is expressed by [30]

$$C_{DL} = A\varepsilon_0\varepsilon_r/\kappa^{-1} \tag{13.4}$$

where κ^{-1} [m] is the Debye length, ε_0 [F/m] is the vacuum permittivity $\varepsilon_0 = 8.854 \times 10^{-12}$ F/m and ε_r is the relative permittivity.

Debye length κ^{-1} [30] is approximately calculated by

$$\kappa^{-1} = \sqrt{\frac{\varepsilon_0\varepsilon_r k_B T}{2 n_e n_z{}^2 e^2}} \tag{13.5}$$

where k_B [J/K] is the Boltzmann constant $k_B = 1.38 \times 10^{-23}$ J/K, e [C] is the elementary charge $e = 1.60 \times 10^{-19}$ C and n_e [m^{-3}] is the ion density.

Moreover, R_Ω [Ω] and X_Ω [Ω] in (13.6) represent the real part and the imaginary part of solution impedance Z_Ω [Ω], respectively, which are expressed as

$$\begin{aligned} Z_\Omega &= R_\Omega + jX_\Omega \\ R_\Omega &= \frac{A d \sigma_c}{A^2\sigma_c{}^2 + 4\pi^2 f^2 \varepsilon_0{}^2 \varepsilon_{rc}{}^2 A^2} \\ X_\Omega &= \frac{2\pi f d \varepsilon_0 \varepsilon_{rc} A}{A^2\sigma_c{}^2 + 4\pi^2 f^2 \varepsilon_0{}^2 \varepsilon_{rc}{}^2 A^2} \end{aligned} \tag{13.6}$$

where d [m] is the distance of electrodes, A [m^2] is the surface of electrode and σ_c is the conductivity of cell solution.

According to the definition of the contact impedance effect ratio ξ, the value of ξ is related to the frequency. Figure 13.2 shows the real part ξ_{Re} and imaginary part ξ_{Im} of the cell solution at different frequencies. According to (13.6), the calculation

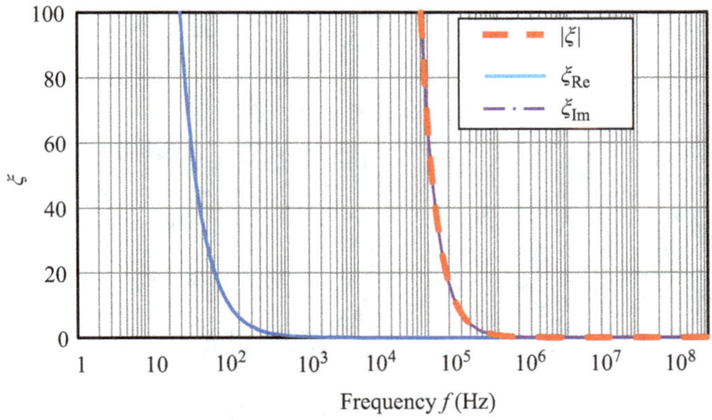

Figure 13.2 Effect of contact impedance by frequency in μEIT sensor

Table 13.1 Symbol and values of calculation in ξ and the comparison of macro- and micro-sensors

Symbol	Quantity	Value	
		Macro EIT sensor	**μEIT sensor**
R_G	Gas constant [J K^{-1} mol^{-1}]	8.31	
T	Absolute temperature [K]	298	
n_z	Valence charge number	1	
F	Farday constant [C/mol]	96,500	
i_0	Exchange current [A]	1×10^{-4}	
c_0	Charge concentration [mol/m^3]	1×10^{-7}	
A	Electrode surface [m^2]	2×10^{-4}	2×10^{-9}
d	Distance of electrode pair	5.5×10^{-4}	
D	Ions diffusion coefficient [m^2/s]	1×10^{-5}	
κ^{-1}	Debye length [m]	1×10^{-6}	
ε_0	Permittivity of vacuum [F/m]	8.85×10^{-12}	
ε_{rw}	Relative permittivity of pure water [−]	78	
ε_{rc}	Relative permittivity of cell solution [−]	100	
σ_w	Conductivity of pure water [S/m]	5.5×10^{-6}	
σ_c	Conductivity of cell solution [S/m]	9.36×10^{-2}	

results are based on the values of the parameters as listed in Table 13.1. As shown in Figure 13.2, frequency has a serious effect on zeta. In low frequencies of $f < 100$ kHz, the values of ξ_{Im} and $|\xi|$ are much larger than those of $f > 100$ kHz. The real part has the same trend. In the low frequency of $f < 100$ kHz, the value of ξ_{Re} is much larger than that of $f > 100$ kHz. The results show that the contact impedance has a great influence on the measurement. When $f = 1$ kHz and $f = 1$ MHz, the values of ξ_{Re} and ξ_{Im} converge to a very small value, respectively. The magnitude $|\xi|$ is almost the same as that of the earthquake. This phenomenon means that the imaginary part of the contact impedance has a great influence on the measurement.

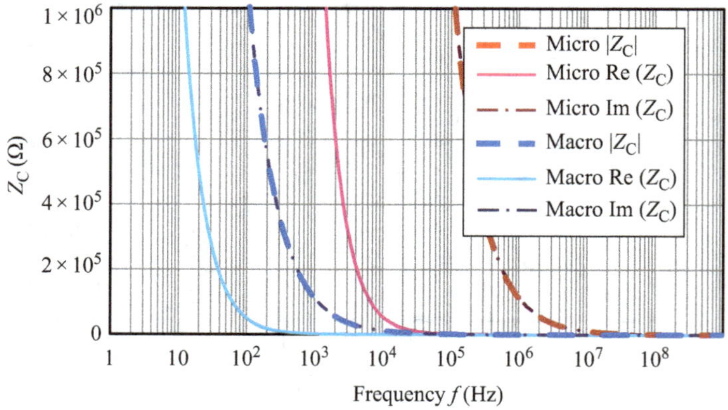

Figure 13.3 Comparison of impedance in μEIT sensor and macro sensor by calculation

Mostly, the obvious difference of micro and macro EIT sensors is the surface area. Figure 13.3 shows the comparison of real part, imaginary part and magnitude of Z_C for micro and macro EIT sensor calculated by (13.6) in a frequency range from $f = 1$ Hz to $f = 100$ MHz. Surface area of micro and macro EIT sensor is assumed as $A_{micro} = 2\times10^{-9}$ m^2 (200 μm × 10 μm) and $A_{macro} = 2 \times 10^{-5}$ m^2 (4 mm × 10 mm), respectively. Table 13.1 shows other relevant parameters used in the calculation. As shown in Figure 13.3, the magnitude of Z_C for macro sensor is decreased rapidly in the frequency range $f < 100$ Hz, but for μEIT sensors, it decreased rapidly in the frequency range $f < 100$ kHz. Through comparative analysis, the influence of frequency on contact impedance of micro-scale impedance sensor is much greater than that of macro-scale impedance sensor, due to the difference of electrode surface area between μEIT sensor and traditional macro-sensor. Therefore, due to the influence of contact impedance, a μEIT sensor produces more noise at low frequency than the sensor at macro scale.

13.4 Cell imaging

13.4.1 Micro-fluidic sensors

In this chapter, a μEIT system is presented to visualise cell concentration distribution in a micro-channel. To overcome the complexity of electrical properties and phenomena induced by the μEIT system, simulation and experiment were conducted to study the optimal image reconstruction algorithm and frequency. In the simulation, three algorithms, which are GVSPM, Tikhonov Regularisation (TK) and PLI, were compared to find the optimal image reconstruction algorithm for the μEIT system. Due to the special electrical properties of μEIT in micro-scale measurement, potential drop induced by the electrical double layer at the electrode-fluid interface is applied to the measured voltage to evaluate the feasibility of the μEIT system. In the experiment, a range of frequencies and the three algorithms were employed to reconstruct images in three cross-sections of the micro-channel to realise 3D cell sedimentation visualisation with the developed μEIT system at an optimal frequency and optimal algorithm. Finally, the developed μEIT system was preliminarily applied to visualise the cross-sectional concentration distribution of cell sedimentation in a micro-channel flow.

With the development of micro-electro-mechanical systems (MEMS), the size of EIT sensor has been reduced from mm to μm. Yao *et al.* developed a multi-layer electrode micro-fluidic EIT sensor for cell sensing and operation as shown in Figure 13.4. The micro-fluidic EIT sensor consists of a micro-channel embedded with platinum electrodes in five cross-sections (see Figure 13.4(a)), a PCB with pin connectors and a frame (see Figure 13.4(b)). The micro electrodes are embedded around a flow channel with a diamond shape (see Figure 13.4(c)). The total length of the channel is $L = 20$ mm, diagonal length $D = 800$ μm and distance of each cross-section is $l = 4.5$ mm. Each cross-section has 16 electrodes (see Figure 13.4(e)). The electrode thickness is $L_t = 10$ μm, the length is $L_w = 200$ μm and the distance of two electrodes is $L_d = 80$ μm (see Figure 13.4(d)).

13.4.2 Simulation of cell distribution

The cell phantom images with the electrical properties are reconstructed by using by the three popular algorithms (i.e. GVSPM, TK and PLI) to evaluate the measurement performance of the μEIT system. The cell phantom images are three different concentrations of cells ($\Phi = 12.5$ vol%, 50 vol% and 87.5 vol%) at the micro-channel bottom. In the phantom, the water conductivity is $\sigma_w = 5.5 \times 10^{-6}$ S/m and the cell

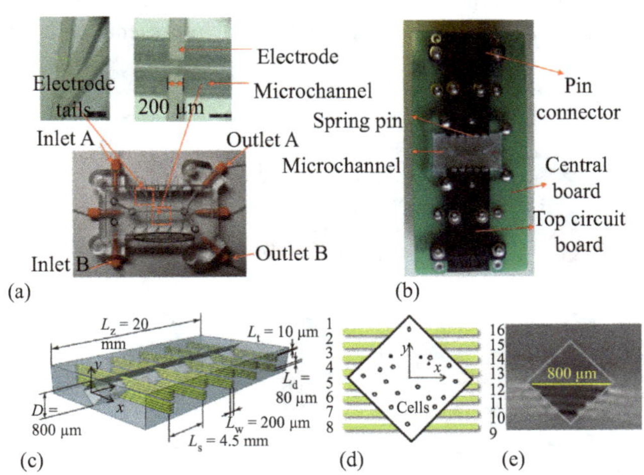

Figure 13.4 Micro-fluidic sensor for cell sensing and manipulation: (a) micro-scope of sensors embedded in micro-channel, (b) sensor board, (c) 3D schematic diagram, (d) one of cross-sectional schematic diagram and (e) micro-scope of one cross-section

conductivity is $\sigma_c = 5.5 \times 10^{-6}$ S/m. In the simulation, 16 electrodes are equally spaced at the periphery of the micro-channel. An AC current $i = 10$ mA is injected through opposite electrode pairs while the voltage differences across the remaining opposite pairs are recorded. For each current injection, the voltages are obtained by solving Poisson equation by a finite-element method (FEM). Each cell phantom image contains $N = 1{,}024$ square pixels based on $n \times n = 32 \times 32$ Cartesian mesh.

When an electrode is brought in contact with a fluid, the ions with opposite charges are separated across the electrode-fluid interface, which is known as electrical double layer (EDL) [31]. Since the charge distribution causes a potential drop δU at the electrode-fluid interface, δU coming from the electrode-fluid interface in the micro-scale must be considered to prove the feasibility of the μEIT system. The potential drop δU_{EDL} induced by EDL at the electrode-fluid interface are $\delta U_{\mathrm{EDL}} <$ 25 mV at the Pt electrode/water interface at room temperature [32]. In the simulation, the average net potential drop from the electrode is $U_{\mathrm{E}} = 84$ mV. Therefore, the total calculated potential drop considering the EDL, U_{T} is expressed as

$$\begin{aligned} U_{\mathrm{T}} &= U_{\mathrm{E}} + \delta U_{\mathrm{EDL}} \\ \delta U_{\mathrm{EDL}} &= \alpha U_{\mathrm{E}} \end{aligned} \tag{13.7}$$

where $\alpha = 5\%, 10\%, 15\%, 20\%, 25\%$ and 30%.

The potential ratio α is used to evaluate the influence of EDL on the accuracy of reconstructed images.

Numerical simulation was conducted to clarify the contact impedance influence on the measurement results. Figure 13.5 shows three reconstructed images of the phantoms by GVSPM, TK and PLI with different cell concentration ($\Phi = 12.5$ vol%, 50 vol% and 87.5 vol%) at the micro-channel bottom. The blue colour represents the fluid with low conductivity σ_w while the red colour represents cell with high conductivity σ_c.

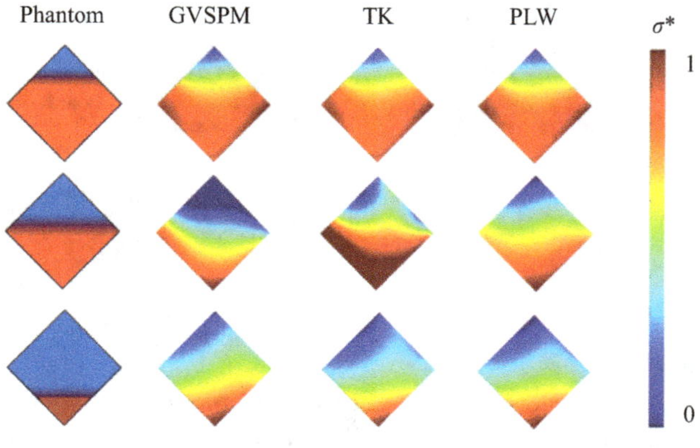

Figure 13.5 Reconstructed images of the phantoms by GVSPM, TK and PLI with different cell concentration ($\Phi = 12.5$ vol%, 50 vol% and 87.5 vol%)

Figure 13.6 shows the relationship between the iteration times (*Iter*) and image error (I_E) of the reconstructed images when $\Phi = 50$ vol%. In Figure 13.6(a), with GVSPM, I_E of the reconstructed image is decreased sharply and then increases as *Iter* is increased. The minimum I_E is $I_E = 0.43$ when $Iter = 2$. In Figure 13.6(b), the images were reconstructed by TK with regularisation factor $\mu = 50$. Similar to GVSPM, I_E was increased and then decreased slowly. The minimum image error is $I_E = 0.63$ when $Iter = 2$. In Figure 13.6(c), with regularisation factor $\mu = 0.005$, I_E is increased as *Iter* is increased. From the comparison results of I_E, GVSPM is much better than TK and PLI due to its lower I_E and faster calculation without regularisation factor μ.

Figure 13.7 shows the relationship between *Iter* and image correlation (I_C) of the reconstructed images when $\Phi = 50$ vol%. In Figure 13.7(a), for the reconstructed images by GVSPM, I_C is increased sharply and then decreased as *Iter* is increased. The optimal iteration time was selected as $Iter = 3$ since the maximum I_C was obtained as $I_C = 0.84$. In Figure 13.7(b), the images were reconstructed by TK with the regularisation factor $\mu = 50$. The maximum I_C was 0.80 when $Iter = 4$. Similar to GVSPM, I_C was increased and then decreased slowly like a parabola. In Figure 13.7(c), with the regularisation factor $\mu = 0.005$, different tendency of I_C was found, which is increased as *Iter* was increased, even until $Iter = 300$. Similar

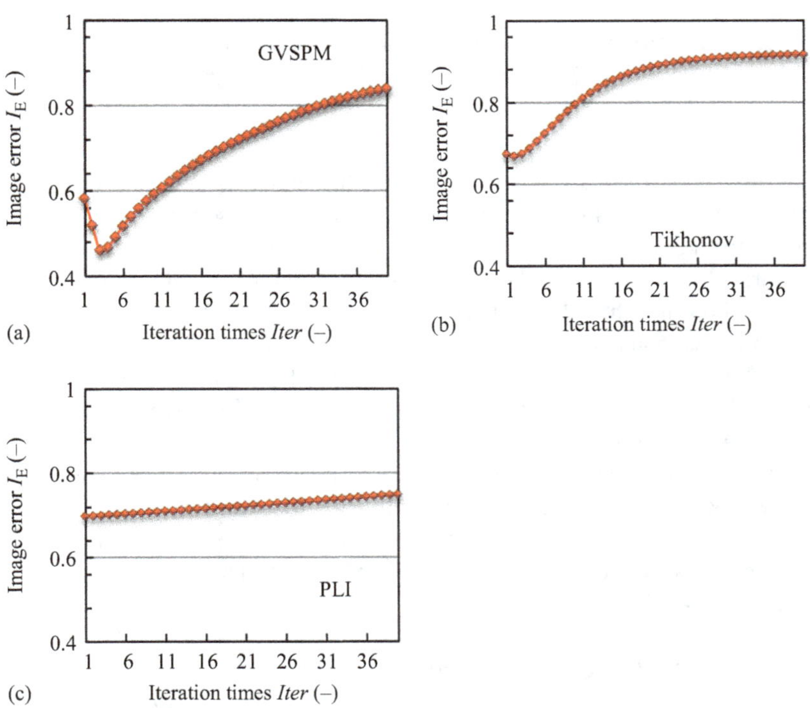

Figure 13.6 Relationship between Iter and I_E of reconstructed images when $\Phi = 50$ vol% by (a) GVSPM, (b) iterative TK and (c) PLI

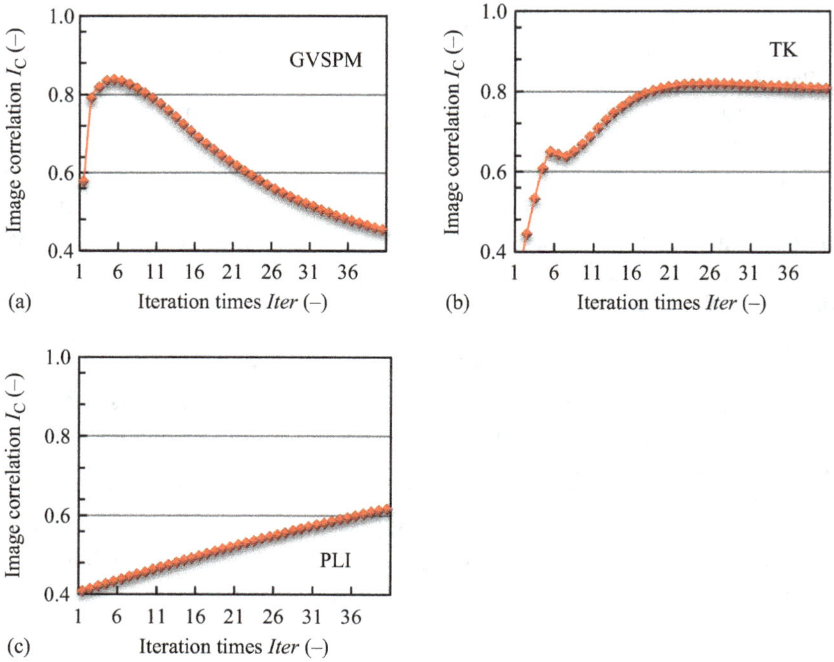

Figure 13.7 Relationship between Iter and I_C from images reconstructed when $\Phi = 50$ vol% by (a) GVSPM, (b) ITK and (c) PLI

to the comparison results of I_E, GVSPM has the higher I_C and faster calculation without regularisation factor μ. Therefore, GVSPM is selected as the optimal algorithm from simulation.

Figure 13.8 shows the quality of the reconstructed images at different potential ratio α by GVSPM when $\mathbf{\Phi} = 50$ vol%.

Quantitative comparison of the reconstructed images is shown in Figure 13.9. As the potential ratio α increases, I_C is increased slowly. At the same time, I_E is decreased slowly except for *Iter* = 10. At *Iter* = 10, I_C and I_E have a little change of 0.26% and 4.7%, respectively.

This comparison demonstrates that the potential drop δU_{EDL} induced by EDL at the electrode-fluid interface does not influence too much on the quality of the reconstructed images, showing that the developed μEIT system is applicable to visualise cell distribution in micro scale with the optimum image reconstruction algorithm of GVSPM at a proper *Iter*.

13.4.3 Cell imaging experiments

Figure 13.10 shows the experimental set-up consisting of a multi-layered micro-channel, a multiplexer, an impedance analyser and a personal computer (PC). Sixteen electrodes embedded in one cross-section of the multi-layered micro-

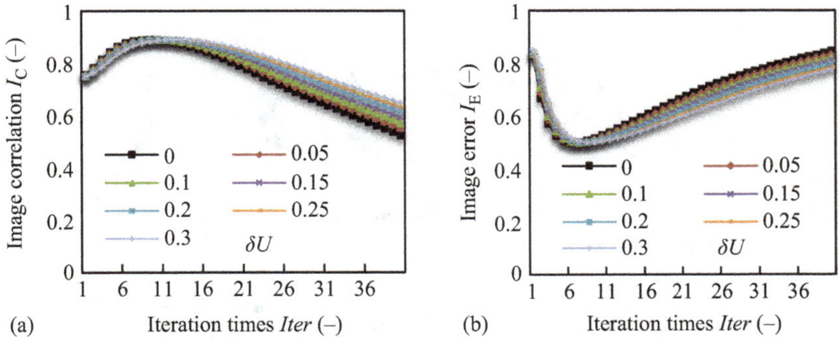

Figure 13.8 Comparison of reconstructed images at different potential ratio α when Φ = 50 vol% by GVSPM

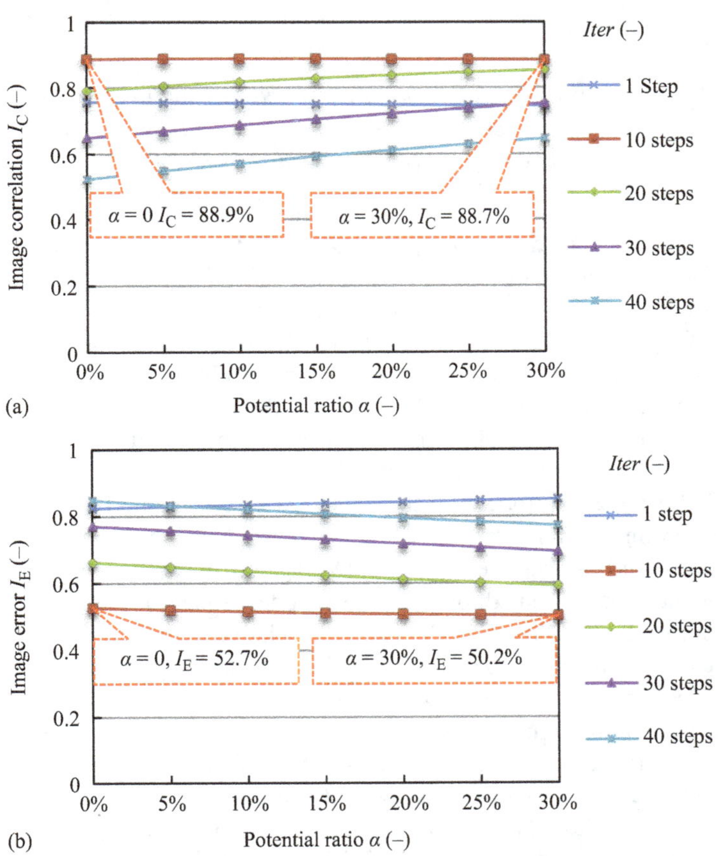

Figure 13.9 Quality comparison of reconstructed images at different potential ratio α by GVSPM

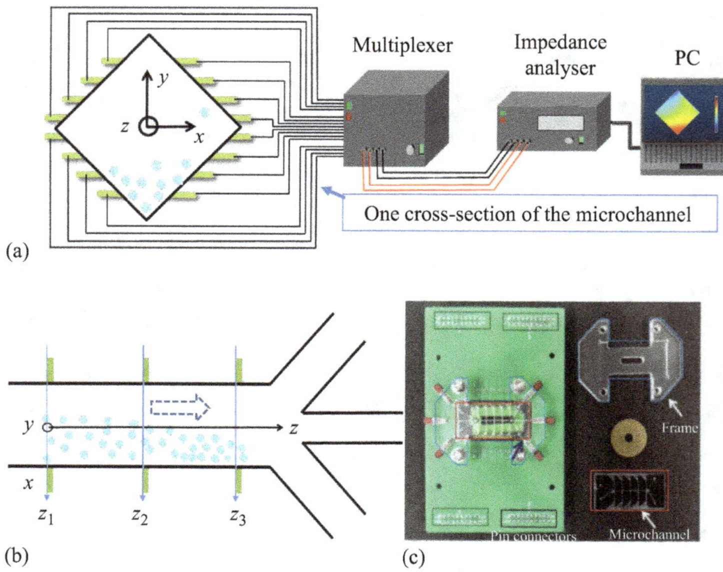

Figure 13.10 (a) Experimental set-up and (b) cell distribution is measured with three cross-sections at $z_1 = 0$, $z_3 = 10$, and $z_5 = 20$ [mm] from inlet and (c) photo of microchannel and connectors

channel are connected to the multiplexer to select the required stimulation electrodes. The dimensions of the micro-channel with integrated electrodes were reported by Yao *et al.* [33]. The length of the main flow channel is $L = 20$ mm, diagonal of the diamond cross-section is $D = 800$ μm and separation distance of the five electrode-integrated positions is $z_0 = 4.5$ mm. In each cross-section, 16 electrodes are integrated, while each electrode has the thickness $k = 10$ μm, length along the channel $w = 200$ μm and inter-electrode distance $d = 80$ μm. Figure 13.10(b) shows the measured cross-sections in the micro-channel flow. In Figure 13.10(c), the electrode tails are connected to a printed circuit board (PCB) via copper terminal connectors. The PCB is equipped with pin connectors to connect a control and measurement system. During this operation, the μEIT system injects a controlled current through one electrode, 'sinks' a second electrode to ground and measures voltages at all electrodes relative to ground.

The current source $i = 10$ mA from the impedance analyser was applied to the 16 electrodes via the multiplexer with an opposite current injection method. The measured electrode-pair number was $M = 104$. The PC was used to collect the data for image reconstruction at the third cross-section z_3. Dried yeast cells (S. cerevisiae, Euroscarf) in living state with concentration of $\Phi = 10$ vol% suspensions were employed as samples. The conductivities of purified water and cells were $\sigma_w = 5.5 \times 10^{-5}$ S/m and $\sigma_c = 2.34 \times 10^{-3}$ S/m, respectively. The AC frequency f was in a range from 100 kHz to 2 MHz. Three algorithms, which are GVSPM, TK and PLI, were used to reconstruct images at the specified frequency of 1 MHz.

The impedance magnitude $|Z|$ of water and yeast cells ($\Phi = 12.5$ vol%) are compared in Figure 13.11 with the electrode pair M at the micro-channel bottom at different frequencies. At low-frequency $f < 100$ kHz, $|Z_{\text{cells}}|$ is much high but decreases sharply below 100 kHz, because the electrical double layer around the electrodes and cells behaves like a capacitor, which prevents the current flow at low frequency [34]. This phenomenon is called interfacial or Maxwell–Wagner polarisation [35]. As the frequency f is increased from 100 kHz, $|Z_{\text{cells}}|$ continues to decease slowly even $|\mathbf{Z}_{\text{cells}}| < |\mathbf{Z}_{\text{water}}|$, due to the frequency-dependent electrical properties of cells, that is the cell membrane behaves like a capacitor, which prevents low-frequency current from entering the cell [36]. This phenomenon is supported by the impedance spectra of water. As the frequency increases, $|\mathbf{Z}_{\text{water}}|$ keeps constant due to the absent of capacitance components brought from cells. According to these results, the image construction of the cell sedimentation is apt to realise at an optimal frequency f according to the electrical properties of the cells.

The reconstructed images are evaluated by voltage error U_{E} at different frequencies with GVSPM as shown in Figure 13.12. The voltage error U_{E} is decreased sharply at low frequencies, then decreases slowly from 100 kHz to 770 kHz. At 1 MHz, the voltage error reaches to the lowest point of $U_{\text{E}} = 0.582$, because at a specific frequency, the measured voltage is more sensitive to the objects, which consist of complex components, such as water, cells and electrical double layer. In particular, the electrical properties of the cells are frequency dependent [37].

The images reconstructed by three algorisms (i.e. GVSPM, TK and PLI) are compared with voltage error U_{E} at 1 MHz as shown in Figure 13.13. The reconstructed image by GVSPM has the lowest image error $U_{\text{E}} = 0.582$ comparing with the other two algorithms. For the reconstructed image by TK and PLI, the regularisation factor μ is an empirical factor greatly influences the results [38], which

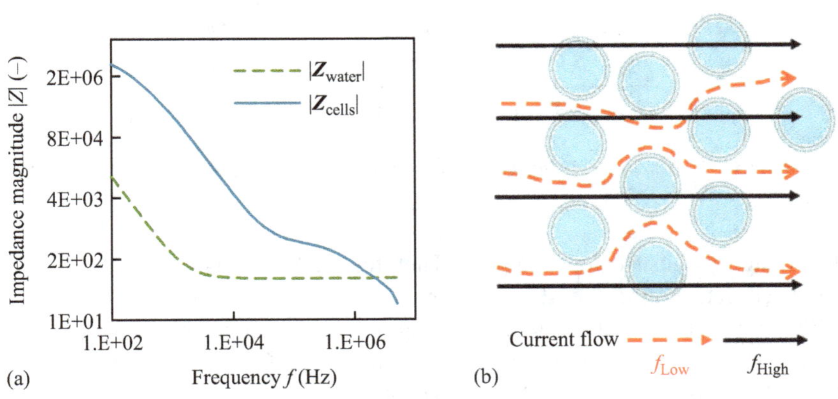

Figure 13.11 Electric characteristics of biological cells: (a) dielectric spectra of yeast cells and (b) frequency dependent of electrical properties of cells, that is cell membrane prevents low-frequency current from entering cell

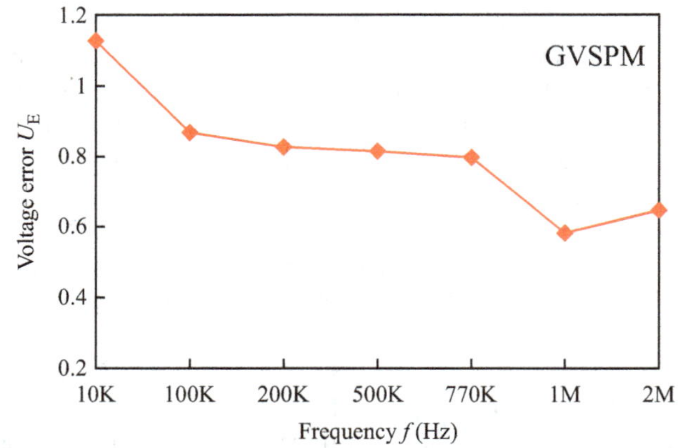

Figure 13.12 Comparison of reconstructed images with GVSPM at different frequencies f by experiments

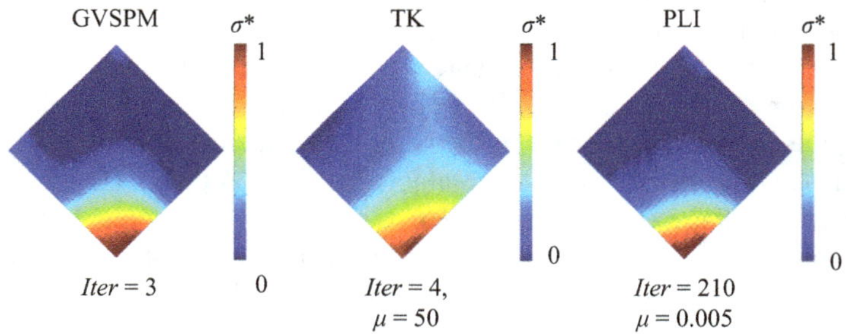

Algorithm	Coefficient	Iteration times *Iter*	Voltage error U_E
GVSPM	*Iter*	3	0.582
TK	*Iter*, μ	4	0.8
PLI	*Iter*, μ	210	0.65

Applied frequency: $f = 1$ MHz

Figure 13.13 Evaluation of image quality among three algorithms by experiments

makes image reconstruction complex. Therefore, GVSPM is superior to TK and PLI for cell detection in the μEIT system.

Figure 13.14 shows reconstructed images of cell sedimentation with GVSPM at 1 MHz at different injection cell concentrations of $\Phi = 5$ vol%, 10 vol% and 15 vol%. Cell suspensions were flowed with a flow rate of $Q = 10$ ml/h pushed by a syringe pump. The red areas distributed at the micro-channel bottom are decreased gradually, which indicates that, cell sediment at the micro-channel bottom in the two-phase flow due to the higher density of the cells than that of the water. The velocity of the cell sedimentation has been analysed in our previous study [33].

Figure 13.14(b) shows reconstructed images of the cell sedimentation in z_1, z_3 and z_5 of the micro-channel. To prove these results quantitatively, the cell concentration distribution measured by EIS method in the previous study is also shown [33]. The cell concentration decreases gradually from $\Phi = 17.5$ vol% to $\Phi = 4.9$ vol%, which visualises the cell accumulation from z_1 to z_5 due to sedimentation. The agreement of the cell sedimentation between μEIT and EIS methods demonstrates that the developed μEIT system can visualise cell sedimentation successfully in the electrode-multi-layered micro-channel.

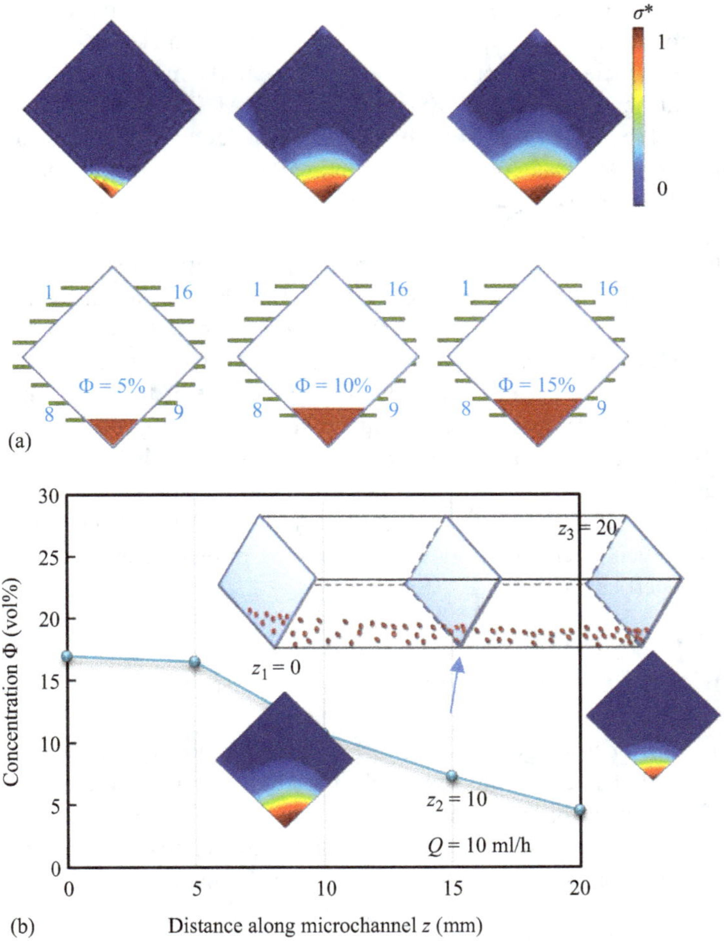

Figure 13.14 Visualisation of cell sedimentation in micro-channel along three cross-sections by μEIT with GVSPM at 1 MHz: (a) reconstructed images at different injection cell concentration $\Phi = 5$ vol%, 10 vol% and 15 vol% at z_3 and (b) reconstructed images along micro-channel at different cross-sections $z_1 = 0$ mm, $z_2 = 10$ mm and $z_3 = 20$ mm

The research results in this section are concluded as follows.

(1) In the simulation, three kinds of phantoms of cell/fluid suspension are established. Three image reconstruction algorithms, which are iterative TK, Landweber iteration and PLI are employed to solve the ill-posed inverse problem. GVSPM was found to be the optimal algorithm for image reconstruction in the present study due to its higher $I_C = 0.84$ and lower $I_E = 0.43$. The potential drop δU_{EDL} induced by the electrical double layer at the electrode-fluid interface was applied to the measured voltage, the I_C and I_E have a little change of 0.26% and 4.7% at a specified $Iter = 10$, which proves the feasibility of μEIT.

(2) In the experiment, yeast cells and purified water were employed as a two-phase flow to measure cell sedimentation in the micro-channel. An AC current with a range of frequencies were applied to the electrode in the micro-channel to find the optimal frequency for the μEIT system with GVSPM, TK and PLI. The optimal frequency for the μEIT system was found as 1 MHz due to its higher measurement sensitivity. GVSPM was found as the optimal image reconstruction algorithm because of its lower voltage error $U_E = 0.582$ and simpler image reconstruction without regularisation factor.

(3) Images of cell sedimentation are reconstructed with GVSPM in three cross-sections of the multi-layered micro-channel at 1 MHz. The reconstructed images show that concentration of cell sedimentation from z_1 to z_5 is decreased along the micro-channel, which are compared with the quantitative results from $\Phi = 17.5$ vol% to $\Phi = 4.9$ vol% in the previous study. The agreement of the cell sedimentation between μEIT and EIS methods demonstrates that the developed μEIT system can realise the cell sedimentation visualisation successfully in the electrode-multi-layered micro-channel.

The developed μEIT system would be useful in visualisation of cell spatial distributions, such as characteristics of cell motion in blood flow according to the electrical properties.

13.5 Blood flow measurement

The electrical properties of blood can reflect the biological characteristics of haemorrhagic fluid. Therefore, the electrical properties of blood have always been a subject of concern in biology and medicine. The study of electrical properties of blood has been applied in biosensors, such as blood viscosity monitor [39] and coagulation monitor [40]. From the multiphase flow dynamics point of view, blood is a suspension consisting of cells and plasma, mainly red blood cells (RBC), in conductive liquid (plasma). The electrical properties of static blood mainly depend on the volume fraction of RBCs (hematocrit) [41], temperature and cell shape [42]. In addition, the electrical characteristics of blood flow are also affected by the flow rate. In the blood, due to the interaction between cells, the distribution of RBCs is not uniform. When the flow rate changes, RBCs aggregate and form clots. In this case, Hanai theory considering the interaction between RBCs can be used for theoretical analysis of cell distribution. Asami *et al.* extended the Hanai formula from the non-spherical particles' point of view [43]. However, this study is only applicable to static flow without considering the changes in the shape and electrical properties of RBCs under flow conditions. Therefore, to evaluate the effect of blood flow velocity on cell electrical properties more effectively, the traditional Hanai mixing formula is revised from the perspective of blood flow related factors (erythrocyte orientation, erythrocyte deformation and double layer thickness).

Furthermore, a single current frequency is used to study the electrical characteristics of blood flow. It is found that the electrical characteristics of blood flow are frequency dependent. Within a specific frequency range, the relationship between resistivity and blood condition is obvious [44]. It is of great significance to determine the appropriate frequency range. At the same time, the double layer formed around the erythrocyte also has a great influence on the blood flow velocity. Therefore, multi-frequency detection is of significance to the study of blood flow characteristics. The purpose of this study is to quantitatively evaluate the electrical parameters of blood flow by applying a certain frequency range of current and explain the effect of blood flow velocity on cell movement.

13.5.1 Experimental setup and method

Figure 13.15 shows the experimental devices, including thermostat (HB-1400; Japan), liquid storage tank (Japan Moriko Medical Device Manufacturing Co., Ltd.), roller pump (RP-PLB; Japan Furue Science KK), flowmeter (VNS-10F, Japan Aichideki Co., Ltd.), impedance analyser (IM 3570, Japan Hioki) and two-electrode sensor. The length of the circulating pipe is 4,500 mm. The blood circulation tube was pre-coated with heparin, with good biocompatibility. The inner diameter $D_1 = 9.5$ mm and the thickness $D_2 = 1.2$ mm. The distance between the two electrodes of the sensor is $l_4 = 60$ mm, the outer diameter of the sensor is

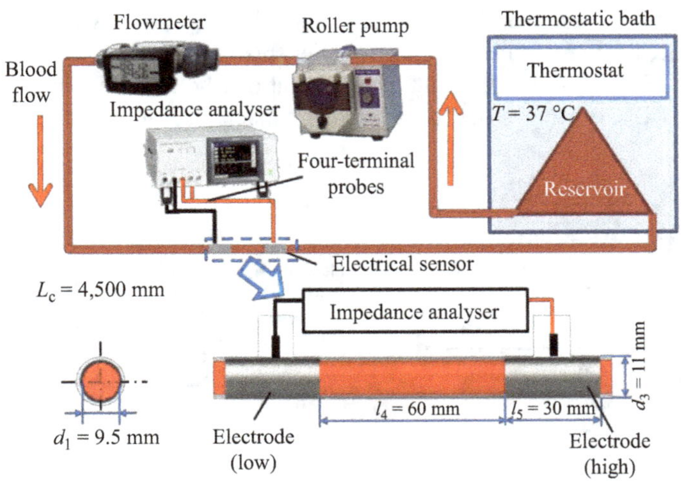

Figure 13.15 Experimental setup for flowing blood

$d_3 = 11$ mm, and the inner diameter is about $d_4 = 9.5$ mm and the length of each electrode is $l_5 = 30$ mm.

Because pig blood is similar to human blood, fresh pig blood is used as a sample and 3.28% trisodium citrate solution of one-tenth volume is used as a flow blood test. The specific volume of whole blood red blood cells is $H = 38\%$. A constant temperature bath is used to keep the blood temperature at 37 °C, which is the normal body temperature of the human body. The applied current value of the sensor's two electrodes is 0.1 mA and the frequency range of AC current is from 0.5 to 5 MHz.

In the experiment, a drum pump was used to produce the flowing blood. The velocity of blood flow ranged from 0 to 2.6 L/min. Pig blood is 800 ml. Before each test, 1 min is waited to stabilise the blood flow. Under these conditions, the change in resistance between the two electrodes is detected by applying an AC current and the change in electrical characteristics of blood flow process is studied according to the change in resistance.

Electrical parameters (characteristic frequency f_c, shape factor α, resistance at zero frequency R_0 and resistance at infinite frequency R_∞) are utilised to evaluate the electrical properties of flowing blood, and they are obtained from the Cole-Cole plot fitted by resistance R and reactance X under multiple-frequency measurement, as is shown in Figure 13.16(a). The locus of Cole-Cole plot is an arc of one semi-circle with its centre depressed below the axis of resistance R, which is described by [45]:

$$Z^* = R + jX = R_\infty + \frac{R_0 - R_\infty}{1 + (jf/f_c)^\alpha} \tag{13.8}$$

where Z^* is the complex impedance of flowing blood as a function of frequency f; characteristic frequency f_c is the frequency where X has the maximum value; α is the shape factor which shows the deviation from pure capacitance ($0 \leq \alpha \leq 1$) and $\alpha = 1$ only in the case that the whole solution is homogeneous and R_0 and R_∞ are the resistances at zero and infinite frequencies, respectively.

Figure 13.16(b) shows the equivalent circuit of whole blood, including plasma resistance, RBC resistance and double-layer capacitance, to explain the physical

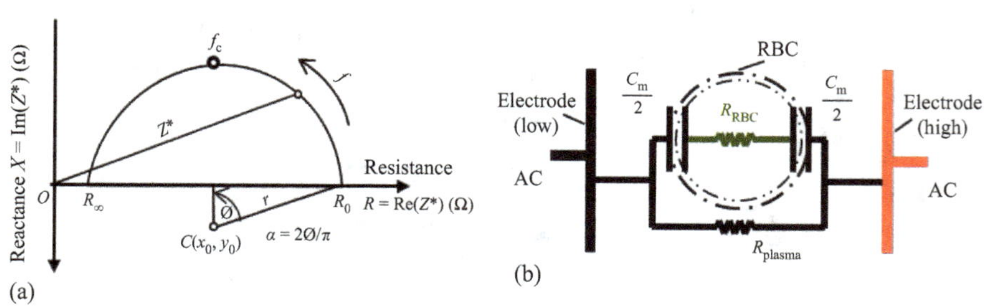

Figure 13.16 Electrical parameters: (a) Cole-Coleplot and (b) equivalent electrical circuit of blood

significance of these electrical parameters. f_c is the characteristic frequency of whole blood, which is affected by plasma resistance, RBC resistance and double-layer capacitance. Because of the changes in the structure and distribution of RBCs, α is less than 1. The thickness of the double layer includes the charged layer around the erythrocyte membrane and erythrocyte. At low frequencies, the current cannot pass through the erythrocyte membrane, so R_0 depends on plasma. At high frequencies, the double layer is considered to be short circuit, so r is determined by the parallel circuits of R_{plasma} and R_{RBC}. The impedance and electrical parameters of the blood equivalent circuit are described as follows:

$$Z^* = \frac{R_{RBC}R_{plasma}}{R_{RBC} + R_{plasma}} + \frac{R_{plasma} - \frac{R_{RBC}R_{plasma}}{R_{RBC}+R_{plasma}}}{1 + j2\pi f C_m \left(R_{RBC} + R_{plasma}\right)} \tag{13.9}$$

$$f_c = \frac{1}{2\pi \cdot \left(R_{RBC} + R_{plasma}\right) \cdot C_m} \tag{13.10}$$

$$R_0 = R_{plasma} \tag{13.11}$$

$$R_\infty = \frac{R_{RBC}R_{plasma}}{R_{RBC} + R_{plasma}} \tag{13.12}$$

13.5.2 Experimental results

To obtain the dielectric parameters of flowing blood and confirm the modified Hanai mixture formula, experiments under various flow rates Q have been carried out. Figure 13.17(a) shows the Cole-Cole plot of flowing blood from the electrical impedance spectroscopy, which is fitted using curve-fitting algorithms (lines are fitted semi-circle) [46]. Actually, twelve Cole-Cole plots with different flow rates were obtained, but only five of them are shown in Figure 13.17(a) to give a view. In the case that Q is increased, the semi-circular radius r_p of Cole plot decreases.

Figure 13.17(b) shows the characteristic frequency f_c and the shape factor α of flowing blood with various flow rates. In the case that Q is increased from $Q = 0$ L/min to $Q = 0.5$ L/min, characteristic frequency f_c increases sharply from 1.73 MHz to 2.39 MHz and the shape factor α is increased from 0.78 to 0.88. After Q is larger than 0.5 L/min, f_c gradually decreases from 2.39 MHz to 2.19 MHz, while shape factor α only changes from 0.88 to 0.93.

Figure 13.17(c) shows the resistance of experimental-based flowing blood at zero-frequency R_0 and at infinite-frequency R_∞. When $Q = 0$ L/min to $Q = 0.5$ L/min, R_0 drops sharply from 996.6 Ω to 804.8 Ω. When the flow rate Q is greater than 0.5 L/min, the resistance R_0 decreases slowly from $R_0 = 804.8$ Ω to 757.0 Ω. The results of Figure 13.17 shows that Q has a great influence on hemo-electronic parameters, especially in the case of $Q < 0.5$ L/min; f_c, α and R_0 are sensitive to the change of blood flow velocity, but r is not. The explanation of this phenomenon will be described quantitatively by the simulation discussion in Figure 13.20.

Based on (13.9)–(13.12), the traditional Hanai formulas are modified to suit the blood flow, taking into account the blood flow related factors (erythrocyte

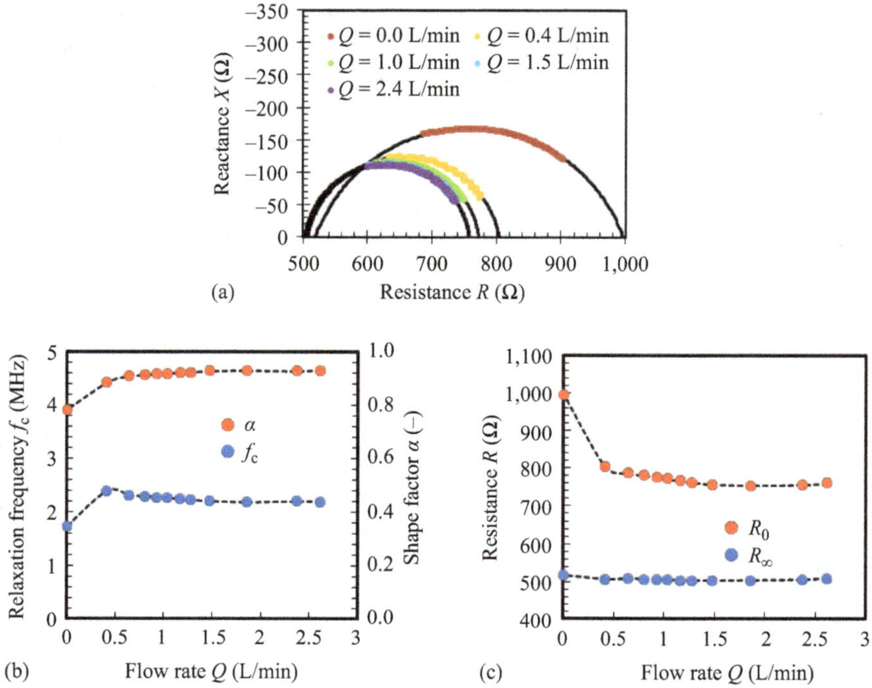

Figure 13.17 Experimental results: (a) Cole-Cole plot of flowing blood, (b) characteristic frequency f_c and the shape factor α and (c) resistance R_0 and R_∞

orientation, erythrocyte deformation and EDL thickness around erythrocyte). The impedance of multi-frequency blood flow is obtained by (13.6) and (13.13). Therefore, electrical parameters (characteristic frequency f_c, shape factor α, resistance at zero frequency R_0 and resistance at infinite frequency R_∞) are obtained by Cole-Cole plot.

$$Z^* = \left(\frac{2\pi}{L} \int_0^{d_1/2} j\omega\varepsilon_0\varepsilon^*_{\text{blood}}(r) r dr \right)^{-1} \tag{13.13}$$

Figure 13.18 shows the simulation results of the distribution of blood flow correlation factors in the cross-section of blood circulation vessels based on the improved Hanai mixture formula. Figure 13.18(a) shows that the low shear rate near the central axis of the tube means that the RBC direction is random ($\alpha_x = \alpha_y = \alpha_z = 1/3$) and the high shear stress near the tube wall means that the RBC direction is longitudinal relative to the flow ($\alpha_x = 0$), which is consistent with the assumption in Figure 13.19(a). In the case of increased flow rate Q, the y-axis length of RBCs D_y rises up, while the x-axis length of RBCs D_x decreases gradually as shown in Figure 13.19(b). The EDL thickness around RBCs reduces in inverse

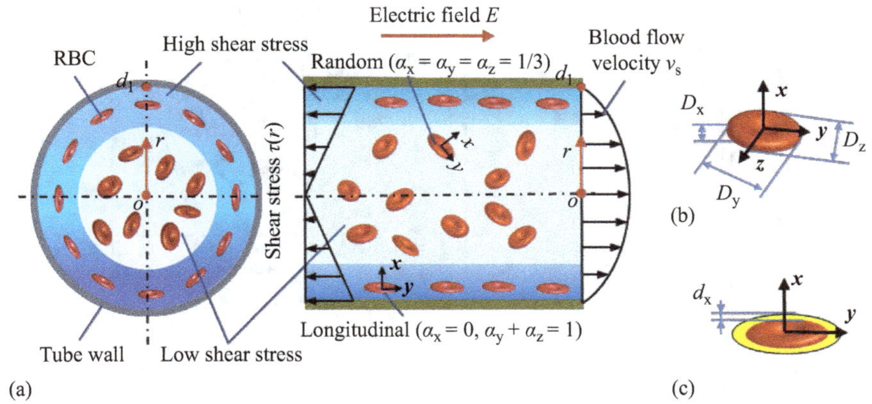

Figure 13.18 Influences of shear stress on RBCs in flowing blood: (a) distribution of RBCs in flowing blood, (b) axis length of RBC and (c) thickness of EDL

proportion to the square root of shear rate as shown in Figure 13.19(c). The variation of these three blood flow-dependent factors surely influences the electrical properties of flowing blood.

The error between the experimental and simulated results e_i is obtained by the following equation:

$$e_i = \frac{T_{Hi} - T_{Ei}}{T_{Ei}} \tag{13.14}$$

where e_i is the error between experimental data and simulated data, and the subscript i represents f_c, α, R_0 and R_1; T_{Hi} is the simulated data from the modified Hanai mixture formula and T_{Ei} is the experimental data.

Figure 13.19 shows the experimental and simulated results of electrical parameters in flowing blood. Figure 13.19(a) shows the characteristic frequency f_c and the shape factor alpha and Figure 13.19(b) shows the resistance at zero frequency R_0 and the resistance at infinite frequency R. The simulation results are obtained by (13.13). Red and blue dots are the results of experiments. Rough solid line (RO+RD+ET) showed the effect of all blood flow related factors (RBC orientation, RBC deformation and EDL thickness) on blood flow velocity Q. The maximum errors between the experimental results and the simulation results are $e_{fc} = 3.4\%$, $e_\alpha = 19.3\%$, $e_{R0} = 3.1\%$ and $e_{R1} = 1.7\%$, respectively. The theoretical evaluation with the improved Hanai mixtures is in good agreement with the experimental characteristics. Regardless of the Q value, the rough dot line (RO+ET) shows the influence of RBC direction and EDL thickness on the blood-dependent factors, that is the influence of RBC deformation is not introduced in the improved Hanai model. The thin dot lines (RD+ET) show the influence of RBC deformation and EDL thickness among the flow-dependent factors, namely the influence of RBC orientation is not introduced into the formula. The thin chain

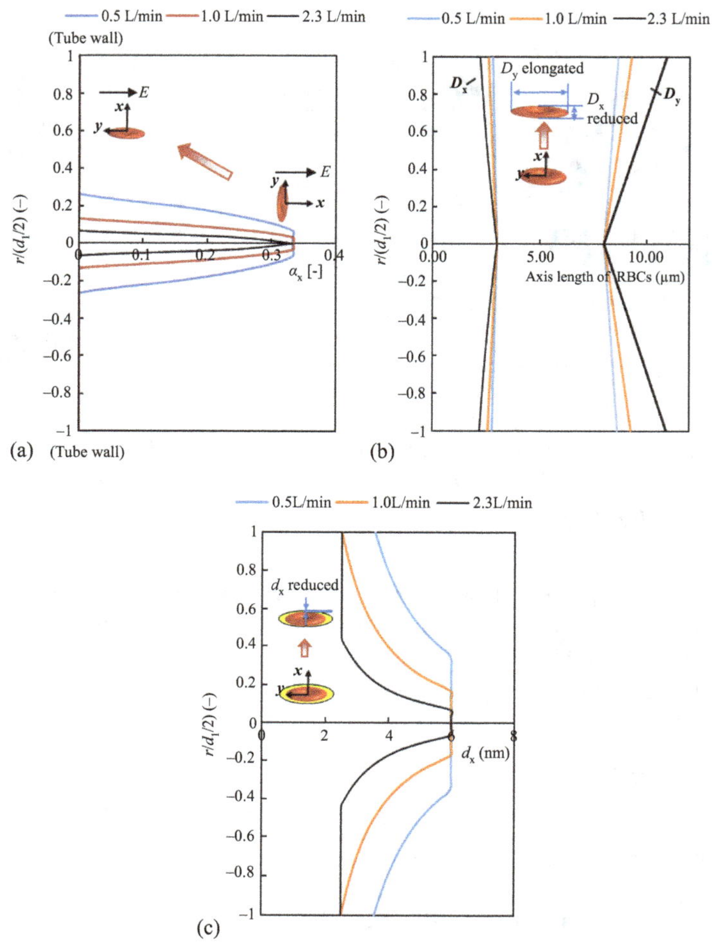

Figure 13.19 Simulation results of blood flow-dependent factors distribution in cross-section of blood circulation tube: (a) RBCs orientation, (b) RBCs deformation and (c) RBCs EDL thickness

lines (RO+RD) show the influence of RBC orientation and RBC deformation among the blood flow-dependent factors, namely the influence of EDL thickness is not introduced into the equation.

The RO+RD+ET line of f_c agrees well with the experimental results of flow Q. The RD+ET line of f_c is shorter, while the RO+RD and RO+ET lines of f_c are higher than Q estimates, which indicates that f_c is greatly influenced by RBC direction, RBC deformation and EDL thickness. During the experiment, the electric field E is consistent with the flow direction. When Q increases, the shear stress near the tube wall makes RBC orientate along the longitudinal direction (same as E direction), making R_{RBC} higher ($D_y < D_x$). In addition, the shear stress extends

RBC along the longitudinal direction, extends D_y and increases RRBC. However, according to Figure 13.19, C_m decreases with Q as D_x decreases. Therefore, f_c and Q rise together, but when Q is high enough, most RBCs are oriented, and deformation and EDL thickness cannot change again, which is the reason why f_c does not change much in the end. Therefore, the electrical parameters f_c measured at multiple frequencies are sufficient to reflect all three blood flow related factors (RBC direction, RBC deformation and EDL thickness).

RD+ET line of α does not show the similar tendency against Q even though RO+RD+ET, RO+ET and RO+RD lines show the same values, which means that α is strongly affected by RBC orientation. According to the Cole-Cole plot, α is influenced by the homogeneity of blood. If the RBCs are in spherical shape, all the RBCs in blood are under the same condition (orientation, shape and EDL thickness), the value of α is nearly 1. However, the RBCs are in dish shape, in the case that Q is small, most of the RBCs are in random orientation, namely, the RBC orientation is not the same (α_x, α_y, α_z), and hence α is small. In the case that Q increases, more RBCs are oriented in the same longitudinal direction, which makes the flowing blood more homogenous and that is considered the reason why α with Q.

Figure 13.20(b) shows that R_0 decreased sharply at first, and then deceased slightly as Q increases. The large drop of R_0 is thought to be caused mainly by the RBC orientation according to the calculation by modified Hanai mixture formula. From Figure 13.20(b), RD+ET line of R_0 does not show the same value against Q even though RO+RD+ET, RO+ET and RO+RD lines show almost the same values. Without considering the RBCs orientation (RO), the error is much high (line RD+ET), which means that R_0 is strongly influenced by RBC orientation. R_0 is mainly depending on R_{plasma} at zero frequency. In the case that RBC orientation changes, the current distribution in plasma and R_{plasma} are influenced as well. Even

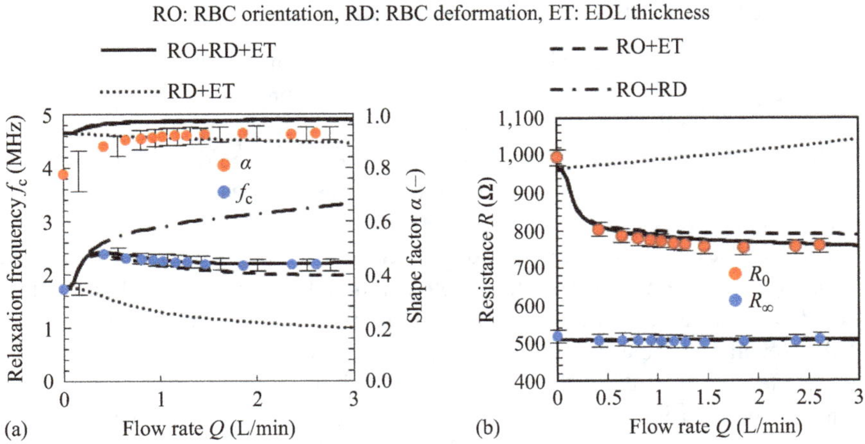

Figure 13.20 Comparison of electrical parameters from experiment and simulation results in flowing blood: (a) characteristic frequency f_c and shape factor α and (b) resistance R_0 and R_∞

though RBC deformation also affects the current distribution in plasma and R_{plasma}, the influence is small. Hence, R_0 is mainly affected by RBC orientation.

All curves of R_∞ show almost the same values and constants, independent of Q, which means that the electrical parameter R_∞ is insensitive to Q. According to (13.5), the value of R_∞ is based on R_{RBC} and R_{plasma}. Although the direction and deformation will increase, the current distribution will also be affected, thus reducing R_{plasma}. Therefore, R_∞ does not change much with Q.

In conclusion, this study clarifies the characteristics of multi-frequency measurement of electrical parameters in blood flow at different flow velocities. In addition, the influence of RBC direction, RBC deformation and EDL thickness on flow rate is also the main reason.

The electrical parameters (i.e., characteristic frequency f_c, shape factor α, resistance at zero frequency R_0 and resistance at infinite frequency R_∞) of the flowing blood were quantitatively evaluated. An improved Hanai mixing formula was proposed. In the flowing blood, the electrical parameters of the blood are greatly affected by the velocity Q. Among them, α and R_0 are insensitive to the velocity of the blood, while R is insensitive to the velocity of the blood. f_c can express all three blood flow related factors (RBC direction, RBC deformation and EDL thickness), while resistance R_0 can only reflect the influence of RBC direction and deformation.

This study quantitatively determined the relationship between blood flow velocity and blood electrical parameters. The results show that the basic state of blood flow can be predicted by monitoring the electrical parameters of blood. This study can help to achieve rapid blood testing, RBC detection.

13.6 Future prospects

The future development of micro-sensors will focus on the following five aspects:

1. In the theoretical and simulation research, the establishment of various models remains to be further improved. The model is not only a cell model but also a model with multiple physical fields coupling.
2. High precision and high efficiency of the cell detection could be realized. Microfluidic chips with different structures and different measurement methods are effective means.
3. The application of the separation of dielectric electrophoresis can be expanded. Other particles can be separated, such as mineral particles and colloidal particles. In addition to solid particles, liquid particles can be considered as one of the separation objects. The separation size from micro-particles to nano-particles can be used as the scope of application.
4. Multi-function integration on the chip is required in many applications. It is necessary to integrate a variety of detection and analysis methods on the chip to observe and analyse the process of cell separation in real time. To eliminate the influence of other substances in cell suspension or to enrich the experimental cells, it is necessary to integrate the sample pre-treatment module. To

realise cell analysis and reuse, it is necessary to integrate the culture after cell separation.

5. Miniaturisation and automation of the experimental system is a tend of the development. At the present stage, the experiment often uses the imaging system with high magnification microscope, and the new microscopic imaging method can be used for experiment without lens shadow imaging technology and the volume and cost are far lower than the former.

The result of artificial counting can be considered in the future experiments. The method of image processing can be taken into account for cell counting, with higher efficiency and higher automation. In addition, improvements are needed in some aspects, such as the survival rate after separation.

References

[1] Pretlow T G, Pretlow T P, 'Velocity sedimentation of cells', *Nature*, 1988, **333** (6168), pp 97–97

[2] Lecrivain G, Barry L, Hampel U, 'Three-dimensional simulation of multilayer particle deposition in an obstructed channel flow', *Powder Technol.*, 2014, **258**, pp 134–143

[3] Brown B, 'Electrical impedance tomography (EIT): A review', *J. Med. Eng. Technol.*, 2003, **27**(3), pp 97–108

[4] Smith R W, Freeston I L, Brown B H, 'A real-time electrical impedance tomography system for clinical use-design and preliminary results', *IEEE Trans. Biomed. Eng.*, 1995, **42** (2), pp 133–140

[5] Bayford R, Tizzard A, 'Bioimpedance imaging: An overview of potential clinical applications', *Analyst*, 2012, **137** (20), pp 4635–4643

[6] Bera T K, Mohamadou Y, Lee K, *et al.*, 'Electrical impedance spectroscopy for electro-mechanical characterization of conductive fabrics', *Sensors*, 2014, **14** (6), pp 9738–9754

[7] Yang W, 'Design of electrical capacitance tomography sensors', *Meas. Sci. Technol.*, 2010, **21** (4), 042001

[8] Tapp H S, Williams R A, 'Status and applications of microelectrical resistance tomography', *Chem. Eng. J.*, 2000, **77** (1–2), pp 119–125

[9] Meir A, Rubinsky B, 'Electrical impedance tomographic imaging of a single cell electroporation', *Biomed. Microdevices*, 2014, **16** (3), pp 427–437

[10] Beebe D J, Mensing G A, Walker G M, 'Physics and applications of microfluidics in biology', *Biomed. Eng.*, 2002, **4** (4), pp 261–286

[11] Asami K, Yonezawa T, Wakamatsu H, Koyanagi N, 'Dielectric spectroscopy of biological cells', *Bioelectrochem. Bioenerg.*, 1996, **40** (2), pp 141–145

[12] Xiao C, Lachance B, Sunahara G, Luong J H, 'An in-depth analysis of electric cell-substrate impedance sensing to study the attachment and spreading of mammalian cells', *Anal. Chem.*, 2002, **74** (6), pp 1333–1339

[13] Wang H, Pilon L, 'Accurate simulations of electric double layer capacitance of ultramicroelectrodes', *J. Phys. Chem. C*, 2011, **115** (33), pp 16711–16719

[14] Yang W, 'Hardware design of electrical capacitance tomography systems', *Meas. Sci. Technol.*, 1996, **7** (3), pp 225–232

[15] Chai K T C, Davies J H, Cumming D R S, 'Electrical impedance tomography for sensing with integrated microelectrodes on a CMOS microchip', *Sens. Actuators B: Chem.*, 2007, **127** (1), pp 97–101

[16] Andía B I, Sauer K D, Bouman C A, 'Nonlinear Backprojection for tomographic reconstruction', *IEEE Trans. Nucl. Sci.*, 2002, **49** (1), pp 61–68

[17] Vauhkonen M, Vadasz D, Karjalainen P A, Somersalo E, Kaipio J P, 'Tikhonov regularisation and prior information in electrical impedance tomography', *IEEE Trans. Med. Imag.*, 1998, **17** (2), pp 285–293

[18] Yang W, Peng L, 'Image reconstruction algorithms for electrical capacitance tomography', *Meas. Sci. Technol.*, 2002, **14** (4), pp 478–484

[19] Zhang H, Cheng L, 'Projected landweber iteration for matrix completion', *J. Comput. Appl. Math.*, 2010, **235** (3), pp 593–601

[20] Caselli F, Bisegna P, Maceri F, 'EIT-inspired microfluidic cytometer for single-cell dielectric spectroscopy', *J. Microelectromech. Syst.*, 2010, **19** (5), pp 1029–1040

[21] Shah P, Zhu X, Zhang X, He J, Li C Z, 'Microelectromechanical system-based sensing arrays for comparative in vitro nanotoxicity assessment at single cell and small cell-population using electrochemical impedance spectroscopy', *ACS Appl. Mater. Interfaces*, 2016, **8** (9), pp 5804–5812

[22] York T A, Phua T N, Reichelt L, Pawlowski A, Kneer R, 'A miniature electrical capacitance tomograph', *Meas. Sci. Technol.*, 2006, **17** (8), pp 2119–2129

[23] Sun T, Tsuda S, Zauner K-P, Morgan H, 'On-chip electrical impedance tomography for imaging biological cells', *Biosens. Bioelectron.*, 2010, **25** (5), pp 1109–1115

[24] Yang Y, Jia J, Smith S, Jamil N, Gamal W, Bagnaninchi P, 'A miniature electrical impedance tomography sensor and 3D image reconstruction for cell imaging', *IEEE Sens. J.*, 2017, **17** (2), pp 514–523

[25] Boyle A, Adler A, 'The impact of electrode area, contact impedance and boundary shape on EIT images', *Physiol. Meas.*, 2011, **32** (7), pp 745–754

[26] Kim B S, Boverman G, Newell J C, Saulnier G J, Isaacson D, 'The complete electrode model for EIT in a mammography geometry', *Physiol. Meas.*, 2007, **28** (7), S57–S69

[27] Peng L, Merkus H, Scarlett B, 'Using regularisation methods for image reconstruction of electrical capacitance tomography', *Part. Part. Syst. Charact.*, 2000, **17** (3), pp 96–104

[28] Takei M, 'GVSPM image reconstruction for capacitance CT images of particles in a vertical pipe and comparison with the conventional method', *Meas. Sci. Technol.*, 2006, **17** (8), pp 2104–2112

[29] Asami K, Yonezawa T, 'Dielectric behavior of wild-type yeast and vacuole-deficient mutant over a frequency range of 10 KHz to 10 GHz', *Biophys. J.*, 1996, **71** (4), pp 2192–2200

[30] Lvovich V F, 'Impedance representation of bulk-material and electrode processes', in *Impedance Spectroscopy*, John Wiley & Sons, Inc., 2012, pp 59–96

[31] Schwartz M, *Electrodes and the Measurement of Bioelectric Events*, Wiley-Interscience, 1972

[32] Sant S B, *Physics and Chemistry of Interfaces*, Wiley-VCH, 2013, pp 1379–1380

[33] Yao J, Kodera T, Obara H, Sugawara M, Takei M, 'Spatial concentration distribution analysis of cells in electrode-multilayered microchannel by dielectric property measurement', *Biomicrofluidics*, 2015, **9** (4), 044129

[34] Yao J, Sapkota A, Konno H, Obara H, Sugawara M, Takei M, 'Noninvasive online measurement of particle size and concentration in liquid–particle mixture by estimating equivalent circuit of electrical double layer', *Particulate Sci. Technol.*, 2015, **34** (5), pp 517–525

[35] Hanai T, 'Theory of the dielectric dispersion due to the interfacial polarisation and its application to emulsions', *Kolloid-Zeitschrift*, 1960, **171** (1), pp 23–31

[36] Davalos R, Rubinsky B, 'Electrical impedance tomography of cell viability in tissue with application to cryosurgery', *J. Biomech. Eng.*, 2004, **126** (2), pp 305–309

[37] Asami K, Sekine K, 'Dielectric modelling of cell division for budding and fission yeast', *J. Phys. D – Appl. Phys.*, 2007, **40** (4), pp 1128–1133

[38] Takei M, Saito Y, 'Application of the generalised vector sampled pattern matching method to reconstruction of electrical capacitance CT images', *Meas. Sci. Technol.*, 2004, **15** (7), pp 1371–1381

[39] Struijk P C, Hoeven J G, Hoedemaekers C, 'On-line blood viscosity monitoring in vivo with a central venous catheter, using electrical impedance technique', *Biosens. Bioelectron.*, 2013, **41** (4), pp 595–601

[40] Chen X, Liang Z, Li D, *et al.*, 'Microfluidic dielectrophoresis device for trapping, counting and detecting shewanella oneidensis at the cell level', *Biosens. Bioelectron.*, 2017, **99**, pp 416–423

[41] Trautman E D, Newbower R S, 'A practical analysis of the electrical conductivity of blood', *IEEE Trans. Biomed. Eng.*, 1983, **30** (3), pp 141–154

[42] Yoshihito H, Yoichi K, Shinji O, Akio Y, Koji A, Makoto K, Isao U, 'Dielectric coagulometry: A new approach to estimate venous thrombosis risk', *Analytical Chemistry*, 2010, **82** (23), pp 9769–9774

[43] Asami K, Yonezawa T, 'Dielectric analysis of yeast-cell growth', *Biochim. Biophys. Acta – Gen. Subjects*, 1995, **1245** (1), pp 99–105

[44] Huu D N, Kikuchi D, Maruyama O, Sapkota A, Takei M, 'Cole-Cole analysis of thrombus formation in an extracorporeal blood flow circulation using electrical measurement', *Flow Meas. Instrum.*, 2017, **53**, pp 172–179

[45] Cole K S, Cole R H, 'Dispersion and absorption in dielectrics I. Alternating current characteristics', *J. Chem. Phys.*, 1941, **9** (4), pp 341–351

[46] Wang Z L, Zhao T, Yao J F, Kishikawa Y, Takei M, 'Evaluation of the electrochemical characterizations of lithium-ion battery (LIB) slurry with 10-parameter electrical equivalent circuit (EEC)', *J. Electrochem. Soc.*, 2017, **164** (2), pp A8–A17

Jiafeng Yao is a Lecturer in the College of Mechanical & Electrical Engineering, Nanjing University of Aeronautics & Astronautics. He received his BEng degree from The First Aviation Academy of PLA, China in 2008, MSc from Nanjing University of Science & Technology, China in 2011 and PhD from Kumamoto University, Japan in 2014. From 2014 to 2016, he was a JSPS (Japan Society for the Promotion of Science) Research Fellow in the Department of Mechanical Engineering, Chiba University in Japan. His honours and awards include Scholarship from China Scholarship Council for doctoral course in Japan from October 2011 until September 2014 and Research Funding for Doctoral Students' Independent Research, supported by Kumamoto University (2013 and 2014). His research interest includes intelligent measurement, electrical impedance measurement and microfluidics.

Masahiro Takei is a Professor and Vice Dean of Graduate School of Engineering in Chiba University, Japan. He has worked in University of Leeds, UK as a guest researcher supported by the Royal Society in 2007. His research interest includes multiphase flow dynamics, visualisation and measurement, process tomography and inverse problem, micro-fluidics and bio application. He was the President of International Society for Industrial Process Tomography, Vice Chairman and board member of Japan Section of American Society of Mechanical Engineers, board member of Visualisation Society of Japan, Councillor of the Society of Powder Technology, Japan, board member of Japanese Society Experimental Mechanics, Guest Editor of *Measurement Science and Technology* and *Flow Measurement and Instrumentation*, editorial board member of *Journal of Visualisation*.

Chapter 14

Ultrasound imaging sensors

Jianguo Ma[1], Xiaoning Jiang[2] and Lijun Xu[1]

List of abbreviations

CMUT	capacitive micro-machined ultrasonic transducer
CW Doppler	continuous wave Doppler
FBG	fibre Bragg grating
FM	frequency modulation
FOV	field-of-view
FP	Fabry–Perot
HIFU	high intensity focused ultrasound
IVUS	intravascular ultrasound
LDV	laser Doppler vibrometer
MPB	morphotropic phase boundary
PW Doppler	pulsed-wave Doppler
PZT	lead zirconium titanate
SNR	signal-to-noise ratio

14.1 Introduction

Ultrasound is cyclic mechanical vibration with frequency higher than 20 kHz, in comparison to audible sound that human being typically hear in a frequency range between 20 Hz and 20 kHz. Ultrasound requires matter or a medium with particles to vibrate to conduct or propagate. In the wave propagation, particles are deformed or displaced and then re-established after the wave passes. Types of waves include shear, surface and longitudinal waves with the latter being one of the most common used in biological applications. The characteristics of the travelling ultrasound energy depend on the medium that it is travelling through.

The ultrasound wave delivers both energy and information. In an ultrasound imaging application, a given ultrasound wave is transmitted from surface into a

[1]School of Instrumentation and Optoelectronic Engineering, Beihang University, China
[2]Department of Mechanical and Aerospace Engineering, North Carolina State University, USA

body of specimen. Non-uniformities inside the specimen generate echoes that return to the surface again, and the echoes are detected by an ultrasonic sensor at the surface. The time-of-arrival of the echoes and the information modulated in the echo waveforms delivers information inside the specimen. Organising the information can lead to an ultrasound imaging.

This chapter includes basic acoustic theories, introduction of multiple ultrasound sensors and typical ultrasound imaging modes. Fundamentals of acoustics will be introduced in Section 14.2, followed by multiple types of ultrasound sensing technology in Section 14.3. Eventually, an ultrasound imaging system and some typical ultrasound imaging modes will be discussed in Section 14.4. Finally, Section 14.5 summaries the chapter.

14.2 Ultrasound wave

This section introduces fundamental knowledge of sound waves. First, basic oscillation systems are introduced to provide basic background of ultrasound theories. Second, theory and characteristics of several typical beam profiles are discussed. Finally, the section provides several useful properties during propagation.

14.2.1 Fundamentals of acoustic wave

Ultrasonic waves are basically mechanical waves. Introduction to ultrasonic waves starts with the longitudinal vibration of bars as a one-dimensional (1D) sound wave. Majority of ultrasonic imaging conditions use the longitudinal vibration mode. Other vibration modes share similar equations, and exhibit similar properties as the longitudinal mode. Therefore, they are not repeated in this chapter. For details of other vibration systems as transverse vibration or torsional vibration, please refer to a book [1].

14.2.1.1 Longitudinal vibration of bars

A mass-spring system is a simplified lamped-element system, which demonstrates the characteristics of oscillation. Conversion from a lamped-element system to a continuous system yields the longitudinal vibration of bars.

Let's consider a bar of length L with an uniform cross-sectional area A and uniform density ρ as shown in Figure 14.1. A longitudinal force causes displacement u at position x of the bar. The goal is to find the displacement u at position x and time t:

$$u = u(x, t) \tag{14.1}$$

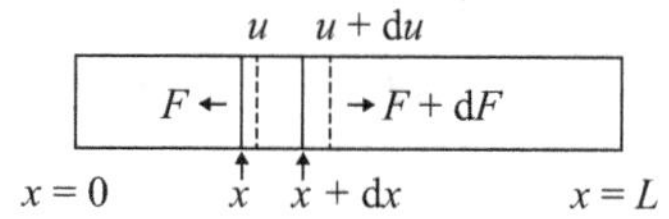

Figure 14.1 A bar with longitudinal vibration

Let's take the cross-section of thin layer (x to $x + \mathrm{d}x$) on the bar as the mass, and take the forces applied on both surfaces of the layer as the spring. The displacements at x and $x + \mathrm{d}x$ are u and $\mathrm{d}u$, respectively. The force at position x is

$$F = -EA\frac{\partial u}{\partial x} \tag{14.2}$$

where E is the Young's modulus and EA is like the spring constant k in the mass-spring system.

Similarly, the force at position $x + \mathrm{d}x$ is

$$F + dF = EA\frac{\partial}{\partial x}\left(u + \frac{\partial u}{\partial x}\right) \tag{14.3}$$

Note that the sign of the force denotes the direction of the force. The net force applied on the thin layer is

$$dF = \mathrm{EA}\left(\frac{\partial^2 u}{\partial x^2}\right)\mathrm{d}x \tag{14.4}$$

With a density ρ, the mass of the layer is

$$m = \rho A\mathrm{d}x \tag{14.5}$$

According to Newton's Second Law,

$$EA\left(\frac{\partial^2 u}{\partial x^2}\right)\mathrm{d}x = \rho A\mathrm{d}x\left(\frac{\partial^2 u}{\partial t^2}\right) \tag{14.6}$$

Simplifying the above equation yields

$$\frac{\partial^2 u}{\partial x^2} = \frac{1}{c^2}\frac{\partial^2 u}{\partial t^2} \tag{14.7}$$

where c is sound speed defined as

$$c = \sqrt{E/\rho} \tag{14.8}$$

The general solution of the second-order differential equation is

$$u(x,t) = Ae^{j(\omega t - kx)} + Be^{j(\omega t + kx)} \tag{14.9}$$

where ω is the angular frequency and k is the wave number defined as $1/\lambda$ (λ is the wavelength).

The general solution is the 1D wave propagation equation consisting both forward (x positive direction) wave and backward (x negative direction) wave with the amplitude of A and B for each wave. The solution is a general form for ultrasonic wave in a certain frequency. There may be more than one frequencies in the medium, in which case the waveform is the superposition of the waves at each frequency.

14.2.1.2 Vibration modes of bars with boundaries

If the bar is terminated at boundaries $x = 0$ and $x = L$ (see Figure 14.2), then the wave propagation is specialised according to the boundary conditions. If $Z_1 = Z_2 = Z_0$, indicating that the boundaries are perfectly matched at $x = 0$ and $x = L$, then the wave propagates without any reflection, performing the same as the plane wave. On the contrary, if $Z_1 \neq Z_0$ and $Z_2 \neq Z_0$, then the specific medium vibrates as a standing wave. Under extreme conditions, the boundaries are considered as free or fixed, with boundary impedances Z_1 and Z_2 equal to zero or infinity. There are three types of typical boundary conditions at the impedance: (1) free-free, (2) fixed-fixed and (3) fixed-free. Three groups of vibration modes exist at these boundary conditions.

For a free-free boundary condition, both sides of the boundaries are free to move so that there is no stress at either boundary:

$$\left.\frac{\partial u}{\partial x}\right|_{x=0} = 0 \tag{14.10}$$

$$\left.\frac{\partial u}{\partial x}\right|_{x=L} = 0 \tag{14.11}$$

Constrained by the boundary condition, the vibration modes are

$$u_n(x,t) = \widetilde{A}_n \cos\left(\frac{n\pi x}{L}\right)e^{j\omega_n t}, (n = 1, 2, 3, ...) \tag{14.12}$$

where n donates the nth harmonic vibration mode.

Equation (14.12) is the general form of the standing wave in the medium with free-free boundary conditions. The first three vibration modes in the group are shown in Figure 14.3. If the boundaries are symmetric ($Z_1 = Z_2$), then the centre surface ($x = L/2$) of the medium stays stationary if not specifically excited. In that case

$$u|_{x=L/2} = 0 \tag{14.13}$$

and the even order harmonics do not show up. Vibration modes in this case are

$$u_n(x,t) = \widetilde{A}_n \cos\left(\frac{n\pi x}{L}\right)e^{j\omega_n t}, (n = 1, 3, 5, ...). \tag{14.14}$$

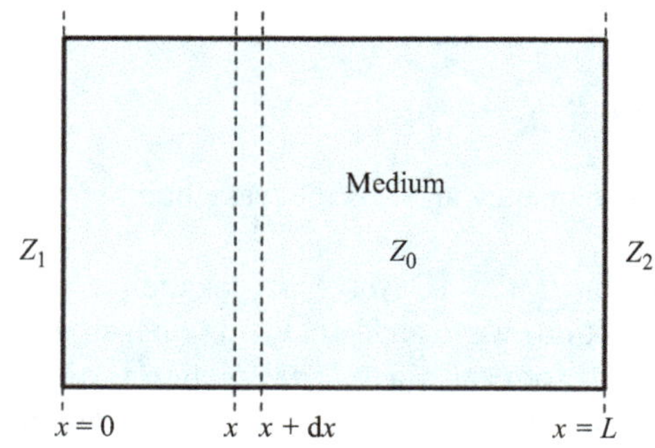

Figure 14.2 1D wave propagation

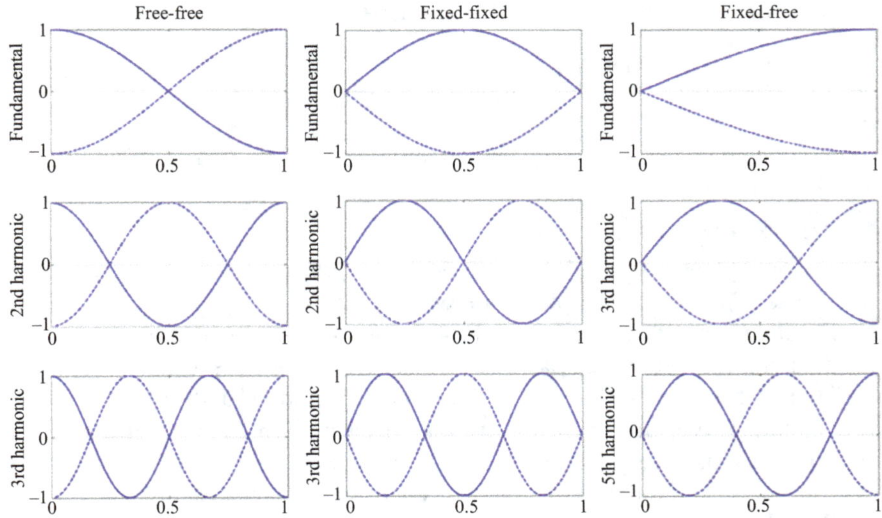

Figure 14.3 Standing wave vibration modes of free-free, fixed-fixed and fixed-free boundary conditions

This group of vibration modes in (14.14) is very close to active piezoelectric materials in piezoelectric transducers, which has low (not exactly free) impedance boundaries at both sides and the two boundaries are almost symmetric. In most cases, the piezoelectric plates work at the fundamental frequency.

On the opposite, a fixed-fixed boundary condition indicates the medium is clamped at both boundaries, resulting in no displacement at the boundaries as

$$u|_{x=0} = 0 \tag{14.15}$$

$$u|_{x=L} = 0 \tag{14.16}$$

Vibration modes at this boundary condition are

$$u_n(x,t) = \widetilde{A}_n \sin\left(\frac{n\pi x}{L}\right) e^{j\omega_n t}, (n = 1, 2, 3, \ldots) \tag{14.17}$$

The first three vibration modes are shown in Figure 14.3.

The third boundary condition is the fixed-free boundary condition, with one boundary of zero displacement and the other of zero stress. It is an equivalent issue either to be fixed at $x = 0$ or $x = L$. In this topic, the fixed boundary is located at $x = 0$ while the free boundary is located at $x = L$. The boundaries are

$$u|_{x=0} = 0 \tag{14.18}$$

$$\left.\frac{\partial u}{\partial x}\right|_{x=L} = 0 \tag{14.19}$$

Vibration modes at this boundary conditions are

$$u_n(x,t) = \widetilde{A}_n \sin\left(\frac{n\pi x}{2L}\right)e^{j\omega_n t}, (n = 1, 3, 5, ...) \tag{14.20}$$

14.2.2 Ultrasound beams

14.2.2.1 Plane wave

An ultrasonic plane wave is 1D vibration, which propagates along one direction with constant phases across planes perpendicular to the propagation direction. The 1D bar vibration discussed in the Section 14.2.1 is a plane wave, which considers vibration along the bar while variations on the cross-section plane are neglected. All the properties of the plane wave, including both travelling wave and standing wave, are the same as that discussed in the bar vibration. In vibration systems, every particle in a vibrations system oscillates at certain frequency or a combination of several frequencies. Displacement was used to describe the oscillation. However, the displacement is not an easily detectable parameter. Acoustic pressure is a typical description of the amplitude of acoustic waves. The pressure p is proportional to the particle velocity v, which is the first derivative of the displacement u. The relationship between the velocity and the displacement is

$$v = \frac{\partial u}{\partial t} = j\omega \cdot u(x,t) = j\omega(Ae^{j(\omega t - kx)} + Be^{j(\omega t + kx)}) \tag{14.21}$$

Note that the particle velocity is not the sound speed in a certain medium. The particle velocity varies over time and varies among particles, which is an instant description of the molecular behaviour. On the contrary, the sound speed is a constant number for a certain medium in a given condition. In short, the particle velocity describes the particle vibration whereas the sound speed describes the wave propagation.

The parameter $j\omega$ describes the phase and amplitude relationships between velocity and displacement. In most cases, particle velocity and pressure are used without referring to the displacement. Therefore, the parameter $j\omega$ is usually omitted and combined to the amplitude parameter A and B. The pressure is the multiplication of particle velocity v and the acoustic impedance Z:

$$p = vZ \tag{14.22}$$

In most cases, the pressure is used to indicate the amplitude of sound waves. The constant value Z is also omitted and combined to the amplitude parameter A and B. For a 1D sound wave

$$p = Ae^{j(\omega t - kx)} + Be^{j(\omega t + kx)} \tag{14.23}$$

Note that the amplitude parameters A and B denotes different amplitude constant versus those for displacement. The acoustic impedance is defined as

$$Z = \rho_0 c \tag{14.24}$$

where ρ_0 and c are the density and sound speed at equilibrium state.

The acoustic impedance plays important roles in many applications, such as wave propagation and reflection, and impedance matching layer design. A specific unit 'Rayl' was associated with the acoustic impedance in honour of John William Strutt, Baron Rayleigh (1842–1919), which is defined as 1 Rayl = 1 kg m^{-2} s^{-1}.

For air at a temperature of 20 °C and atmospheric pressure, the density is 1.21 kg/m^3 and the sound speed is 343 m/s. The acoustic impedance is 415 Rayl. In distilled water at 20 °C and 1 atm, the density is 998 kg/m^3 and the sound speed is 1,482 m/s. Accordingly, the acoustic impedance is 1.48 MRayl (Mega Rayl).

The instantaneous intensity $I(t)$ of a sound wave is the instantaneous rate per unit area, at which work is done by one element of fluid on an adjacent element. It is given by

$$I(t) = pv \tag{14.25}$$

with the unit of watts per square meter (W/m^2).

The time-average intensity I of $I(t)$ is the time-average rate of energy transmission through a unit area normal to the direction of propagation, which is defined as

$$I = \frac{1}{T}\int_0^T pv\mathrm{d}t \tag{14.26}$$

For a plane wave, the time-averaged intensity is

$$I = \frac{P^2}{2Z} \tag{14.27}$$

where P is the peak amplitude of the pressure for a sinusoidal wave.

A plane wave is theoretical simplification of waves, which approximates several cases. First, if the acoustic source is very large and every point on the source plane vibrates in the same phase, then the wave in front of the source is approximately a plane wave. Second, at positions very far from an acoustic source, the wave is constant phase across a large sphere surface, and the curvature radius is large enough to consider the wave as a plane at a local areas.

14.2.2.2 Three dimensional wave

Extended from 1D wave, the three-dimensional (3D) acoustic wave differential equation is

$$\nabla^2 p = \frac{1}{c^2}\frac{\partial^2 p}{\partial t^2} \tag{14.28}$$

where the Laplacian operator is

$$\nabla = \frac{\partial}{\partial x} + \frac{\partial}{\partial y} + \frac{\partial}{\partial z} \tag{14.29}$$

A solution to this partial differential equation requires information of specific beam patterns. For a plane wave traveling in some arbitrary direction, the solution can be

$$p = Ae^{j(\omega t - k_x x - k_y y - k_z z)} \tag{14.30}$$

Several more types of beam patterns with certain symmetries are discussed in the Section 14.2.2.

14.2.2.3 Spherical wave

Expressed in spherical coordinates, the Laplacian operator is

$$\nabla^2 = \frac{\partial^2}{\partial r^2} + \frac{2}{r}\frac{\partial}{\partial r} + \frac{1}{r^2 \sin\theta}\frac{\partial}{\partial \theta}\left(\sin\theta \frac{\theta}{\partial \theta}\right) + \frac{1}{r^2 \sin^2\theta}\frac{\partial^2}{\partial \phi^2} \tag{14.31}$$

If the waves have spherical symmetry, the acoustic pressure p is a function of radial distance and time but not of the angular coordinates, and this equation simplifies to

$$\nabla^2 = \frac{\partial^2}{\partial r^2} + \frac{2}{r}\frac{\partial}{\partial r} \tag{14.32}$$

The wave equation for pressure reduces to

$$\frac{\partial^2 p}{\partial r^2} + \frac{2}{r}\frac{\partial p}{\partial r} = \frac{1}{c^2}\frac{\partial^2 p}{\partial t^2} \tag{14.33}$$

The general solution of the differential equation is

$$p = \frac{1}{r}\left(Ae^{j(\omega t - kr)} + Be^{j(\omega t + kr)}\right) \tag{14.34}$$

The first term represents an outgoing wave diverging from the origin, whereas the second term represents an incoming wave, converging to the origin. The solution fails at the origin because the pressure approaches to infinity as $r \rightarrow 0$. However, this condition never happens for either outgoing wave or incoming waves. For the outgoing wave, the acoustic source cannot infinitesimally small while non-zero acoustic power is emitted. For the incoming wave, significant nonlinearity causes strong losses that limits the pressure amplitude.

In actual cases, the outgoing wave represents a wave with point source whereas the incoming wave represents a focusing wave. Although point sources with non-zero emission do not exist, any acoustic source can be considered as the integration of point sources. An example of this application will be illustrated in Section 14.3.2. Spherical wave is also the approximation if the point is far away from the source (far field in Section 14.2.2.4). In many cases, although the source is not infinitesimally small, it can be generally considered as spherical wave if it is much smaller than the wavelength. Scattering speckles (in Section 14.2.3) and elements of arrays (in Section 14.3.2) are all considered as a point source and emits spherical waves.

The incoming wave means focusing, which generates high intensity at the focal point and specifies the spatial point of interest. The high intensity focused ultrasound (HIFU) can generate high temperature that ablates biology tissues at the focal point, which is used as non-invasive therapy of tumours. In imaging applications, the focused ultrasound generates high spatial resolution (i.e. localised) with high signal-to-noise ratio (SNR) due to the strong intensity, which will be discussed in Section 14.3.2.

14.2.2.4 Piston-shaped source

An acoustic source of practical interest is the plane circular piston, which is the model for a number of sources, including loudspeakers, open-ended organ pipes and ventilation ducts. Let's consider a piston of radius a mounted on a flat rigid baffle of infinite extent. Let the radiating surface of the piston move uniformly with speed $v_0 e^{j\omega t}$ normal to the baffle. The geometry and coordinates are shown in Figure 14.4. The pressure at any field point can be obtained by dividing the surface of the piston into infinitesimal elements, each of which acts as a baffled simple source of strength. The total pressure at a certain point is

$$p(r,\theta,t) = j\rho_0 c \frac{v_0}{\lambda} \int_S \frac{1}{r'} e^{j(\omega t - kr')} dS \tag{14.35}$$

Integration of the pressure is not included in this chapter. Characteristic properties of the acoustic field will be provided. For the points on the z-axis (axial direction), the pressure oscillates dramatically as positions close to the baffle, and decreases gradually at positions farther than the natural focal point ($r = a^2/\lambda$), where a is the radius of the baffle. The relative pressure along the axial direction is shown in Figure 14.5(a).

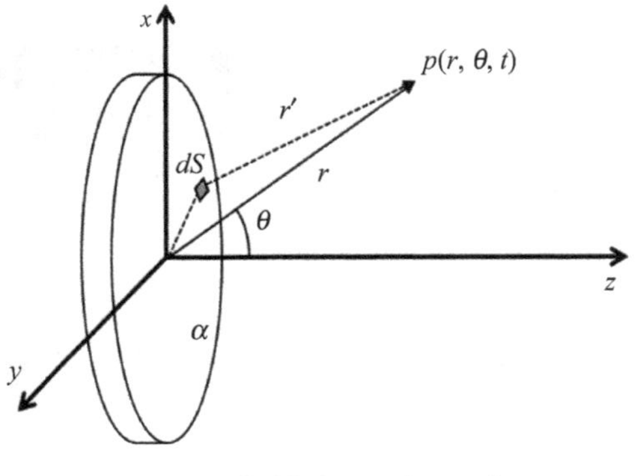

Figure 14.4 Pressure in acoustic field depends on distance and angle deviated from axial direction

The piston baffle is axial-symmetric. Therefore, the acoustic field is symmetric as well. As a consequence, the acoustic pressure is a function of the distance r and the angle θ from the axial direction, but not a function of the azimuth. The acoustic field at the plane across the z-axis is shown in Figure 14.5(b). The natural focal point generally separate the acoustic field as near-field and far-field. In the near-field, the amplitude of the acoustic pressure varies dramatically. On the contrary, the pressure in the far-field exhibits clear patterns, one main lobe is along the axial direction and side lobes at both sides of the main lobe.

14.2.3 Reflection and scattering

Acoustic energy is partially reflected and partially transmitted at mismatched boundaries (see Figure 14.6(a)). A matched boundary indicates the fact that the equivalent impedances across the boundary are in complex conjugate with each other ($Z_1 = Z_2^*$, with the star standing for complex conjugate operation). If Z_1 and Z_2 are real values, then $Z_1 = Z_2$. If the boundary is not matched ($Z_1 \neq Z_2^*$), the reflection coefficient R_P and transmission coefficient T_P are given as

$$R_P = \frac{Z_2 - Z_1^*}{Z_2 + Z_1^*} \tag{14.36}$$

$$T_P = \frac{2Z_2}{Z_2 + Z_1^*} \tag{14.37}$$

where the subscript $()_P$ stands for the pressure.

If the reflecting boundary is not infinitely (or sufficiently) large plane but a small particle with smaller dimension compared with the wavelength, then it acts as a scattering point source (see Figure 14.6(b)). The diverging signal is measurable at a wide range, which is an important feature for array sensing because each element has to receive the signal. Detailed array sensing will be discussed in Section 14.3.2.

14.2.4 Attenuation

The above calculations are based on an assumption that the loss is negligible in the wave propagation. As the acoustic wave propagates in a lossy medium, a portion of

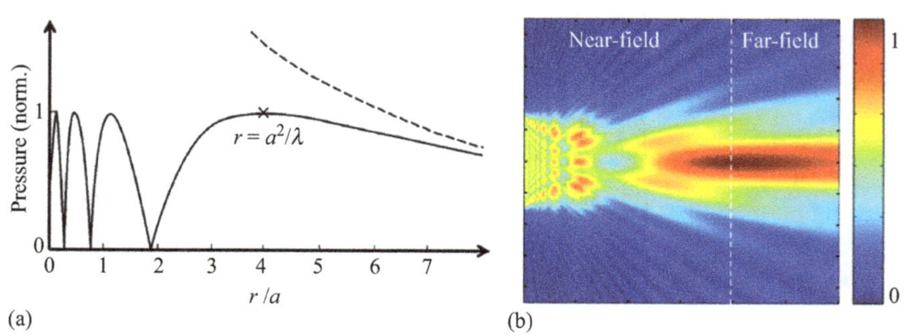

Figure 14.5 (a) Pressure amplitude at z-axis (axial direction) and (b) pressure amplitude in a two-dimensional acoustic field

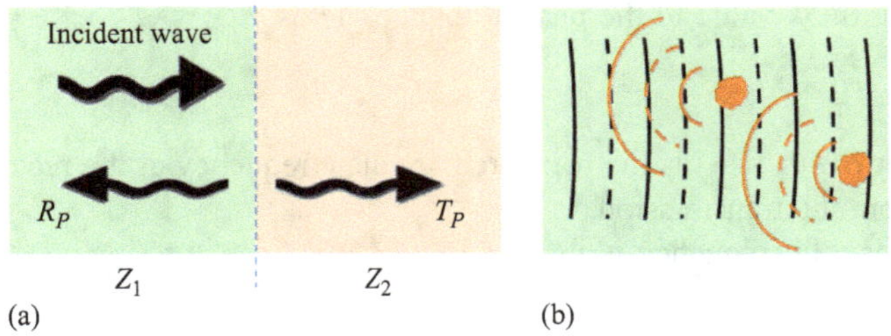

Figure 14.6 *(a) Reflection and transmission of ultrasonic wave at impedance boundaries and (b) scattering of acoustic wave at particles smaller than wavelength*

the mechanical energy is converted into thermal energy and dissipated. For the plane wave, the output pressure amplitude P is

$$P = P_0 e^{-\alpha z} \tag{14.38}$$

where P is the pressure amplitude at depth z, P_0 is the original pressure with $z = 0$, and α is the attenuation coefficient.

In biomedical applications, the attenuation is frequency dependent, which is expressed as $\alpha \propto f^b$, where f is the frequency and the exponential b is typically slightly larger than 1. In diagnostic ultrasound applications with relatively low scattering and frequencies below 50 MHz, the exponential b is approximately considered as 1, and the unit of attenuation coefficient is dB/cm/MHz.

A precise value of b is slightly larger than 1, which is dependent on the frequency, the bio-tissue, and the amplitude of the pressure. Depending on the precision requirement, higher order frequency dependence can be considered

$$\alpha = \alpha_1 f + \alpha_2 f^2 + \alpha_3 f^3 + \text{residual} \tag{14.39}$$

14.2.5 Doppler shift

If the acoustic source is stationary while the observer travels during the wave reception, then the frequency detected by the observer deviates from the transmitted frequency. Take a right-traveling plane wave as an example

$$p = Ae^{j(\omega t - kx)} \tag{14.40}$$

If x in the equation varies with time, then the phasor change within time dt is

$$p = Ae^{j(\omega \mathrm{d}t - k\frac{\mathrm{d}x}{\mathrm{d}t}\mathrm{d}t)} = Ae^{j(\omega - kv)\mathrm{d}t} \tag{14.41}$$

The wave number k is defined as

$$k = \frac{2\pi}{\lambda} = 2\pi f \frac{v_o}{c} = \omega \frac{v_o}{c} \tag{14.42}$$

where v_o denotes the velocity of the observer.

Inserting the k value to the phasor change yields

$$p = Ae^{j\omega(1-\frac{v_o}{c})dt} \tag{14.43}$$

where the term $\omega(1 - \frac{v_o}{c})$ is the measured angular frequency at the receiver, which is shifted from the transmission.

The measured frequency is

$$f = \left(1 - \frac{v_o}{c}\right)f_0 \tag{14.44}$$

Similarly, if the acoustic source travels while emitting acoustic wave, then it will cause another frequency shift term and the measured frequency becomes

$$f = \left(\frac{c - v_o}{c - v_s}\right)f_0 \tag{14.45}$$

Note that the signs of velocity for both the source and the observer refer to the wave propagation direction. They are positive values if they are the same as the wave propagation direction, and negative values if they are opposite. In high dimensional space, the motion direction of the source/observer modifies the frequency shift as

$$f = \left(\frac{c - v_o \cos\theta_o}{c - v_s \cos\theta_s}\right)f_0 \tag{14.46}$$

where θ_s and θ_o denote the angle between the source/observer motion and the wave propagation, respectively.

The frequency shift is named as Doppler shift. Unlike Specific Theory of Relativity that describes electromagnetic waves in vacuum, ultrasound travels in medium. As a result, all the velocities are absolute velocities referred to the medium. The sound speed c varies in different inertial frames of reference. Travelling along the wave propagation direction at the sound speed results in measurement of constant pressure.

14.2.6 Nonlinearities in ultrasound imaging

All acoustic equations derived so far are based on linear ultrasound. The oscillation equations used Hook's Law that requires the retrieving force to be proportional to the particle displacement from equilibrium state $f = -kx$. In actual cases, however, the force is not exactly proportional to the displacement, especially if strong acoustic intensity induces large particle displacement. Many parameters, like sound speed and densities, are dependent on the acoustic intensities as well. Therefore, linear ultrasound wave is an approximation for a weak vibration, whereas nonlinear responses exist in every ultrasound system, especially with strong excitations.

There are many sophisticated equations and solutions for nonlinear ultrasound, which are out of the range for this chapter, which focuses on imaging applications instead of theoretical acoustics, and provides some descriptive results without sophisticated equations. This section briefly introduces some properties of nonlinearity, and detailed application of the properties will be discussed in detail Section 14.3 regarding to the specific imaging modalities.

In linear systems, by definition, the ultrasound echo is the linear combination of the transmitted ultrasound signal, where the linear combination indicates delay and amplitude modulation of the signal. Linear combination of the signal does not alter the frequency of the signal, that is the ultrasound echo remains the same frequency as the transmitted excitation. In nonlinear systems, on the contrary, the reflected or scattered signal is distorted in a manner that it is no longer a combination of the excitation signal, and consists signals in other frequencies than the excitation frequency. The additional frequencies are denominated as the harmonics, and the original frequency is the fundamental frequency. Because the excitation wave at fundamental frequency is periodic, the harmonics are typically periodic as well. The harmonic frequencies are typically integer times as the fundamental frequency or half of the fundamental frequency [2]. The harmonic signals reveal nonlinear information of tissues or micro-bubble contrast agents, leading to an effective imaging modality named as harmonic imaging [3,4]. Detailed harmonic imaging techniques and examples will be discussed in Section 14.4.7.

14.3 Ultrasound sensors

An ultrasound sensor is used to generate ultrasonic waves and detects the echoes. The sound wave is typically produced by a piezoelectric transducer encased in a plastic housing. Strong, short electrical pulses from the ultrasound machine drive the transducer at the desired frequency. The frequencies can be anywhere between 1 and 18 MHz, though frequencies up to 50–100 MHz have been used experimentally in a technique known as bio-microscopy in special regions, such as the anterior chamber of the eye [5]. Transducers based on old technology focus their beam with physical lenses. Transducers of new technology use phased array techniques to enable the ultrasound machine to change the direction and depth of focus [6]. Sound is focused either by the shape of the transducer, a lens in front of the transducer, or a complex set of control pulses from the ultrasound scanner, in the beamforming technique. This focusing produces an arc-shaped sound wave from the face of the transducer. The wave travels into the body and comes into focus at a desired depth.

The return of the sound wave to the transducer results in the same process as sending the sound wave, except in reverse. The returned sound wave vibrates the transducer and the transducer turns the vibration into electrical pulses that travel to the ultrasonic scanner, where they are processed and transformed into a digital image.

14.3.1 Piezoelectric ultrasound transducers

14.3.1.1 Piezoelectricity

An electrical insulator can be polarised by an applied electric field $\vec{E}$. Electric charges inside the material do not flow freely or accumulate at the surfaces as a conductor, but just shift slightly by the external electric field from their equilibrium positions. As the positive and negative charges shift towards opposite directions, the material shows net negative charge at the high electric potential side and positive charge at the low electric potential side. This phenomenon is dielectric polarisation, which is defined as $\vec{P} = \varepsilon_0 \chi \vec{E}$, where ε_0 is the dielectric constant in vacuum and χ is the electric susceptibility. The dielectric polarisation generates an internal electric dipoles that compensate the external electric field and result in a lower electric field inside the material. If the material is originally centro-symmetric or isotropic, then the material can be deformed by the external electrical field because of polarisation, and this effect is denominated as the electro-strictivity, as shown in Figure 14.7. The dipoles are generated by external force, so the material is always elongated by the external electric field (see Figure 14.7(b) and (c)). Mechanical compression could not generate charges on the surfaces because there are no inherent dipole if external electric field is absent (see Figure 14.7(d)). A classification of the dielectric materials is shown in Figure 14.8.

On the contrary if the material is non-centrosymmetric (see Figure 14.9(a)), the external electric field can either expand (see Figure 14.9(b)) or compress (see Figure 14.9(c)) the material with a relatively large deformation compared to electro-strictive effect. The non-centrosymmetry of the material indicates the fact that the positive charge centre does not overlap the negative one. An applied external source pushes the positive centre to the low potential direction while

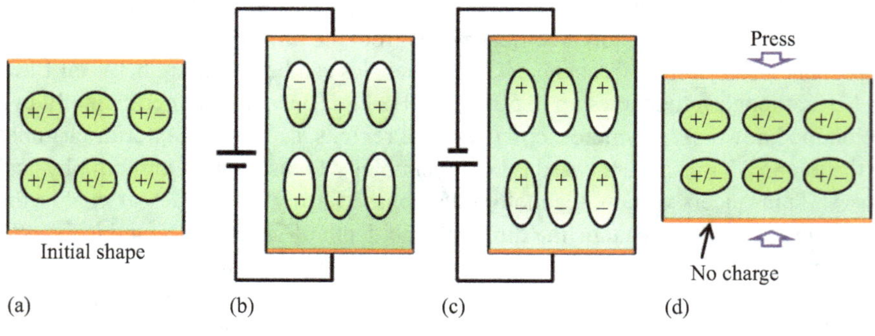

Figure 14.7 Electro-strictive material: (a) at initial shape, (b) and (c) with external electric field and (d) with external mechanical compression

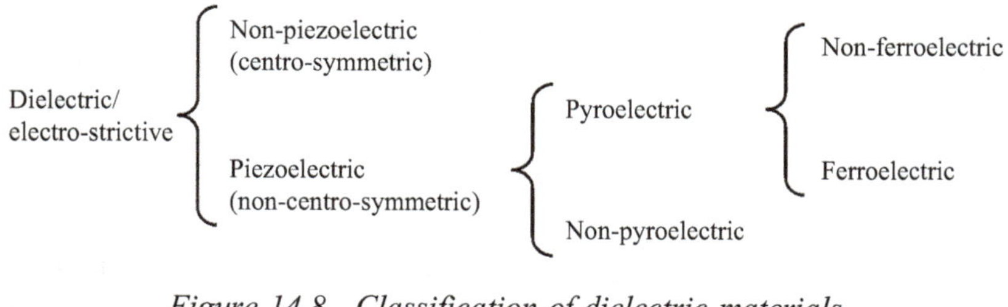

Figure 14.8 Classification of dielectric materials

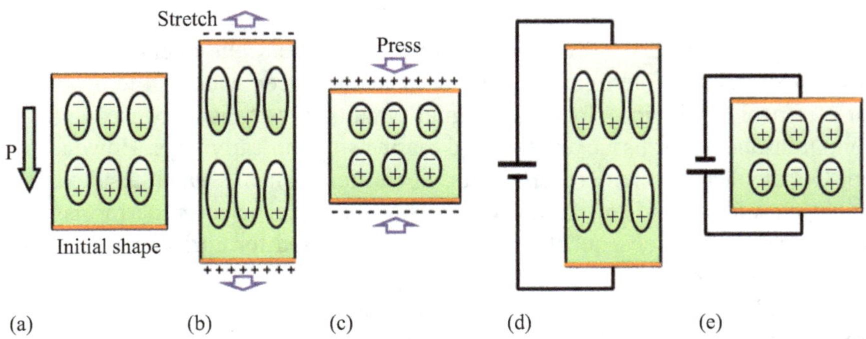

Figure 14.9 *Reverse piezoelectricity: (a) at initial shape without external electric field, (b) expanded by external forward electrical field, (c) compressed by reverse electric field, (d) being stretched by external force and (e) being compressed by external force*

pushing the negative centre to the positive direction. As a result, the material expands if the external electric field is the same as the polarisation of the material and shrinks if it is reversed. Such electric field-induced deformation is denominated as reverse piezoelectric effect and such material is denominated as piezoelectric material. If the polarisation is sensitive to temperature, then the material is pyroelectric material. Some pyroelectric materials are ferro-electric materials, the characteristics of which are that the polarisation can be reversed by external electric field, and residual polarisation exists after the external electric field is removed. Most high performance piezoelectric materials are ferroelectric materials, such as lead zirconium titanate ($Pb[Zr_xTi_{1-x}]O_3$, PZT) or relaxor-based lead titanate ($PbTiO_3$, relaxor-PT) such as lead magnesium niobate-lead titanate ($Pb[Mg_{1/3}Nb_{2/3}]O_3$–$PbTiO_3$, PMN-PT). Relationships among these materials are shown in Figure 14.8.

Applying force on the piezoelectric material in the polarisation direction, electric charges are generated. This is the direct piezoelectric effect found by Curie brothers earlier than the reverse piezoelectric effect [7]. In steady state (see Figure 14.9(a)), the internal diversion of positive and negative charge centre is neutralised by the ambient charges so that there are no net charges on the surfaces. The applied external force pushes the two charge centres closer (see Figure 14.9(e)) or pull them farther (see Figure 14.9(d)), breaking the neutralisation and generating net charges at surfaces of the piezoelectric material. The direct and reverse piezoelectric effects are the opposite scenarios that transduce the energy between electrical and mechanical form.

14.3.1.2 Piezoelectric materials

Piezoelectric materials consist of many forms of materials, including crystal, ceramic, polymer and composites of piezoelectric material with passive polymers. Quartz, a piezoelectric crystal, is the first material found with piezoelectricity, which is one of the few naturally available piezoelectric materials. Quartz exhibits extremely high (~10^4 order) quality factor (Q-factor), together with good reliability, long life and very low dependency on temperature. Consequently, quartz can provide extremely stable frequency output, which has been widely used in watches and electrical oscillation clocks for digital circuits. Quartz is also used as ultrasound transducers for tissue ablation and marine sonar in early days. However, the high Q-factor leads to long reverberation in the time domain, which results in low imaging resolution and low bandwidth. Besides, the piezoelectric performance of quartz is very low. Consequently, quartz is seldom used for ultrasound transducer any more after PZT was developed around 1952.

PZT, $Pb[Zr_{1-x}Ti_x]O_3$ ($0 \leq x \leq 1$), is a ceramic with perovskite crystal structure (see Figure 14.10(a) and (b)) that exhibits much higher piezoelectric properties than natural piezoelectric materials [8] and therefore is the most widely used as active material for ultrasound transducers after being developed. Structure of PZT depends on both the percentage of Ti ($\frac{\mathrm{Ti}}{\mathrm{Zr + Ti}}$ in mol) and the temperature of the material (see Figure 14.10(c)) [9]. At temperature higher than the Curie temperature T_C ($\sim$ 350 °C), PZT is a cubic structure (see Figure 14.10(a)) without polarisation and therefore no piezoelectricity. Below Curie temperature, PZT exhibits polarisation (see Figure 14.10(b)) and it performs the best piezoelectricity at a Ti percentage of 48%–52%, close to the morphotropic phase boundary (MPB). For different applications, PZT materials are modified to induce domain wall motions for different properties. If donors are doped, metal (cation) vacancies are generated to produce high piezoelectric properties, with the expense of larger loss due to the internal friction. The high piezoelectric coefficients, high loss, and donors doped PZTs are usually named as 'soft PZT', which are used in low power applications, such as imaging transducers to take advantage of the performance without being damaged by the high loss.

Relaxor-based ferro-electric single crystal is a group of piezoelectric material possessing extraordinary piezoelectric properties [10]. This group of relaxor-PT-based

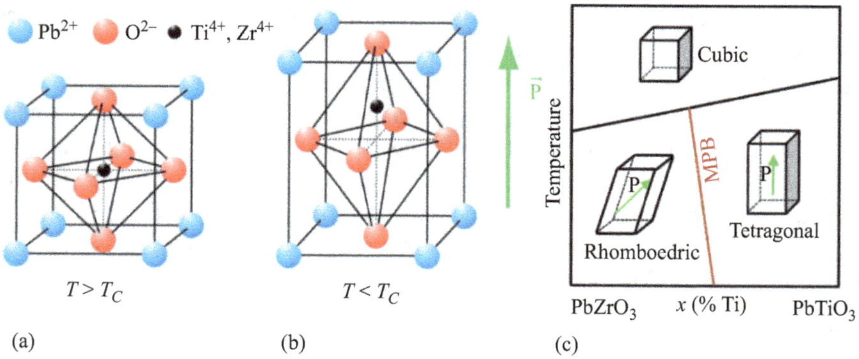

Figure 14.10 Perovskite structure of PZT material: (a) in cubic shape at $T > T_C$, (b) in polarized shape at $T < T_C$ and (c) phase transformation of PZT dependent on composition and temperature

crystals include $Pb[Mg_{1/3}Nb_{2/3}]O_3$–$PbTiO_3$ (PMN-PT), $Pb[Zn_{1/3}Nb_{2/3}]O_3$–$PbTiO_3$ (PZN-PT), $Pb[Yb_{1/2}Nb_{1/2}]O_3$–$PbTiO_3$ (PYN-PT), $Pb[In_{1/2}Nb_1/2]O_3$–$Pb[Mg_{1/3}Nb_{2/3}]O_3$–$PbTiO_3$ (PIN-PMN-PT) and so on [11,12]. PMN-PT exhibits superior performances in the relaxor-PT family and is consequently one of the most widely used material for high end piezoelectric transducers. The drawback of PMN-PT material is the relatively low operation temperature (phase transformation temperature T_{rt} = 60 °C–95 °C), so that other alternative relaxor-PT crystals are also widely researched for high temperature applications, although the piezoelectric properties are slightly lower.

The semi-crystalline polymer poly-vinylidene fluoride PVDF, ($[C_2H_2F_2]_n$) and the copolymer poly-vinyledenedifluoride–trifluoroethylene (PVDF-TrFE) exhibit polarisation because of the asymmetry alignment of hydrogen and fluorine atoms (PVDF in Figure 14.11) [13]. As a polymer, PVDF exhibits very different properties from the piezoelectric ceramics or crystals [14]. Characteristic acoustic impedance and stiffness of PVDF are very low, making it an ideal material for acoustic wave reception. Besides, its flexibility and mechanical stability assured that it can be easily manufactured to various shapes. However, its low piezoelectric properties limit its utilisation where high transmission performances are required.

14.3.1.3 Transduction between electrical and mechanical waves

An ultrasound transducer makes use of both direct and reverse piezoelectric effects, and transduces between electrical and mechanical energy. For the acoustic wave transmission, an electrical signal, usually a short and high voltage impulse, is applied on the piezoelectric material. Due to the reverse piezoelectric effect, a short package of acoustic wave is generated, the waveform of which is the convolution of the impulse and the transmission response of the transducer. The acoustic wave propagates forward and is reflected or scattered by impedance non-uniformities and propagates back to the transducer. The backward acoustic wave, commonly named as an echo, excites the transducer and generates electrical signal by direct piezoelectric effect. The electric signal is digitised by a data acquisition system. With the sound speed given, the round loop distance that the sound travels, the sound speed multiples the time-of-flight is twice the distance of the target from the transducer.

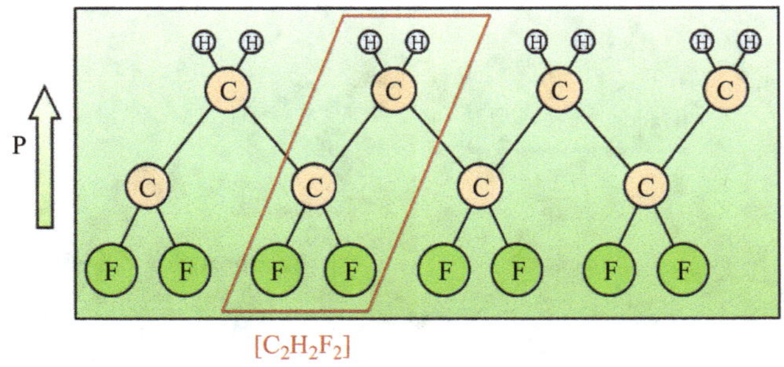

Figure 14.11 Carbon chain of PVDF polymer

The detected signals are the convolution of the impulse excitation, the transmission response of the transducer, the scatter response of the target and the reception response of the transducer. The transmission and the reception response of the transducer is combined as a pulse-echo response, which is an important feature of the transducer performance. Because the convolution is multiplication in frequency domain, a broadband pulse-echo response is beneficial for the acquisition of broadband signal in frequency domain and short impulse in time domain, resulting in high resolution imaging.

14.3.1.4 Matching layer

As discussed in Section 14.2.3, the reflection at a boundary between two materials depends on the impedance of the two materials. However, the high-performance piezoelectric materials, for example those in the PZT and relaxor-PT families discussed in Section 14.3.1, exhibit very high acoustic impedance ($Z > 30$ MRayl), whereas the impedance of human soft tissue is about 1.5 MRayl. The significant impedance mismatch causes an energy reflection of >80% and <20% of energy transmits from the transducer to the soft tissues. The low transmission efficiency causes low resolution and low SNR.

An impedance matching layer sandwiched between the piezoelectric material and soft tissue would promote the transmission efficiency. If an intermediate layer with impedance Z_2 is sandwiched between two media Z_1 and Z_3 (see Figure 14.12), the transmission coefficient for ultrasound intensity T_I is given as

$$T_I = \frac{I_3}{I_1} = \frac{4Z_1Z_3}{(Z_1 + Z_3)^2 \cos^2 kD + (Z_2 + \frac{Z_1Z_3}{Z_2})^2 \sin^2 kD} \tag{14.47}$$

where Z_1, Z_2, Z_3 are the impedances at the three media, k is the wave number $k = 2\pi/\lambda$ and D is the thickness of the intermediate layer with Z_2.

If the thickness of the intermediate is designed as quarter wavelength ($D = \lambda/4$), then the transmission coefficient is given by

$$T_I = \frac{4Z_1Z_3}{(Z_2 + \frac{Z_1Z_3}{Z_2})^2} \tag{14.48}$$

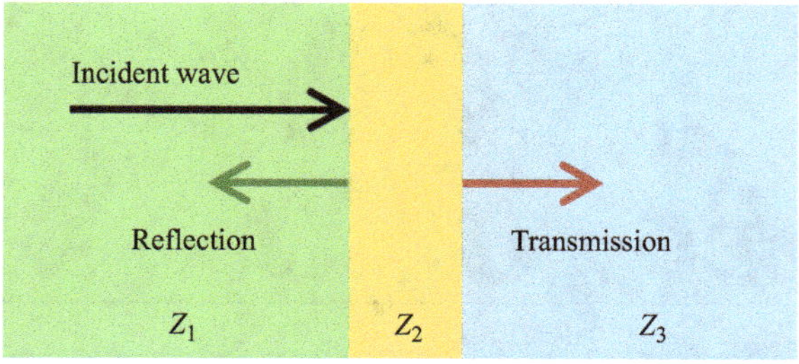

Figure 14.12 Wave propagation in a one-dimensional three-media system

Furthermore, if the impedance of the intermediate layer is designed as $Z_2 = \sqrt{Z_1 Z_3}$, then the transmission coefficient becomes

$$T_I = 1 \tag{14.49}$$

It means that all the energy from the incident wave in the first medium is delivered to the third medium. In the ultrasound transducer design, adding a thin layer with specific thickness and specific acoustic impedance will substantially promote the transmission efficiency. This layer is denominated as matching layers in transducer.

In brief, the design of the matching layer requests the thickness of the intermediate layer as quarter wavelength, and the acoustic impedance as the geometric mean of the piezoelectric material and the from medium. This matching mechanism is not the only design [15]. Multiple or gradient matching layers are used for a broader bandwidth [16]. New matching method was used for very high frequency (>100 MHz) ultrasound transducers [17,18].

14.3.1.5 Backing layer

For an ultrasonic imaging, the field-of-view (FOV) is at one side of the piezoelectric layer with transmission efficiency promoted by the matching layer. However, the wave generated by the piezoelectric layer propagates to the other direction, the back direction, as well. If the backward wave is reflected at certain types of targets, then the piezoelectric layer receives that echo as well, which is indiscernible from the signals from FOV. To remove the 'wrong' signal from the 'wrong' direction, the backward wave has to be eliminated.

Highly absorptive materials are typically attached to the piezoelectric layer at the back side, which is denominated as backing layer. The backing layer is relatively thick to ensure that almost no echoes propagate through the layer back to the piezoelectric layers.

14.3.1.6 Typical structure of a piezoelectric ultrasound transducer

As discussed above, the typical piezoelectric ultrasound transducer consists of a piezoelectric layer to transduce the electrical energy to mechanical ultrasound wave and vice versa, a matching layer to promote the energy transmitting efficiency, and a backing layer to absorb the backward wave [6]. The typical structure is shown in Figure 14.13. Electrodes are coated at both surface of the piezoelectric layer to deliver and receive electric signals. The coating electrode is usually very thin, typically 100 nm, thin enough to be neglected in the mechanical vibration for up to 100 MHz ultrasound transducers.

14.3.2 Piezoelectric array

14.3.2.1 Typical array structures

To generate a two-dimensional (2D) image, the ultrasonic beam is swept. A transducer may be swept mechanically by rotating or swinging. An 1D-phased array transducer (see Figure 14.14(a)) may be used to sweep the beam

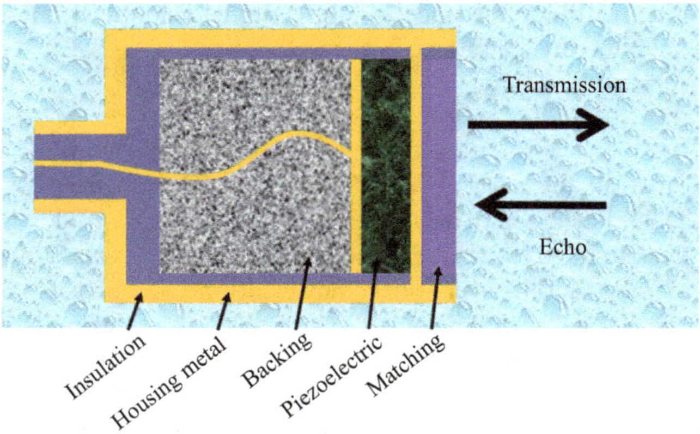

Figure 14.13 Typical structure of a piezoelectric ultrasound transducer

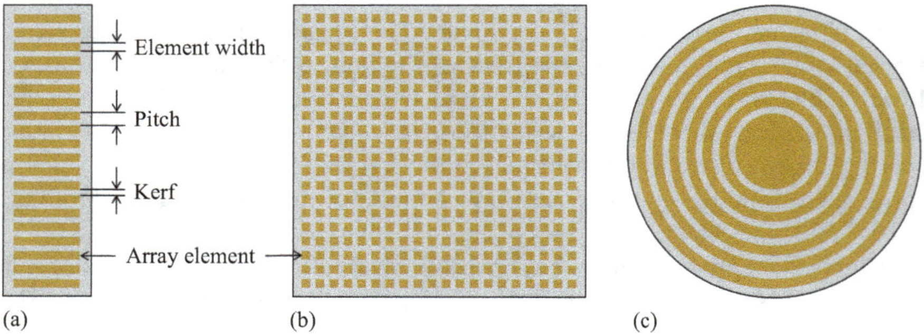

Figure 14.14 (a) 1D array, (b) 2D array, and (c) annular array

electronically. The received data are processed to construct the image. The image is then a 2D representation of the slice into the body. 1D linear array is the most commonly used array in ultrasound imaging.

3D images can be generated by acquiring a series of adjacent 2D images. Commonly, a specialised probe that mechanically scans a conventional 2D-image transducer is used. However, because the mechanical scanning is slow, it is difficult to generate 3D images of moving tissues. Recently, 2D-phased array transducers (see Figure 14.14(b)) that can sweep the beam in 3D have been developed. These can image faster and can even be used to generate real-time 3D images of a beating heart. Because of significantly increased number of elements, it is difficult to make a 2D-phased array. The electrical system is also very complex.

For a relatively simplified 1D imaging with variation of focusing depth, ring-shaped array (see Figure 14.14(c)) is designed with concentric arrangement, which is named as annular array. Because of the dynamic focusing capability and relatively low cost, annular array has some advantage in therapeutic ultrasound, for example high intensity focused ultrasound.

14.3.2.2 Array excitation modes

An ultrasonic beam generated by a phased array can be focused and steered by properly delaying the signals going to the elements for transmission or arriving at the elements for receiving (see Figure 14.15(a)). Phased arrays allow dynamic focusing and beam steering. Dynamic focusing can be achieved in transmission and in reception. However, multiple transmissions of pulses are needed for dynamic focusing during transmission for slowing down the frame rate. Transmission dynamic focusing is usually done in discrete zones, whereas receiving dynamic focusing can be done in many more zones or almost continuously.

A linear switched array (sometimes called a linear sequenced or simply a linear array) is operated by applying voltage pulses to groups of elements in succession (see Figure 14.15(b)). In this way, a sound beam is moved across the face of the transducer and electronically produces a picture similar to that obtained by scanning a single-element transducer manually. The amplitude of the voltage pulses can be uniform or varied across the aperture as shown in Figure 14.15(b) by arrows of varying length. Amplitude apodisation or varying the input pulse amplitude across the aperture is sometimes used to suppress side lobes at the expense of worsening the lateral resolution. If the electronic sequencing or scanning is repeated fast enough (e.g. 30 frames per second), a real-time image can be generated.

14.3.3 Capacitive micro-machined ultrasound transducers

A capacitive micro-machined ultrasonic transducer (CMUT) is a relatively new concept in the field of ultrasonic transducers. Most of the commercial ultrasonic transducers today are based on piezoelectricity. CMUTs are the transducers where the energy transduction is due to change in capacitance [19]. CMUTs are constructed on silicon using micro-machining techniques. A cavity is formed in a

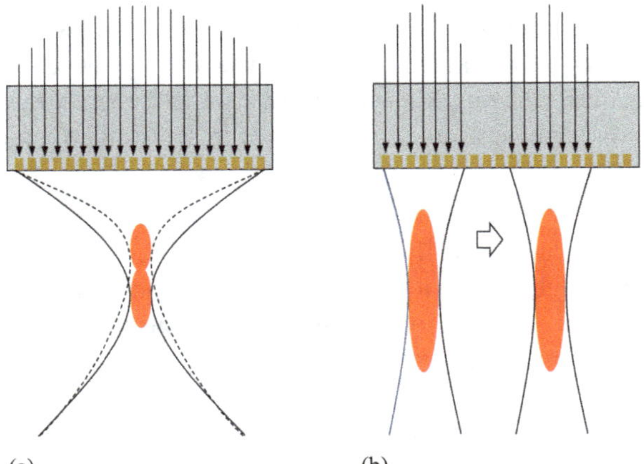

Figure 14.15 (a) Excitation of a phased array and (b) excitation of a linear array. Red ellipse areas indicate the focusing zone for imaging

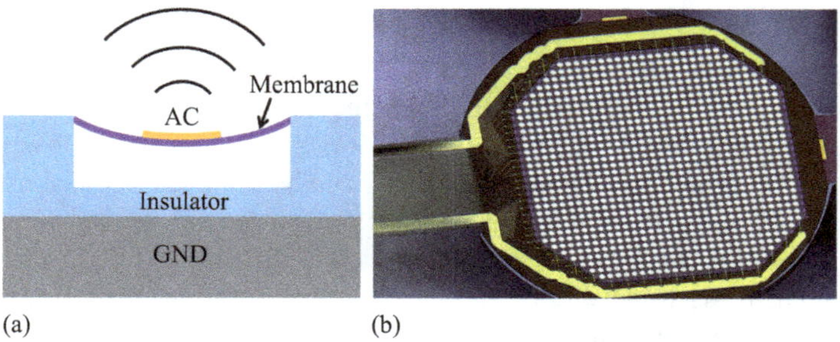

Figure 14.16 (a) CMUT structure and (b) example of a CMUT transducer with large number of elements (white round areas). Courtesy of Phillips Innovation Services

silicon substrate, and a thin layer suspended on the top of the cavity serves as a membrane, on which a metallized layer acts an electrode, together with the silicon substrate, serving as a bottom electrode (see Figure 14.16(a)).

If an AC signal is applied across the biased electrodes, the vibrating membrane will produce ultrasonic waves in the medium of interest. In this way, it works as a transmitter. On the other hand, if ultrasonic waves are applied on the membrane of a biased CMUT, it will generate alternating signal as the capacitance of the CMUT is varied. In this way, it works as a receiver of ultrasonic waves.

As CMUTs are micro-machined devices, it is easier to construct 2D arrays of transducers using this technology (see Figure 14.16(b)). This means a large number of CMUTs can be included in a transducer array providing larger bandwidth compared to other transducer technologies. To achieve a high-frequency operation using CMUTs is easier due to its smaller dimensions. The frequency of operation depends on the cell size (cavity of membrane) and on the stiffness of the material used as a membrane. As it is built on silicon, the integration of electronics would be easier for the CMUTs compared to other transducer technologies. The properties to use in high frequency with large bandwidth makes it a good choice to use as a transducer in medical imaging, especially in an intravascular ultrasound (IVUS). Because of its broader bandwidth, it can be used in second-harmonic imaging.

14.3.4 Optical ultrasound sensors

Besides piezoelectric and capacitive transduction of ultrasound, optical sensing of ultrasound exhibits high sensitivity, broad bandwidth and immunity from electromagnetic interference. Typical optical ultrasound sensors include fibre Bragg grating, Fabry–Perot interferometer and Laser Doppler vibrometer.

14.3.4.1 Fibre Bragg grating

A fibre Bragg grating (FBG) is a type of distributed Bragg reflector constructed in a short segment of optical fibre that reflects particular wavelengths of light and transmits all others [20]. This is achieved by creating a periodic variation in the refractive index of the fibre core, which generates a wavelength-specific dielectric

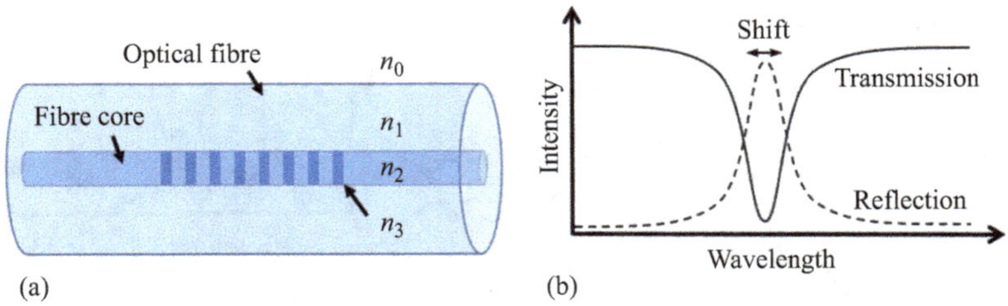

Figure 14.17 (a) FBG-based ultrasound sensor and (b) wavelength of reflected by FBG is dependent on pressure applied by ultrasound wave

mirror (see Figure 14.17(a)). An FBG can therefore be used as an inline optical filter to block certain wavelengths, or as a wavelength-specific reflector.

The reflection at the FBG is highly dependent on the wavelength of the light, and there is a very sharp peak at specific wavelength (see Figure 14.17(b)). Excited by acoustic compression or expansion, the position of the reflection peak shifts slightly, which causes significant variation on the light intensities at specific wavelength. Detection of the light intensities leads to the detection of ultrasonic waves.

Because only the light with specific wavelength is reflected, many FBGs can be fabricated on one fibre, leading to an array sensing capability with only one fibre. Furthermore, optical fibre is immune to electromagnetic interference. These features make FBG very suitable for distributed sensing, such as structural health monitoring of bridges or buildings.

14.3.4.2 Fabry–Perot interferometer

In a Fabry–Perot (FP) interferometer, a light signal is reflected backward and forward within the cavity many times [21]. The output will be a maximum when all the exit beams are in phase and drop quickly when they are slightly out of phase. Thus, a small phase change introduced in the cavity can result in a large change in the output intensity.

An FP interferometer system consists of a wavelength tuneable Laser, a fibre circulator, a photo-detector and the FP cavity (see Figure 14.18(a)). The wavelength of the light is tuned at the largest slope of the light intensity for a maximum sensitivity and linearity. As the ultrasound wave compresses or expands the cavity of the FP interferometer, the thickness varies in response of the wave. The thickness variation causes significant changes in the light intensity, which can be detected by photo-detector (see Figure 14.18(b)).

The FP interferometer has two major advantages: (1) the high sensitivity due to the sharp response of multi-beam interferences, and (2) the compactness because the sensing cavity is typically a thin membrane at the fibre tip. The major disadvantage of the FP interferometer is the small dynamic range.

14.3.4.3 Laser Doppler vibrometer

If a light beam is reflected by a vibrating surface, then the frequency of the light is shifted proportional to its velocity, a phenomenon referred to as the Doppler shift

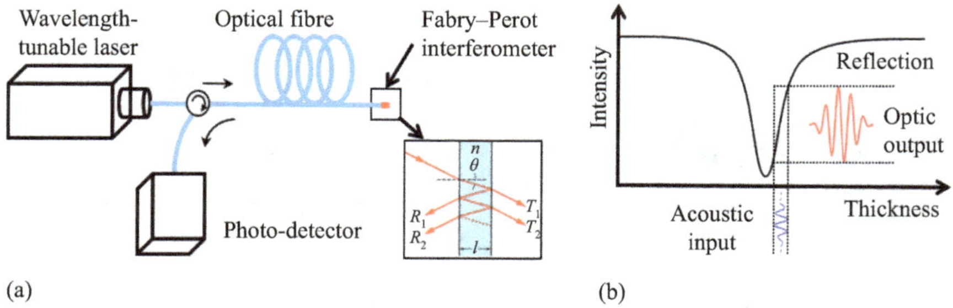

Figure 14.18 (a) Ultrasound sensor based on Fabry–Perot interferometer and (b) at largest slope of light intensity, optic output is very sensitive to thickness, which turns acoustic signal to optic output

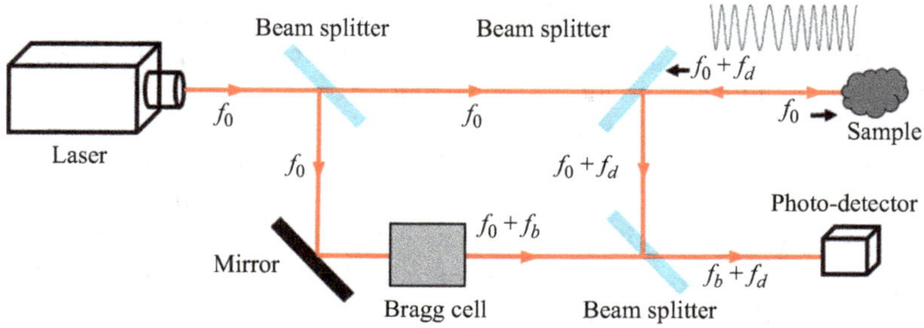

Figure 14.19 Laser Doppler vibrometer

[22]. A Laser Doppler vibrometer (LDV) can measure the vibration of the surface without contact. The laser beam from the LDV is directed at the surface of interest, and the vibration amplitude and frequency are extracted from the Doppler shift of the reflected laser beam frequency due to the motion of the surface (see Figure 14.19). After phase demodulation, the output of an LDV is generally a continuous analogue voltage that is directly proportional to the target velocity component along the direction of the laser beam.

A vibrometer is generally a two-beam Laser interferometer that measures the frequency (or phase) difference between an internal reference beam and a test beam. The most common type of laser in an LDV is the helium–neon laser, although Laser diodes, fibre Lasers and Nd:YAG Lasers are also used. The test beam is directed to the target, and scattered light from the target is collected and interfered with the reference beam on a photodetector, typically a photodiode. Most commercial vibrometers work in a heterodyne regime by adding a known frequency shift (typically 30–40 MHz) to one of the beams. This frequency shift is usually generated by a Bragg cell or acousto-optic modulator.

The beam from the Laser, which has a frequency f_0, is divided into a reference beam and a test beam with a beam splitter. The test beam then passes through the

Bragg cell, which adds a frequency shift f_b. This frequency shifted beam then is directed to the target. The motion of the target adds a Doppler shift to the beam given by $f_d = 2v(t)\cos(\alpha)/\lambda$, where $v(t)$ is the velocity of the target as a function of time, α is the angle between the laser beam and the velocity vector and λ is the wavelength of the light.

Light scatters from the target in all directions, but some portion of the light is collected by LDV and reflected by the beam splitter to the photodetector. This light has a frequency equal to $f_0 + f_d$. This scattered light is combined with the reference beam at the photodetector. The initial frequency of the laser is very high ($>10^{14}$ Hz), which is higher than the response of the detector. However, the detector does respond to the beat frequency between the two beams, which is at $f_b + f_d$, typically in the tens of MHz range.

The output of the photodetector is a standard frequency modulated (FM) signal, with the Bragg cell frequency as the carrier frequency, and the Doppler shift as the modulation frequency. This signal can be demodulated to derive the velocity versus time of the vibrating target.

14.4 Ultrasound imaging

14.4.1 Ultrasound imaging system

The transducer and the arrays described in Section 14.3 are used to transmit and receive ultrasound signals and generate the ultrasound images. In addition to the ultrasound probes, other electrical components are required in the imaging system, which generate, process, and display the ultrasound signals into an image. Typically, the electrical system consists of a central processing unit, a pulse generator, a signal amplifier, a data acquisition component and a monitor to display the ultrasound images (see Figure 14.20).

The central processing unit controls the whole system, defining the imaging mode, synchronising transmission and reception, processing the results to display and so on. The pulse generator produces high-intensity short pulses to excite the ultrasound transducer and transmit the ultrasonic wave. The ultrasound is scattered by the non-uniformities at interfaces between organs or small inhomogeneities in

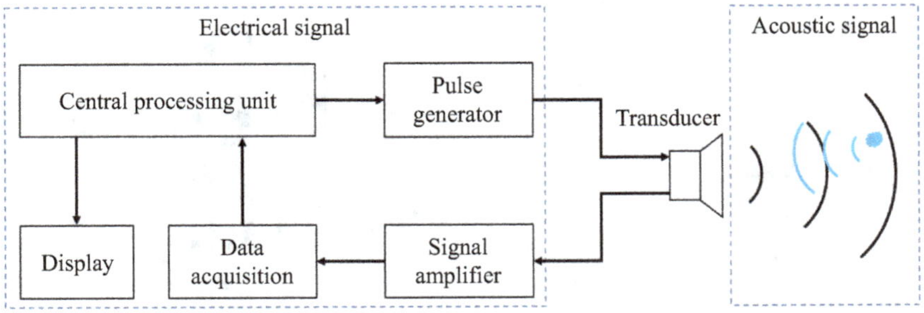

Figure 14.20 Typical components of an ultrasound imaging system

the organ, such as cells, small blood vessels and ducts. The returned echoes from the tissues are converted to electrical signals by the same transducer. As the echoes are relatively weak, amplification of the signal is required. The amplified signals are then digitised and collected by the central processing unit. After post-processing, ultrasound images are displayed in a monitor according to the ultrasound echo signals.

If the ultrasound transducer is a phase array, then the pulse generator would be an array with precise delay lines on each element. Multiplexers are required to selectively active certain channels for transmitting and receiving signals. There are many imaging modes:

- A-mode – amplitude mode
- B-mode – brightness mode
- M-mode – motion mode
- IVUS – intravascular ultrasound
- PW Doppler – pulsed-wave Doppler
- CW Doppler – continuous wave Doppler

14.4.2 Amplitude mode (A-mode) ultrasound imaging

Traditional ultrasound imaging technique is developed based on the scattering at positions with non-uniform acoustic impedance. An impulse is sent towards the non-uniform medium and the time-of-flight of the echoes are used to calculate the distances of the scattering points from the transducer (see Figure 14.21(a)). The amplitude of the echo ultrasound envelope reveals the boundary information, including the boundary position, impedance difference, size of the scattering point and so on. Consequently, 1D imaging in time (axial direction) domain is named as A-mode (i.e. amplitude-mode) imaging. Each line of signal is usually called as an 'A-line'. Sound velocity can be assumed to be a constant because it varies very

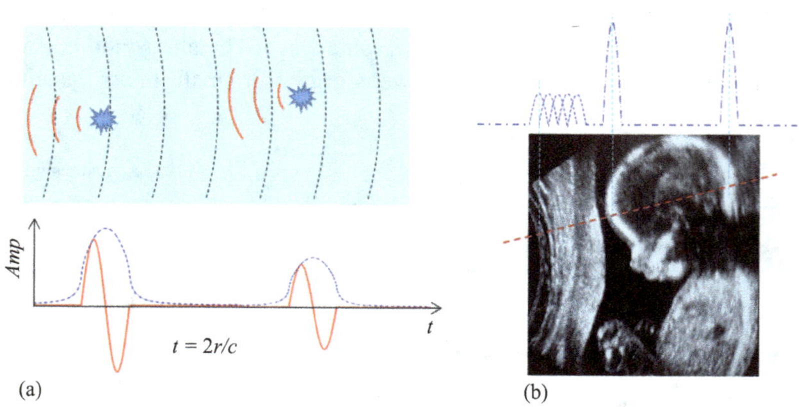

Figure 14.21 Ultrasound imaging: (a) A-mode and (b) B-mode. Courtesy of GE Healthcare

little in tissues, typically <5% [23]. This assumption may cause errors in distance, area or volume measurements and image distortion sometimes.

The A-mode imaging provides information on how far a scattering point is. If a wide beam is used, then the field-of-view (FOV) is large, but there is no information on the direction or the position of the target. On the contrary, a narrow beam provides information on the direction and position of the target because only the points within the beam generate echoes, and a mechanical scan is needed to enlarge the FOV.

14.4.3 Brightness mode (B-mode) ultrasound imaging

If a series of A-mode scans from different angles or different positions are collected together, then the position of scattering point can be located in a 2D map [24]. The echo amplitude is used to modulate the brightness of each pixel in the image (see Figure 14.21(b)). As a consequence, the imaging mode is denominated as B-mode (i.e. brightness mode) imaging.

The modulation of echo amplitude to grey level is made adjustable according to specific clinical application (see Figure 14.22). The most straightforward method is linear modulation, that is the brightness of each pixel is proportional to the amplitude of the echo. However, the amplitude of ultrasound signals typically covers a large dynamic range, making small signals invisible in B-mode images. To solve this problem, most B-mode imaging uses log-compressed modulation, which makes good illumination for both large and small signals. For specific applications, windowed grey-scale mapping may be used to enhance the image contrast of tissues in regions where there are no strong reflectors or strong echoes need to be suppressed. These options are available in commercial scanners.

With the modulation of the amplitude, B-mode images are 2D ultrasound images composed of bright dots representing the ultrasound echoes (see Figure 14.21(b) and Figure 14.23). The brightness of each dot is determined by the amplitude of the returned echo signal. This allows for visualisation and

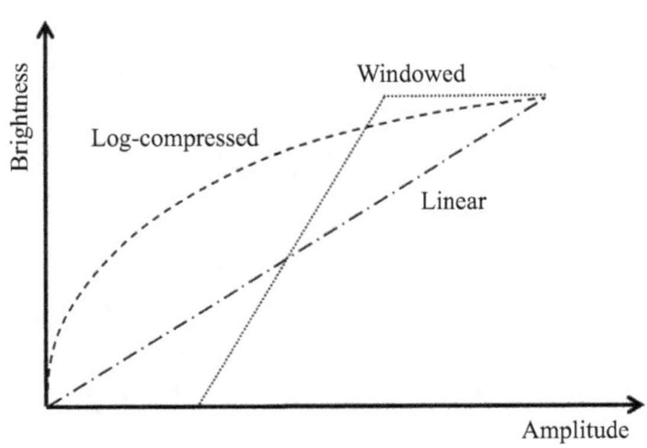

Figure 14.22 Typical brightness modulation forms

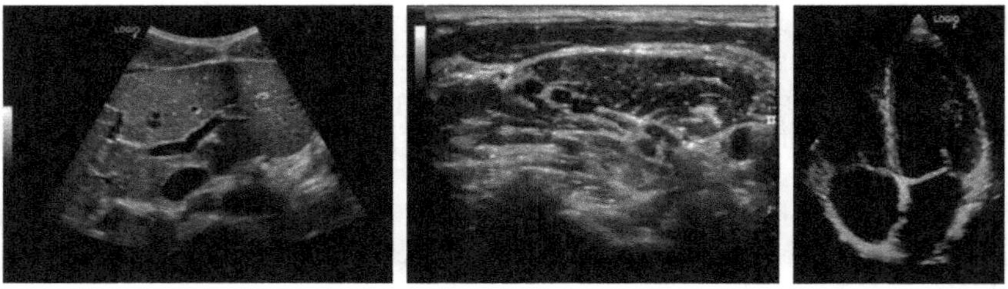

(a) (b) (c)

Figure 14.23 B-mode ultrasound images: (a) liver, (b) brachial plexus nerve, and (c) heart with four chambers. Courtesy of GE Healthcare

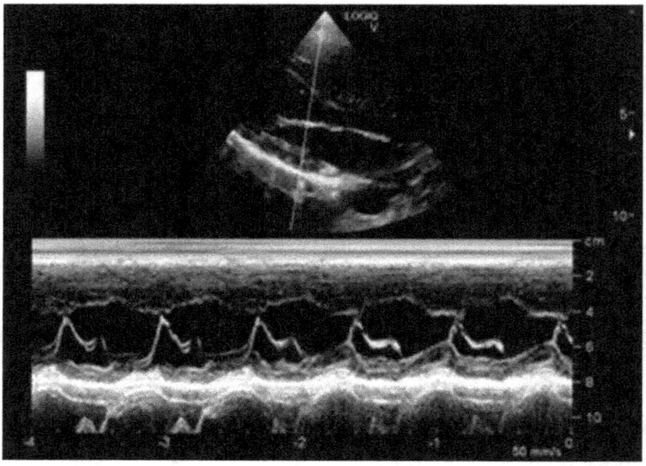

Figure 14.24 M-mode image of the heart. Courtesy of GE Healthcare

quantification of anatomical structures, as well as for the visualisation of diagnostic and therapeutic procedures for small animal studies.

B-mode ultrasound imaging is the most common used ultrasound imaging. Usually, the B-mode image bears a close resemblance to the anatomy, which may be seen by eye, if the body could be cut through in the same plane. Abnormal anatomical boundaries and alterations in the scattering behaviour of tissues can be used to indicate pathology.

14.4.4 Motion mode (M-mode) ultrasound imaging

An A-line displays the amplitude of a line at a certain position and angle. If the echoes from the specific position and angle are constantly recorded and aligned together, then the image illustrates the motion of the tissue in a certain direction. In Figure 14.24, the top is a B-mode display of the heart. The white line indicates the ultrasound beam direction, along which the M-mode image in the bottom is acquired. Imaging of the specific line and aligning them in horizontal direction shows the motion along the line. Motion caused by the heart beat results in the variation of echo positions, and the motion is clearly visible [6].

The M-mode was the preferred imaging modality in the early days of ultrasound. M-mode is defined as time motion display of the ultrasound wave along a chosen ultrasound line. It provides a mono-dimensional view of the heart. All of the reflectors along this line are displayed along the time axis. The advantage of the M-mode is its very high sampling rate, resulting in high time resolution, so that even very rapid motions can be recorded, displayed and measured. The disadvantage is that the ultrasound line is fixed to the tip of the ultrasound sector. It is therefore difficult to align the M-mode perpendicular to the structures, which are displayed (i.e. the septum), thus leading to false measurements. Anatomical M-mode circumvents this limitation by reconstructing the M-mode from the 2D image by post-processing. The anatomical M-mode permits free positioning of the cursor line. However, the time resolution is significantly less than that of the conventional M-mode.

14.4.5 Intravascular ultrasound (IVUS) imaging

Based on the sound wave delivering method, the ultrasound applications can be classified as transcutaneous and percutaneous. The transcutaneous ultrasound delivering method is to generate ultrasonic wave outside of the body and deliver the wave through the skin. The advantage of this imaging method is the non-invasive feature and no much limitation on the dimension of the transducers. It is the most well developed and widely used for real-time 3D imaging. This transcutaneous imaging mechanism is mainly limited by the penetration depth in applications, in which high frequency ultrasound is needed for high resolution imaging. In applications, such as blood vessel wall imaging and vasa vasorum angiography [3,25,26], percutaneous ultrasound method is usually used, which delivers ultrasonic wave by an IVUS transducer inside a catheter (see Figure 14.25(a)).

IVUS transducer is a miniaturised ultrasound probe attached to the distal end of a catheter for vessel wall and lumen imaging [27]. The proximal end of the catheter is attached to computerised ultrasound equipment. It allows the ultrasound transducer to see from inside blood vessels out through the surrounding blood column, visualising the endothelium on the inner wall of blood vessels in living individuals.

The arteries of the heart (the coronary arteries) are the most frequent imaging target for IVUS. IVUS is used in the coronary arteries to determine the amount of atheromatous plaque built up at any particular point in the epicardial coronary artery. The progressive accumulation of plaque within the artery wall over decades is the setup for vulnerable plaque, which in turn leads to heart attack and stenosis (i.e. narrowing) of the artery (known as coronary artery lesions). IVUS is used to determine both plaque volume within the wall of the artery and/or the degree of stenosis of the artery lumen.

First, angiographic techniques are used and the physician positions the tip of a guidewire, usually 0.36 mm diameter with a very soft and pliable tip and about

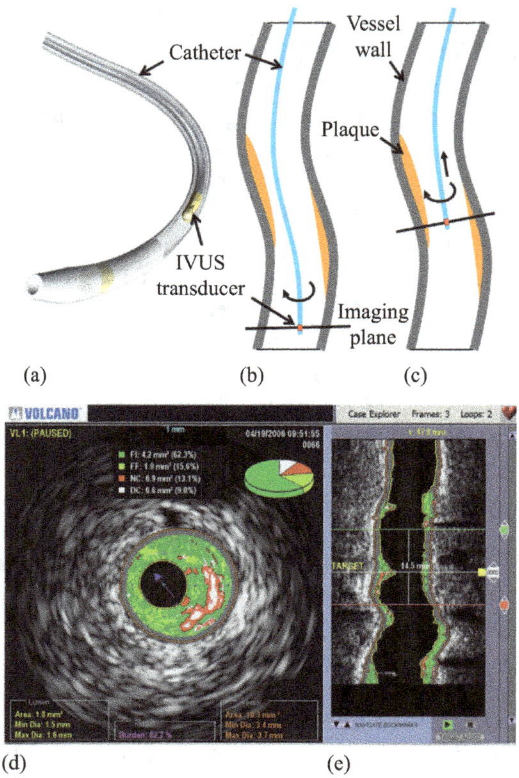

Figure 14.25 IVUS imaging. (a) Distal end of IVUS catheter with ultrasound transducer. (b&c) In an ultrasound imaging, the transducer rotates to image a slice, and pulls back for 3D imaging. (d) Cross-sectional and (e) side view of a section of vessel. Courtesy of GE Healthcare

200 cm long. The physician steers the guidewire from outside the body, through angiography catheters and into the blood vessel branch to be imaged. Second, the ultrasound catheter tip is slid in over the guidewire and positioned, using angiography techniques so that the tip is at the farthest away position to be imaged. The sound waves, emitted from the catheter tip, are usually in the 20–40 MHz range, and the catheter also receives and conducts the return echo information out to the external computerised ultrasound equipment, which constructs and displays a real time ultrasound image of a thin section of the blood vessel currently surrounding the catheter tip, usually displaying images at 30 frames/second. Third, the guide wire is kept stationary and the ultrasound catheter tip is slid backwards, usually under motorised control at a pullback speed of 0.5 mm/s (see Figure 14.25).

The blood itself and the healthy muscular tissue portion of the blood vessel wall is relatively low-scattering, which is just black circular spaces in the images.

On the contrary, heavy calcium deposits in the blood vessel wall both heavily reflect sound, that is are very echogenic, but also distinguishable by shadowing [28]. Heavy calcification blocks sound transmission beyond and so in the echo images, are seen as both very bright areas but with black shadows behind.

14.4.6 Doppler ultrasonography

Doppler ultrasonography is medical ultrasonography that employs the Doppler effect to generate imaging of the movement of tissues and body fluids (usually blood), and their relative velocity to the probe [24]. By calculating the frequency shift of a particular sample volume, for example flow in an artery or a jet of blood flow over a heart valve, its speed and direction can be determined and visualised. Colour Doppler or colour flow Doppler is the presentation of the velocity by colour scale. Colour Doppler images are generally combined with grayscale (i.e. B-mode) images to display duplex ultrasonography images, allowing for simultaneous visualisation of the anatomy of the area. This is particularly useful in cardiovascular studies (sonography of the vascular system and heart) and essential in many areas, such as determining reverse blood flow in the liver vasculature in portal hypertension.

The Doppler data are displayed graphically using spectral Doppler, or as an image using colour Doppler (i.e. directional Doppler, see Figure 14.26(a)) or power Doppler (i.e. non-directional Doppler, see Figure 14.26(b)). This Doppler shift falls in the audible range and is often presented audibly using stereo speakers, producing a very distinctive, although synthetic, pulsating sound.

All modern ultrasound scanners use pulsed Doppler to measure velocity. Pulsed wave instruments transmit and receive series of pulses. The frequency shift of each pulse is ignored. However, the relative phase changes of the pulses are used to obtain the frequency shift (since frequency is the rate of change of phase). The major advantage of pulsed wave Doppler (PW Doppler) over continuous wave (CW Doppler) is that distance information is obtained (i.e. time between

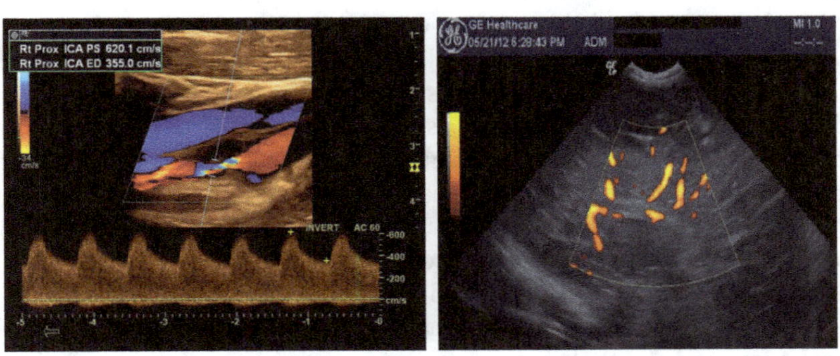

(a) (b)

Figure 14.26 Doppler ultrasound imaging: (a) colour Doppler imaging and (b) power Doppler imaging. Courtesy of GE Healthcare transmitted and received pulses multiplied by sound velocity equals distance) and gain correction is applied. The disadvantage of pulsed Doppler is that the measurements can suffer from aliasing.

There are no standards for the displaying colour Doppler. Some laboratories show arteries as red and veins as blue, as medical illustrators usually show them, even though some vessels may have portions flowing towards and portions flowing away from the transducer. This results in the illogical appearance of a vessel being partly a vein and partly an artery. Other laboratories use red to indicate flow towards the transducer and blue away from the transducer. Still other laboratories display the Doppler colour map in accord with published data, with red shift representing longer wavelengths (scattered) from blood flowing away from the transducer, and with blue representing the shorter wavelengths from blood flowing towards the transducer. Because of this confusion and lack of standards, the sonographer must understand the underlying physics of colour Doppler and the physiology of normal and abnormal blood flow in the human body.

14.4.7 Harmonic imaging

Harmonic imaging exploits nonlinear propagation of ultrasound through body tissues. The high-pressure portion of the wave travels faster than low pressure resulting in distortion of the shape of the wave. This change in waveform leads to the generation of harmonics (i.e. multiples of the fundamental or transmitted frequency) from a tissue.

As images produced with harmonic imaging have a higher resolution and are associated with fewer artefacts than conventional fundamental imaging, it is now the principal technique used in echocardiography. When an ultrasound wave passes through a body, the tissue generates 'harmonic ultrasound waves' because the tissue resonates. The resonance frequency is typically a multiple of the original frequency (i.e. the transmitted frequency), at which the tissue was insonated. Harmonic wave generation is similar to the overtones of a musical note. The harmonic frequencies are reflected back to the transducer, where they are used to create the image. Harmonic imaging typically leads to decreased reverberation and side lobe artefacts, increased axial and lateral resolution, cyst clearing, increased SNR and improved resolution in patients with large body habitus.

The second harmonic frequency is used most commonly [2]. The first reason is the bandwidth limitation of ultrasound transducers, which cannot cover high-order harmonics. The second reason is that signal at the second harmonic is much stronger than that at high-order harmonics. In the second harmonic imaging, all other harmonic frequencies and the fundamental component are filtered out. The receive frequency of modern transducers is two-fold higher than the sending frequency.

High order harmonic imaging, usually named as super-harmonic imaging, requires a transmission at relatively low frequency, and a reception of high order harmonics at high frequencies. Single frequency ultrasound transducers cannot cover both frequency bands. Dual frequency ultrasound transducers are designed to transmit and receive ultrasound waves individually (see Figure 14.27(a)) [3,25,29,30].

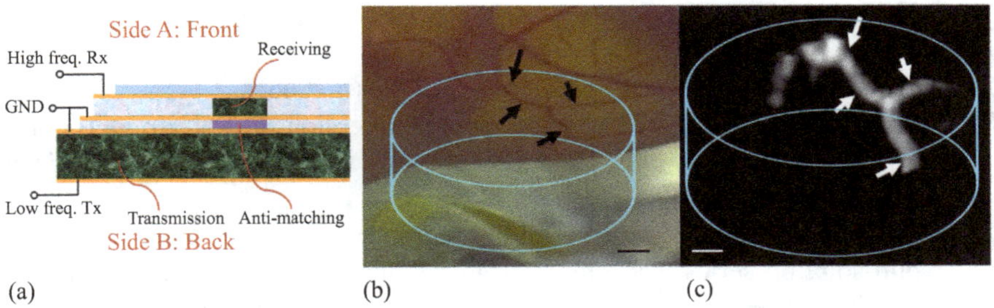

Figure 14.27 (a) Dual frequency ultrasound transducer for super-harmonic imaging and (b and c) contrast enhanced imaging of vessels in a chick embryo [25]

Because of significant difference from fundamental frequency scattering, the high order harmonic signals results in very clear ultrasonic images (see Figure 14.27(b) and (c)) [4,25,31].

14.5 Summary

This chapter describes theories and technologies about ultrasound imaging. As a mechanical wave, ultrasound propagates in the imaging zone, and the echoes deliver information about the position and properties of non-uniformities inside the specimen. The most commonly used device for ultrasound transmission and reception is piezoelectric ultrasound transducers and their arrays. Vibration modes and design principles were discussed in detail. Other ultrasound transmission and/or reception technology include CMUT and optical-based acoustic sensing. Multiple imaging modes are available in ultrasound imaging. Finally, typical ultrasound imaging modes are described, including A-mode, B-mode, M-mode, IUVS, Doppler imaging and harmonic imaging modes.

References

[1] Kinsler L E, Frey A R, Coppens A B, Sanders J V, *Fundamentals of Acoustics*, John Wiley & Sons, Inc., 2000

[2] Goertz D E, Frijlink M E, Tempel D, *et al.*,'Contrast harmonic intravascular ultrasound: A feasibility study for vasa vasorum imaging', *Invest. Radiol.*, 2006, **41** (8), pp 631–638

[3] Ma J, Martin K H, Dayton P A, Jiang X, 'A preliminary engineering design of intravascular dual-frequency transducers for contrast-enhanced acoustic angiography and molecular imaging', *IEEE Trans. Ultrason., Ferroelect., Freq. Contr.*, 2014, **61** (5), pp 870–880

[4] Gessner R C, Frederick C B, Foster F S, Dayton P A, 'Acoustic angiography: A new imaging modality for assessing microvasculature architecture', *Int. J. Biomed. Imag.*, 2013, 936593

[5] Szabo T L, *Diagnostic Ultrasound Imaging: Inside Out*, Academic Press, 2014

[6] Shung K K, *Diagnostic Ultrasound*, CRC Press, 2006

[7] Curie P, Curie J, 'Développement par compression de l'électricité polaire dans les cristaux hémidres à face inclinés', *Bulletin de la Societe ineralogique de Fran*, 1880, **3** (4), pp 90–93

[8] Jaffe B, Roth R, Marzullo S, 'Properties of piezoelectric ceramics in the solid-solution series lead titanate-lead zirconate-lead oxide: Tin oxide and lead titanate-lead hafnate', *J. Res. Nat. Bur. Stand*, 1955, **55** (5), pp 239–254

[9] Liu W, Ren X, 'Large piezoelectric effect in Pb-free ceramics', *Phys. Rev. Lett.*, 2009, **103** (25), 257602

[10] Cross L E, 'Relaxor ferroelectrics', *Ferroelectrics*, 1987, **76** (1), pp 241–267

[11] Zhang S, Luo J, Hackenberger W, Shrout T R, 'Characterization of Pb (In1/2Nb1/2) O3–Pb (Mg1/3Nb2/3) O3–PbTiO3 ferroelectric crystal with enhanced phase transition temperatures', *J. Appl. Phys.*, 2008, **104** (6), 064106

[12] Geng X, Ritter T A, Shung K K, 'Optimization of 1-3 relaxor-PT single-crystal/polymer composites', in *Proceedings of SPIE 3664, Medical Imaging 1999: Ultrasonic Transducer Engineering*, 21 June 1999, pp 24–30

[13] Pfister G, Abkowitz M, Crystal R, 'Pyroelectricity in polyvinylidene fluoride', *J. Appl. Phys.*, 1973, **44** (5), pp 2064–2071

[14] Ohigashi H, 'Electromechanical properties of polarized polyvinylidene fluoride films as studied by the piezoelectric resonance method', *J. Appl. Phys.*, 1976, **47** (3), pp 949–955

[15] Zhu J, 'Optimization of matching layer design for medical ultrasonic transducer', *PhD Dissertation*, Pennsylvania State University, 2008

[16] Li Z, Yang D-Q, Liu S-L, *et al.*, 'Broadband gradient impedance matching using an acoustic metamaterial for ultrasonic transducers', *Sci. Rep.*, 2017, **7**, 42863

[17] Fei C, Chiu C T, Chen X, *et al.*, 'Ultrahigh frequency (100 MHz-300 MHz) ultrasonic transducers for optical resolution Medical Imagining', *Sci. Rep.*, 2016, **6**, 28360

[18] Fei C, Ma J, Chiu C T, *et al.*, 'Design of matching layers for high-frequency ultrasonic transducers', *Appl. Phys. Lett.*, 2015, **107** (12), 123505

[19] Oralkan O, Ergun A S, Cheng C-H, *et al.*, 'Volumetric ultrasound imaging using 2-D CMUT arrays', *IEEE Trans. Ultrason., Ferroelect., Freq. Contr.*, 2003, **50** (11), pp 1581–1594

[20] Tsuda H, Kumakura K, Ogihara S, 'Ultrasonic sensitivity of strain-insensitive fiber bragg grating sensors and evaluation of ultrasound-induced strain', *Sensors*, 2010, **10** (12), pp 11248–11258

[21] Ismail N, Kores C C, Geskus D, Pollnau M, 'Fabry-Pérot resonator: Spectral line shapes, generic and related Airy distributions, linewidths, finesses, and performance at low or frequency-dependent reflectivity', *Opt. Exp.*, 2016, **24** (15), pp 16366–16389

[22] Castellini P, Martarelli M, Tomasini E P, 'Laser Doppler Vibrometry: Development of advanced solutions answering to technology's needs', *Mech. Syst. Signal Process.*, 2006, **20** (6), pp 1265–1285

[23] Dunn F, Tanaka M, Ohtsuki S, Saijo Y, *Ultrasonic Tissue Characterization*, Tokyo, Springer, 1996

[24] Hoskins P R, Martin K, Thrush A, *Diagnostic Ultrasound: Physics and Equipment*, Cambridge University Press, 2010

[25] Ma J, Martin K H, Li Y, *et al.*, 'Design factors of intravascular dual frequency transducers for super-harmonic contrast imaging and acoustic angiography', *Phys. Med. Biol.*, 2015, **60** (9), pp 3441–3457

[26] Li Y, Ma J, Martin K H, *et al.*, 'An integrated system for superharmonic contrast-enhanced ultrasound imaging: Design and Intravascular Phantom Imaging Study', *IEEE Trans. Biomed. Eng.*, 2016, **63** (9), pp 1933–1943

[27] Honye J, Mahon D J, Tobis J M, 'Intravascular ultrasound imaging', *Trends Cardiovasc. Med.*, 1991, **1** (7), pp 305–311

[28] Schoenhagen P, Nissen S E, *An Atlas and Manual of Coronary Intravascular Ultrasound Imaging*, London, UK, CRC Press, 2004

[29] Martin K H, Lindsey B D, Ma J, Nichols T C, Jiang X, Dayton P A, 'Ex vivo porcine arterial and chorioallantoic membrane acoustic angiography using dual-frequency intravascular ultrasound probes', *Ultrasound Med. Biol.*, 2016, **42** (9), pp 2294–2307

[30] Ma J, Steer M B, Jiang X, 'An acoustic filter based on layered structure', *Appl. Phys. Lett.*, 2015, **106** (11), 111903

[31] Bouakaz A, Krenning B J, Vletter W B, Cate F J ten, Jong N de, 'Contrast superharmonic imaging: A feasibility study', *Ultrasound Med. Biol.*, 2003, **29** (4), pp 547–553

Jianguo Ma is an Associate Professor in School of Instrumentation and Optoelectronic Engineering at Beihang University, China. He is also an Adjunct Researcher in Beijing Advanced Innovation Center for Big Data-Based Precision Medicine. Dr Ma received his BSc (2008) and MSc (2011) degrees from School of Physics, Shandong University, and received his PhD (2014) degree from North Carolina State University. He received his Postdoctoral training from the University of California, Los Angeles (2015–2016). His research includes multi-frequency piezoelectric ultrasound transducer design for a variety of diagnostic and therapeutic applications, optic sensing of ultrasonic vibrations for high sensitivity and broad band ultrasound detection and spectral pattern recognition of ultrasonic echoes for tissue characterisation. He has published 18 journal papers, 8 conference papers, and 2 book chapters. Dr Ma's research was supported by Young Elite Scientists Sponsorship Program by China Association for Science and Technology since 2017. He served as a Guest Editor for special issues 'Ultrasound Transducer' of the journal '*Sensors*' since 2018.

Xiaoning Jiang is a Distinguished Professor of Mechanical and Aerospace Engineeing at North Carolina State University, USA. He is also an Adjunct Professor of Biomedical Engineering at North Carolina State University and University of North Carolina, Chapel Hill. Dr Jiang received his BSc, MSc and PhD degrees from Shanghai Jiaotong University (1990), Tianjin University (1992) and Tsinghua University (1997), respectively. He received his postdoctoral training from Nanyang Technological University (1996–1997) and Pennsylvania State University (1997–2001). He was the Chief Scientist and Vice President of TRS Technologies, Inc. prior to joining North Carolina State University in 2009. Dr Jiang is an author and co-author of 2 books, 5 book chapters, 9 US Patents, 102 peer reviewed journal papers and over 100 conference papers on piezoelectric ultrasound transducers, ultrasound for medical imaging and therapy, drug delivery, ultrasound NDT/NDE, smart materials and structures and M/NEMS. He is a member of the technical program committee for a few international conferences including *IEEE Ultrasonics Symposium* (TPC-5), *SPIE Smart Structures and NDE*, *ASME IMECE*, *IEEE Nano* and *IEEE NMDC*. He is the Chair of NanoAcoustics Technical Committee for IEEE NTC, IEEE NTC Distinguished Lecturer (2018–), an editorial board member for the journal '*Sensors*' and a Senior Associate Editor for the *ASME Journal of Engineering and Science in Medical Diagnostics and Therapy*. Dr Jiang is an ASME Fellow.

Lijun Xu is a Senior Member of the IEEE. He received his PhD from Tianjin University, China in 1996. From 1997 to 2001, he was an Associate Professor in the School of Electrical Engineering and Automation, Tianjin University. From 2002 to 2006, he was a Research Fellow at the University of Greenwich, and the University of Kent, and a Higher Scientific Officer at the Institute of Cancer Research, University of London, UK. Currently, he is a Professor and the Dean of the School of Instrument Science and Opto-Electronic Engineering, Beihang University, China. His research interests include tomographic imaging, digital imaging and dynamic process monitoring. He has authored and co-authored more than 200 publications. He was awarded the National Science Fund for Distinguished Young Scholars of China in 2012, elected into the Ministry of Science and Technology Innovation Talents Promotion Plan in 2013, chosen as a Chang-Jiang Scholars Program Professor by the Ministry of Education, China in 2014 and elected into a high-level talents special support plan: '10,000 Plan' in 2016.

Index

www.ingramcontent.com/pod-product-compliance
Lightning Source LLC
LaVergne TN
LVHW011547200926
848484LV00001B/295